Before Building: Site Planning in the Digital Age

SECOND EDITION

R. Gene Brooks

David W. Lestage

Prentice Hall

Boston Columbus Indianapolis New York San Francisco Upper Saddle River
Amsterdam Cape Town Dubai London Madrid Milan Munich Paris Montreal Toronto
Delhi Mexico City Sao Paulo Sydney Hong Kong Seoul Singapore Taipei Tokyo

Editorial Director: Vernon R. Anthony
Acquisitions Editor: David Ploskonka
Editorial Assistant: Nancy Kesterson
Director of Marketing: David Gesell
Executive Marketing Manager: Derril Trakalo
Senior Marketing Coordinator: Alicia Wozniak
Marketing Assistant: Les Roberts
Project Manager: Maren L. Miller
Senior Managing Editor: JoEllen Gohr
Associate Managing Editor: Alexandrina Benedicto Wolf
Senior Operations Supervisor: Pat Tonneman
Senior Art Director: Diane Y. Ernsberger
Cover Designer: Bruce Kenselaar
Cover Image: R. Gene Brooks and David W. Lestage
AV Project Manager: Janet Portisch
Full-Service Project Management: Heidi Allgair, Element-Thomson North America
Composition: Element-Thomson North America
Printer/Binder: Edwards Brothers
Cover Printer: Lehigh-Phoenix Color/Hagerstown
Text Font: Minion

Credits and acknowledgments borrowed from other sources and reproduced, with permission, in this textbook appear on the appropriate page within the text. Unless otherwise stated, all artwork has been provided by the author.

Library of Congress Cataloging-in-Publication Data

Brooks, R. Gene,
Before building : site planning in the digital age / R. Gene Brooks, David Lestage. -- 2nd ed.
p. cm.
Rev. ed. of: Site planning, 1988.
Includes index.
ISBN-13: 978-0-13-508069-6
ISBN-10: 0-13-508069-X
1. Building sites--Planning. I. Lestage, David. II. Brooks, R. Gene, Site planning. III. Title. IV. Title: Site planning in the digital age.
NA2540.5.B76 2012
720.28--dc22

2011014750

10 9 8 7 6 5 4 3 2 1

Prentice Hall
is an imprint of

www.pearsonhighered.com

ISBN-10: 0-13-508069-X
ISBN-13: 978-0-13-508069-6

Contents

Preface vii

Acknowledgments ix

PART I

SITE PLANNING IN HISTORY 1

CHAPTER 1 Historical References 2

A Context of Religion and Authority 2

Nature, Place, and Form 3

Conventions and Inventions 5

Plan, Order, and Site 7

Order in the Wilderness 12

Reconciling Vision with Reality 14

CHAPTER 2 Environmental Ethics: A Value System in Change 17

The Judeo-Christian Ethic in the New World 17

Manifest Destiny 19

Sustainable Communities 20

A Question of Values 21

CHAPTER 3 New Ethics, New Law, and New Process 23

NEPA Purpose and Policies 23

Progress of a Different Color 24

PART II

THE EVOLUTION AND IMPACT OF COMPUTERS ON THE PROFESSIONAL OFFICE 25

CHAPTER 4 Until Recently 26

A New Understanding of Roles and Responsibilities in Practice 26

A Glance in the Rearview Mirror 27

The Digital Office 28

Perceptions and Expectations 29

Lay Expectations of the Profession 29

Intern Expectations of the Profession 30

Professional Expectations 30

More than a Digital Pencil 30

CHAPTER 5 From Graphite to CPU 32

Computer Platforms 32

Graphical User Interfaces 32

Raster Graphics 33

3D Modeling 33

Vector Graphics and CAD 33

The Internet 34

Google Earth 34

CHAPTER 6 Organizing and Creating Drawings 35

File Sharing 35

Computer File Organization 35

CAD Organization 36

An Overview of the AIA Guideline for Layer Names 37

Creating a Digital Drawing 37

Simple Drawings 38

AutoCAD and MicroStation 38

Geographic Information Systems (GIS) 38

ArcInfo: Using Data to Create Graphics 38

Property or Cadestral Maps 49

Building Information Modeling (BIM) 49

PART III

AN ENVIRONMENTAL INVENTORY AND ASSESSMENT 51

CHAPTER 7 Mapping: The Method of Inventory 52

Due Diligence 52

The Method of Inventory 53

Documentation 58

CHAPTER 8 Geology and Soils 70

Subsurface Geology 70

Surface Geology and Natural Properties 70

Structural and Drainage Characteristics 71

Performance Standards 72

CHAPTER 9 Vegetation 73

Landscape Character 73

Evaluating Existing Vegetation 73

The Role of Plant Materials 74

Wind Control 75

Erosion Control 75

Energy Conservation 75

Wildlife Habitat 75

The Planting Plan 75

Native Plants and Drought-Tolerant Plants 76

CHAPTER 10 Hydrology 78

The Hydrological Cycle 78

Resource Distribution and Conflicts 80

Surface Water: Streams, Creeks, Rivers, and Wetlands 81
Surface Water as a Process 82
Wetlands 83

CHAPTER 11 Climate and Site 86
Comfort 86
The Climatic Elements 87
Solar Radiation 88
Air Temperature 94
Humidity 96
Air Movement/Wind 97
Information Sources and Interpretation 102
Solar Geometry 102
Climatological Data 104
General Climatic Analysis 107
Site-Microclimate Mapping 110

PART IV
THE PLAN FOR CHANGE 117

CHAPTER 12 Earthworks: Shaping the Site 118
Concepts and Vocabulary 118
The Contour Line 118
Topography 120
Slope Analysis 122

CHAPTER 13 Grading 128
The Grading Process 131
Grading and Drainage 136
Grading Principles 136
Drainage: Considerations and Plan 138
Grading for Streets and Roads 141
Estimating Cut and Fill 148

CHAPTER 14 Retaining Walls 153
Structural Design Review 153
Precast Concrete Retaining Wall Systems 158
Mechanically Stabilized Earth (MSE) Walls 161
The Mechanically Stabilized Earth Process 161

CHAPTER 15 Topography and Surveys 164

CHAPTER 16 Storm Water Management and Erosion Control 170
Floods: Description and Regulation 171
Floodplains 172
Storm Water Runoff Analysis 178
Erosion: Types and Process 180
Erosion and Sedimentation Control 182
Erosion Reduction and Abatement Methods 185
Open Paving Systems 187

PART V
STREETSCAPE AND SITE IMPROVEMENTS 189

CHAPTER 17 Streets and Roads 190
Street Systems 190
Thoroughfare Classifications and Criteria 192
Intersections 193
Traffic Circles, Rotaries, and Modern Roundabouts 195

CHAPTER 18 Parking and Signage 197
Parking Configurations 197
Pedestrian Transition and Access for People with Disabilities 199
Disabled Person Parking 202
Drop-Off Zones 203
Signage 204
Lighting 206

CHAPTER 19 Pedestrian and Bicycle Circulation 208
Pedestrian Systems 208
Connectors and Linkages 209
Scale 210
Pedestrian and Living Streets 212
Accessibility 214
ADA Regulations 215
Design Guidelines 215
Bicycles and Bikeways 218
Bicycle Facilities Planning 218
Planning Criteria 219

CHAPTER 20 The Building and Energy 226
Building Metabolism 226
Predesign Thermal Load and Building Energy Analysis 229

CHAPTER 21 Land Use Controls 234
History and Background 234
Eminent Domain and *Berman* v. *Parker* 237
Hawaii Housing Authority v. *Midkiff* 238
Penn Central v. *The City of New York* 238
Kelo v. *The City of New London* 239
Conclusion and Counsel 239
Euclidean Zoning: The Basic Ordinance 240
Cumulative Zoning 240

Special Exceptions and Special Uses 241
Performance Standards 241
The Parking Schedule 244
The Decision Makers 245
The Local Legislative Authority 245
Planning and Zoning Commissions 245
The Board of Zoning Appeals or Adjustment 246
The Hearing Examiner 247
Neighborhood Zoning 248
The Rezoning Process 248
Rezoning Alternatives 249
Performance Zones 249
Down-Zoning and Back-Zoning 251

PART VI

APPLYING THE THEORY 253

CHAPTER 22 The Site-Planning Process 254

Goals and Objectives 254
Research and Data Collection 255
Inventory 255
Site R 258

Appendix A Stations Providing Local Climatological Data 275
Appendix B Sun Pegs for Various Latitudes 277
Appendix C Model CAD File Organization 280
Glossary 282
Index 285

Preface

This text was written to help both students and practitioners understand the issues and technology supporting the process of site planning. It was not composed as a surrogate course guide or outline. The intent was to provide a broad resource for faculty teaching site planning, regardless of their philosophical perspectives or design profession. They can customize their courses by choosing those aspects of the text that are key to their course and pass on the rest.

A genuine effort has been made to bring an understanding of the role of the digital world into the site-planning process. The computer is a tool, a means to an end, and little else. Its superior capabilities are limited to speed in processing, storing, duplicating, and sharing data in a variety of digital forms. To reinforce the idea that a computer does not possess the power of cognitive thinking, it has been observed that the computer only reduces the time from when a poor decision is made to the time when the results of the decision will occur.

Understand that site planning is a process in which assumptions and anything short of due diligence invite litigation. The digital world assists a site planner in retrieving information, processing the salient components, and forwarding those components to other members of the team involved in the process. Here's where the PC is superior to paper and pen: Documentation, sharing, and storage are unmatched. But a site planner must still manually process the data. We do not want to diminish the value of the PC but to understand its limitations. We recognize that the professional world requires that you have skilled competency in CAD. But we also want you to have the insight to understand how the PC world works in the context of paper and pen. This is not an either/or condition but one where you use a hammer to drive a nail and a deep-well socket wrench to change a spark plug. Both tools (i.e., skills) are required.

The overarching intent of this work is to cover the principles, issues, and interrelationships that are considered in the site-planning process. Since application is the acid test of most theory, this text leans on the application side. Aside from brief chapters on history, ethics, and environment, the majority of the book examines the basic components, issues, and methods that make up site planning and development.

The assumptions are:

1. There is a cluster of common issues that must be addressed and resolved in the development of any site plan.
2. The specifics of reality are far more complex than is academic doctrine.
3. In the final analysis the critical question in the design of any site relates to its *reason*. Long after the scholarly discussions of design juries are only a vague memory, the questions posed by review boards, clients, legislative authorities, and citizens will be: "Why?" "What was the rationale for that decision?" "What were the alternatives to this proposal?" Some of the questions will be polite and a smile will follow any but the most nonsensical answer. There will also be circumstances in which every word, sentence, and punctuation of the answer will be scrutinized for flaws in a defendant's competency.

Reason or rationale are often found in the two elemental bases of design: form and function. The argument as to which of these two principles should enjoy the dominant position is ongoing. There is a segment of the design professions that supports the position that the creative processes are diluted when functional or technical issues are anything more than benign agents in a problem. Advocates of this philosophy suggest that the creative mind works at its highest degree of potential when the constraints of the problems are minimal. This view holds that form must be the principal consideration since function left to its own devices rarely creates a design of any aesthetic achievement, and more often than not results in solutions devoid of spirit. The opposing opinion suggests that functional reasoning must be the primary consideration; otherwise, anything of any size can be any place. The functionalist view holds that beauty, in form, is a very subjective consideration, and points to the passing of numerous design styles as evidence of its capricious nature.

Most practioners, regardless of their philosphical persuasion, are of the opinion that there are few benign agents in the problem-solving process. This more holistic alternative holds that the more we know about a problem, the more responsible our solution will be. This does not impugn the value of form, but that every professional knows his or her creative judgment has become more precise with experience and research.

Although we know and recognize that the site is a common design problem with a set of common community-related issues, there is some question as to whether the site-planning process is always activated with common sense. Common sense is that collection of resources that each person knows will emerge, when all else fails, to act as a mental safety net. What is often forgotten is that the first hope is not for stellar innovation but for good judgment. Other good things follow.

Given the mix of philosophical differences and professional biases today toward environment, design, and structure, the concept of site planning enlists an array of responses. The pervasive dilemma is that many approach the process of site planning as though their work is isolated, independent, and immune to the impacts of nature, legislation, and economics. While each of these will influence the disposition of any man-made change to the site, the challenge is to develop a plan of aesthetic achievement that reflects the natural attributes of the site in concert with the law and fiscal responsibility.

Prior to applying any theory to site planning, it is critical to understand that like other design issues, site planning is a process of interrelated investigations. Our common sense tells us that questionable soils, impossible slopes, and drainage swales are all conditions that need to be approached with some degree of caution. Since the extent or range of a site's problems are dependent on research and analysis, these are the obvious first steps in the process. Problems emerge most often when intuitive decisions, keyed to form, are based on optimistic conclusions from partial data. Almost every professional has had at least secondhand knowledge of a situation in which a "brief analysis" would suffice until the financing became firm; then things really took a turn for the worse.

An understanding of the influence of climate on a site is another aspect of site planning that requires the use of common sense. Similar to the site's physiography, these issues are critical enough that one neither assumes nor guesses as to the seasonal paths of the sun. If a garden court's success is dependent on direct sunlight, would not the documentation of that illumination be an essential part of the early research? If we recognize that the east and west sides of a building will receive the highest heat load in the summer, would not orienting the building to minimize the heavier heat loads have been more cost-effective than the sun screens and additional air-conditioning requirement for the east-west alternative? If we understand that the sun can provide supplemental heat and light to a room, would not it have been more responsive to have oriented the building in such a way as to exploit this energy?

It must be added, that, in reality, site planning problems are seldom black or white. The contour lines may run perpendicular to the preferred orientation of a structure; or the zoning envelope forces the building into a broad swale that bisects the property; or the flattest area of the site is shadowed from the winter sun by a multistory structure. Since almost every problem has conflicting objectives, the process dictates the development of criteria on which alternative action might be based; otherwise, anything is possible. The site planner must have accrued sufficient acumen to recognize when a subjective decision creates a condition that violates common sense. In a professional context this is called *judgment.*

The National Architectural Accrediting Board (NAAB) has four general categories on which they review the design component of architectural programs for accreditaion. They are *analysis, synthesis, communication,* and *judgment;*[1] judgment is without question the key property, as it integrates both technical and artistic components in its process. Other educational institutions have similar criteria that require some aspects of the educational process address this critical element. Almost regardless of the issue or test, judgment is the critical ingredient, as it provides the direction to the other three skills. What is easy to forget is that judgment rests on a foundation of common sense.

Sooner or later your judgment will be put to the test. That is the nature of the design profession as all seek to demonstrate their understanding and creativity. Those who enjoy a flourishing practice know that relative success is based on their command of fundamentals, which permit them to reach beyond those basics. They will also testify that failure to recognize and resolve basic problems can jeopardize both project and practice.

We believe this book is different from the numerous other publications listed in your library's database to the degree that it seeks to examine the three forces at work on every site in their most basic terms:

1. The natural site conditions; the significance of the site's natural context and critical resources
2. The process and methods of altering the site to accommodate man-made components
3. The legal controls that delineate the extent to which the site can be changed, and their collective interrelationships in site planning.

Each of these three topics must be understood in a comprehensive way, as they both support and encumber each other in a dynamic process. Not only are they operationally inseparable, but a working knowledge of their finer parts is critical to diagnosis and design. Ignoring any one of these constituents invites criticism and solutions of dubious results. While it is valuable to understand "how to do it," it is equally important to have some insight into "why" something is done. The intent of this larger perspective was to provide the reader with a grasp of the past as the basis upon which to reckon with change.

The bias maintained throughout the book is that change is, for most sites, a requirement. But accepting change and improvements does not imply that radical change is the best under any circumstances. The site planner's role is a continuing process of balance and reconciliation. It begins with an intellect that can recognize those qualities of a site that make it unique and continues as a process that integrates change in a way that respects those attributes, thereby enriching the architecture and the lives of its occupants.

R. GENE BROOKS
DAVID W. LESTAGE

[1]National Architectural Accrediting Board. *NAAB Criteria and Procedures* (Washington, D.C.: NAAB, December 1984). p. 14.

Acknowledgments

Every author has a group of friends and colleagues who have supported his or her publishing efforts, and I am no exception. This major revision to an earlier text would have been significantly more difficult without the critique of Phillip Neely, and to him I am indebted. I have benefited from Phillip's experience as a practicing landscape architect, as he provided me with direction that I could not have found alone. The contributions of Truett James, architect, and Michael Ostdick, landscape architect, are still key to presenting, what I believe, is the most complete text available on the topic of site planning. Their perspectives and insights reflect professional acumen that sets their work apart. I also want to express my appreciation for the contribution of the late Dr. Ernest Buckley's work to this text. And to John McDermott: I want to thank him for his friendship, elastic encouragement, and counsel over the past 30 years. John always been always there to help me open the can of worms.

A special thanks to my co-author David, for his practical knowledge and understanding of the digital world in site planning. This text would not have seen the light of day without David's hands-on savvy of the hardware, software, and processes. We both acknowledge dial technology will continue to exert an ever-expanding role in our professional practice. David is unique in that he has the experience to critique what we do and why, with the balanced understanding that there is no technical substitute for some time-honored skills.

And finally, my heartfelt appreciation to my wife, Betty, for her patience, tolerance, and steadfast support during the preparation of this work.

R. Gene Brooks

I would like to express my appreciation and thanks to Tom Skinfill, a landscape architect in Orange County, California, and to Max Conrad, professor of landscape architecture at Louisiana State University in Baton Rouge, for sharing with me their perspectives and insights. Also, my thanks to Charles Patout, for his technical contributions and computer graphics.

It has been a privilege for me to work alongside Gene Brooks on this book. His passion and knowledge of the subject of site planning is both significant and inspiring. My sincere thanks to Gene for extending the invitation to me to co-author, and gratitude for his constant encouragement to see it through.

And finally, thank you to my wife, Cindy, for her constant affection, motivation, and encouragement.

David W. Lestage

We are grateful to the staff at Element-Thompson North America for providing the editing necessary to make our work more complete, cohesive, and easier to comprehend. Their queries and suggestions always provoked us to write better and often by saying less. If there is but one positive comment regarding this book, it must be shared with these outstanding wordsmiths.

PART ONE

SITE PLANNING IN HISTORY

- **CHAPTER 1:** Historical References
- **CHAPTER 2:** Environmental Ethics: A Value System in Change
- **CHAPTER 3:** New Ethics, New Law, and New Process

CHAPTER ONE

HISTORICAL REFERENCES

A CONTEXT OF RELIGION AND AUTHORITY

History is a principal resource in developing an understanding of any design discipline, as it permits an analysis of a particular set of events relative to their context and subsequent implications. History is the enviable 20–20 hindsight. Aside from the opportunity to separate the myth from the reality, history provides the opportunity to identify primary models and scrutinize their applicability to current conditions. One of the problems with an examination of the early history of site planning is that it was a natural extension of architecture and city planning. These two disciplines have been the principal sources of literature and practice, which provided their common theory, form, and order.

For many, city planning is an active component in the review of site planning. Leon Battista Alberti's suggestion that a house was a small city oversimplified the complex nature of cities and their real applicability to site planning and architecture. While studying the evolution of towns and villages is an engaging endeavor, the commonalities between a city's form and site-specific planning problems can have marginal corelationships. Although the plan for a city may have been the subject of some deliberation relative to its shape and physical development, a site in that city may or may not have been influenced by the same or similar criteria. The best example of this point is that although there were plans for ideal cities (Fig. 1-1), for good reasons we do not see plans of "ideal" sites. Although site planning has relied on town planning for theory and formal models, scrutiny of the real implications of that process and analogies can become obviously thin. What might be relevant in the relationship between town and site planning may not be the form as much as the prevailing social or ethical philosophy. The perception of physical plans also creates some problems in translating the city to the site. The form of a city is a reflection of the social, political, cultural, and economic forces at work in that society over an extended period. The quaint medieval streetscape we visit and enjoy today may have been a steaming sewer in the fourteenth century. Our inclination is to interpret the success or failure of history through our own values of good, bad, order, chaos, and so on. Transferring forms and physical models of the past to the twenty-first century is one of the prevailing reasons for the sea of differences between the visions of planners and the reality of today. If the visual and environmental results of the twenty-first century city are dubious (for which there is some supporting evidence), they may be based, at least in part, on a superficial understanding of history.

The scale and authorities controlling site planning in history are also worth noting. The large land ownerships controlled by king, pharaoh, or maharaja did not lend themselves to the scale of site planning in the sense that we use it today. Properties were often administered by noblemen and leased to smaller tenants, who built structures, raised livestock, and/or farmed. Since each tenant was a part of a larger organizational structure, in urban European history there appears to have been less emphasis on individual land ownership and building and more on the plan or development of a community. Most of the models taken from architectural history are of major monuments created for monarchs or religious institutions, wherein an expression of a culture could be articulated. The significance of the smaller site-planning problem (vernacular architecture) has been considered less profound,

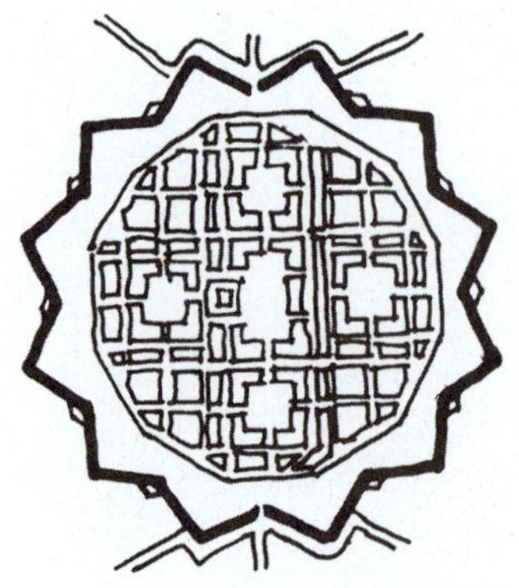

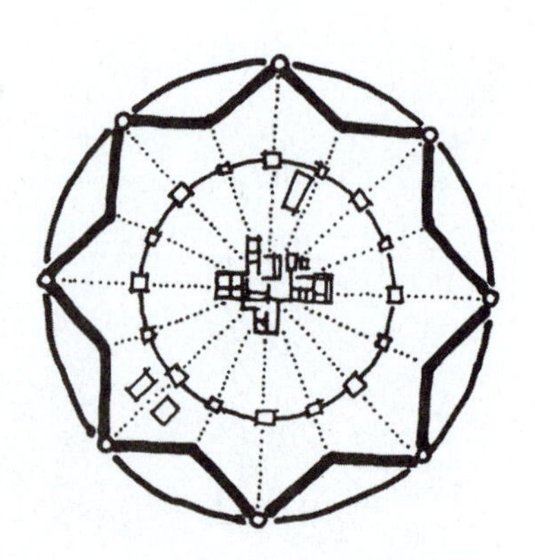

FIGURE 1-1. Plans for ideal cities.

as it relied on conventions and ethics as opposed to formal intellectual ideas. What may be learned from site planning in history is an appreciation for the relationships that the ancients structured between humanity, earth, and deity, and the physical implications of those alliances to site and settlement. Although the relationships vary, decisions were nonetheless based on conventions that focused to define place, to separate the sacred from the secular, and in the process, to survive.

To the ancients, existence and well-being had a direct functional relationship with the environment. All the elements that provided them with food, clothing, and shelter were directly attributed to the recognition of earth as deity. To the American Indian, "Father Sky and Mother Earth" were a union that gave meaning to life beyond the scope of the daily essentials. As such, the reverence held for the relationship between humanity and earth was one of survival, ethics, and religion. Amos Rapoport's interpretation of Deffontaines in *Geographie et Religions* is: "The founder [of the site] chooses the location not from geographical conditions but by seeking to conform to the decisions of the Gods."[1]

Numerous cultures believed the decisions associated with the location of a building, or the placement of a structure or community, to be a process that obligated one to consult the gods before a stone was turned or a field plowed. Since the earth was one of the few "known" quantities, it is easy to understand the importance attributed to nature and the cosmos. Finding the place for a community settlement, house, or field was no casual decision, particularly for the religious person. If human beings were to establish any sense of order in a world of chaos, the place must be fixed, or to quote Mircea Eliade: "If the world is to be lived in, it must be founded."[2] Some definition between that place which was sacred and that which was secular was of prime importance. Sometimes, "signs" or visions provided one level of decision making, and when those did not occur, a reasoning process was employed.

[1]Amos Rapoport, *House, Form and Culture* (Englewood Cliffs, N.J.: Prentice Hall, Inc., 1969), p. 75.

Nature, Place, and Form

Among the myths and rites associated with the definition of sacred places, the physical qualities of nature were an ever-present resource. Springs, trees, and particularly hills and mountains were often identified as sacred precincts. Stones and rocks have also held a special reverence with people throughout time. The most holy shrine of Islam at Mecca contains a stone that is believed to have fallen from heaven. Although Americans feel the compelling urge to carve faces in mountains, the ancients considered rock outcroppings to mark places "where the essence of the earth could still be perceived."[3] These were considered to be holy places in which dwelled the earth's spirit and were often visited to "obtain states of heightened vision."[4] Many religions have references in their liturgy to holy mountains or hills. In an effort to integrate theology with the daily activities of the practitioner, numerous mountain/hill forms became sites for convents, monasteries, mosques, and churches (Fig. 1-2). Dominant earth forms obviously provided excellent visual metaphors which could easily be integrated into those of the prevailing religion.

[2]Mircea Eliade, *The Sacred and the Profane* (New York: Harcourt Brace Jovanovich, Inc., 1957), p. 2.

[3]Nigel Pennick, *The Art and Science of Geomancy* (London: Thames and Hudson Ltd., 1979), p. 22.

[4]Ibid.

FIGURE 1-2. Mont St. Michael.

As an example, the Hebrews consider Palestine to be holy high ground, as it was not submerged by the great flood described in the Old Testament.[5] The Islamic faith believes the *Ka'aba* to be the highest place on earth, as "the pole star bears witness that it faces the center of Heaven."[6] The ancient Greeks sought to identify sacred precincts by pronouncing Mt. Olympus to be the home of the Greek gods and Delphi to be the center (navel) of the earth. The center of the world was also a feature to which many cultures attached religious significance. In the twelfth century, the alignment of the sun on Jerusalem on the summer solstice was such that it prompted declarations that the city was the center of the earth. Although the perception of the center of the earth might vary from one religion to the next, the concept that embraced the heavens was common to many. The rationale relative to the summer solstice was also the criterion used by the Chinese in locating the center of the world and the capital of the perfect Chinese sovereign.

Although there was doubt among many ancient cultures relative to an earth-born deity, the physical power and visual spectacle of the heavens were sufficient to create an abundant number of myths associated with the sky. The heavens became the home of gods for many cultures, as its infinite quality created concepts of eternity and omnipresence. Although human beings enlisted various conventions to locate the center of the world, the corollary to that was often an "axis mundi," or vertical axis, through the center connecting the earth to the heavens. That concept became manifest in the construction of human-made mountains (*ziggurats*) or temples whose design attempted either to connect the secular to the sacred or to imitate the cosmos in its design (Fig. 1-3).

[5]Mircea Eliade, *The Myth of the Eternal Return.* Bollingen Series XLVI (New York: Pantheon Books, Inc., 1954), p. 10.

[6]Ibid., p. 15.

In the temple of Queen Hetshepsut near Deirel Bahari, a mountain has been utilized as a surrogate for the pyramid. This decision not only embodied the notions associated with the sacred quality of hills but successfully shadowed attempts to place the temple in time.

The physically prominent was only a segment of the significance of nature to the ancients. To the religious, nature was never something that transpired by virtue of its own processes. "Nature is never natural; it is always fraught with a religious value."[7] Nature was a daily reminder of an inexplicable power that left only evidence of its meaning. The vegetation cults were not based on the appearance of spring as an annual birth as much as on the "prophetic sign of the mystery."[8] Although the secular view of trees may have been no more than an expression of continual birth and death, to many ancients the tree was the symbol of cosmos whose fruit could impart infinite wisdom, power, and eternal life. Trees were simple expressions of the same perceptual phenomenon: a mystery of inexhaustible life and energy. Trees were attributed particular importance in antiquity, for they were links between heaven and earth and believed to be the residence of specific spirits. Europeans had identified numerous religious and ethical meanings with trees. In Germany, it was believed that elves and spirits lived under oaks, elders, and firs. Care was taken to avoid damaging or uprooting these particular species in case one should, in the process, disturb their mythological tenants.[9] In German law, trees were the meeting places of the secret vigilant Vehmic Courts.[10] Scandinavian tribesmen believed that trees housed the spirits of deceased family members. The "old religion" did not die with Christianity. As a part of their land-clearing process, Finns

[7]Eliade, *The Sacred and the Profane,* p. 116.

[8]Ibid., p. 151.

[9]John R. Stilgoe, *Common Landscape of America. 1580–1845* (New Haven, Conn.: Yale University Press, 1982), p. 8.

[10]Pennick, p. 19.

FIGURE 1-3. Ziggurat.

FIGURE 1-4. Stonehenge.

and Swedes in the New World would recite incantations to the tree spirits prior to burning the fallen timber.[11]

An ancient Irish law stipulated that the destruction of any one of the seven holy trees (apple, alder, birch, hazel, holly, oak, and willow) was punishable by the fine of a cow, an extremely valuable property at that time.[12] The attitude of reverence for trees was not limited to ancient Europeans but continues today. Nigel Pennick reports in *The Art and Science of Geomancy* that construction work on a Singapore site was halted by the local workers' refusal to cut a holy tree, until labor from outside the Malaysian community was hired to remove it.[13]

Conventions and Inventions

Since survival was based on the delicate balance between the whims of nature and individual needs, primitive cultures invented various strategies to explain the world, creation, and cosmos. Aside from the nature-related forms discussed earlier, some cultures sought to develop principles that would serve to guide a change in the earth's form. The effort to provide some structure to the unknown, earth or cosmos, was the essence of countless settlements, all of which sought answers to a series of questions. The first was: "Where is the right place? How do we locate the appropriate site for a settlement that maintains the balance among earth, sky, and deity?" The second might have been: "Where is the holy ground? Where are those sacred zones that it is important to recognize? What are the limits of the settlement, or what is its form?" Although the first two questions might be reversed, the question of form and direction was always one that required the use of various rituals.

The art of *geomancy* provided guidance and direction for site planning of the great pyramids of the Giza in Egypt, Stonehenge in Britain, and the Forbidden City of Peking. "In geomancy, the world was conceived as a continuum in which all acts, natural and supernatural, conscious and unconscious, were linked . . . one with the next."[14] In respecting that concept, one exercised great caution in any act which sought to modify that symbiotic relationship.

Probably the most notable example of the ancient geomancy is Stonehenge at Salisbury (Fig. 1-4). The significance of the site continues to evolve; recent research indicates that the location was a burial site in addition to functioning as a calendar. The latest discovery on human burial activity in the circular ditches underscores the significance of the site to primitive people as early as 3100 B.C. The 1965 publication *Stonehenge Decoded* suggested that the monumental trilithons were most likely used to forecast changes in the seasons as well as the lunar eclipse phenomenon.[15] Imagine the range and scale of unknown quantities in the thirteenth century B.C. and the impact of a structure that could predict the alignment and arrangement of the sun and moon. The structure provided a partial insight into the tenuous sequence of planting, cultivation, and harvest, upon which survival depended. The significance of Stonehenge to a community surrounded by an environment that was the source of both life and death was obviously profound. The impact of early geomancy is probably more obvious when we consider its contribution to our collective knowledge in terms of something as simple as direction. Although the concept of direction is attributed to humanity's individual orientation—front, back, left, and right—the division is also reflected in the geometry of the globe: two poles and east–west. The division of direction into corresponding colors was considered and adopted by cultures throughout the world and is illustrated in Table 1-1.

As to the shape of the settlement, the Roman model was a circle. Using the cardinal (four) points, the circle was then quartered.[16] The form of Roman cities reflected the criteria of Vitruvius, who theorized that the shape of the city was created from the crossing of two roads, dividing the world as a compass, and surrounding the region of that intersection with walls.[17] A similar quartered system was also employed by ancient cultures in Bali and Asia, which followed a ritual of separating settlements into four sections that collectively resembled a square. The construction of the Sioux's sacred lodge also reflects the cardinal points, with four doors, four windows, and four colors.[18]

---

[11]Stilgoe, pp. 7 and 176.

[12]Pennick, p. 20.

[13]Ibid.

[14]Ibid., p. 8.

[15]Ibid., pp. 31–35.

[16]Christian Norberg-Schultz, *Existence, Space and Architecture* (New York: Praeger Publishers, 1971), p. 23.

[17]Ibid., p. 22.

[18]Eliade, *The Sacred and the Profane,* pp. 45 and 46.

Table 1-1 Division of Direction and Color

Culture	Direction				
	Center	East	North	West	South
Aztec		Red	White	Black	Blue
Chinese	Yellow	Blue	Black	White	Red
Irish		Red	Black	Green	White
Hindu		White	Brown	Yellow	Red
Maya/Hopi	Blue-green	Red	White	Black	Yellow
Tibetan	White	Blue	Green	Red	Yellow

Source: Nigel Pennick, *The Art and Science of Geomancy* (London: Thames and Hudson Ltd., 1979), p. 96.

The concept of direction and orientation provided a basic mechanism for the development of various geometric typologies (squares, circles, crosses, pentagons). These techniques and methods were employed in the development of specific sites as well as ideal cities. All were based on the belief that humanity was inextricably linked to the planet. As Tao Te Ching summarized: "The world is a holy vessel . . . let him that would tamper with it, beware."[19] Although each culture nurtured a philosophy that was unique to a region or a continent, the common quality was one that integrated the physical, natural environment with the supernatural, religious one; and it was humanity's obligation to maintain a balance between them.

The Chinese geomancy, feng shui, provides a number of excellent examples in which mountains correspond with planets. In feng shui there are functional analogies concerning the ancient landscape that provide guidance toward any change in the relationship between humans and earth. An established set of complex rules governs the practice of feng shui. One particular theory was that evil spirits traveled only in a straight line and could not turn corners.[20] Buildings were sited and designed with the specific purpose of providing an indirect access that would "trap" or shield a home from intrusion by an evil spirit. Although a person unaware of this principle might interpret the plan to be a classic "linking" or "transition zone," the reality of the rationale is quite different.

The geomancer believed that each site was vested with natural aura or forces. The "place" could be fixed when the forces were enclosed for a purpose rather than allowed to seep away into the external continuum. An eastern Indian society employed an astronomer to fix the forces in the preparation of a building site. The astronomer identified the location for the building cornerstone, which was believed to be the location of the snake's head supporting the world. A sharpened stake was driven in the earth to "fix" the symbolic snake of chaos, and the cornerstone was set on the stake. The ritual was repeated at each construction site to create symbolic form from the amorphous.[21]

Although to some cultures "fixing" the place was a site-specific convention, to the Romans, there was a relationship between sites and people. *Genius loci,* or the "spirit of place," was a Roman concept in which every person had his or her own guardian spirit, genius, or guiding force that was trying to shape his or her character. The concept of *genius loci* goes back to the *diamon* of the Greeks and reflects ancient humanity's perception of life and destiny. "In particular he recognized that it is of great existential importance to come to terms with the genius of the locality in which his life takes place."[22] *Genius loci,* like geomancy, emerged to give meaning to a relationship between places and people. Since at any one time, many people may inhabit a site, *genius loci* must be multivalent. Although *genius loci* relies on the physical quality of a site and reflects a distinct character, that character is laced with "complexities and contradictions."[23]

Most people understand that our individual worlds are neither common nor perceived with common values from a background of common experiences. To complicate an already questionable perceptual problem, there is another view which suggests that the substance of urban life emerges from the interaction of people, not from the physical place. It is within this multifaceted context of place occupied by people of various cultural backgrounds that site planning functions.

If, as Christian Norberg-Schultz maintains, "man's existence is dependent upon the establishment of meaningful and coherent environmental image of existential space,"[24] then the methods by which "meaningful and coherent" emerge become critical to the process. Over time, many early conventions conceived to assist in the process of giving meaning to sites and settlements fell into disrepute. Successes were either extremely limited or perceptually biased, as few rituals endured that can be applied to any site. Although myths and rites dissolved in the light of knowledge, what we can glean from the centuries-long process of change is the evolution of four basic forms used to fix a place or form space.

1. The image and form of the object building are capable of fixing a place (Fig. 1-5a).

[19]Pennick, p. 162.

[20]Ibid., p. 66.

[21]Eliade, *The Sacred and the Profane,* p. 54.

[22]Christian Norberg-Schultz, *Genius Loci* (New York: Rizzoli International Publication, Inc., 1980), p. 18.

[23]Norberg-Schultz, *Existence, Space and Architecture,* p. 69.

[24]Norberg-Schultz, p. 114.

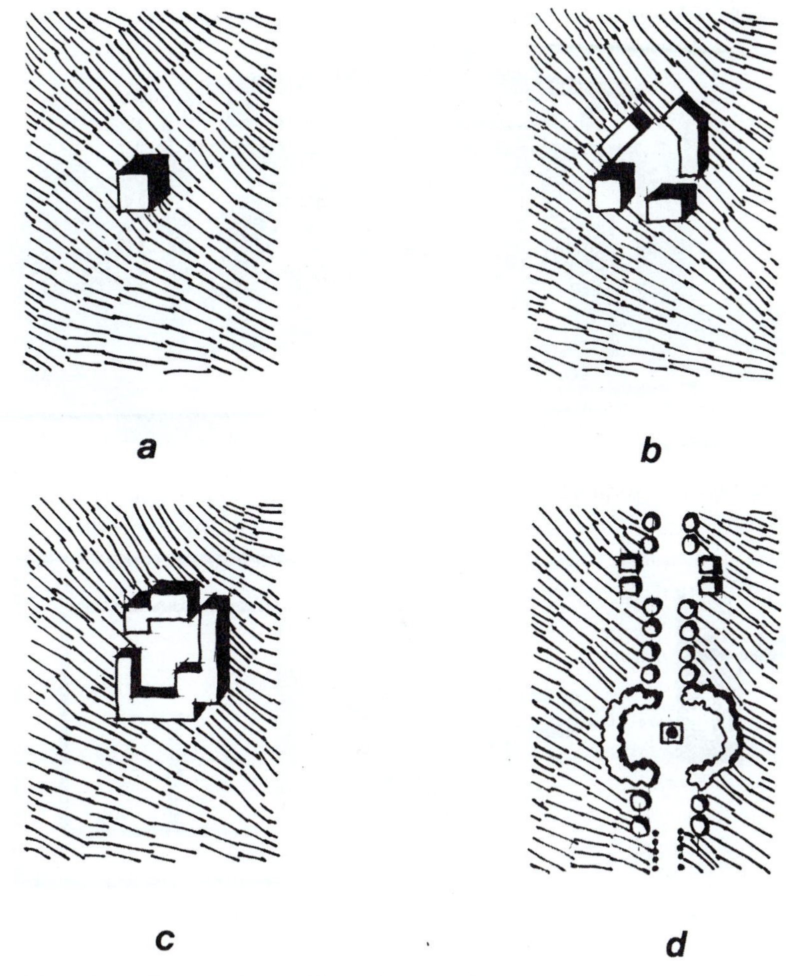

FIGURE 1-5. Four basic models of site planning.

2. A collection of independent structures, which although unattached, create a coherent image of place (Fig. 1-5b).
3. The form of a building can be such that a place may be fixed by the enclosure of its facades (Fig. 1-5c).
4. Although the kinetic implications of the word "path" are somewhat contradictory, paths are nonetheless capable of forming coherent, meaningful images[25] (Fig. 1-5d).

These four concepts and their numerous variations make up the basic human-made models. All are common in the built environment. There are, on occasion, similar qualities in the natural environment. Each of us has been in places where the view, treeline, landform, or water's edge was in its own way a coherent identifiable place without the introduction of human-made invention. The one concept that has only recently become more clear is the relationship between humanity and the natural environment. Fewer resources and an expanding population are easily overshadowed by a technology that has taken humanity into a realm once considered to be the home of gods. This larger perspective has made society more conscious that the balance between earth and humanity is as acute as ever.

The challenge of creating place thus has two interlocking parts: to form space that permits a plurality of interpretations, but to do it in such a way as to recognize the existing "place" and integrate the identity of humanity with the identity of earth.

PLAN, ORDER, AND SITE

The development of ancient conventions and rites framed the parameters of site planning for numerous cultures as each sought to bring order to a world of chaos. Although methods to fix or define a place varied, the most common form of human-made invention was the building. History is rich with examples of this site planning model. Some are more successful than others at creating an image of place, as opposed to an object, as concepts of structure easily dominate context and landscape (Fig. 1-6).

Another site planning model was mastered by the Egyptians, who were expert at the development of paths acting as "place" (Fig. 1-7). An extensive number of their monuments used the path (sequential images) in conjunction with buildings to magnify the funerary ritual. Large-scale land ownership, stratified social orders, and well-articulated religious principles provided a context and structure through which various cultures functioned to create place.

[25]Ibid., pp. 39–53.

FIGURE 1-6. Villa Rotunda.

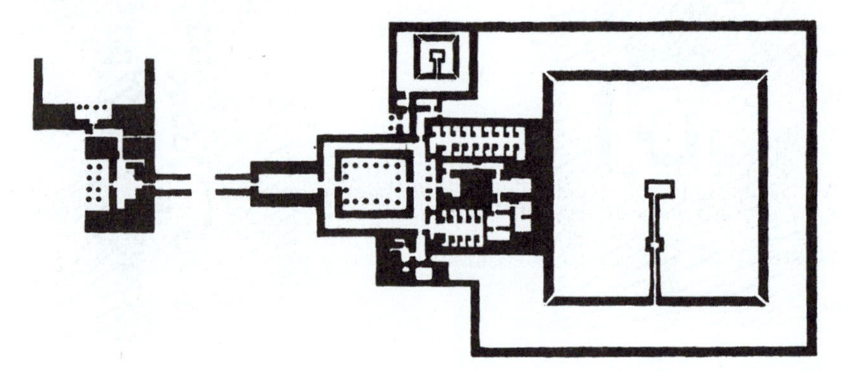

FIGURE 1-7. Plan of pyramid complex of Sahura.

Hellenic Greece, like other societies, sought to create some human-made order in the field of nature (the cosmos). The prevailing theory was that "the ordering of space on earth would mirror the order of the universe."[26] In pursuit of that order, the Greeks evolved two additional basic site planning models: a place defined by a collection of buildings and a place defined by an enclosure of structures.

The Acropolis is an outstanding example of place given identity by an arrangement of unattached buildings. The site and composition of structures convey a clear image of the precinct separate from the individual achievements of classical architecture. The Greeks, as did others, ascribed a value to high, physically prominent areas (Fig. 1-8). It naturally followed that the sacred site for the city of Athens would be located on a visibly prominent hill with the mortal complement of the community at the base elevation. The physical quality of historic Athens and its districts relies heavily on the Ionic order's preoccupation with the finite. This predisposition to develop plans of fixed, forever-controlled shapes influenced decisions associated with both town and site planning. Since the Acropolis was initially a fortress, its site planning incorporated vestiges of that earlier role, which may partially account for its barren quality. An "end state plan," as such, did not apparently exist until sometime between 447 and 437 B.C.[27] Temples were built, disassembled, and moved over a period of years to accommodate new structures and additions and to identify the correct relationship between structure and site. Fig. 1-9 illustrates the changes in the Acropolis in the period between 530 and 450 B.C.[28]

Whereas the Acropolis was sacred, an agora was completely secular, and plans often reflected the differences. The agora at Corinth is an example of a site plan that develops an identity from the enclosure of building facades (Fig. 1-10, page 10). Whereas the Acropolis was conceived as a group of individual building *volumes* designed and positioned to create a specific spatial quality, the Corinth agora consisted of a series of structures designed to control the quality of exterior space with *facades*. The agora at Miletus (Fig. 1-11, page 10) is a reminder that although different models were employed in different regions, each was a deliberate, conscientious effort to create an exterior spatial identity.

Although there is a marked difference in the concept of the buildings' arrangement, there is a similarity in the treatment of the ground plane and vegetation. The sites for various agoras and their planning almost totally exclude vegetation, or integrate natural landforms. Although the

[26] C. A. Doxiadis, *Architectural Space in Ancient Greece* (Cambridge, Mass.: The MIT Press, 1972), p. 20.

[27] Ibid., pp. 36–38.

[28] Ibid., p. 29.

FIGURE 1-8. The Acropolis.

Greeks' attitude toward the role of vegetation in the context of building is not clear, the absence or inclusion of living plants as a part of the site did not occur by chance.

Numerous public spaces throughout Hellenic Greece were developed without a network of geometric systems (radial, orthogonal, hexagonal, etc.), but C. A. Doxiadis maintains that many utilized an organizational structure based on geometry and the visual quality of space. His research supports the notion that the siting of structures in numerous locations throughout Greece was based on a system of "polar coordinates." This concept builds on the premise that with humanity as the observer, each object (building) should be located on a site relative to humanity's view.[29] *Architectural Space in Ancient Greece* contains extensive documentation of 15 sites that were developed on the basis of a strict set of Euclidean principles. Doxiadis's hypothesis is that numerous structures in major public spaces were oriented relative to a fixed view from a propylon in accordance with a predetermined set of geometric angles.

The Altis at Olympia (Fig. 1-12) incorporated aspects of place defined by building facades, a collection of structures, an object building, and a path. When multiple concepts are used to reinforce the role of others, the potential for creating an enduring image is increased substantially. It should be noted that the Roman Forum of Trajan and the Imperial Forum are excellent examples of building facades used to

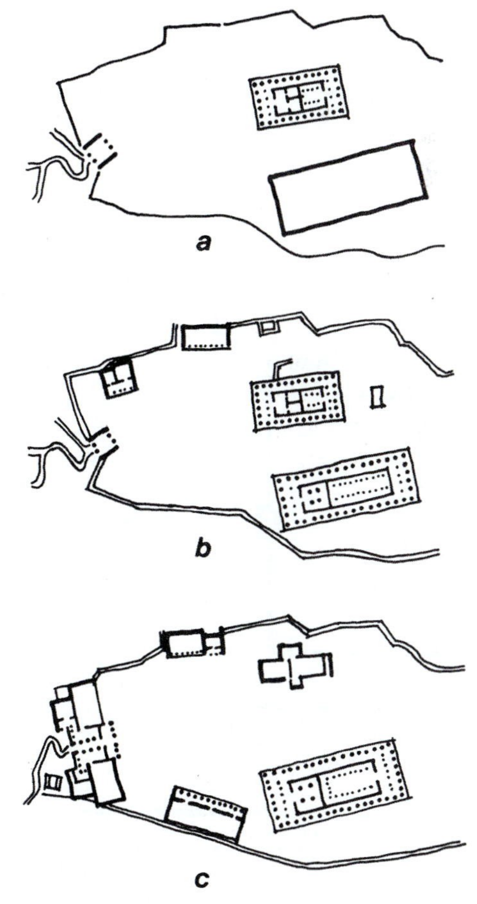

FIGURE 1-9. The Acropolis; plan 530 to 450 B.C.

[29]Ibid., pp. 4 and 5.

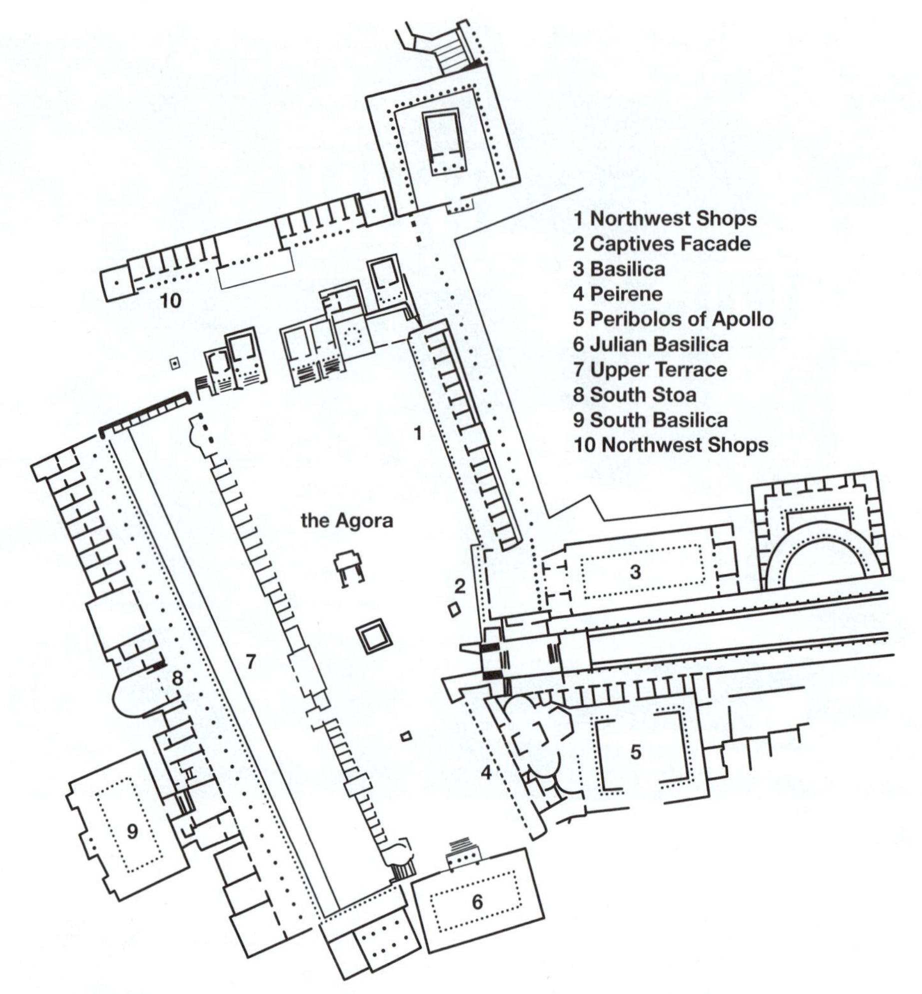

FIGURE 1-10. The agora at Corinth.

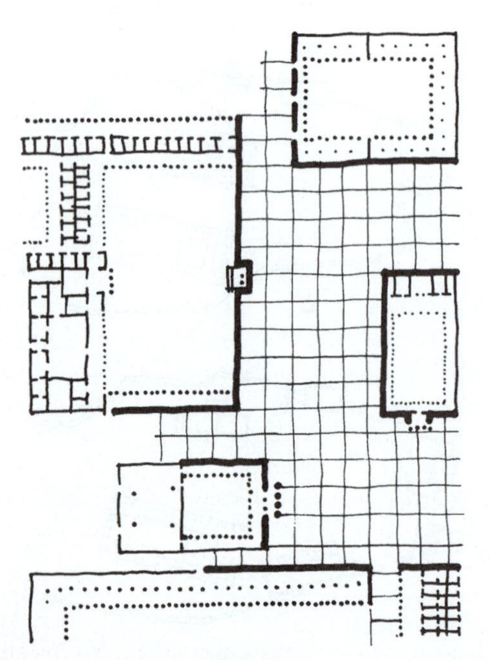

FIGURE 1-11. The agora at Miletus.

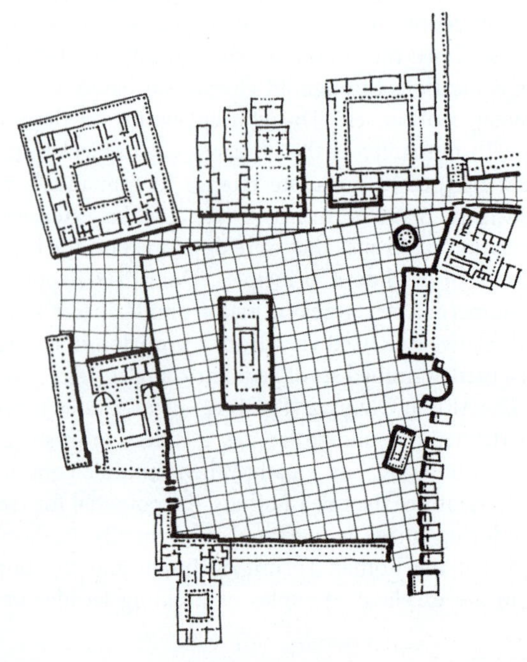

FIGURE 1-12. The Altis at Olympia.

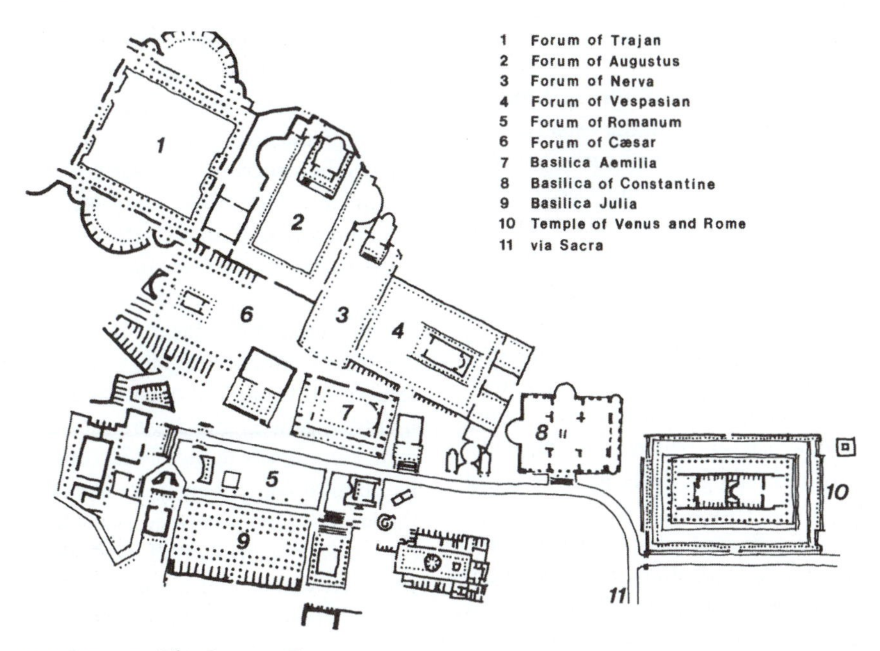

FIGURE 1-13. The Roman Forums.

give form and image to a place (Fig. 1-13). Whereas the Forum of Trajan evolved as an assembly of monuments and temples around a common open zone, the Imperial Forum was designed to enclose and define space with a building form. Both contain easily identifiable spaces which are visibly coherent and rich in variety.

Although neither the Acropolis nor many agoras were developed with an orthogonal grid, the concept had been in use in urban societies of the Mediterranean and the Middle East from the third millennium. The fact that the grid had been the organizational structure for Greek cities with high topographical relief implies that the decision to employ the system had little to do with a site's natural quality but relied more on some grand-design decision. The city of Priene was developed with an orthogonal grid on topography so steep that many streets took the form of stairs. Others have attested to the Greeks' concern for topography in site planning, and their preoccupation with creating physical order in a natural context is clear. This is consistent with the decisions of various cultures throughout history which relied on the geometry of the grid to create order in an unstructured environment.

The proliferation of the grid street system has been attributed to the Greek planner Hippodamus. His work is visible in numerous Greek colonial towns, which have qualities similar to that of his home, Miletus (Fig. 1-14). Although the orthogonal system was not a new concept, it was continually under refinement. Dinocrates, the town planner for Alexander the Great, universalized the grid pattern plan as he sought to give a common physical order and/or form to the conqueror's world. But in spite of the proliferation of the orthogonal plan popularized under the reign of Alexander, the form survived the test of time only in those societies that saw a value in the geometry beyond its functional capabilities.

By the second century before Christ, war and corruption had seriously affected Greece's former Golden Age, and

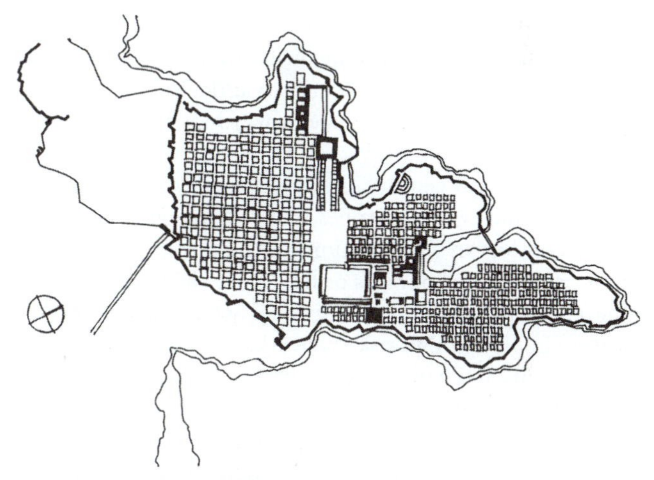

FIGURE 1-14. Plan of Miletus.

Greece was little more than another Roman province. Unlike Greece, Imperial Rome was never a democracy and the previous Greek attitudes toward the finite were never embraced by the Romans. The philosophy of the Roman Empire may best be expressed as "colonial expansionist," with Rome at the center of the universe. The grid was immensely successful, as it was simple to establish and easy to expand. It was easy to manipulate the size of the township, and the regularity of the grid facilitated control and access. Numerous Roman colonial cities established throughout Italy and Europe were based on the orthogonal grid. Long after the collapse of the political empire, various cities, including Paris, London, Vienna, and Cologne, flourished on the sites of former Roman cities and included remnants of the original grid plans.[30]

[30]Leonardo Benevolo, *History of the City* (Cambridge, Mass.: The MIT Press, 1980), p. 221.

Order in the Wilderness

The grid became a common organizational system throughout Europe and was particularly useful in the expansion of colonial settlements. Although the grid had variations that reflected the nuances of each culture, the model successfully accomplished two objectives: It provided a clear organizational structure, as it had in the past, for the development of settlements, and it facilitated the subdivision of land.

For tenant farmers in Europe, where land was a scarce resource, cooperation and community were physical and economic necessities. The word "community" was most accurately defined in the concept of *landschaft*. According to John Stilgoe, this word, which is German in origin, traditionally implied a cluster of dwellings, fields, meadows, and homes for about 300 people. This was a collective enterprise that objectified a self-sufficient agricultural community of tradition, purpose, order, and strength.[31] More than the notion of land as an area or property, *landschaft* was thought of as a space defined and ordered by humanity. Similar to other concepts, *landschaft* implied a common bond or obligation between men and the land. When resources were limited, cooperation and community were paramount concerns to individuals. *Landschaft* was apparently a concept applied primarily to agricultural communities, as the building-dominated townships related little to either land or community.

Seventeenth-century European society was extremely stratified into classes, families, and social orders. There was some comfort in the traditions of the European class system, as everyone knew his or her "place" in society. But for the energetic, ambitious individual who sought a higher station in life through risk and work, the New World was an opportunity to change that predetermined place. Any immigrant, of whatever craft or profession, who sought a new life and identity knew the potential in colonial settlement. Although the European hierarchy focused on the community first and the individual second, the American model reversed that relationship and codified the concept in the Bill of Rights. The significance of the relationship between property and status was not a phenomenon of colonialism in Europe. Historical precedents favoring those who own property date to the Roman Senate, which was "restricted by the *Lex Claudia* of B.C. 218 to Land owners."[32] The ownership of property in the new world meant social, economic, and political change, all of which were eagerly sought by indentured servants and free men alike.

Settlement of the New World occurred at a time of technological innovation and change in agricultural methods. This change shifted the role of the farm from that of a self-supporting family enterprise to that of a market-oriented establishment.[33] In addition to the change in economic conditions, there was a change in the "value" of community. Many settlers had misgivings about any collective controls that resembled Old World methods. Although colonial governments had policies for land assignments that supported the structure of village or township, few took shape.

In the New World, land was plentiful. The ethical and socioeconomic structure that was the foundation of the *landschaft* in Europe was not visible in America. But in spite of the disappearance of common pastures and collective maintenance, the concept of *landschaft* did not disappear overnight. Until the middle of the eighteenth century, recollections of its physical quality were the scale by which settlements in New England were measured. As attention shifted from community to individual concerns, the social fabric and its physical component were also undergoing change. Since the larger, market-oriented farms did not lend themselves to the spatial organization of the *landschaft*, the need for a surrogate structure emerged in the form of neighborhoods.[34]

Whereas rural communities struggled with social, economic, and physical transitions, urban townships were somewhat more secure, with their own identity. The prevailing text on city planning was "The Ordering of Towns," published in 1683.[35] Among the various recommendations of this broadly read manuscript was the postulate that an ideal town should be 6 miles square. Although there may have been disagreement on the site of a town, various shapes, particularly the orthogonal form, were generally accepted and followed. The grid plan enjoyed widespread use in colonial America. The organizational structure was well known and implied that the community had established a preconceived plan and order for its growth. The system was obviously an intellectual concept and as powerful a statement of humanity and order as could be made in an environment of nature and disorder. Unlike the ancients, who relied on conventions or myths to separate secular from sacred or provide an order for a site or settlement, the colonists were intellectually well prepared. With a few notable exceptions (Annapolis, Boston, and Washington), the grid was the prevailing planning strategy. Visiting Europeans often commented on the extensive use of the grid in American cities. An Englishman, Francis Baily, noted on his arrival in America:

> That perfect regularity in which [Philadelphia] is built, is not to be approved by some; but it is what I most admire; indeed it accords so much with the ideas of the Americans in general that it is a practice which is almost universally adopted in laying out their new towns and improving their old.[36]

The grid as a matrix provided numerous variations relative to the siting of municipal buildings and public parks, and in view of its widespread success, it became the conceptual model of the unexplored territories, although in his

[31]Stilgoe, pp. 11–13.

[32]William Alexander McClung, "The Mediating Structure of the Small Town," *Journal of Architectural Education*, Spring 1985, p. 4.

[33]Stilgoe, pp. 80–81.

[34]Ibid., p. 82.

[35]Ibid., p. 99.

[36]John W. Reps, *Cities of the American West: A History of Frontier Urban Planning* (Princeton, N.J.: Princeton University Press, 1979), p. 3.

journal, Francis Baily would later recant on his enthusiasm for the grid:

> I have taken on occasion to express my approbation of the American mode of laying out their new towns . . . but I think that often times it is sacrifice of beauty to prejudice, particularly when they persevere in making all their streets cross each other at right angles without regard to the situation of the ground.[37]

The fact is that throughout history the grid was the plan form used by various civilizations when the objectives were to establish colonial townships or to control empires. The orthogonal plan was simple, easy to comprehend, embraced a rational (intellectual) quality, and expedited the survey and allocation of land parcels. Although the grid was a widely accepted form for many urban areas, expanding the concept to include territories larger than most European nations had some dubious implications. But the Continental Congress of the 1780s found that the frontier was as insecure as the nation's treasury. In an effort to resolve both problems, Congress began to consider liquidating its extensive land holdings west of the Appalachians. The idea was to generate income from the sale of the land while encouraging settlement and stabilization of the western frontier.

It was one task to develop a set of policies for the political structure of a democratic society but quite another to formulate strategies for the physical development of that society in an uncharted land. Accepting the fact that formulation of the Declaration of Independence and Constitution was no small achievement, the articulation of a mechanism for the actual settlement of the frontier was also a major challenge.

[37]Ibid., p. 13.

The strategy that was ultimately adopted by Congress was based, in part, on two developments. The first was a collection of surveying constants promulgated by an English inventor/surveyor, Edmund Gunter. Gunter developed a surveying "chain" of 100 links with a total length of 66 feet. The chain's value was in the multiple uses of its lengths. Eighty lengths of Gunter's chain on either side of a square enclosed a square mile, which had an area of exactly 640 acres (Fig. 1-15). The second development was the publication "The Ordering of Towns," mentioned earlier. These two marginally related events provided the regimen for the 1785 law to survey the territories west of the Appalachians.[38]

In spite of Thomas Jefferson's presence as chairman of the committee responsible for the development of a survey strategy, the Congress did not totally embrace the concepts developed by his committee. It did, however, seize on two important recommendations. One was the development of a survey that would be comprehensive and lock the uncharted territories into an orthogonal grid. Second, the orientation of the grid was to be based on lines of longitude and latitude. The legislation mandated townships 6 miles square, with subdivisions of 36 parcels of 640 acres throughout the territories. This was in complete accord with Jefferson, who visualized a democratic society as a nation of free men on individual farms with a minimal number of cities.

The implementation of a grid that was locked into a north–south longitude (meridian) was, however, doomed to failure. Congress did not have the presence of mind necessary to understand that there would be substantial errors in a format which required that a survey be based on true north–south alignments that converged at the poles. In spite

[38]Stilgoe, p. 100.

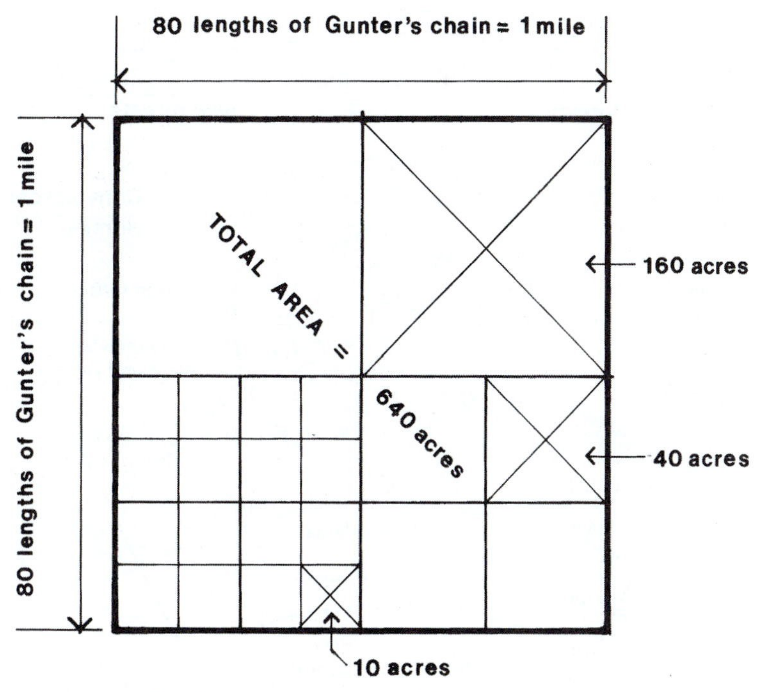

FIGURE 1-15. Depicting area measure of Gunter's chain.

of vocal criticism of the methodology, Congress in its "wisdom" prevailed. Although that august body would spend the next three congressional sessions creating legislation to correct the flaws inherent in the original act, the conceptual power of the grid was very pervasive. Although urban in origin, the indelible field of the grid was transformed into fence, plowed field, and road. Although the philosophical base of the idea had nothing to do with the environment, there is nothing that compares with the impact of the orthogonal grid on the American landscape.

RECONCILING VISION WITH REALITY

The fact that the grid worked only part of the time did not appear to matter. Like other dogma, we used it where it met little resistance and modified it to accommodate difficult conditions. Even though the construction of roads was relegated to alignments that were on the grid and not on the diagonals, people struggled through. It was almost as though we did not control the system. The myth of the solution endured despite its incredible flaws.

The power of an image or idea can sometimes take on a larger-than-life or mythical state and dupe otherwise rational people into trading a perfectly good cow for a handful of magic beans. History is resplendent with examples wherein the more preposterous the proposal, the broader the appeal.

Revolutions are often blinded by the light of change to the implications of their solutions, and the modern movement of the 1920s was no exception. The issues that gave rise to the revolution in art, architecture, and city planning in the second decade of the twentieth century had various roots. The "City Beautiful" movement, which began in the 1890s, had provided the initial impetus for a credible change in city planning in the United States. This had been followed closely by the "Garden Cities" movement conceived by an Englishman, Ebenezer Howard. Howard, like other social reformers of the era, believed the city to be the antithesis of everything humane and worked to develop alternative strategies to resolve various English social problems. Howard's solution was a series of low-density, satellite new towns in the rural hinterlands of London (Fig. 1-16). Lewis Mumford, Sir Patrick Geddes, and Clarence Stein were all anti-city, and all actively supported the development of these new communities in Europe and the United States, with varying degrees of success.

The acceptance of the Garden Cities concept by architects, landscape architects, and city planners on both sides of the Atlantic provided an intellectual setting for Le Corbusier's "Radiant City." Unlike the low-density settlements of Howard and Stein, Le Corbusier's vision was of a city of high-rise towers set in a tablet-like park. Le Corbusier explained:

> Suppose we are entering the city by way of the Great Park. Our fast car takes the special elevated motor track between the majestic skyscrapers: as we approach nearer, there is seen the repetition against the sky of the twenty-four skyscrapers; to our left and right on the outskirts of each particular area are the administrative buildings, and enclosing the space are the museums and university buildings. The whole city is a Park [Fig. 1-17].[39]

Although the plan of the Radiant City was physical, it was also an attempt to structure a social utopia based on

[39]Jane Jacobs, *The Death and Life of Great American Cities* (New York: Vintage Books, 1961), p. 21.

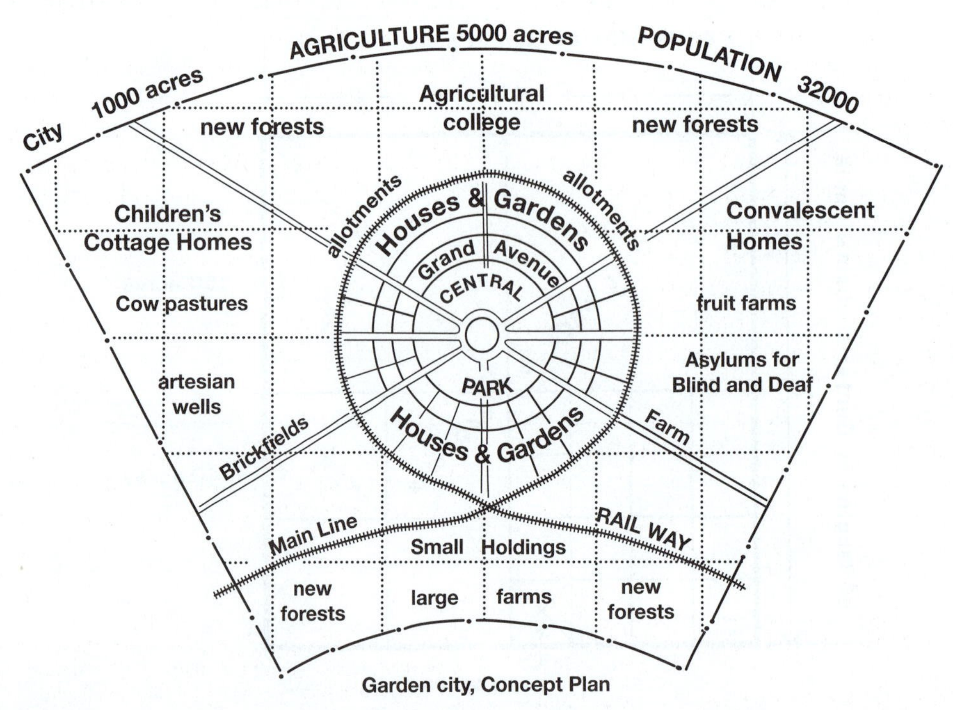

FIGURE 1-16. Garden Cities concept plan.

FIGURE 1-17. Radiant City.

what he called "maximum individual liberty." Although Le Corbusier accepted the concept of Garden Cities, he maintained that "the solution was to be found in the *vertical* Garden City."[40]

The city as a physical and social problem was first on the agenda of the Congress Internationaux d'Architecture Moderne (CIAM). Le Corbusier had been instrumental in organizing this prestigious group of architects and city planners, who in 1931 pledged their skills to solving urban problems. Le Corbusier had spent years giving shape to his invention. Although part of it was due to the success of the Garden City, part must also be attributed to the physical magnitude and simplicity of the proposal. The visual and aesthetic clarity of the site planning of Greece was to become visible qualities of his large-scale design proposals. He attributed importance to the role of the "shaper of the city," which came to be an implicit part of his work in South America and Algiers. Le Corbusier held the opinion that the architect would give shape, order, and stature to the landscape with architectural achievement. In these projects Le Corbusier approached the plan as though "the ground is clear, even if it is occupied."[41] This conceptual bias was consistent with earlier comments in which he expressed the need of clear and level ground to receive a building; the "aesthetic of a carpet." This absence of any natural qualities provided an unencumbered canvas for art. To the revolutionary, the meaning of culture, land, environment, and society mattered little. To Le Corbusier the city was similar to what A. Choisy saw at the Acropolis: "a site which fire had cleared."[42]

Redevelopment proposals for older European cities were constantly thwarted when the planner was obligated to address the problems of small individual property owners. Unlike the Acropolis, the city was a patchwork of small land parcels which created a serious impediment to large-scale planning proposals. With funding from various French industrialists, Le Corbusier worked to resolve the problems of urban Paris. When the need to resolve the problems associated with individual properties distorted the plan, Le Corbusier simply suggested dissolving the concept of individual properties. In defense of his 1930 plan for that city, he concluded: "The unification of the land, a true revolution in the sacred concept of property . . . we must demonstrate that there can be no modern city planning without the unification of the land."[43]

The concept of the Radiant City, the Voisin plan for Paris, and Le Plan de Paris enamored architects and city planners alike. Although supporters of the Garden Cities movement were appalled at the densities suggested by Le Corbusier, the simplicity with which the visionary resolved the problems of low-income housing, traffic, and office, commercial, and retail space was greeted with genuine enthusiasm. He brought order to the chaos of the city.

On the other hand, Jane Jacobs, author of *The Death and Life of Great American Cities,* suggests that there is nothing undesirable in something that appears to be visually chaotic.[44] There is nothing fundamentally wrong with a city because it cannot be interpreted by a simple key. She contends that the nature of the city is complexity, that the structure of the city is a mixture of uses, and that understanding

[40]Ibid., p. 22.

[41]Francois de Pierrefeu and Le Corbusier, *La Maison des Hommes* (Paris: n.p. 1942), p. 73. Cited from "The Terrain of Architecture," *36 Lotus International,* 1982.

[42]A. Choisy, *Histoire de L'Architecture,* Tome I, p. 238. Cited from "The Terrain of Architecture," *36 Lotus International,* 1982.

[43]Le Corbusier, "Rapport sur le parcellement du sol des villes," *L'Architecture Vivante,* Spring–Summer 1931, pp. 21–22. Cited from "The Terrain of Architecture," *36 Lotus International,* 1982.

[44]Jacobs, pp. 223–226.

the city as a total organism requires understanding its smaller parts first.

What we see and understand better now is that in the rush to rebuild and resolve new social challenges, the modern movement discounted everything that gave meaning to the culture and natural aesthetics of landscape. Working from the premise that the industrial revolution could provide the technology to resolve the housing problems of the world, variations in property, community, context, topography, vegetation, and geomorphology restricted the development of solutions appropriate for mass production and consumption. Since the machine was the solution, the problem had to be scaled to accommodate the limitations of the machine. This limited view of problems and solutions is not unusual, as revolutions in general are often skewed by polemics and segmented views of reality.

The postwar audience of two continents sifted through the precepts and proposals of the visionaries, then extracted and applied the physical components where the need seemed to fit. Although the social rationale for the movement was critical to developing a supporting constituency in sufficient numbers to make the products credible, some simply accepted and applied the form and image products.

The redevelopment policies and programs of the postwar U.S. government were not incongruous with the physical concepts promulgated by the international school of the modern movement. Housing and slum clearance programs were features of the 1949 Housing Act, which promised "a decent and safe living environment for every American family." The theory was to turn the wartime U.S. industrial and technological force into a peacetime power that would transform cities and suburbs into utopian communities. The will of the people, expressed through its Congress, created a cornucopia of development programs that focused on housing and urban renewal. Subdivisions, inner-city neighborhoods, and sites were developed by legions of entrepreneurs who leveled hills and filled valleys to make way for millions of acres of imageless housing. The grid provided the guidelines for the front- and side-yard setbacks. Each house repeated the same interval, prescribed by the geometry of the streets. All the site planning, as such, was done in the context of the design of the streets and the establishment of the subdivision regulations and zoning ordinances.

It was not that Americans lacked the technology to approach the problem with more sensitivity; it was almost as though we did not have the time. It was easier to plan and redevelop the cities with the "carpet aesthetic" and the structure of the grid. The grid provided the order, the revolution, the new architecture, and the planning. There were few who challenged the despoilment of the landscape. It was simply viewed as an unfortunate, albeit acceptable cost of "progress." And, of course, the myth endures.

The indelible nature of the grid and the simplistic ideas of early twentieth-century modernists left a legacy of sites and communities devoid of social life and natural landscape. Although the intent of all was a built environment of "better" quality, "better" is a subjective value judgment. Some contend that "better" is not necessarily neat and orderly. On occasions, "better" might have qualities of disorder or a wider variety, where the site and structure become a backdrop for the daily activities of people. One of the many common oversights associated with site planning is accepting the way in which people will, or will not, use a site, regardless of the site planner's intent. Where the visionary is oblivious to the "warts and all" of society, the experienced site planner understands and accepts the varieties of preferences expressed by people, culture, and history on a site. Jane Jacobs's message to planners was: "A city cannot be a work of art."[45] The search for visual order has very definite limits. Nature has an order of its own. Plant communities, soils, topography, and microclimate are all part of an interrelated system. Although these components may appear to be without visual order, it does not mean that there is an absence of order. Making the site plan a work of art means understanding the limits of those things that should not be changed and having the creativity to bring those components together into a unified whole. Although art is of value, and important in making something of life and reality, it is not a substitute for either.

[45]Ibid., p. 372.

CHAPTER TWO

ENVIRONMENTAL ETHICS: A VALUE SYSTEM IN CHANGE

The fact that a site must accommodate change is, in most instances, an acceptable condition. How that change will take place and under what circumstances is in large part influenced by the philosophical attitudes of society. Since the values of society find a way of materializing themselves into laws and ordinances, or the lack thereof, the impact of a society's value system cannot be casually dismissed. Although the foundation of a culture's values may have various roots, probably one of the most profound is religion. The guidelines for acceptable behavior in society are often corollary with those believed to be important theocracy. Remember Prohibition?

Recall from Chapter 1 the influences of early religions on changes to the natural landscape. The dominant religious perspectives in America today, however, are generally moderate in their view relative to the environment. Accepting the fact that various members of the clergy may take a vocal position concerning a specific issue of environmental quality, a collective religious consensus on the environment appears to be some time away.

Although religions exert potent and enduring pressure on the ethics of society, they realistically share that influence with the social, economic, and political character of a culture. It appears that when all four forces can be brought together into a single movement, a society's ethics are more articulate and visible. Such was the case in the colonial era in this country.

THE JUDEO-CHRISTIAN ETHIC IN THE NEW WORLD

Land ownership in Europe was rigorously controlled by the aristocracy. The prevalent arrangement was one in which the common people leased land and, as tenants, worked to produce an income that was shared with the landlord. People's lives were tied to the land socially, economically, and culturally. The parcel of property they were permitted to till was often allocated year after year to the same family, yet no one that worked a "strip," as it was called, fenced that property, as there was a common strength in the collection of farms. The order of the cultivated fields in the wilderness provided an image of collective strength and reinforced the social compact among individuals, community, and nobleman.[1]

A permanent house or structure was an equally valuable commodity. Although the cotter or laborer had a crude shelter, it was temporary and was considered a dwelling, not a house. A house was different not only in its more permanent quality, but also because it was one of the components in what the English considered a "stead." John Stilgoe writes in *Common Landscape of America* that the "stead" held a meaning which integrated the building and the land, recalled today in our concepts of homestead and farmstead.[2] So important were land, homestead, and their meaning in the community that punishment of a serious crime such as heresy would result in the destruction of one's home and the plowing of the yard. The stead provided and ensured a family of status and position in the community for which there was no substitute.

Immigration to the New World provided an opportunity for land ownership that was, for most people, unattainable in Europe. However, many found the cost of their land to be terribly severe. Of the few true wildernesses remaining in Europe at the end of the fifteenth century, none compared to those of North America. The New World was almost incomprehensible. The dimensions of the land, savages, and climate comprised a very hostile set of conditions. The fact that the survival rate of the Jamestown colony was only 20% provides grim documentation of the impending hostility of the New World.

To the settlers the dream of economic independence in America far outweighed the cost in hardships. To the mass of European and Asian immigrants to follow, land and property meant the right to vote and, as such, gave one a role in controlling his destiny. The relationship between land ownership and voting privilege was a part of the ideology of the architects of the new republic. Compared to the political structure of the colonies, a democratic government did not exist in most European countries in the seventeenth century. Despite later reforms, which permitted voting rights by those other than the propertied class, the concept of equating the ownership of land with the right to vote died hard. As recently as the early twentieth century, when women's suffrage was a topic of national debate, the city of New York permitted women to vote only when they owned property and only when taxes were an issue.

If the fruits of economic and social gain were not enough incentive, the permissive attitudes of colonial governments

[1]John R. Stilgoe, *Common Landscape of America, 1580–1845* (New Haven, Conn.: Yale University Press, 1982), p. 13.

[2]Ibid., p. 14.

associated with religion were very desirable. Pocahontas's spouse, Captain John Smith, suggested to his followers that settlement of the New World was a holy war. One only had to recall God's commandment to Adam and Eve, in Genesis 1:28:

> And God blessed them, and God said unto them, be fruitful and multiply and replenish the earth and subdue it and have dominion over the fish of the sea and over the fowl of the air, and over every living thing that moveth upon the earth.[3]

Captain Smith and many who followed visualized their efforts to be in complete accord with those of Jehovah. They maintained an anthropocentric view which held that *humanity* was the central aim of creation and, as such, enjoyed the right of dominion over the earth. The pervasive philosophy was one of dominance over all things, with the caveat that the results would be for the glory of God. This anthropocentric view was Judeo-Christian in origin and obviously had little respect or tolerance for other religious perspectives. It was sympathetic to the political and economic goals of New World settlement and, as such, provided religious reinforcement for other motives.

The interpretation of Genesis 1:28 has been a topic of debate by religious scholars for some time. In Hebrew, the key word in the text is *khibshuah,* which translates as "fill the earth and subdue it." Other references in the Old Testament using the word are very harsh, as it is based on the root "to trample, tread underfoot, to force humble, ravish."[4] The Greek translation is not as caustic, and where it is used elsewhere in the Old Testament expresses simple domination.

Josephus, a Jewish historian writing around the first century A.D., stated that God commanded human beings "to take care of the plants."[5] An ancient commentary on this passage of the Old Testament claims the following: "If he merits it, he is to have dominion, if not he descends to the lower level with other animals."[6] Josephus's text implies that the dominion over creation depends on the responsibility that human beings display in exercising that dominion.

In this light, the concept of covenant between man and God maintains man as the agent of the creator. As Robert Gordis, a modern Jewish scholar, has put it: "The Hebrew and Jewish interpreters would prohibit exploitation." Judaism goes much further and insists that man has the obligation not only to conserve, but also to enhance, because man is the "co-partner of God in the creation."[7]

The interpretation of the same scripture by St. Thomas Aquinas rendered a similar picture. The largely ignored work of the thirteenth-century Italian priest and scholar suggests a more pluralistic attitude. Thomas Aquinas, an outstanding theologian of the Catholic Church, believed that the Biblical text implied that man's obligation to the earth was as a steward, with God at the center of the universe. This stewardship role implied further that man was a link in the chain of the environment, and as the most "intelligent," the role of dominion was more that of "keeper" with an obligation of preservation. Whether Thomas Aquinas was influenced by Josephus is open to speculation, but the similarity of their conclusions suggests that the literal interpretation of "dominion" had disputed theological support.

John Stilgoe is of the opinion that "Christianity destroyed the ancient oneness of man and nature."[8] To early Christians, the wilderness was pagan and the place of idolaters. Even the word "wilderness" has as its roots "the nest of the wild [beast]" and was considered to be beyond the realm of man's control.[9] Christianity, as a religion, was objectified in settled land, not woods. The farmhouse in a setting of tilled soil and flourishing pasture was all a part of man's abstract concept of order, harmony, and peace.

John Stilgoe was not by himself in his views regarding the impact of Christianity on an emerging nation. Ian McHarg's unambiguous view was that the Judeo-Christian ethic gave religious credence to the Western man's sense of dominion over nature. While his book *Design with Nature* was one of the most significant books on environmental consciousness in city and regional planning, some hold the view that the polemic lacked a complete review of dominion in Christian stewardship. To be fair, McHarg and Stilgoe should not be singled out as the only authors to promote the Christianity-as-culprit theme. Nancy Watkins Denig[10] notes that Lynn White, Jr.'s *The Historical Roots of Our Ecological Crisis,* Science CLV, 1967, and Arnold Toynbee's *The Genesis of Pollution* are both examples supporting the view that the scriptures drew a straight line between a human-centered universe and the environmental ethics of our founding fathers.

But how can one accept an explanation for the taming of the New World and the Judeo-Christian ethic as the theoretical justification when there is evidence citing ancient Egyptian agricultural practices and post-Sung and Chinese deforestations as examples of humanity's careless intervention into nature thousands of years before the Bible? Doesn't the fact that cultures through the ages conjured theoretical rationale for actions contrary to their best interest create a disconnect between the Judeo-Christian ethic and where we are today? Some suggest that while there were cultures that understood the link between their collective well-being and that of the environment, others were focused on simply survival. Although scholars will

[3] *Holy Bible,* King James Version (New York: Abradale Press, 1969), p. 2.

[4] A. B. Davidson, *Analytical Hebrew and Chaldee Lexicon of the Old Testament* (MacDill A.F.B. Fla.: MacDonald Publishing Company), p. 369.

[5] Josephus, quoted in *Antiquities,* Vol. I (Grand Rapids, Mich.: Baker Book House, 1974), pp. 1–3.

[6] Midrash, *Genesis Rabbah,* Vol. VIII (New York: Bloch Publishing Company, Inc.), p. 12.

[7] W. Gunther Plaut and Bernard J. Bamberger, *The Torah; A Modern Commentary* (New York: Union of American Hebrew Congregations, 1981), p. 25.

[8] Stilgoe, p. 8.

[9] *Oxford English Dictionary,* Vol. II, pp. 3778–3779; Kersey, *A New English Dictionary,* under "Wilderness"; W. G. Hoskins, *The Making of the English Landscape* (London: Hodder and Stoughton, 1973), pp. 95–103; Yi-Fu Tuan, *Landscapes of Fear* (St. Paul, Minn.: University of Minnesota Press, 1980), pp. 73–86.

[10] "'On Values' Revisited: A Judeo-Christian Theology of Man and Nature," *Landscape Journal* Vol 4, No. 2,1985 0277-2426/85/00002-096 copyright 1985 by the Board of Regents of the University of Wisconsin System.

re-review both sides of the question of survival versus religion and parse a variety of languages for the most accurate definition of *dominion*, it is very likely that our forefathers did neither.

The attitudes of dominance and concepts of ownership were unknown to many of the Native Americans, to whom the earth and environment were an extension of the family unit. The land was not owned; it belonged to everyone. This was reinforced by the religious belief that deceased family members were reincarnated as parts of the landscape—"my brother the tree, my father the pond." The absence of a land ownership concept by many Native American Tribes was another stroke of luck for the colonists.

Armed with the commandment of God in a land of incredibly abundant natural resources, they embarked on an enterprise to subdue an entire continent and, in the process, bring Christianity and civilization to its inhabitants. To support the exploitation process, the few regulations that did exist in the New World became fewer as one moved toward the frontier. This collection of circumstances propelled the frontier farther to the West, but more important, it crystallized the national land ethic: growth. Growth materialized the essence of America as a land in which an ambitious person could find economic independence, with minimal governmental interference, through the development of the land and its resources. This perception of the Americas, which began with the landing of Sir Walter Raleigh in 1584, is visible today in the massive urban/rural migration fueled by regional economic change.

Manifest Destiny

The extremely mobile nineteenth-century population found new frontiers regardless of whether they followed the timber industry from Illinois to Oregon, the cattle drives from Texas to Kansas, or gold and silver from Colorado to California. This mobility encouraged an exploitive view of the environment and galvanized the concept of land as a commodity. The successful entrepreneurs who converted natural resources into economic fortunes became models for others to follow. In light of the immense hardships encountered by the colonists, the ability to successfully convert an adversary environment into one of financial gain only improved the balance sheet. In the process of forging a democracy with a willingness to accept all comers, progress became a creed.

The birth of the nation spawned an expansionist philosophy that enjoined the imagination and aspirations of millions. Economic progress and the industrial revolution meant jobs and increased wealth for the entire population. The growth achievements and opportunities of the young nation could be measured by the multitudes of emigrants who left families and homelands to begin again. The risk of starting over in America was small considering the tales of success, wealth, status, and progress carried by those who survived the hardships.

The vision of progress was the engine of the future. But visions, like myths, not only ignore and shed the flaws of reality but survive to create utopian images. Currier & Ives engravings of colonial America portray a bucolic setting extrapolating the picturesque from the genuine and, in the minds of many, is still that of "New England" today. We forget that the working conditions of the urban poor at the turn of the century were essentially voluntary slavery. Separating myth from reality poses a problem for a society, as it involves the examination of larger-than-life personalities or events in real, accountable terms. What exacerbates the problem is that it requires scrutiny of society's own values in the process.

Others have observed that we keep two sets of books: one myth, one reality. This is a much more convenient arrangement, as it permits a society to present one set of ideals while living with another. Such was the situation when seventeenth-century references to the settlement of the New World as a "Holy War" were reincarnated in the policy of "Manifest Destiny." In an editorial supporting the annexation of Texas in 1845, journalist and diplomat John Louis O'Sullivan suggested the tenet that "territorial expansion of the nation is not only inevitable but divinely ordained."[11] In his detailed account of the Native Americans' struggle with the U.S. government, *Bury My Heart at Wounded Knee,* Dee Brown suggests that Manifest Destiny "lifted land hunger to a lofty plane." To the Native American, the policy implied that the white man was the "dominant race and, therefore, responsible for the Indians—along with their lands, forests, and mineral wealth."[12]

Rationalization of the numerous Native American treaty infractions was extensive and almost always based on the view that land and its inherent resources were commodities. Furthermore, without federal intervention and control, the expertise to manage the resources adequately would not be available. The policies and development philosophies of our political leaders of the early nineteenth century were heavily influenced by the British economist Adam Smith (1723–1790). Smith's publication *An Inquiry into the Nature and Causes of the Wealth of Nations* was an extensive examination of the distribution of industry and commerce in Europe. Smith's thesis was that a minimum of governmental interference promoted the best use (production) and distribution of capital. This concept of noninterference, or laissez-faire, was based on Smith's theory that each person, permitted to act in his or her own best interest, would create a condition that was in the best interest of all. Obviously, any meddling by government with free competition in the market would create negative effects.[13]

Even if we accept the concept that land is a commodity and essential to the long-term welfare of the nation as a whole, why do we not care for the resource better than we do? Knowing that land resources are finite, should we seriously allow the extraction of resources or the deposit of waste to occur without review, control, or comment by local, state, or federal government?

[11] *United States Magazine and Democratic Review,* July–August 1845.

[12] Dee Brown, *Bury My Heart at Wounded Knee* (New York: Holt, Rinehart and Winston, 1970), p. 8.

[13] *Funk and Wagnalls Encyclopedia,* Vol. 21 (New York: Funk & Wagnalls, Inc., 1976), p. 422.

The "land as commodity" philosophy maintains that whatever resource imbalance might exist can be corrected by the fundamental entrepreneurial spirit that built this country. The opposing opinion maintains that the land is a resource, not a commodity. This philosophical position suggests that land is not necessarily "owned" by anyone, as the land lives on while ownerships change over time. These two opposing positions are not always as polarized as presented here, but the basis for disagreement is operationally close.

The "resource" philosophy suggests that the solution to various environmental problems requires an examination of the limited resources of the planet and the interdependencies of one resource on another. This conservation ethic holds the view that we have squandered our resources for decades and it is time for another accounting. The political bias of the conservation ethic is diametrically opposed to that of Adam Smith and suggests that the laissez-faire market is essentially responsible for the massive water and air quality degradation which has resulted from industrial practices that have treated the environment as a public toilet. Well, how did we change?

By the first decade of the twenty-first century, almost everyone recognized that there was no free lunch; we knew we needed to just keep the costs to a minimum. We were willing to pay something, but please, hold the real pain. The petroleum industry was doing its part by making an effort to provide the "blue-plate special" in a barrel, and the public was willing to burn it all without a clue about the risks the provider was taking. Federal agencies tiptoed along a fine line between the legal extent of their authority and becoming a drag-chute in the eyes of the public. "Drill here–drill now" bumper stickers said it all, as the oil industry had its collective foot on the floor, in an effort to provide cheap energy at an acceptable risk.

But if ever there were an event that crystallized the arguments of ethics, vested interests, and political philosophies, the Deepwater Horizon debacle in the Gulf of Mexico in May 2010 was that event. The explosion of the Deepwater well head and failure of the blow-out preventer at 5,000 feet below sea level created an environmental disaster of monumental proportions. The ruptured well head spewed millions of gallons of crude daily into the Gulf, as the environmental, economic, and social costs of the cleanup of the estuaries and beaches of the Louisiana, Mississippi, and Florida coastlines grew exponentially.

Talking heads droned endlessly, embellishing how their particular perspective could have avoided the calamity or, conversely, justifying the efforts of the deep-water oil exploration. The Deepwater Horizon tragedy was a massive failure of a technically complex endeavor, in one of the most environmentally sensitive locations in North America. At one time the philosophy was that an unfettered, unregulated free market would create performance standards that would serve a larger society. However, the Deepwater Horizon disaster was evidence that industry was unable to police itself.

Sustainable Communities

It may be that the pursuit of "sustainable community standards" was well out ahead of the essential change in our ethics. There is the view that if 10 people were asked for a definition of *sustainable community*, you would get 20 different definitions. Although "sustainability" for our way of life was what the oil industry was trying to facilitate, how could that ever be possible under even the most ideal circumstances? How could we sustain what we have today as the planet's resources shrink and our population explodes?

On occasion, we behave as though we are bipolar. In our efforts to create a broader source of energy sources, we convert food (such as corn) to fuel for internal combustion engines, but we disregard the fact that there are areas of the globe that have suffered for decades from malnutrition. But in a casual audit of the books, how can we use the food as fuel and call the result a contribution to sustainability? There is the observation that if the United States were genuinely interested in sustainability, some effort would be made to reduce fuel consumption by reducing interstate highway speeds. This was a successful strategy in the oil crisis of the mid-1970s, when OPEC placed a moratorium on crude imports to the United States. Was the forced "rationing" a nuisance? Yes. But we suffered through it and reduced the number of highway deaths in the process. But as soon as OPEC changed its policies, we saw no benefit to our conservation and returned to the mind-set of 70 mph speed limits.

Sustainability for many Americans is closer to "freeze my lifestyle in time (that's the sustain part) but make it cheaper and available more quickly." What we fail to recognize is that real *sustainability* is inextricably linked to another *s* word: *sacrifice*. We shouldn't be too quick to judge; the efforts by many to successfully develop sustainable cities recognizes that although resources can be limited, *change* can be good.

It wasn't long ago that America came to realize that historic preservation is a valuable, viable community objective. The word *sustainability* was not a part of the popular lexicon in the early 1960s. But during that time, and gaining momentum in the following decades, we discovered that there was something genuinely important about those derelict, historic buildings. We found that those resources linked us to our heritage and history while helping to create lasting identity for our communities. Aside from a growing appreciation for the design of period architecture, we found that the structures and materials could be refurbished and, when required, become home to activities and uses with dramatically different purposes from those of the original structure. We discovered a pride in their restoration and celebrated the new openings. Historic preservation is now a tested, visible component of sustainable communities, but it was preceded by a change in our *ethical* standards as we changed our understanding and appreciation of their contribution to creating a sense of place and history. Developing sustainable communities, parallel to the successes of historic preservation, without a change in the ethics and vision, is an objective that neither rhetoric nor money can bridge.

We can take a lesson from the past: It has been suggested that the operative word that transports *sacrifice* to *sustainability* is *change.* Rationing during World War II was a federal edict that found broad public support. We were at war, and everything we could do to help the war effort was on the mind of every adult. There was a change in our ethics that did not require a law. Ethics stems from a sense of right and wrong. While we create small successes that serve as models of what could be, until there is a bedrock change in our ethical standards, it is unlikely that we can achieve an environmentally conscientious society capable of creating genuinely sustainable communities. We always look to the emergence of a leader, a discovery, or a major achievement to create a remedy to our condition. In the final analysis, it's more likely we'll find that Mahatma Gandhi's simple directive to his countrymen regarding their behavior has application to this age as well: "We must be the change we wish to see."

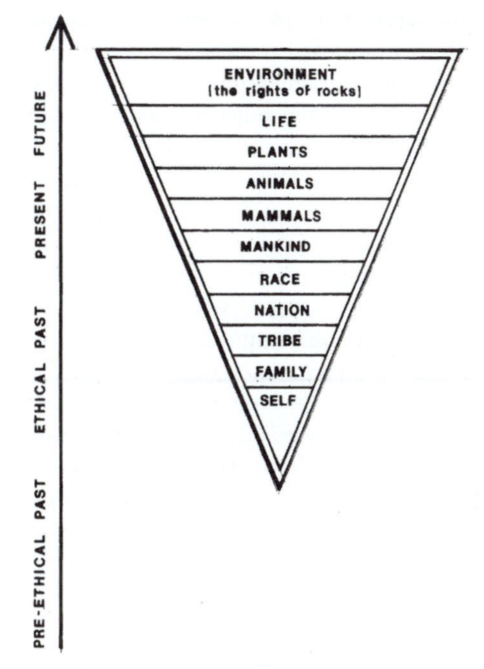

FIGURE 2-1. Diagram of an ethical relationship.

A QUESTION OF VALUES

The conflict between land as a resource and land as a commodity has been a topic of extensive debate and consternation. The model of this confrontation between the land and ethical standards is probably best illustrated in Garrett Hardin's treatise "The Tragedy of the Commons." Hardin's rejoiner to Adam Smith's theory was based on an accepted practice in fourteenth century England that permitted neighboring property owners the right to graze in the "wastes" or moors belonging to a manoral lord. These common property rights were often multiple in any one field, as both cultivation and grazing might occur simultaneously. This was a difficult situation at best as the quality of the commons was dependent upon a *quid-pro-quo* relationship between those who shared the rights to these fields. When population densities were low and the natural processes imposed certain limits on the number of livestock, the common pasture was easily capable of supporting the demands of both man and beast. Time and circumstances had worked to create and maintain a balance between the commons and livestock as a resource. Although each herder recognized the importance of the commons as a resource to all, each also decided that his own best interest would be better served by adding one animal to the commons. The rationale was that the economic value of one animal was well worth the "fractional" impact on the quality of the commons. Considering the fact that the costs would have been paid for by a marginal degradation to the commons and the rest of the community, every herdsman elected to increase his herd by one. The results: an overgrazed commons and poor herds. Livestock that could survive in a healthy environment at limited levels failed to produce in comparable quantity or quality. The net effect of uses that attempted to extend the holding capacity of the land beyond its realistic potential was the generation of serious long-term economic and environmental losses. This absence of a quid-pro-quo relationship with each person, as well as the land, produced predictable results.

Observations on this relationship that exists among man, environment, and an ethical standard have been made by many. In an essay by Roderick Nash on Aldo Leopold, Nash developed a model of the ethical relationships, illustrated in Fig. 2-1.[14] The model suggests that the first quid-pro-quo relationship existed when man adopted the attitude that someone else's life was more important than his or her own. When "survival of the fittest" was no longer the creed on which man chose to exist, a change in our ethics became plausible. Over time, this change has generally expanded from the individual to include family, tribe, and nation. Although some have adopted an attitude that some mammals (other than human beings) have "rights," extending the concept to include plants and rocks creates serious debate.

There have been proposals put forth which support the argument that the environment does have rights. Christopher Stone's *Should Trees Have Standing?* discusses the principles behind legal "standing," which he suggests exists when we establish national forests and parks. The Grand Canyon is a collection of rocks and is, furthermore, considered a national treasure. The decision as to the extent of those rights is based on values, priorities, and our ethics at the time.

The "man dominant" position implied in Genesis has had a definitive effect on the philosophy behind our government and legal system. That philosophy had been practiced in the United States for over 200 years without caveat or

[14]Roderick Nash, "Do Rocks Have Rights?" *Center Magazine*, November–December 1977, p. 6.

question—until recently. It may be time for a reevaluation of the canons which support that philosophy in view of the range and scale of environmental problems that confront the United States today. Even if the more moderate attitude of human-steward is wrong, how much worse could the impending results be?

The primary concepts associated with developing an understanding of man's relationship with the environment can best be summarized in the four basic laws of ecology formulated by the renowned environmentalist Barry Commoner.[15]

The first law is: "Everything is connected to everything else." The development and preparation of life itself is dependent on interrelated processes. Commoner's second law is: "Everything must go somewhere." This is really the basic law of physics—matter is indestructible. There is no "waste" in ecology or nature; nothing is discarded. Principal among the reasons for the environmental crisis today is that we have extracted extraordinary amounts of natural resources, converted them into other forms, and discharged the residuals without realizing that everything must go somewhere.

The third law is: "Nature knows best." This axiom is sometimes difficult for us to accept, for in many instances we look to technology to solve our problems. Even the concept of recycling is concomitant with the natural processes. One of the characteristics of living systems is that for every organic substance produced by a living organism, there exists somewhere in nature an enzyme capable of breaking down that substance. The essence of this is that no organic substance is synthesized without provision for its degradation. The fourth law is: "There is no such thing as a free lunch." In ecology, just as in economics, this law warns that every gain is won at some cost. The tenets of the three previous laws are incorporated in this precept. Because the global ecosystem is a connected whole in which nothing is gained or lost without an imperative for improvement, anything that is extracted by human effort must be replaced. Payment cannot be avoided; it can only be delayed.

[15]Barry Commoner, *The Closing Circle* (New York: Alfred A. Knopf, Inc., 1971), pp. 33–45.

CHAPTER THREE

NEW ETHICS, NEW LAW, AND NEW PROCESS

With an understanding of why a "man dominate" ethic was so common and widely accepted, it is easy to understand why it was so pervasive for so long. Although growth and change fueled the most envied economy in the world, both qualities were unsuspecting harbingers of costs that would be paid by other generations. While the economy enjoyed a flourishing growth, it was time for introspection. The 1960s was a time of wholesale inquiry into the ethics and values of government and industry. The sentiment of Congress was that something was wrong with the conduct of various agencies and that their actions had convoluted the trust of the nation's natural resources. Congress was aware of a 1954 Supreme Court decision, which expressed in unambiguous terms that there were valid issues, beyond monetary, that were valid qualities of every community. Based on the premise that the environment has an intrinsic value to society, Congress reinforced the Court's perspective and adopted the National Environmental Policy Act (NEPA) of 1969. As important as the policies that the Environmental Protection Agency (EPA) set forth was the establishment of a process that provided the opportunity for neighborhoods to challenge the decisions of local, state, and national agencies. Signs with the slogan *Power to the People* could be painted, used, and passed on for use by a variety of activist groups. America had indeed found its voice.

Although it is not imperative that every site planner understand the intricacies of NEPA, an overview of the law's intent and the process is worthwhile. Since the passage of the legislation, many state and local governments have enacted laws. The proliferation of additional legislation reflects the fact that the concept of environmental quality is not an abstract idea but rather an issue that individuals and governments can, and do, control. Thorough research and due diligence suggest a call to state and local governments to determine the existence of any relevant EPA requirements or procedures as a part of the inventory process.

NEPA PURPOSE AND POLICIES

NEPA is probably one of the most controversial and significant laws to be passed in recent history. For the first time in this act, Congress mandated an active federal role in the maintenance and control of the environmental quality of the nation. The legislation had several purposes:

1. Encourage productive and enjoyable harmony between humanity and the environment
2. Promote efforts to prevent or eliminate damage to the environment and biosphere and stimulate the health and welfare of man
3. Enrich the understanding of the ecological systems and natural resources important to the nation

The thrust of the NEPA was to engage every federal agency in a balanced decision-making process that considered the total public interest. This carried with it the obligation of each federal agency to critique, through an Environmental Impact Statement (EIS), its projects and to provide other agencies or private parties with that evaluation. The hope was that this would ensure a more continued collective effort on programs and projects throughout federal, state, and local entities. Prior to the enactment of NEPA, interagency conflicts were the standard bill-of-fare. The Federal Highway Administration was supporting the construction of highways in national parks, and the Army Corps of Engineers was financing projects that would eliminate millions of acres of farm and park land.

We know now that the effect of this legislation on the environmental design professions has been significant beyond the scope of the law itself. First, the legislation welded into prominence the concept that the physical environment was something of value to be considered in terms other than economic. Second, the issue of environmental quality has become something of importance from a national perspective, and it has also filtered down to town hall meetings and grade-school classrooms. The growth of awareness and consciousness of the environment as something that requires stewardship has been nothing short of remarkable.

Third, regardless of any other accomplishments, the legislation delineated a process that bound each federal agency to a role of inventory, analysis, and examination of alternative solutions. The process has proven to be valuable to the federal establishment in the scrutiny of feasible options, but it has also become a model for state and local governments. The review process and citizen participation in the law have also had a major impact on those federal agencies that had an inclination for translating a community's best interest into agency projects. One observation related to the scope and implications of NEPA is that it was the single most important change in land use regulation in the United States since zoning.

Looking back, it is easy to see that NEPA created a new plateau for the legal consideration of environmental issues—at least where federal government dollars are used. Many assume the positions taken today by the federal government and private citizens regarding environmental issues to be routine. However, the movement has taken four decades to evolve to the place where it is today. The fact that we have acknowledged that we hear an alarm bell regarding climate change represents a sea change in attitudes. Indeed, the acceptance of the new "green" building codes, among others, is remarkable. It is also sinking in that the future will only have more stringent rules, not more lax ones. We *may* have realized that the world is not only smaller today than yesterday but that our failure to recognize the difference between what we need and what we want invites a tide of repercussions.

PROGRESS OF A DIFFERENT COLOR

The phrase "going green" was so overused in 2008 that one national newspaper flagged "green" as a term that was used more for marketing products than as a shift to a pro-environment ethic. *Going* implies change. Without much fanfare, we may have witnessed the broad-based shift in the public's ethical standards necessary to support environmentally responsible development. But first things first. Change in a mind-set is fundamental to a change in behavior. What we need to know is what aspects require change—and in what ways. Generally speaking, society resists change. It is therefore a watershed event when society expresses support for any reform regarding development. It is even more significant when that support is reflected in laws and the process through which development is permitted.

If there is to be a change in our view of site planning, similar to what we have seen in the built environment, we must first change the vision regarding what we want to achieve. As an example, if the vision is a completed site plan that includes a manicured lawn over 70% of the site, leaving the remaining 30% to impervious surfaces, then it is not likely that the goal of a "green" site plan will be achieved. It would be difficult to eradicate 100% of the natural vegetation for a site and develop a plan that sets as its goal one that would retain as many resources as are plausible. The vision is inconsistent with the design criteria.

However, if the goal is to retain as much of the green that exists on the site as possible, then a change in the policies set a standard for clearing the site. The word *clearing* has some implications; many builders believe that "clearing the site" is both the goal and vision. If, however, the process called for "cautious site preparation," then careful removal of resources would more accurately reflect a policy of conservation. Changing what we consider to be the image of the end-state plan is one of the most powerful changes that can be made.

It is easy to see that if the vision for every residential site is a lawn, that image carries some significant implications. While we once planted grass to keep from walking in the mud when it rained, we now see grass planted in areas where we almost never walk—except to mow the lawn and fertilize the lawn and water the lawn and. . . . But if the vision for the site was to be "green," why would we not be willing to accept the existence of the scrub brush as components of a garden that we would supplement with other vegetation for their contribution to a garden? It's easy to see that the trees, large and small, are worthy contributors, so why would those other plants in the under story not be part of the composition of a green site plan? Can we save everything? Not likely, but we can develop a "green site preparation" plan that would focus on minimizing damage made during the construction process.

It has been suggested that the definition of *weed* is "an unwanted plant." If that standard can be applied to every site and the policy is to "go green," could that imply that there are few plants that would be considered weeds? When the need for site clearing is reduced, the use of motorized equipment, from bulldozers to chain saws, is reduced. Almost all vegetation that can grow and mature with minimal maintenance is worthy of being considered as part of a garden in a commercial activity. In addition, the use of irrigation can be reduced because vegetation that was retained matured with what precipitation fell on the site prior to the development of the site. Whereas residential land uses almost always have adequate care and attention, commercial uses can become victims of landscape plans that have relied on well-funded maintenance programs. We should consider looking at all the maintenance costs and extending those costs beyond the occupancy date when evaluating how "green" a plan should be.

Various regions of the United States today are experiencing longer periods of drought and are drilling deeper water wells at an alarming rate. Lakes and reservoirs are shrinking to show more of shallow edges and boats resting aground than ever before in our history. We are rapidly depleting what seemed to be inexhaustible. Periodically, we need to be reminded that those who cleared the immense forests of the Midwest simply migrated across the northern tier of states, clearing the land of all timber, until they arrived in Washington and Oregon, with intent to clear that as well. What we often fail to recognize is the need for conservation before a resource is depleted. Moving "green" site planning from unique to commonplace can be accelerated only if policies facilitating conservation become more commonplace.

A green site implies that the site is also cognizant of the value of water conservation and cost-effective site drainage. Decisions associated with water conservation begin with retaining water on a site and using the captured precipitation to be dispersed to support a green site. We need to avoid collecting precipitation into downspouts and drains and piping all onsite runoff into storm sewers. That process is not only a dated and costly public policy, but it does nothing to support either water conservation or green site planning. The physician's oath that begins "First, do no harm" has application to site planning. The premise of the oath is that no action be taken that would result in exacerbating a patient's illness. If that policy has currency when applied to site planning, then we must begin with well-documented inventory and proceed cautiously. Think of treating every change as though it must be done with a pick and shovel.

PART TWO

THE EVOLUTION AND IMPACT OF COMPUTERS ON THE PROFESSIONAL OFFICE

- **CHAPTER 4:** Until Recently
- **CHAPTER 5:** From Graphite to CPU
- **CHAPTER 6:** Organizing and Creating Drawings

CHAPTER FOUR

UNTIL RECENTLY

Depending on your level of education and profession, you may have little appreciation for the trek that design professionals have traversed over the past 20 to 30 years, with the advance of the digital age. You may have been caught in this transition. Or you may just be hanging up your cap and gown and wondering "what transition?" Because this book will find different people in various stages of their education, we assume that you have been watching the evolution unfold for a while and are just getting up to speed on the technology.

If you feel like you were born with a flash drive in your hand, do not skip over the chapters in Part II. Our predecessors who drew every line by hand on a site plan used a different thinking process. Today, importing topographical files from other time zones takes as much time as it used to take to sharpen a pencil. Understanding those laborious thought processes has value.

It is not our intention to create an application users' manual or computer system review. Rather, the objective here is to focus on issues that surround the computer technology as they relate to the professional office environment and experience.

We believe that the critical components to working with digital applications are found primarily in two areas: the design approach and organization. First, as best as possible, an office should marry the design process to the digital approach, setting standards related to software, hardware, networking, and filing. Second, organization of the digital information itself is crucial. Modern computer technology helps us organize site information, proposed concepts, construction details, and other technical information. When this information is organized into a logical and practical system, it supplies the designer with an array of tools.

To help maintain clarity in this text, we use the following standards for nomenclature of site-planning professionals. *Interns* are recent college graduates and/or practitioners-in-training. *Professionals* are usually licensed practitioners and often project managers. We use the term *principals* for the most seasoned practitioners and managers with firm ownership responsibilities.

A NEW UNDERSTANDING OF ROLES AND RESPONSIBILITIES IN PRACTICE

In many ways, resolving the issues that marry structure and site has not changed since humans considered something other than a cave as a permanent structure. Although the past 30 years have witnessed the most dramatic technological changes to date in how we convey, describe, and document our intent to others, our thinking about how these changes reflect the traditional design process has not kept pace. We must examine how the integration of technology into a traditionally hand-drawn medium either enhances or aggravates the overall design process. First, we will examine the implications of technology on the roles that professionals and interns perform.

It is fundamentally appropriate for a site planner to use a pencil and paper to record and visualize the design process. A sketch can represent the visual thinking of the site planner—as a picture of a thought. The more complex the problem, the more likely the need for additional pictures. The computer can extend our thinking, as it allows us to add or create additional layers of analysis and documentation. Both hand-drawn and computer sketching are valuable and should be exploited for their intrinsic contributions. The key is knowing the limitations of each.

When skillfully utilized, a computer can become an extension of a planner's intuition by enhancing his or her ability to organize, analyze, and visualize. However, as computer use in professional offices has increased, we have observed that a generational gap has developed. This breach has led in some situations to misunderstandings between those who are used to completing drawings by hand and those who are facile with computer-based drawing. Often, these misunderstandings are rooted in the unrealistic expectations of seasoned practitioners of the capabilities of computers and computer-aided design (CAD). In addition, interns may mistakenly believe that computers open doors to *design* solutions, to which only time and experience hold the key.

In the past, project leadership and problem solving were tasks commonly performed by one individual. As in the past, today we still find that the varied roles and responsibilities in the professional office and job market are yet to converge. Recent graduates may desire roles as designers, but due to their relative lack of experience, many are relegated to the role of CAD techs. Oftentimes, larger design issues are left in the hands of those with more design experience, though many of those professionals have only a casual understanding of the technology involved. They "stand on the shoulders" of those with the CAD skills and talents.

As colleges and universities introduce more new graduates into the realm of working professional offices, principals look with hope and optimism to the fresh infusion of ideas and idealism. But as seasoned professionals, they remember the difficulties they addressed in their first real design jobs. These principals might remember those tasks as tedious; however, the hours spent drafting, running prints, compiling project books and specifications, and doing a variety of other tasks were rewarding and educational. Universities and colleges continue to do their best to provide broad overviews of the wide variety of professional office practices in preparing their students for graduation and professional careers. Still, many students and interns harbor illusions and expectations of the professional office that fail to match the reality they encountered with their first job experiences.

At one time, a site planner needed to have the ability to be creative and express ideas with freehand drawings. Designers wished they had computers that could help them prepare the cumbersome base information at the start—in a better, faster, more accurate way. Today, principals wish interns could hand draw or sketch rather than prematurely jumping into the computer work.

Most employment listings over the past 10 years have included requirements such as "AutoCAD, Photoshop, Microsoft Office proficiency." Today, professionals are presumed to possess those skills, and the requirements often also include proficiency in 3D modeling and digital analysis modeling. It seems that the days of "good hand graphic skills required" are long past. Upon graduation, most interns do not possess high-quality graphic skills. Regardless of how the technology has come to dominate offices, it is common for interns who persevere in developing drawing skills to be given preferential treatment in hiring. A potential employer is likely to translate examples of an applicant's superior graphic skills to be equivalent to the skills of a designer.

It is important that young professionals prepare themselves in school with the tools they will use every day in the preparation of design drawings. Those include both hand-drawn graphics and the digital tools of CAD and illustration software packages. They also include business software generally found in suites, such as Microsoft Office, with Word, Excel, and PowerPoint.

A GLANCE IN THE REARVIEW MIRROR

To better assess the change from hand drawings to digital drawings, let's look back a few decades. This was no mere transition; it was a change the magnitude of which site planners had never experienced before. It was the equivalent of doors opening and closing, not some sort of ethereal morphing of events. Recall that the personal computer (PC) was popularized in the 1980s. Prior to the creative efforts of those who brought CAD into existence, all drawings had been hand drawn just as they had been for centuries. The only changes were in the drawing utensils and media. Leonardo da Vinci would have been comfortable with pens, pencils, and Mylar. The design professions were also comfortable with these media and materials. Reproduction was accomplished through a direct printing process at a relatively low cost. No layers could be inadvertently turned "off" when a copy was made. And what you saw was what you got; if it was not on the print, it did not exist.

When digital drafting hardware and AutoCAD software emerged in the mid-1980s, only a few could see the profound changes on the horizon. Many forward thinkers thought that digital drawing would finally and forever address the issues of accuracy, interdisciplinary coordination, and storage. It was also hoped that it would resolve the many other goblins of pen-and-ink drawings.

Although the costs of both hardware and software were extremely high, professional offices experienced pressure from the market to pay in order to remain competitive and practice on the cutting edge of the new technology. Of course, this took a toll on the profitability of many. Those with budget management responsibilities found themselves on the "bleeding edge." For professional practitioners, it was difficult to predict the future by examining the past.

As if all the confusing and complicated changes were not enough, academic institutions, which had historically provided educational leadership in technology, had few faculty with either academic education or practical experience in the digital environment. While accreditation boards agreed with the objectives of professional programs and subject material that would be taught, the specifics of the detailed curriculum to teach a particular topic were left to each individual institution. This permitted each institution to create curricula that emphasized its unique vision. There are therefore legitimate variations in accredited undergraduate and graduate degree programs with degrees in the same topic. Variations in curricula involving digital application were less rational.

Early introductory computer education in design schools was often hit-or-miss. Some institutions benefited, with plenty of funding for academics as well as the then-current hardware and software. Others proceeded cautiously and tended to be led less by academic professionals than by those with expertise in the technology, particularly programmers. Few could have predicted the variety of teaching methodologies that would be developed. Even fewer could have conceived the complexity of hardware and software solutions that progressed along parallel paths toward consensus.

Intellectual arguments were particularly splintered regarding the role of computers overall. The expertise of one faculty at one institution was not equal to that of another. It was not that there was disagreement per se; there was little to no effort to coalesce agreement on the subject. But if there is one characteristic common to all academic settings, it is *independent* thinking, for which we all share the burden and benefits.

And, of course, the wants of those supporting additional PCs and software soaked up every stray dollar that all other faculty members coveted to support their own area interests and research. Deans begged for more money to acquire more and newer hardware, software, and reproduction capabilities. It was hard to keep up. Every year, there was something new and different. There seemed to be a limitless number of hardware and software alternatives in which institutions could invest. The technology topic became a huge divisive wedge for many design faculties.

And if all the previous fractures were not enough, the differences between school and professional practice were another prism that scattered opinions. Aside from the pedagogical differences between academic institutions, the space between schools and the working office included a wall that students were required to scale with or without help from school. Many traditional design academics scorned the development of the technology, saying that it was not drawing. For students trying to learn the technology, it mattered little: "Okay, it isn't really drawing, but so what? I need a job when I graduate, and I need to know what the profession requires." And in many cases, the profession was out in front of educators. Some offices took another view and were reluctant to invest in what many thought was the purview of engineers alone. It was plausible for new interns to find themselves as experts in a professional office absent the design skills and leadership acumen necessary to function as "experts."

Many once thought the digital environment would clean up the ubiquitous hangnails of pen-and-ink drawings and substitute a system of powerful accuracy with robust coordinating strategies. But the explosion of digital graphic innovation has produced many designers with distinct levels of expertise and interests. The proliferation of software has helped the design professions create digital structures capable of virtual walk-through tours that give clients an understanding of a proposal that could not have been imagined three decades ago. We realize now that the simplicity that digital graphics were once expected to bring remains just out of reach. Many found themselves caught in the whipsaw of change. Those trying to draw while using both sides of the brain created digital graphics after doing all the thinking with a pencil. And why not? In the final analysis, few care as long as trades and consultants understand what is wanted and the job is completed within budget and on time.

Despite the fact that we have more technology today than ever before, we demand to know more, and we want it now. But even when we have everything we want, there is still a struggle to separate the relevant data from that which can physically fill the hard drive. Yes, it may be there, but do we need it all?

THE DIGITAL OFFICE

A number of digital graphics programs are commercially available. A site planner's office is fortunate in that it is not limited to a single choice of CAD or illustration software. From the myriad packages of design applications, firms have the opportunity to select those applications that best fit the type of work they perform. An intern, on the other hand, must be prepared for a professional world without standards and that appears to move in multiple directions, with random commitments. To an intern, it may appear that business judgments are made oblivious to state-of-the-art hardware and software. Very often, a professional office has not acquired the advanced cutting-edge computing systems that interns likely used at college or university.

The decisions a site-planning office makes regarding productivity, profitability, and the work environment in general have significant effects on what CAD platform and illustration software the firm uses. Regardless of a firm's size, the decision to purchase PCs, Macs, workstations, or hybrids is of critical importance. The office communication system, computer network, and compatibility are issues that must be documented, evaluated, and reconciled. A firm's principals will have to make final decisions regarding an office's digital network. Because few have the time or inclination to consider the issues, principals should consider appointing a committee composed of staff members representing the office's various user groups to develop the fundamental organization and CAD criteria. While this process is slow and requires genuine participation by all stakeholders, it helps get good results and reduces the potential for major oversights and omissions.

An intern will very likely find that many decisions have been made by principals who lack a full understanding of the most current or most capable hardware or software. Final decisions are commonly based on affordability and are often made by professionals who may have only a limited knowledge of the real options that exist. In addition, professionals often lack sufficient understanding of how their decisions affect productivity on a day-to-day basis. Interns may have to solve technical difficulties associated with those decisions. For these interns, this can provide a real opportunity or great frustration. It is critically important that an intern be prepared to be flexible and patient. Just as in another time, when interns had to practice to adopt the drawing styles of their employers, so today an intern must adjust to the varied digital approaches (the hardware and software processes) used in offices.

With the development of the PC in the early 1980s, along with affordable computer hardware, many principals were willing to accept interns with limited design experience, as long as their computer skills were credible. In the past, design principals had a command of all the skills necessary to prepare drawings for presentations as well as for construction contracts. Most offices today have few, if any, totally CAD-illiterate professionals. But some senior professionals may have been either slow or reluctant (or both) to acquire the computer skills necessary to function in their own office environments. This rather common condition has resulted in a type of hybrid office that relies on both hand-drawn and computer-generated graphics. Few offices today either rely exclusively on graphics prepared by hand or prepare every graphic using digital technology.

PERCEPTIONS AND EXPECTATIONS

The issues pertaining to CAD and computer technology in an office are complex. However, it is important to keep the big picture concerning site planning at the forefront. Understanding digital technology allows us to create documents and details with a degree of precision that simply cannot be built on many sites. Used correctly, the technology helps see, analyze, and develop solutions that are appropriate in scale, time, and economy.

The natural gaps in generational thinking and viewpoints between principals and interns are amplified by the technology. Because interns often possess CAD skills exceeding those of principals and more senior professionals, the latter tend to believe that CAD tech interns are more capable than their actual experience indicates. Professionals must manage the divide that is prevalent between the expectations of principals and intern designers. Modern offices must endeavor to merge the tech skills of the junior to the design experience of the senior. Failure to properly manage this merger can lead to disconnected thinking—not to mention wasted time and ruptured budgets in failed attempts to solve real project challenges while pursuing solutions to issues that are unrelated to the project at hand.

Principals and professional managers sometimes take a hands-off approach to this issue of disconnection. They imprudently leave younger, inexperienced staff the tasks of solving design problems. In the past, it would have been far more unlikely that a principal would have left such problem solving to a draftsperson. Today, however, due to the fact that some senior professionals are intimidated by technology or computers, interns may be left to develop design solutions beyond the scope of their experience. The results of these efforts may fall short of real problem solving and creativity.

Principals may have difficulty offering criticism to interns on their design solutions due to a lack of common ground. In decades past, interns were not asked to provide final presentation sketches, construction drawings, or reports until they had demonstrated proficiency in graphics and organizational skills. Today, however, some principals must be reminded that knowledge of digital applications and hardware does not give interns the skills principals have acquired over careers of many years.

Managing the integration of design capability with technical know-how is paramount in today's professional world. Amassing and activating all the newest gadgetry and technology alone does not equate to savant knowledge and skill to perceive and plan the future. Professional offices can achieve new levels of sophistication in analysis, production, and presentation only with well-trained staff who have a base of solid design skill and experience. This includes the experience of modern computing and digital techniques unavailable to site planners of a few decades past.

From the beginning of the digital revolution, principals have been challenged by constantly changing technology. Unfortunately, unlike the graphic techniques of pen and ink in the past, the continued evolution of computers, applications, and digital techniques renders obsolete the purchases, hiring, and training of yesterday. Advances and innovations in digital software have exploded in the past three decades. And although the trend has moved the site planner's office ever closer to being an all-CAD environment, there is a growing sense that the inevitable event is tied directly to the emergence of an affordable touch-screen technology. In the meantime, the market will continue to rely on new interns and young professionals who have acquired expertise in school. While this presents young professionals with a great opportunity in the computing realm, it could fragment their efforts to develop their design skills. For most professionals, combining technology and design skill today is significantly more complex than it was in years past, and the complexity grows with each promise that a digital solution will make it easier.

Professional offices must confront the issues of perception and expectations as they are related to the practical application of CAD. For professionals, the expectations fall into two categories: the expectations of the interns, who believe that the basic computer knowledge learned in school will provide a professional boost, and the expectations of the principals, who may believe that hiring an intern with basic computer knowledge will provide the firm with a marketable advantage over competitors.

Lay Expectations of the Profession

Among lay clientele, there is a broad misconception regarding the plausible capabilities of computer-savvy design professionals. The general public is constantly bombarded by images of high-tech solutions. Walk into any movie theater today, and you will be dazzled by the amazing effects created by computer-generated imagery. The prevalence and ease of access to digital information have led some to conclude that slick computer-generated graphics and presentations can be produced instantaneously and easily. Little or no regard is given to budget constraints, schedules, and, notably, the skill of the artist. While it is true that some professional firms produce extraordinary computer imagery, rivaling that of Hollywood, most design offices lack both the commitment and budget to do so.

Technology is an ever-changing and evolving process. It calls professionals to constantly update their skill and experiment. Practitioners, educators, and professional societies related to site design must continue to educate the general public about the actual possibilities and limitations afforded by technology. Regardless of how or whether the public gets a clear understanding of the matter, design professionals will have to come to terms with the complex issues presented by this technological evolution.

Intern Expectations of the Profession

Site planners must build on their education and abilities, and they must deal with ever-changing technology. While possessing the ability to separate existing topography from proposed water lines may be a very useful exercise in the development of a parking lot plan, it does not particularly help a planner in determining an appropriate parking lot plan or where and how deep the water line should be placed.

Technology as a tool does not reveal any decision to the designer that the designer has not already made. CAD in particular should be thought of and used as if it were a pencil. (Photoshop could therefore be thought of as a colored pencil.) Accurate work in a CAD environment is as important as accurate work in simple mechanical drafting. The computer will not, however, resolve or solve input errors. In fact, the ability to draft in CAD well beyond the limits of field verification can become a burden.

College students are naturally drawn to the use of computers. Whereas older generations were intrigued by the possibilities of digital applications, a modern student's world is immersed in technology. Some students and interns are attracted to computers as a way to compensate for deficiencies in freehand drawing skills. Many students continue to believe that computer knowledge learned in school will be a fast track to employment and management. Students and interns should be careful not to rely on computers as a crutch they believe they can use to compensate for an incomplete education. Students should seek a solid working understanding of CAD and illustration software, as these are the tools they will use every day in an office. The traditional design skills of hand drawing and an understanding of the design principles will always be of paramount importance to the design professionals and managers who employ them.

Professional Expectations

Professional practitioners expect that incoming students and interns will possess a solid working knowledge of CAD and illustration software. They expect interns to understand the basic use of word processing and spreadsheet software, and in some cases, database and 3D modeling software. But while computer knowledge is important, professionals will be looking for potential employees who can demonstrate design skills and execute drawing production. In evaluating interns, principals would be prudent to keep in mind that quite often students have access to computing power unavailable to most professional firms. The inability of some interns to re-create portfolio-level work with the limited resources of most firms can be frustrating for both professionals and interns. Students who possess a very high level of skill with computers are as rare as those who possess a very high level of skill in freehand drawing.

MORE THAN A DIGITAL PENCIL

It may well be that all designers have something in their DNA that "hardwires" the connection between their eyes and their brain. This is not necessarily a good or bad condition, but one that needs to be recognized. Some of us are easily smitten by the simple beauty of something, regardless of how structurally flawed or inherently toxic the object may be. As such, we are often prone to accept, without thinking or caveat, the accuracy or significance of an image. And the more elegant the image, the more willing we may be to embrace its message.

When designers had to physically draw every line, they had some time to think about the role or functionality of each line. But when lines appear on a PC's monitor, we are sometimes slow to digest the significance of the lines in the context of an actual site. A visit to the site provides background information that helps us create a mental graphic. While the line work in a drawing is important, the topography of a site can be experienced only via a site visit. But until a planner stands atop a hill and sees 200 yards beyond the property line, the value of the line work is of somewhat limited value. When the line work becomes a visual vista, it is because it has been interpreted by other data and has meaning.

It is easy to click a mouse button and pull up all the associated data integrated within an AutoCAD drawing file you just picked up at the local government's mapping or geographic information system (GIS) division. But mentally processing all the information is something else again. Synthesis is the deliberate process of our minds "turning off" the layers of extraneous data and focusing on the interaction of the components that are relevant to the design issue. Synthesis is the process that directs us to determine the composition of the problem.

We are not suggesting that the PC is capable of synthesis. This fact, however, is often forgotten in the midst of the constant use of digital resources. The most insidious aspect of the computer is that we insist on attributing to it capabilities it does not have. This is partly due to the fact that we are barraged with more information than we can mentally process, and we are prone to shove into the breach anything that appears to be helpful. We are often duped into believing that when we have entered all the data, the problems will become clear. This is not necessarily so. Some things may appear to be shortcuts to the wellspring of knowledge, and the computer is a particularly enticing example of this. The PC's capabilities, and particularly the speed with which it performs, can beguile a professional. We may find ourselves using a very powerful tool in the wrong place and time, not because of a stated promise by the manufacturer but because of the deference we give to it.

Site planners who worked before PCs had an advantage in that they created a scaled drawing of a site. Today, a CAD operator works at 1:1 scale. Consider doing the creative thinking in the before-PCs frame of mind (that is, with a pencil) and moving to the PC when more detail is required. That is, think with a pencil in your hand before moving to the digital pencil. When using a PC for design tasks, think as though half your brain functions exclusively as a pencil, while the other half as a PC. Recognize the limitations of each side of the brain and understand that site-planning problems require the cooperative participation of both sides. Specifically, use the PC particularly when you require documentation (photographic and line work images) and when you need to store data, send and receive information, and duplicate information.

Pen and paper are tools, and so are PCs; each provides unique opportunities. Just because a PC is capable of providing, storing, and reproducing all the data associated with a site does not mean that having all that data on one plot is the best way to address a problem.

It is important for designers to be thorough and not move so quickly into the details that they neglect the bigger picture. This principle holds particularly true in dealing with digital applications. In many applications, the software enables a user to render at a level of detail that cannot be constructed. Moving into detailed work earlier in the process than is appropriate can lead to time wasted later in editing and to reworking of solutions.

Learning anything is a process of creating a foundation of concepts and facts on which to place more complex theory and proofs. It is important to make notes about the challenges and issues of each project, both in design and execution. Intellectual growth is based on feeding the brain. We have all heard that we learn by making mistakes. It is not necessary, however, to actually make every mistake. We can learn from the mistakes of others. If you see something that does not appear to be correct, ask "Is that right? Was that the best way to solve the problem?" These questions are not reserved for apprentices. It is healthy to question the process in a professional way, asking questions such as "Did we get all the information we needed? Is there a more effective way to conduct project review? Is our problem-solving process working as well as it could?"

Many enthusiastic undergraduates launch theories and predictions from a platform of two years in college. Exercise some restraint. Understand who you are and where you are; think about where you want to be. Technology does not make a designer; many great designers existed before PCs. But having a keen knowledge of hardware and software is also a skill that not everyone can claim. The same can be said for drawing talent. We know that the CAD environment will continue to evolve, and every designer with a talent for freehand drawing will be sought.

The language of site planning involves having an understanding of the components to be considered in the planning process, including line work, images, and quantifiable data. One of the best lessons in the design process lies in the review of the differences between initial design expectations and a completed project on the ground. Take the time to look at and understand the differences.

CHAPTER FIVE

FROM GRAPHITE TO CPU

Computer organization involves more than just files and folders. How information is stored within files should be standardized. Without an intra-file system, a designer can forget the *art* of design. One can become bogged down, left to decipher and interpret the countless possibilities of layers and acronyms. This chapter provides examples of methods for standardizing and coordinating this information.

COMPUTER PLATFORMS

One of the most important decisions made in a professional firm is selecting the computer platform to be used in the office. The selection of a platform–PC versus Mac–influences hardware architecture and software operating systems and applications. Typical decisions involve computer architecture, operating system, programming languages, and related runtime libraries and graphical user interfaces (GUIs). Determining factors include the type of work commonly performed, the current experience and anticipated training of personnel, the budget, and system maintenance requirements.

Because most professionals are computer literate in only one platform, many offices opt for a single-platform configuration. Single-platform configurations are attractive to project managers and principals for two main reasons. First, a single-platform configuration generally has simpler networking and printing requirements, and personnel training is less demanding and costly. Second, single-platform offices are relatively easy to set up and maintain, as the system usually comes preconfigured from the manufacturer so that computers readily communicate with each other, with printers, and with the Internet.

Despite the operational simplicity of a single-platform configuration, such a configuration is not perfect. One main criticism is that a single-platform configuration limits the selection of software options. Many CAD and illustration packages are so restrictive that they are compatible with only one platform. Some may suggest that thanks to emulation software, these barriers don't really exist, and this is true as a workaround. However, the practicality of running multiple operating systems (Linux, Mac, Windows, and so on) on a single platform begins to negate the point of the simplicity of the single-platform configuration. Oftentimes these emulators come at the price of overall speed, and in the final analysis, they are rarely viable options.

Some offices, however, require a wider variety of options in computing. These offices may employ cross-platform configurations–that is, multiple platforms on a single office-wide network. While the setup, management, and training required for cross-platform configurations can be complex, the benefits include much more diverse and flexible access to software applications. In addition, principals and project managers may find that cross-platform configurations are more cost-effective because the platform architecture can be tailored to the standards and protocols of specific software.

Today, developing technology is beginning to free offices from difficult platform decisions. The emerging cloud computing presents a distinct option that is not subject to all the idiosyncrasies of single and multiple platforms. More and more Internet-based computing will provide the software resources designers and mangers require. Cloud computing represents a paradigm shift in computer functionality much like the shift from mainframe to client/server computing in the early 1980s. Details are abstracted from the users, who no longer need expertise in or control over the technology infrastructure "in the cloud." Capital expenditures on hardware and software are reduced for the professional office, and computer services are treated like subscriptions or utilities with a monthly service fee.

GRAPHICAL USER INTERFACES

The development of the graphical user interface (GUI; pronounced "goo-ey") dramatically changed the use of computers in homes, schools, and every office that uses digital graphics. Initially introduced in the late 1960s, the GUI underwent extensive software modification by Xerox and was popularized by Apple in the late 1980s. Today, most professionals are familiar with multiple GUI formats, such as Microsoft Windows, Mac OS X, and Linux. All these platforms use a WIMP (window, icon, menu, pointing device) style of interaction. By using a physical input device, such as a mouse or pen stylus, the user controls the position of a cursor or other tools on a "desktop," allowing access to information organized in windows and represented with icons.

The evolution of computer interfaces is ongoing. Just as WIMP-style interfaces changed how we interact with computers, "post-WIMP" devices will lead the way to new types of digital devices. The development of touch-screen technology has already exploded in handheld devices and smartphones. Gaming systems from companies such as Microsoft and Sony use NUIs (natural user interfaces), and the technology used to develop these NUIs is very likely to spill over into CAD and illustration/animation software in the future. Such technological innovations will continue to bring opportunities and challenges to universities and professionals using digital graphics.

RASTER GRAPHICS

A raster graphic, or bitmap, is a graphic medium that differs from vector graphics (used in AutoCAD and MicroStation). With raster graphics, images are formed using a rectangular grid of colored pixels rather than vectorized lines. The quality of a raster graphic image depends on its resolution. Unlike a vector graphic, a raster graphic cannot scale up in resolution without diminishing image quality. For designers and artists, raster graphics provide a more practical format for digital images and photographs, whereas vector graphics provide a better platform for precision drafting. Both AutoCAD and MicroStation allow for the importation of raster files.

Photoshop, developed and published by Adobe Systems, is one of the most valued tools for raster graphics used in professional offices. Originally developed in the late 1980s, Photoshop is a raster graphic–based software package that has become an industry standard for image editing, color illustration, and digital art. Photoshop allows for the creation of object libraries and standardized drawing palettes. It also has functions to separate objects onto multiple layers, much like its CAD counterparts. These basic functions are only a small part of its dizzying variety of powerful editing tools.

As with CAD, the power of Photoshop lies in its assortment of tools and functions. A basic criticism of such platforms is that they can be difficult to master without proper instruction. However, it is vital that interns get up to speed with Photoshop, as most professional firms that work with illustrative drawings use a raster-based image editor to create presentation graphics. Firms should create standardized palettes and tools to assist their employees in the use of the software.

3D MODELING

3D software platforms are slowly becoming more commonplace in professional offices. Though these tools have been available and affordable since the mid-1980s, too often they have been difficult to learn and ill suited for site planners.

In the late 1990s, new approaches to 3D were developed, with an emphasis on user friendliness. One such package is SketchUp, developed by At Last Software and sold to Google in 2006. With its patented push/pull tool, which can be used to quickly create geometric extrusions, SketchUp appealed to architects, landscape architects, and engineers. Its low cost and shallow learning curve make it an attractive choice for both novices and those who are CAD savvy. It is a powerful tool for quickly setting up 3D models for illustration and analysis.

Both AutoCAD and MicroStation have robust integrated 3D capabilities that are useful to site planners in the areas of slope analysis, cut and fill, and animation. Both platforms provide for simple integration of SketchUp models into a CAD model file. Often these simple models created by SketchUp are preferred for planning purposes to those created with more robust 3D platforms.

Vector Graphics and CAD

Vector graphics provide the primary format for designers using CAD platforms. They are composed of basic drawing elements called *primitives.* Primitives are composed of lines, points or nodes, curves and arcs, and regular and irregular shapes. Vector graphics have an advantage in that they can be translated to a smaller file size and provide indefinite zooming without a loss of resolution; element modification via scaling, relocation, rotating and filling; and a basis for 3D perspective.

AutoCAD, developed by Autodesk, is by far the leading computer drafting software used by designers around the world. It was originally released in 1982 as the first comprehensive computer drafting software for PCs. Though other drafting and design software existed at the time, the shift from mainframes to PCs would begin to reshape how designers would approach the development of important graphic documents that had previously always been prepared by hand drafting.

Compared to today's software, the early releases of AutoCAD had very modest functionality and usability. It is difficult to remember those first years of typing in commands, saving files to floppy disks, and printing to dot-matrix printers. AutoCAD has from the start utilized a series of primitive entities, such as lines, ellipses, multi-lines (called poly-lines), and text, as building blocks for more intricate and complex shapes and forms. This method has become the standard format for most CAD platforms following in the AutoCAD model.

Autodesk released versions for UNIX and Mac, but it eventually settled exclusively on the Microsoft Windows platform. AutoCAD is a powerhouse software package that delivers all that a designer could desire and more in the domain of computer drafting. AutoCAD was one of the first applications to offer unlimited layers, millions of colors, multi-grouped elements called blocks, 3D parametrics (that is, parts and assemblies that allow universal modification at a later point in the design process), and an internal programming feature called AutoLISP.

Because AutoCAD is a vector-based software application, its initial use for site planners was limited. Using AutoCAD to create free-flowing arcs, spirals, and curves was nearly impossible. Maps and plans were drafted by hand, as they had always been done, and they were then handed off to a CAD technician who would redraft the plans into a vectorized coordinate-based system. This interface did not allow for great flexibility or creativity. For some, the desire for digitized information came with an unwelcome "dumbing down" of the shapes and forms due to either the limitations of the software or drafting skill.

It is not unusual to see a hand-drawn sketch scanned as a raster file (that is, a file containing bitmapped drawing information) imported into AutoCAD and then "traced" by an intern. Though today's software is extremely capable and allows for considerable flexibility, many professionals prefer to hand-draw ideas first. Even with all the capability that AutoCAD offers, many offices continue to use it exclusively as a drafting tool.

In the early 1980s, Bentley Systems developed MicroStation as a head-to-head competitor to AutoCAD. Bentley ultimately released MicroStation versions for multiple platforms, including PCs and Macs. Though taking a separate development path, MicroStation incorporates many features and tools that are identical to those of AutoCAD in function. Users of MicroStation use virtually the same drafting process as AutoCAD users.

AutoCAD and MicroStation are the most robust CAD platforms available for design offices using PCs. Both suites offer the end user multiple add-ons that deliver more specific working templates and tools for architecture, civil engineering, and landscape architecture. Both platforms are available in lower-end, less-expensive, yet powerful drafting versions of the full software. In addition, Autodesk and Bentley are working toward interoperability of AutoCAD and MicroStation. This will ultimately result in seamless crossover between platforms for users, and it represents a giant step forward in cooperation and productivity. Cloud computing versions of AutoCAD enhance the interoperability and usability of both platforms.

THE INTERNET

Precisely how the Internet functions as a digital resource in a design office may require a brief explanation. The Internet is a global system of interconnected computer networks that use a standardized Internet protocol suite (called TCP/IP) that serves a multitude of users worldwide. This computer interconnectedness between users gives designers access to a panoply of information quickly and reliably in a manner completely unknown just a generation ago.

The Internet has become a standard research, marketing, recruitment, and networking tool for professional offices and universities. Powerful search engines allow site planners to quickly gain access to site-related information, including site history, vegetation, soil, and regulatory ordinances.

In a design office, everyone is expected to be able to access information from Internet sources quickly and reliably. One critical negative aspect of the Internet is the reliability of the information sources. Cross-checking data and developing a library of reliable resources is fundamental to having a credible database. While information from some sites requires verification, many local, state, and federal agencies support Web sites that are both reliable and reasonably current sources of public information. Design staff are expected to understand how to organize and maintain bookmarks or favorite sites in order to more effectively build a library of reliable sources.

Google Earth

Google Earth is a virtual globe that utilizes satellite imagery, aerial photography, ground-level photography, and computer modeling (via Google SketchUp) superimposed over a map of the earth. It comes in both a free version and the commercial Google Earth Pro. Google Earth can be an extraordinarily valuable resource for remote sensing, providing a designer with aerial views and, in some areas, street views of sites, neighborhoods, or cities. It also provides information about existing terrain, buildings, and other physical features, as well as, to a limited degree, the location of existing vegetation.

Google Earth provides interoperability with some software packages, including AutoCAD, MicroStation, and SketchUp. This interoperability permits the integration of Google Earth images directly into a project file via input of geographic coordinate data. Though this can be very useful, remember that the free version of Google Earth is not necessarily regulated or maintained to a specific design standard; the information it provides can be verified only through site reconnaissance and surveys. The commercial package Google Earth Pro is a more robust version of the software, with more powerful search tools and enhanced graphic capability, and as such it is preferred by many professional offices.

CHAPTER SIX

ORGANIZING AND CREATING DRAWINGS

FILE SHARING

Thanks to Web sites, File Transfer Protocol (FTP) sites, e-mail, DVDs, and flash drives, sharing information has never been easier than it is today. Computer files can be easily shared, stored, and organized, using a firm's or an organization's filing system. Whether files are uploaded and stored for sharing within an FTP site or in the company server for sharing over a local network, users can access stored information easily and quickly. Today's methods represent a dramatic improvement over the space-consuming filing and storing of large and irregularly sized and shaped construction drawing sets, base maps, aerial photos, and so on.

An effective way to share large design files is via an FTP portal. An FTP site is part of a Web site with limited accessibility. The owner of a file can upload that file to the FTP site, and a designated user (another consultant, the client, or someone else) can then gain access to and download the file by entering the username and password supplied by the file owner. The designated user can in turn upload files to the FTP site. This method allows the transfer of very large files very quickly. It is also useful because it provides a secure method for transferring confidential information.

Perhaps the most popular method for sharing information is via e-mail. The advantages of e-mail are that files can be searched and sorted, providing an easy method of documenting communications. This method also has the advantage of providing the sender with notification if a problem has occurred during file sharing. Information can also be sent to smartphones and other handheld devices, which has proven to be exceedingly useful. This is a particularly efficient way to stay in contact with other professionals on a team who may be away from the office. One limitation of sharing files via e-mail is that some Internet carriers have implemented file size limits for e-mail.

Most firms use some type of local network with a central server for file storage. Local networks may be wired or wireless. Creating a standardized file storage system is among the responsibilities of principals in an office. They need to develop and deploy the system, and others must learn to use it. The particular details of the workings of a local network are the responsibility of a network administrator, who may or may not be a firm employee.

CDs and DVDs are also used for file sharing. They are typically shared via mail or courier, particularly when file size or Internet access is an issue.

It is important to always verify with the end user the format in which he or she wishes to receive information. Even different versions of the same software can cause problems in files. If the information to be shared is for information only and intended to be unedited, using portable document format (PDF) files is recommended. Adobe Systems provides PDF viewing software for free as an Internet download.

A final consideration is file security. Sharing information with clients or other consultants is often required. However, sharing digital file information in a format that permits editing by other users exposes a firm to liability and potential litigation. When sharing files that can be edited or altered, it is prudent and recommended to request a signed release of liability from the user.

COMPUTER FILE ORGANIZATION

Organizing a computer filing system is a key step in creating a comprehensive office management system. While most offices have developed a standard filing system, it is not unusual for such systems to be seriously flawed and fall well short of their user-friendliness and security targets. Correcting a poorly conceived system is often as demanding and difficult as starting over. It is critical to an office's functionality that a system be developed, implemented, and respected. A well-constructed filing system that supports adaptability and storage will, over time, serve well in the maintenance of the office's records and data.

Most offices expect some level of confusion on the part of any new staff member, intern, or professional in learning the computer and computer-aided design (CAD) protocol of a new office. It is important from an operational viewpoint that firms develop clear standards for both file management and intra-file management. Not all firms are expert at establishing standards. Many outsource the task of protocols to third-party providers. Many small firms lack the budget to finance a consultant to establish a file management plan. Because planners and landscape architects provide services inconsistent with design projects, their needs can often require a format that reflects the specific nature of the firm or services. Therefore, using an outside source may not be the best decision. For example, many companies do not outsource their business filing but simply treat the management of computer files as another form of filing and file management.

It is interesting to observe that rarely do professionals, particularly new employees, provide criticism or even comment on an office's central file methodology. This can be attributed to the filing system's nearly sacrosanct status and what would be considered "common" knowledge as to how it is organized and functions. In contrast, computer file management in a design office can be a source of constant complaint and criticism. Criticism commonly stems from a sense of unease and unfamiliarity with the file setup and the frequent requirements for access to keep a project on schedule.

The simplicity and familiarity of an office's central file must be married to the accessibility and functionality of the computer server. In order to avoid chaos and reduce confusion and complaints, two objectives must be met: First, the firm must develop an organized computer filing system based on the overall firm filing scheme. In other words, a firm should not try to reinvent the wheel. Second, it is important to get wide user acceptance and establish a disciplined approach to the system.

A correlation of how an office's central file is organized and how the computer server is organized is one option. It is not unusual for design firms to have a filing system uniquely tailored to the specific type of work or the idiosyncrasies of the firm's principals. For example, central filing systems can be organized alphabetically by client, chronologically by date, or using another unique method. Computer files should follow the adopted system. The specific details of the filing organization an office implements are not as important as the consistency of the computer scheme to the office central file. Until the time comes (if ever) that design offices are totally paperless, much of what is filed in the computer is simply a duplicate of what is filed as a hard copy. Thus, matching the filing schemes creates, in essence, a single filing system to be learned and used.

In creating a filing scheme, remember that simplicity is powerful. Do not be lured into creating unending folders within folders. Most operating systems can display items in the order in which they were created or modified or in alphabetical order by filename. Burying files deep inside folders is usually unnecessary. Also keep in mind that modern operating systems allow for well over 30 characters in filenames and folder names. Therefore, avoid using abbreviations to identify folders and files and instead use descriptive filenames.

Strict adherence to a firm's filing system is vital to the proper functioning of that system. Users should be prohibited from deviating from established protocols because doing so can result in chaos in the short term and loss of data in the long term. Modern CAD systems are woefully inflexible when it comes to moving or renaming files that have been defined within a drawing file. Once reference files (often called *xrefs*) are defined, they should generally not be moved or renamed. Deviation from an original filename or location will cause the user to search for and reattach reference files. With large, complex projects, simply renaming a folder containing multiple files can require many hours to reattach reference files. Professional acceptance of a developed system is critical. Because all the professionals in an office have some level of access to the central computer filing system, everyone must respect the office standard and avoid corruption due to personal bias or laziness.

A final consideration in office filing is whether to limit access to the central filing system. If a firm decides to do so in order to limit corruption or shield sensitive data, it is a common practice to control access to computer files in specific categories. For example, standards and common templates might be accessible only to senior management, while interns may be limited to access only the base-level information, data, and correspondence. See Appendix C for an example of a graph that illustrates this standard.

CAD ORGANIZATION

No CAD organizational standards are so flexible that they will fit every site-planning problem or situation without adjustment. Therefore, a site planner needs to be cognizant of the prevailing industry standards and what works for each design problem. Although one size does not fit all, understanding what will be the best fit is the first step in getting a project's digital layers organized sensibly. This requires having a comprehensive understanding of the general area of practice and expertise of a given office.

It is important to keep in mind the challenges related to sharing CAD files. For instance, not all firms use AutoCAD. Most CAD software packages today recognize the ubiquity of AutoCAD and provide for the ability to save information in the .dwg AutoCAD native file format. However, when using other CAD platforms to save and transfer information in the AutoCAD format, one must be aware that there is a risk of data corruption and/or loss of information. This is particularly true when transferring information related to attributes defined in layers (for example, line thickness, line weights, colors, styles, override settings).

The task of creating CAD standards from scratch is daunting, but help is available. The American Institute of Architects (AIA) has continued to release *Layering Guidelines* since it first addressed the issue in 1996. While many firms decide to develop in-house standards that suit their specific requirements, the AIA guidelines offer a sound starting point. Other sources include the *Uniform Drawing System* provided by the Construction Specifications Institute (CSI) and the *Tri-Service Plotting Guidelines* from the U.S. Department of Defense. All these guidelines are available for a nominal fee. Some CAD platforms, such as those released by Autodesk and Bentley, now include several different standards. Many local agencies and institutions such as cities, highway departments, and universities have also developed layering guidelines that private practitioners can use when developing documents for specific agencies and institutions.

Whereas some offices may find that the AIA guidelines offer too few layers related to a particular area of practice, they may find that certain other agency standards are too specific and offer hundreds of layers more than are really required for a private-sector project.

In developing independent office standards, a professional must keep in mind several factors. The first, and perhaps the most critical, involves the sharing of CAD information. Multiple professional disciplines may work together on any given project, and they must all be accommodated. Another factor to consider is the specific development challenges unique to each site-planning project. A professional must also anticipate and allow for an array of site conditions, both existing and proposed, that can develop over the course of a project's development. Finally, one must create a system that is sufficiently simple that a CAD intern can adapt to its structure. At the same time, the system must be sufficiently robust to produce clear and coherent project drawings for presentation and construction documents.

An Overview of the AIA Guideline for Layer Names

Recognizing that CAD would become commonplace and the sharing of information routine, in 1990 the AIA appointed a task force to address a range of issues associated with the emerging technology. Creating the *Layer Guidelines*, organized as a hierarchy, was the first step in establishing protocols for expansion and user-defined extensions to the layer list.

The AIA *Layer Guidelines* provide an excellent example of the process and understanding of the complex issues involved in developing CAD standards. In developing a unique interoffice standard, a CAD professional would do well to develop a system that incorporates the thinking of these guidelines, if not their actual layers. Also, keep in mind that standardization need not stop at layer naming but is also applied to attributes such as color, thickness, width, line style, and printing characteristics of the lines.

The *Layer Guidelines* state that layer names are to be alphanumeric, which permits easy-to-remember abbreviations. Two layer name formats are also defined. The original short format has been revised to be consistent with the ISO CAD standard and is recommended. It is less cryptic than the original short format and is compatible with new international CAD standards. The long format is compatible with the 1990 *Layer Guidelines* long format, and some users may prefer it.

Layer Names. Format Schematic—Character Fields:

AB CDEF GHIJ KLMN

A and B are discipline designators.

AB **CDEF** GHIJ KLMN

C, D, E, and F are major group designators.

AB CDEF **GHIJ** KLMN

G, H, I, and J are minor group designators.

AB CDEF GHIJ **KL**

K and L are status designators.

Consider the following example:

A—WALL FULL—D

In this example, discipline is A (architectural), major group is wall, minor group is full, and status is D (demolition).

Discipline Designators. The discipline designator, or agent responsible code, is used to indicate the origin of the graphic information, such as architect or landscape architect. It provides a means for updating information. The field is a two-character field, with the second character either a hyphen or a user-defined modifier.

Major Group. Major groups identify the building system. Although major groups are logically grouped according to the specific agents responsible, it is possible to combine major group codes with any of the agent codes. Examples are WALL for wall, DOOR for door, and LITE for light fixtures.

Minor Group. Minor group is an optional four-character field for further differentiation of major groups. For example, A—WALL FULL indicates architecture, wall, full height.

The following common modifiers are defined for use in the minor group field:

IDEN—Identification

PATT—Pattern

PART—Partial-height partitions

Status. Status is an optional two-character field that starts with a hyphen for further differentiation of new construction, remodeling, existing to remain, and so on.

The following common modifiers are defined for use in the status field:

N—New work

E—Existing to remain

R—Relocated items

T—Temporary work

Annotation. Annotation is made up of text, sheet titles, and borders. Annotation is designated by the major group ANNO.

Consider the following example:

A—ANNODIMS

In this example, discipline is A (architectural), major group is annotation, and minor group is dimensions.

CREATING A DIGITAL DRAWING

If we have learned anything in the era since the invention of the PC, we have learned that nothing is forever. Somewhere someone is working on new software that will, at least in theory, make things simpler, faster, and cheaper—for a price. Why are these bright people working on the new software? Because we are not content with what we have today. We want more, quicker, and we are willing to pay for it.

Obligations to clients are wide ranging, and many professionals are unclear about what the character of graphics needs to be. For some, graphics are parts of a PowerPoint presentation to a client; for others, they are parts of publications or portions of contract documents. Let's take the contract documents issue first because it is the easiest to address. We don't want to be the only one out of step with the other consultants on a team, so where there is consensus, we set our preferences aside to get the job out. If there are ongoing problems with the consensus software, we work with others to find common ground. The environment around the development of construction drawings is a place for consensus building, not prima donnas. This is not to say that you should never raise a question regarding decisions with broad implications. But you should never find yourself responding to the question "Why are you doing that?" by replying "Because we did it that way yesterday." If there is a better way, find it; but do so after the job at hand is out of the picture.

But what if a PowerPoint presentation needs to be plotted? For offices that have an array of digital graphics software, a site planner must understand the potential of each type of software in order to make an intelligent match between needs and what is available to create a graphic product. Most often, large-scale presentations focus on large-scale issues. Regardless of the size of a presentation, remember to keep it simple and include no more than three topics on one graphic. Include only graphic elements that provide information necessary to the issue. Extraneous graphic information that clutters an image is not helpful to the reader or audience. Not every client understands technical drawings or maps. Not everyone is visually oriented.

Simple Drawings

Simple drawings are graphics that have no intelligence; their line work has no data other than distance, and their closed polygons have no data describing the polygon other than area. AutoCAD, MicroStation, and Map Maker can create this variety of graphic line work. Although AutoCAD creates simple graphics, the lines do contain simple information such as line length or area information in closed polygons. The software does not, however, permit a CAD operator to attach data to the polygon or other attributes to the line work, and there is no database of the line work or polygons. The AutoCAD drafting software is significantly more powerful than Map Maker, but it is less robust than ArcInfo. To modify an image for use in Map Maker, one must change the data or line work in ArcInfo.

Map Maker is Environmental Systems Research Institute (ESRI) software that also uses the data in a geographic information system (GIS) to create a graphic. ArcView software is less sophisticated than ArcInfo, and Map Maker is even more limited regarding graphic options. On the plus side, Map Maker has an even shallower learning curve than ArcView. It is extremely popular for "front desk" operations, where general information and large area maps meet most needs. This is not to say that a general base map that describes all the roads, physical features, and major vegetation is of little value to a professional site planner—because it is. It may be one of the few maps that everyone in a community would recognize. Unlike either a GIS database or an AutoCAD file, a Map Maker data file is not something that most jurisdictions sell over the counter.

AUTOCAD AND MICROSTATION

AutoCAD and MicroStation are the two software applications most commonly used by those who need to create digital drawings. Architects, landscape architects, surveyors, cartographers, and those from every other engineering discipline involved in the building and construction industry use AutoCAD and MicroStation extensively. Bentley's MicroStation and AutoCAD were designed for CAD operators. This ubiquitous software is used to create 2D and 3D graphics. Community colleges and universities have offered classes in such software for decades. A working knowledge of this software is almost a requirement for employment.

If we want a physical drawing, it must be created using the drafting software and then image plotted, unless it is hand drawn on paper. Among the advantages of a software-drafted solution is that the file size is smaller than is the file size with images that are hand drawn and then digitized in ArcInfo. But remember that whereas there is "intelligence" associated with a drawing created in ArcInfo, this is not the case in AutoCAD.

GEOGRAPHIC INFORMATION SYSTEMS (GIS)

Almost 50 years ago, a group of geographers began creating the fundamentals for the technology we know today as GIS. By the early 1990s, the concept of a system capable of integrating graphic and alphanumeric data had broad professional interest, and many local and state governments were involved in the development of such a system. For the software ArcInfo, the adjective *powerful* was not used as indiscriminate hyperbole. Just as the name implies, ArcInfo combines graphics (arcs) and data (information) into a system that permits the user to query software and retrieve answers as a map. The widespread use of ArcInfo around the world today testifies to its acceptance as an exceptionally powerful software-mapping technology with an array of applications.

ARCINFO: USING DATA TO CREATE GRAPHICS

ArcInfo has broad appeal. It supports the development of digital maps as well as a database. Based on database inquiries, ArcInfo can create graphic responses. But unless you are enrolled in a professional degree program where an introduction to GIS is part of the curriculum, it is likely that you will not come in contact with the technology in an academic setting. Most people engaged in site planning are focused on applications of CAD software.

Although ArcInfo software provides a graphic product, in its truest sense, it is not graphics software with the robust character of MicroStation, AutoCAD, or Adobe Photoshop. You can create a digital export (.dxf) file from ArcInfo and download a digital "layer," or coverage, from ArcInfo to AutoCAD, but that file contains only the data for the line work graphic, not the alphanumeric data from the ArcInfo file. At this time, no simple graphics programs are capable of providing information and graphics that are incorporated into ArcInfo.

The name *ArcInfo* is a clue to the capability of the software as it creates a drawing (arc) with information (info) attached. When a series of lines creates a closed polygon, the software calculates and creates the centroid. Graphics using this software are most commonly created by asking the software a question or query; the response or answer is in graphic form. The information embedded in the ArcInfo database is the key to the creation of the graphic. For example, many public planning agencies use ArcInfo to create their zoning, land use, and street maps. Figs. 6-1 and 6-2 are illustrations of property and topographic maps created using ArcInfo. Communities that have the resources to do so create property maps and attach the data from their tax maps in each parcel polygon in the jurisdiction of the municipality.

ArcInfo represents a powerful union of data and graphics. Queries can be made of the ArcInfo software, using the data in tax files, such as zoning, land area, property value, and so on. ArcInfo is popular software that many public agencies use to develop their GIS mapping programs. The term *GIS map* is often used interchangeably with the term *ArcInfo map*. Remember that *GIS* is a *system* whereby statistical data about a specific geography is stored in a database and used to respond to inquiries using *ArcInfo software*.

Because a GIS map consists of a series of layers (Fig. 6-3) or map themes, it is possible to create graphics that exclude some layers. This provides an opportunity to use only the specific layers of the GIS database necessary for the creation of a graphic. Knowing what you need and what you can do without makes for a potentially clearer picture than one with extraneous information. The key is to plan ahead and know what you want, as well as what obscures your message.

Similarly to CAD drawings, ArcInfo maps have three feature types: points, lines, and polygons. For example, a polygon could describe a land use or body of water (Fig. 6-4), a line could be used to describe a stream or road centerline (Fig. 6-5), and a point could be used to locate a water well or a surveyed benchmark (Fig. 6-6). And data can be attached to each feature type. A point could have geodetic coordinates, a line could have a precise length with geodetic coordinates at each end, and a polygon might have an area and a centroid.

A CAD file drawing of a structure on a site is likely to have multiple layers. The same format is employed in ArcInfo. Whereas an AutoCAD file or a MicroStation file drawing may have the foundation plan on one layer and the floor plan on another, property line work in an ArcInfo drawing is one layer, or coverage, and topography is another coverage. While the property and topo coverages consist of line work, the difference between AutoCAD layers and an ArcInfo coverage is data. In ArcInfo, data can be attached to points, line work, or closed polygons of line work.

Although ArcInfo has an extensive number of users in urban settings, its applications are not limited to urban conditions. Whereas CAD drawings are graphics, the GIS is a system in ArcInfo capable of creating graphic answers to a query relying on its database. For example, a digital layer of a tree farm depicting different tree species would be considered a coverage of line work that forms polygons (Fig. 6-7). Alphanumeric data identifying a tree species and the date of a tree's planting could be attached to individual polygons, creating a tree species coverage (Fig. 6-8). The same line work used to form the tree species coverage could also be used to create timber yield coverage with data from the timber yield database (Fig. 6-9). If a soil type coverage of the same geography were created, it would be possible to answer a query about the highest yield of each tree species from each soil type. Using the three coverages and the timber yield data, the software could provide the user with a graphic answer to the geographic location of the yield of each tree species on each soil type.

Some agencies have funded the development of digital aerial orthophotography (that is, rectified aerial photography in digital form) as a component of a GIS. These are often excellent graphics as they are aerial photography rather than drawings, and some find them less challenging to understand than maps. Inquire about their availability when you request a plot from a GIS database. On occasion, plots can be created using digital ortho images as the background and line work from the GIS database. Some jurisdictions have developed policies that permit the sale of a limited amount of data from a GIS file. For obvious reasons, selling data is often accompanied by caveats and legal agreements, which must be respected.

ArcView is a software program that uses a menu set of commands to create graphics from a GIS database. This software was developed to facilitate access and applications of GIS with preprogrammed applications. This menu-driven software was developed to shorten the time needed to use GIS. Like ArcInfo, it prepares graphics in response to inquiries of a database. It is less complex than ArcInfo, but more training in its use is required than for Map Maker.

It is important to know the difference between what you want and what you need. Approach the preparation of graphics as though every layer has a cost associated with it. If there is information on a graphic that does not contribute to the visual message, turn off the layer. Finally, do not be afraid to hand draw on plotted graphics if necessary. Just be sure that the pen or pencil used to do so is suited to the paper used for the plot. Sometimes it is quicker and easier to create a new layer of information by hand. Often, graphics are created using multiple digital layers or overlays. The problem with this is that it can result in confusing images,

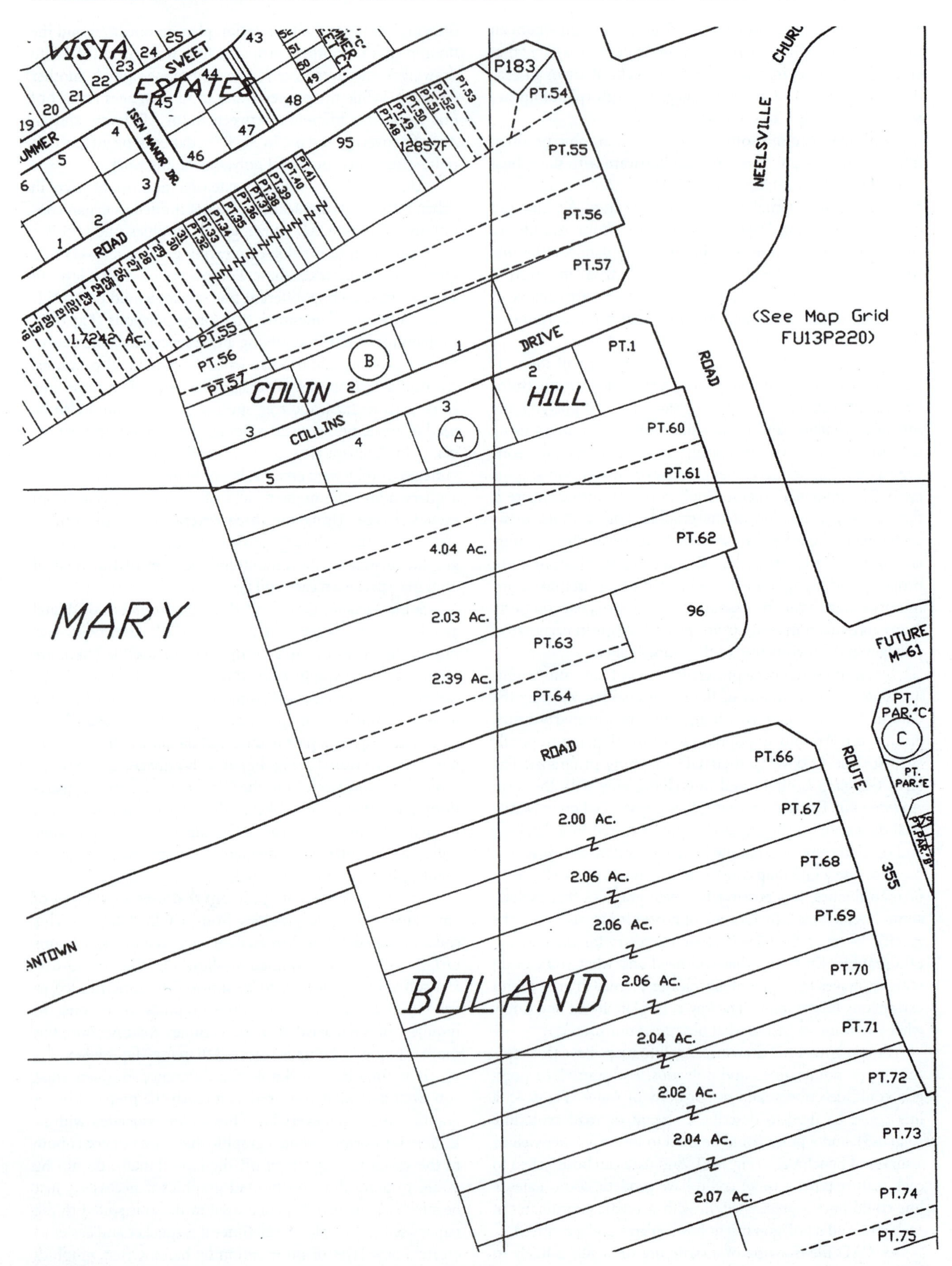

FIGURE 6-1. An example of a property map created using ArcInfo.

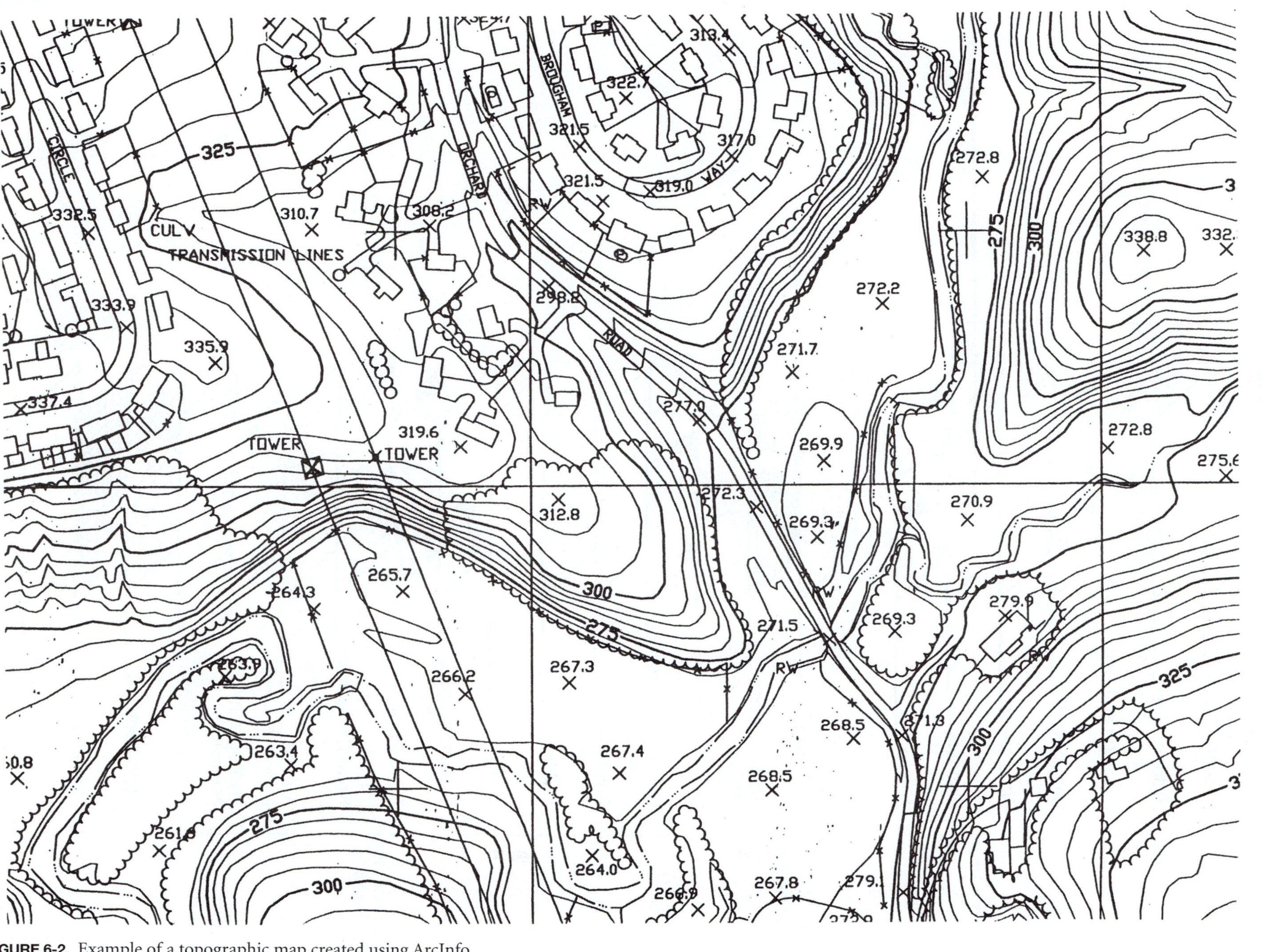

FIGURE 6-2. Example of a topographic map created using ArcInfo.

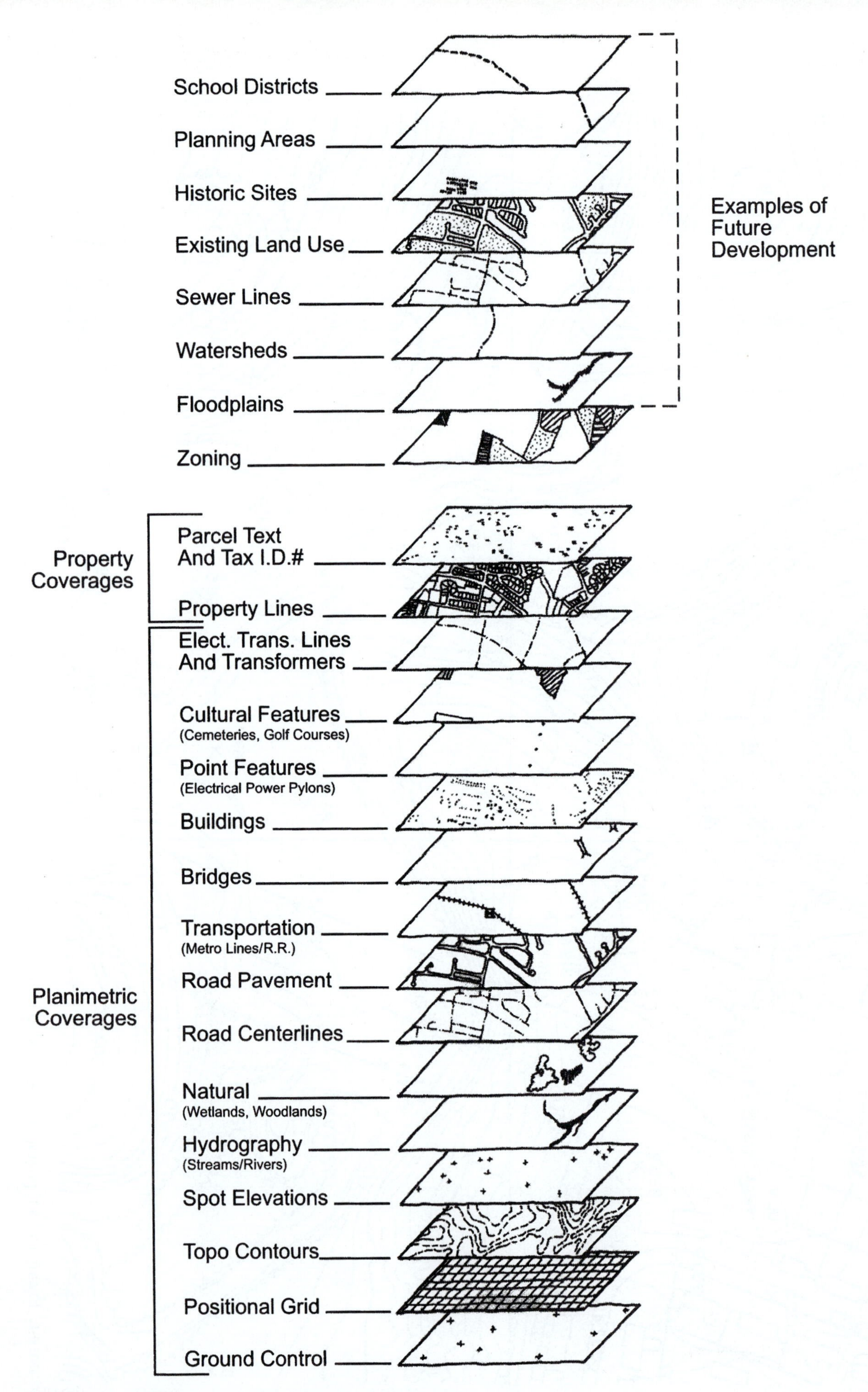

School Districts
Planning Areas
Historic Sites
Existing Land Use
Sewer Lines
Watersheds
Floodplains
Zoning
Examples of
Future
Development
Property
Coverages
Parcel Text
And Tax I.D.#
Property Lines
Elect. Trans. Lines
And Transformers
Cultural Features
(Cemeteries, Golf Courses)
Point Features
(Electrical Power Pylons)
Buildings
Bridges
Transportation
(Metro Lines/R.R.)
Road Pavement
Road Centerlines
Planimetric
Coverages
Natural
(Wetlands, Woodlands)
Hydrography
(Streams/Rivers)
Spot Elevations
Topo Contours
Positional Grid
Ground Control

FIGURE 6-3. A GIS map consisting of a series of layers.

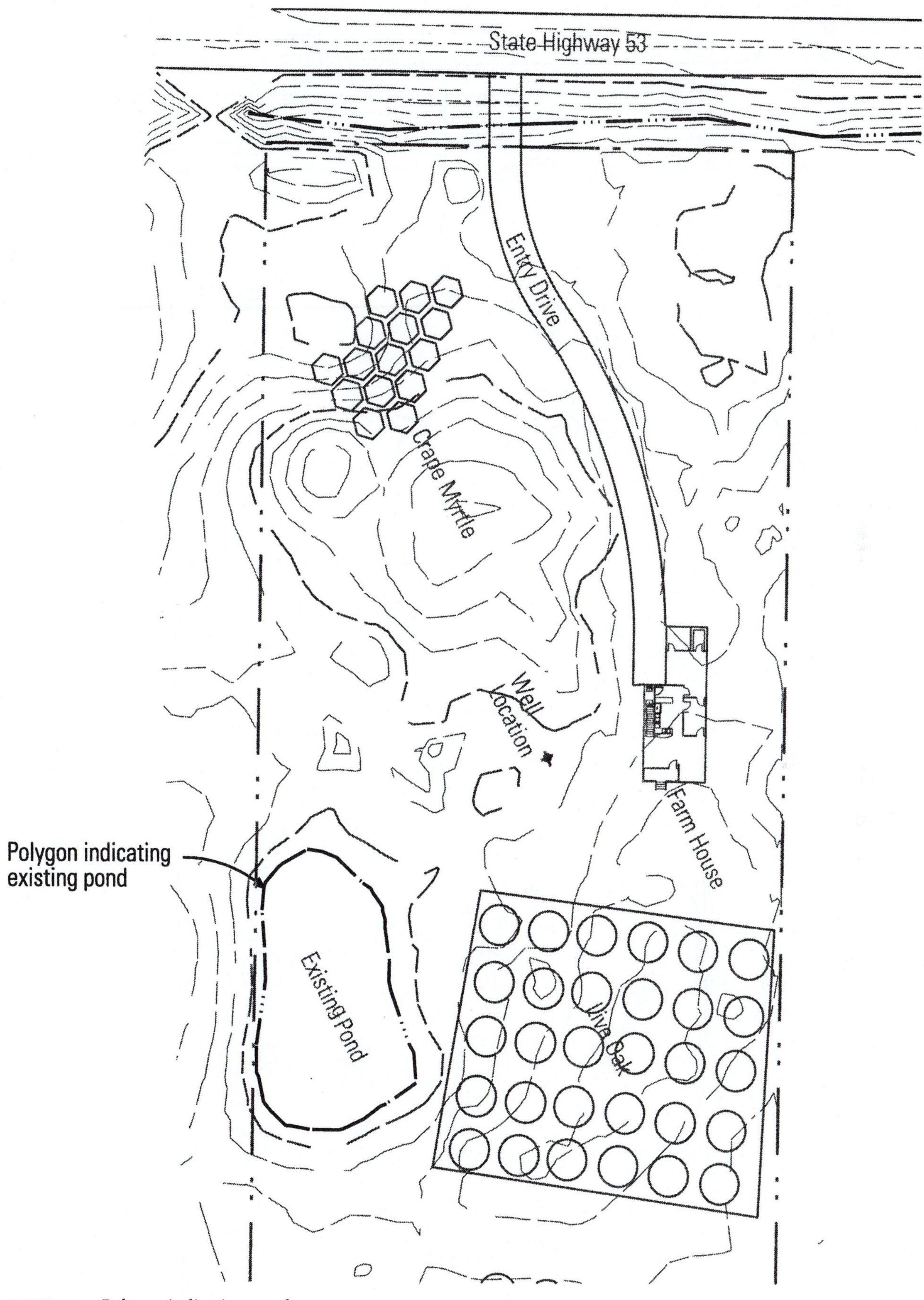

FIGURE 6-4. Polygon indicating ponds.

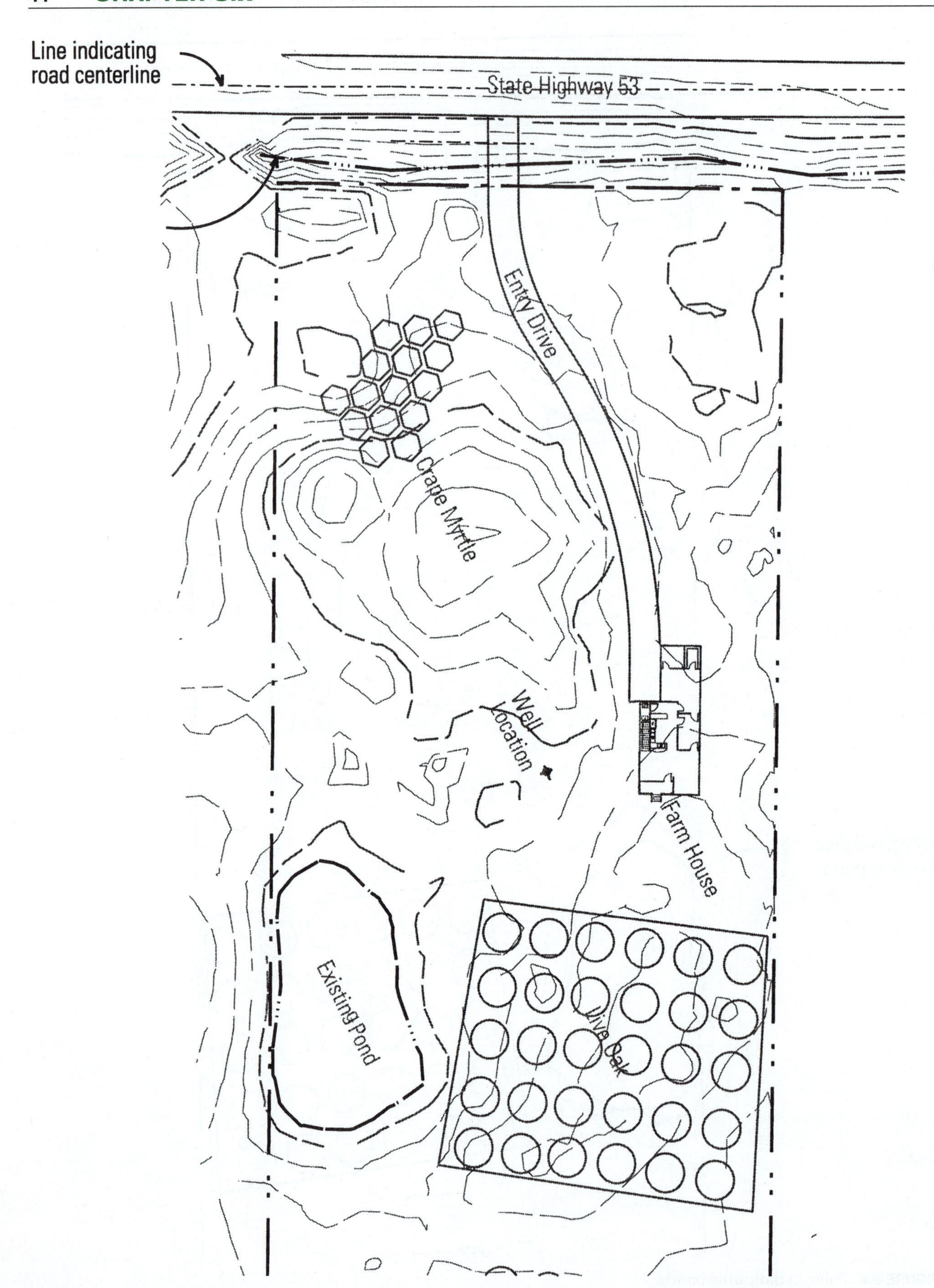

FIGURE 6-5. Lines indicating roadway centerlines and swale centerlines.

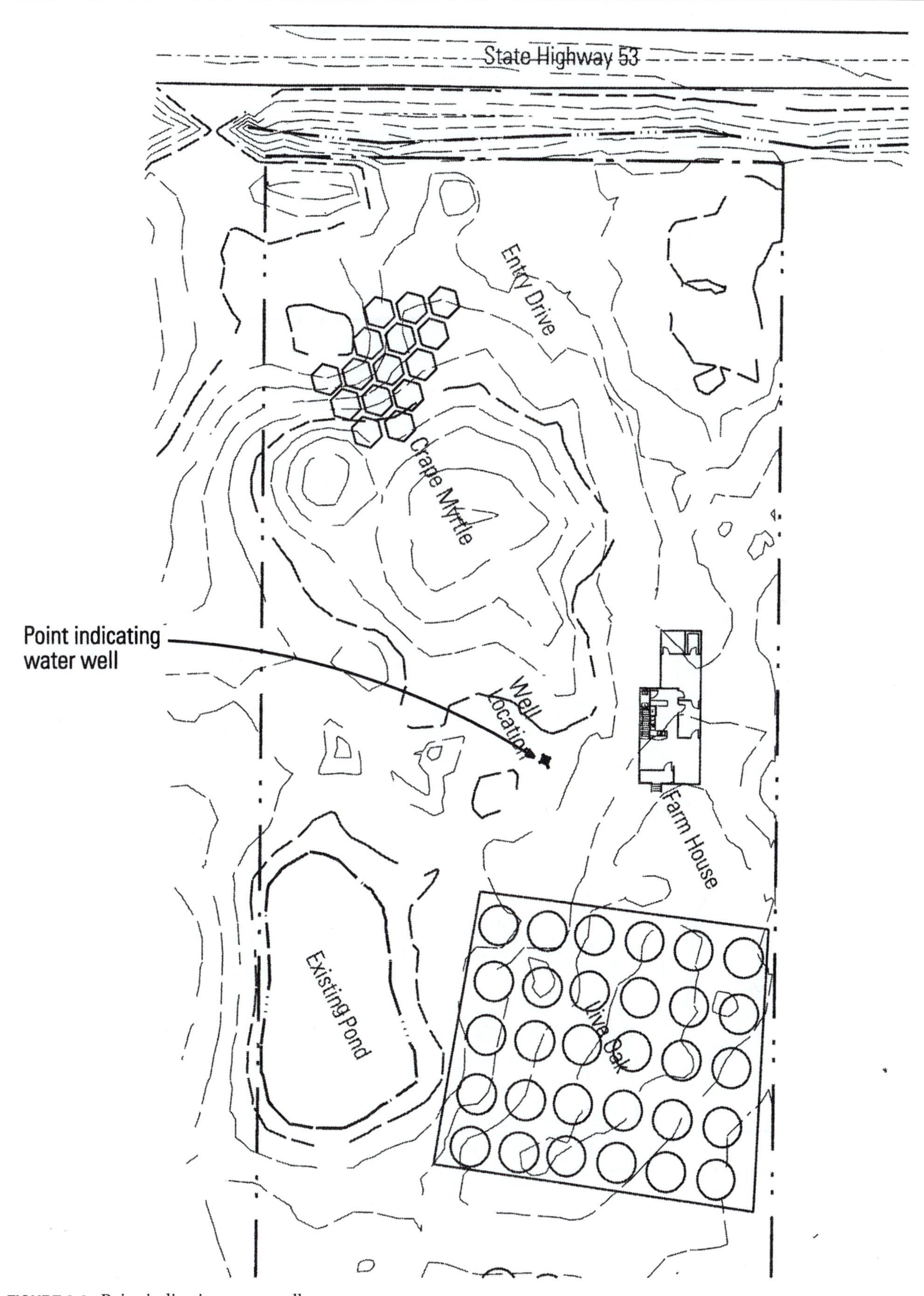

FIGURE 6-6. Point indicating water well.

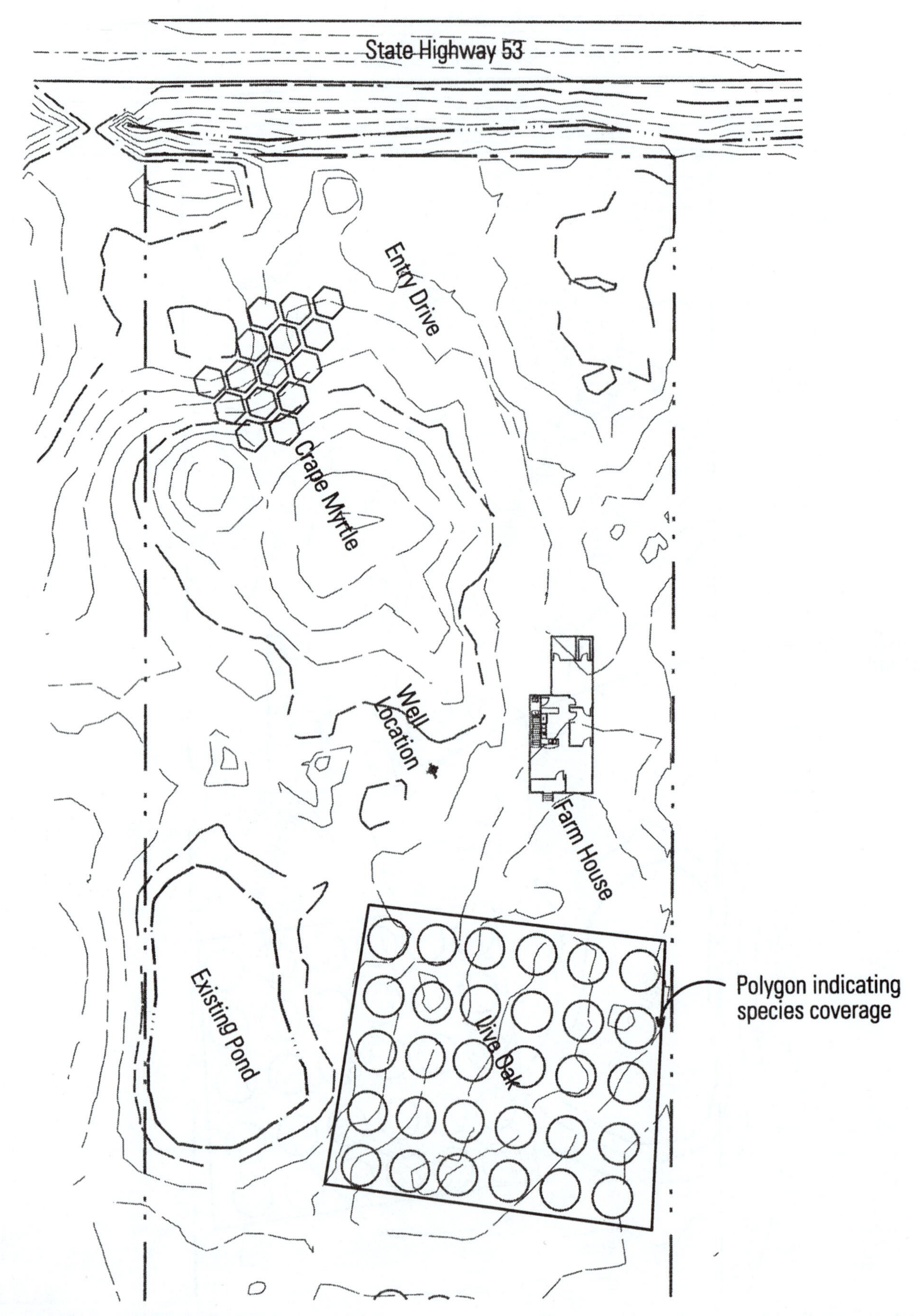

FIGURE 6-7. Polygon indicating species coverage.

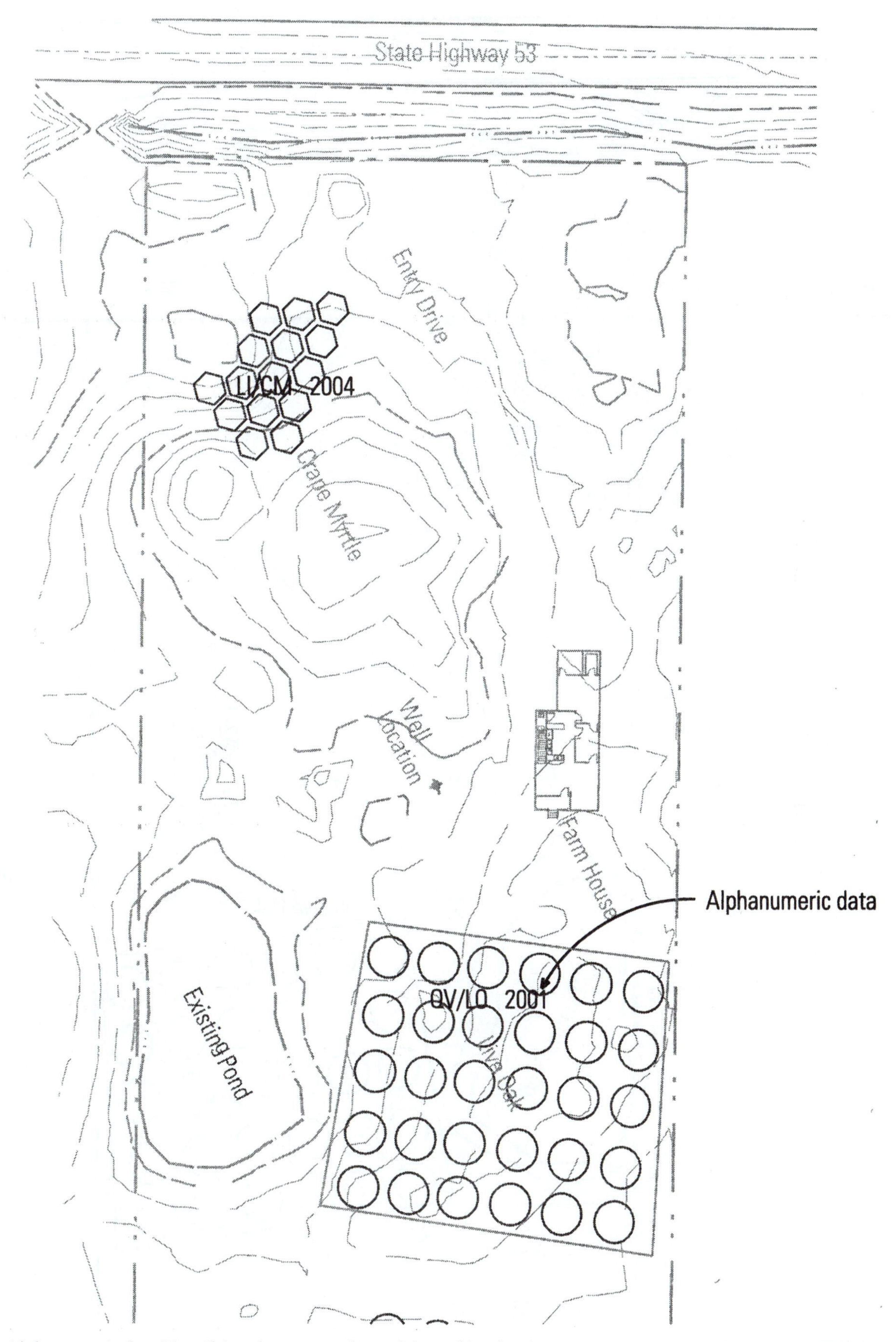

FIGURE 6-8. Alphanumeric data identifying the tree species and date of its planting.

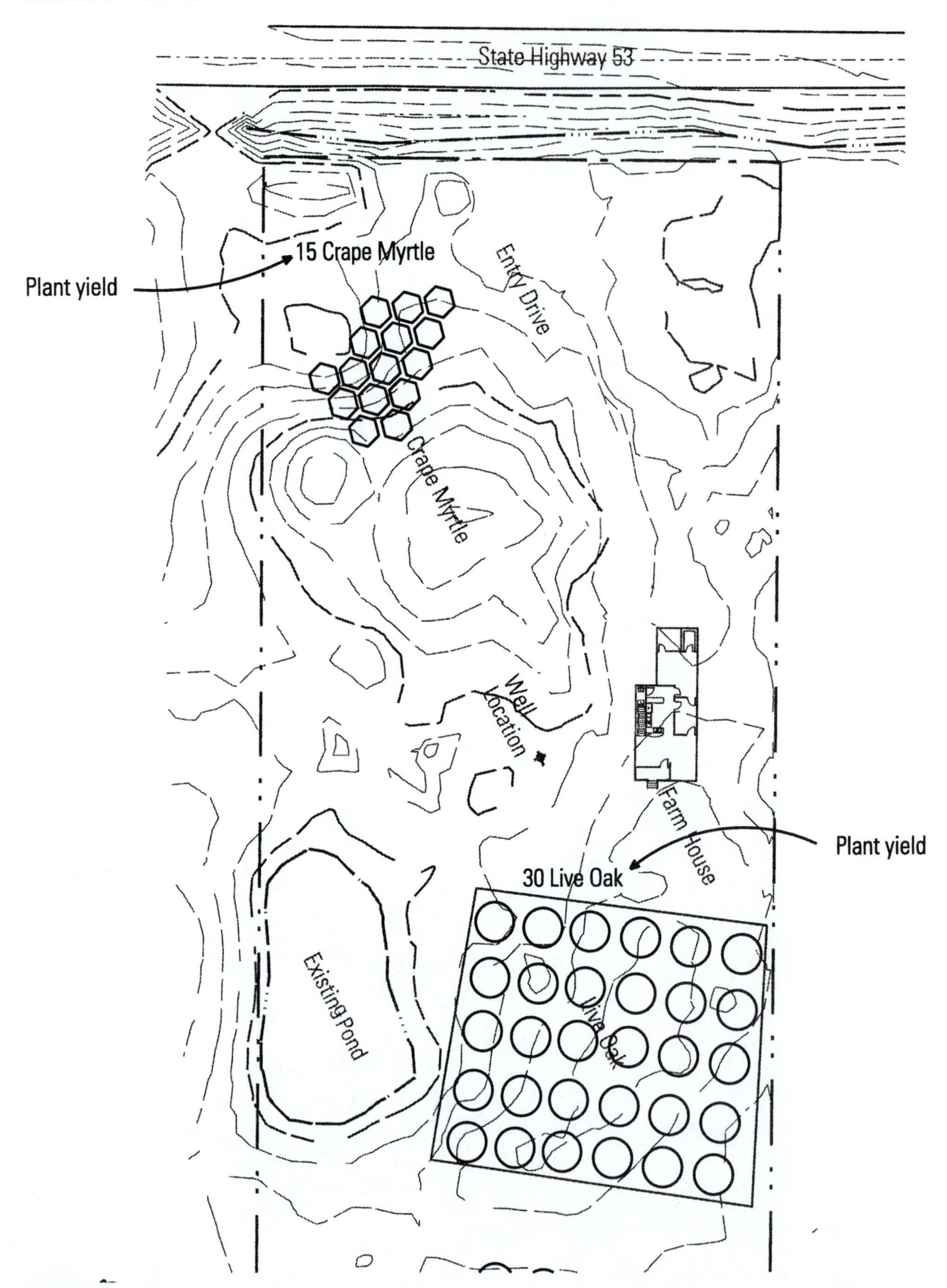

FIGURE 6-9. Timber yield coverage.

huge file sizes, and graphics that can never realistically be used in publications. Avoid requesting a map by using the name of the software used to create the map. And always request a digital ortho to be used alongside a map that displays the same information with line work. To begin, ask and answer these questions:

Project Data

1. Is this graphic for analysis only?
2. Is this part of a base mapping of information and data?
3. Is the graphic available in digital form? If so, what software and what release or version was used to create it?

Presentation

1. What are the messages you want to convey in the graphics?
2. Who is the audience, and what is the size of the event?
3. Does this have a long-term life, or is this a one-time meeting?
4. Would this graphic be used in a final report or publication?

The range of GIS data available from local, state, and federal governments grows yearly. The more comfortable agencies have become with the digital world and the ability of their staff to serve the public's demands, the easier the access has become. Furthermore, what we know to be available today could easily change before the ink is dry on the first printing of this book. Visit the Web site of every agency you think could be a resource or of assistance and make a record in the project file for future reference.

Property or Cadestral Maps

The property (line work) coverage and companion property tax database file are the foundation of most urban GIS mapping. While each GIS should be designed to reflect the problems and issues to be considered through mapping, most urban GIS efforts supplement the property coverage with either topography or planimetric line work. *Planimetric line work* describes building and utility structures, road paving, railroads, and so on—that is, the features that are visible on a digital orthographic aerial photograph and can be digitized from a digital ortho photograph.

The premises address data in a property tax file permits linking the site address to the property map and other characteristics in the property database. This linking creates a base for the GIS using the ArcInfo software. The common thread here is the database, which provides an ArcInfo user with an array of information that can be queried and displayed in graphic form. And because the premises address and tax identification number of a property is unique to that property, it is one of the most common links to other characteristics in a GIS.

There are at least two different philosophies regarding the creation of a property base map. Some jurisdictions with limited funds opt to create GIS property base maps by scanning existing hand-drafted maps. Additional subdivisions are simply scanned and attached on a best-fit-possible basis. A point, usually near the centroid of each parcel, is selected as the point to which the database information will be attached. This method is common because many political entities recognize the value of GIS but do not have the financial resources to commit to the creation of a property map complete with surveyed benchmarks and companion ground control.

There are at least two different methods jurisdictions use to prepare property maps. First, some jurisdictions create hand-drawn base maps from record plats prepared by registered land surveyors that have used surveyed benchmarks as ground control. Others use a less accurate method that involves simply splicing a new record plat into an existing base map of the same scale. While the initial scanning of the new record plat is the same, the rigor of surveyed benchmarks keeps the ground control for all subsequent subdivisions and digital orthogonal photography closer to reality. Jurisdictions that have the finances to invest in more accurate property line work will more likely use the centroid (point) of the closed property polygons for GIS applications. Be aware of the differences between the two methods; the first is certainly not the second, and an assumption to that effect can be expensive to correct.

Because each coverage is a separate layer, all the coverages can be combined in any way necessary to execute analysis, as long as all share the same geography and a common database. The database is likely to be the most limiting aspect of a GIS, as the database creates the limits of GIS-related questions. If a database contains the related statistics, a graphic answer can likely be plotted. Remember that it is common to find that some property maps are pictures of property maps and surveys. In those cases, the geography of one coverage may not be consistent with that of another. Making assumptions and judgments in these cases should be recognized, along with the risk in the assumption.

Finally, creating a GIS is not a simple exercise. The best way to understand all the applications is to see if your local government has an operational GIS. Seeing is not only believing, but it can open intellectual doors you may not have known existed. For more information and details regarding the applications of GIS, see *Understanding GIS: The Arc/Info Method* by the editors of ESRI Press.

BUILDING INFORMATION MODELING (BIM)

Using the proven concepts employed in GIS software, engineers are well on the way to completing a GIS for building construction. The data aspect is attached, for example, to a building component such as a window, where the window is considered to be the "geography." The data describing that window unit becomes part of the database for the structure and includes information on the window framing material, the glazing, type, thickness, dimension, and so on. This will likely become an exceedingly powerful tool for architects, construction managers, and contractors.

BIM is a developing digital standard for construction and facility management that goes well beyond the capabilities of CAD. Typically modeling sites and buildings in 3D, BIM gives users real-time information about a project. A project developed using BIM can track it from conception, design development, and construction through to the ongoing use and maintenance of facilities. Though its application is most powerful for the building industry, BIM's ability to model spatial relationships, geographic information, and material quantity extractions makes it a potentially powerful tool for site planners.

Functionally, BIM represents a major step forward from CAD. The advantage of this platform is that it bridges the gaps of information loss between designers and general contractors and between contractors and facility owners and managers. A BIM system provides a mechanism to attach data to nearly every aspect of a project. This data includes materials, suppliers, quantities, costs, construction methods, schedules, specifications, and regulatory information.

The significance of BIM lies in the project coordination aspects of the technology. Construction industry leaders believe that widespread use of BIM will have a profound positive effect on project management as its potential for improved coordination is better understood. For example, owners, contractors, and facility managers can be much more involved with a project during the initial design phase and can therefore better maintain cost-effectiveness of the design decisions. In comparison, today the interaction between contractors and designers might be viewed as impractical or even as an intrusion into the process. The future of BIM and its broad application will likely make the interaction between designers and users more commonplace.

PART THREE

AN ENVIRONMENTAL INVENTORY AND ASSESSMENT

- **CHAPTER 7:** Mapping: The Method of Inventory
- **CHAPTER 8:** Geology and Soils
- **CHAPTER 9:** Vegetation
- **CHAPTER 10:** Hydrology
- **CHAPTER 11:** Climate and Site

CHAPTER SEVEN

MAPPING: THE METHOD OF INVENTORY

Each of us at one time or another has become exasperated with the response we got from a verbal description and has exclaimed: "Do I need to draw you a picture?" or, as a rhetorical question in an explanation, asked "Do you get the picture?" The implication is that "you may not understand what I have said, so I will resort to a more remedial form of communication: pictures." We have heard that "a picture is worth a thousand words" and "a model is worth a thousand pictures." Both expressions reinforce the fact that pictures provide us with a method of communication that permits us to understand conditions that words cannot describe. This is not to say that words are less important than pictures, but drawings do permit us to delineate some things that are better portrayed in drawings or maps.

It would be outrageous for us to think of design without drawings. It is equally absurd for us to design without analysis, to analyze without documentation, or to attempt documentation without drawings. Without documentation of a site's character, the assessment of a site is relegated to memory and perceptual bias.

Many professions understand the importance of good documentation and the translation of something into visual evidence for clarity or comparison. The physician uses x-rays, an electrocardiogram (EKG), brain waves, and CAT scans, all of which help provide a better "picture" of an individual's physiographic character than can be obtained otherwise. The social scientist uses pie charts and graphs. The importance of the translation of information from numerical or statistical status is not only for the purpose of clarity but also to provide us with the ability to compare and examine visually the relative significance of each constituent.

If each of us recognizes the value of inventory, documentation, and analysis, why is it not employed more rigorously? The explanation may vary from case to case, but the theme is common. Someone—architect, engineer, landscape architect, or client—is applying predetermined solutions to a site without regard for the site's environmental character or context. It is likely that "solutions" of similar origin in an operating room would provide grounds for a malpractice suit.

DUE DILIGENCE

Due diligence is a term that is commonly used by individuals considering the purchase of a property, a business, or an investment. The term is used to refer to an inquiry regarding the assets and liabilities of an economic venture or acquisition, as in "I have completed the *due diligence* necessary to further consider this project." While the process is one that traditionally focuses on financial issues, the term has broad application also to professionals engaged in the multifaceted aspects of site planning. In today's litigious society, *due diligence* is a seriously relevant exercise.

There are almost no projects that are so fat with fees that they are not on a clock. This is a simple fact of private practice. If one is asked to abbreviate the scope of work to fit a budget, one should consider consulting an attorney and have contractual language crafted that abbreviates the designer's liability. There are projects or commissions that you do *not* want to have—at almost any fee. A client may be unwilling to pay for a topo or a boundary line survey yet want you to proceed as though the boundary line sketch of the site is as accurate as one prepared by a licensed land surveyor. A site is rich with opportunity for litigation. The intent here is to identify those aspects of the site-planning process where omissions and assumptions are common.

The word *caution* is worth repeating as it is similar to the concept of restraint. Many designers are so eager to begin the creative leap that they leap before they have defined precisely where they will land, along with the character of that landing. This is particularly true when every site on paper, absent topography, is flat—not that the site actually is flat. Remember that a physician's methodology is to check a patient's vital signs and health history, along with the histories of the patient's parents and siblings. The reason, of course, is that the physician wants to get it right. The mistakes that are common are not so much errors in math or programming as failures to take the time to get complete accurate facts to begin with.

Since no site exists within a static, pristine set of conditions, it is common for an activity or circumstance offsite to affect or limit the long-term use and development of a site. The physician's inquiry as to a family's health history has the same intent. Human-made or natural, conditions and qualities need to be mapped before a complete picture of a site's assets and liabilities can be clearly evaluated.

Mapping often provides us with the basis for some clear decisions, or at least some intelligent questions: "If these are the circumstances, why do I have this property?" This brings into the process a most important person: the owner and the client. Your documentation, analysis, and recommendations for a site create the first level of understanding upon which the client will participate in the site-planning process. Your credibility will be based on the accuracy and objectivity of the research and your perception as to:

1. The conditions of a site's environmental quality and context
2. The opportunity for change as it relates to your client's program

THE METHOD OF INVENTORY

Since the quality of each site will vary, the scope of the inventory needs to be sufficiently inclusive to consider all reasonable variables. The role, value, and contribution of a site's various natural systems are in part based on their species, condition, maturity, stability, and diversity. Although a higher level of detail in the site inventory provides for a more ecological response, the following are qualities common to most sites that require documentation.

Topography: Topography is a collection of contour lines, a two-dimensional graphic representing a three-dimensional character. The documentation of a site begins with topography as contour lines describe the form and shape of the site's floor (Fig. 7-1). Reading and understanding topography are important, as any change in contour or slope will either reduce or increase water runoff patterns, increase or reduce erosion and/or sedimentation, and potentially influence a site's entire stability.

Collectively, contour lines describe another characteristic: slope. Slopes portray the relative incline of a site and are documented in percentages. A slope is expressed as a percentage of vertical rise or fall over a horizontal distance of 100 ft. For example, if the topographical change is 20 ft across 100 ft of horizontal distance, the area has a 20% slope. Site-specific slope analysis is often broken into smaller percentages: for example, 0 to 4%, 4 to 8%, 8 to 16%, 16 to 20%, and above 20%. Slope analysis is by far one of the best techniques for documenting and quantifying a site's physical topographical character. The resulting picture provides an instant understanding of steep slopes as well as more accommodating grades. Fig. 7-2 is an example of a slope analysis of a 4.6-acre site. See Chapter 13 for a detailed description of slope analysis.

Soils: We look but rarely see soils. Most of the time we see only examples of our misuse and misunderstanding of their capabilities. Soils can be considered for their engineering qualities: depth, permeability, shrink/swell characteristics, chemicals, slope, stability, or resource potential. Although the identification of soil types is good information, when coupled with slopes, a definitively more complete picture of a site's tolerances becomes visible. Soils and topography are essential indices used in evaluating the most appropriate and inappropriate locations for structures. The shape of the site (topography) is the basis for drainage analysis. Soils and slopes collectively identify areas that are unstable, susceptible to erosion, or suitable for a building site and tolerance to regrading.

Subsurface Geology: Subsurface geology and geomorphology provide a site planner with a historical perspective of the site's natural evolution. Of all the documentation relative to a site's ability to support buildings or structure, none is more crucial than subsurface geology. Although bedrock provides an excellent base for foundations, rock outcroppings are a sure sign of rock near the surface and a warning of potential problems in subsurface excavation (Fig. 7-3).

Rock exists in numerous categories, and it's important to identify the depth of rock strata, the depth to bedrock, and the rock's density and permeability. The existence of rock on a site can mean extensive construction costs. Its removal can change a site's stability, settlement potential, and dispersal of groundwater and have an impact on local wells, wetlands, bogs, and swamps. See Chapter 8 for a more detailed review of geology, soils, and their common characteristics.

Hydrology: Water in one form or another is a part of every site. The surface water is easy to identify and can often become an amenity for a site. Water affects a site as:

1. Precipitation that becomes runoff
2. Groundwater or standing surface water in a pond
3. A water course traversing the site

Since water can change its "contribution" to a site from amenity to a serious problem, the experienced site planner recognizes the potential for both and develops plans to accommodate that change. Chapter 10 provides a comprehensive overview of hydrology and surface water as they influence site development.

Marsh and Wetlands: Probably one of the most misunderstood roles and functions of water for many years was the marsh. For years, developers, architects, landscape architects, engineers, and planning commissions expended enormous amounts of energy filling marshes and swamps. What many failed to understand is that a marsh is a most sensitive environment and absolutely crucial to some fish in the food-chain cycle (Fig. 7-4). Various states have passed legislation that prohibits the filling of wetlands. Probably one of the best examples of wetland protection was expressed in the 1972 decision by the Wisconsin Supreme Court in *Just* v. *Marinette County.* In sustaining the county's prohibition against wetland filling, the court stated:

> An owner of land has no absolute and unlimited right to change the essential natural character of his land so as to use it for a purpose for which it was unsuited in its natural state and which injuries the rights of others. . . . The changing of wetlands and swamps to the damage of the general public by upsetting the natural environment and the natural relationship is not a reasonable use of that land which is protected from

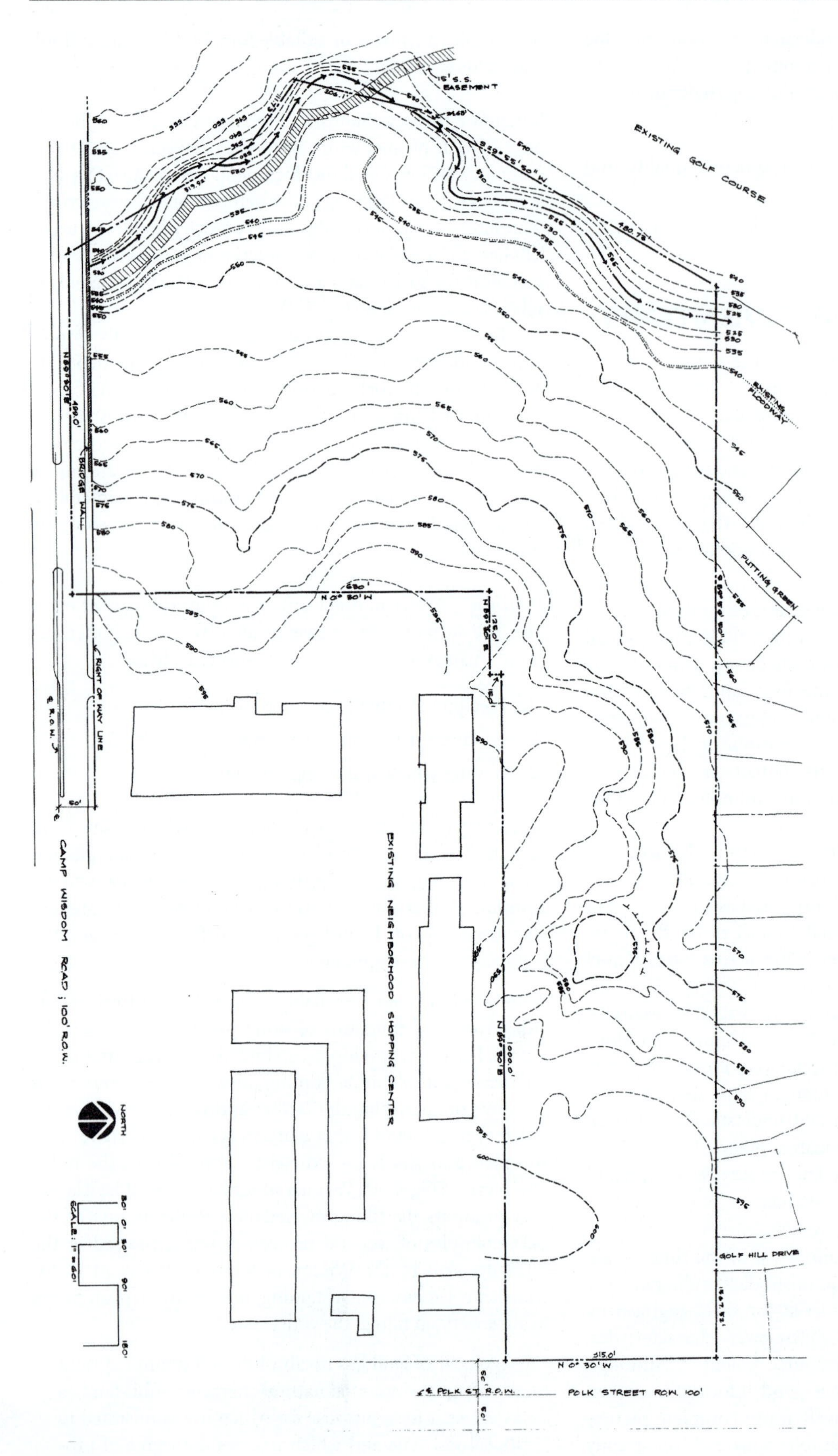

FIGURE 7-1. Topography.

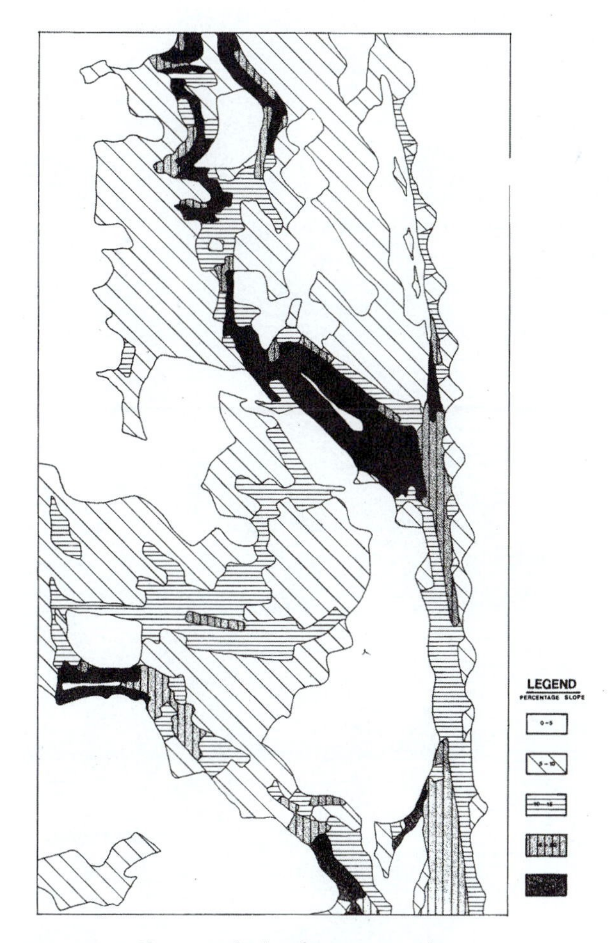

FIGURE 7-2. Slope analysis of a 4.6-acre site.

police power regulation. . . . [N]othing this court has said or held in prior cases indicate [sic] that destroying the natural character of a swamp or a wetland so as to make that location available for human habitation is a reasonable use of that land when the new use, although of a more economical value to the owner, causes a harm to the general public.[1]

The Clean Water Act (CWA) of 1972 set the cornerstone for reducing the degradation of surface water in the United States by establishing the National Pollution Discharge Elimination System (NPDES). The NPDES was expanded in 1987, in Sections 301 and 502, to prohibit any discharge of dredged or fill material into the water of the United States, "including wetlands unless authorized by a permit from the Army Corps of Engineers." If, however, through any on-site analysis or photographic review, any wetlands are found, it is imperative that the perimeter be documented, along with the character of both flora and fauna. The geographic location of any wetlands and photographic documentation will be the primary information that the Army Corps of Engineers will want to review if there is any development near or in a wetland.

Aside from their contribution to the environment, swamps and marshes are inherently poor building sites, as they have high shrink/swell levels and collapsing soils; in addition, by their nature, they drain poorly. There are two other areas that one must approach with caution:

1. Development in areas of high water table
2. Development in areas that have been locations of filling or dumps

Almost any project developed in an area of high water table invites a myriad of problems that only begin with the structure. In the development of any site with a high water table, one should ask "Is this an intelligent place for this project?"

[1] 201 N.W.2d at 768.

FIGURE 7-3. Rock outcropping.

FIGURE 7-4. Common wetland character.

Sites of prior filling or dumps are always rather obscure and sinister. Unless there is a caveat in the land purchase contract that permits the buyer to void the contract if any unusual subsurface conditions arise, in the language of the vernacular, "what you see is what you get." There is nothing quite as demoralizing as core samples that yield bits of plastic milk cartons and domestic tin. The implications are, hopefully, obvious. The additional costs associated with a foundation on a landfill can turn an inexpensive property and profitable development into a financial debacle. It is not that foundations cannot be designed for the minimal bearing capacities of a landfill, but unless the cost of the property was very low, the foundation costs can be prohibitive.

Vegetation: Plant cover is the most sensitive index of soil and weather conditions. For example, willow, hemlock, and red maples mark areas that are poorly drained; oak and hickory associations grow on warm, dry land; and spruce and fir grow in cold, moist places. Each locality has a matrix of soil, water, and temperature characteristics for trees and ground cover.

Each site has a range of levels of plant life that needs to be considered, from the dominant canopy down through the trees of the understory, shrubs, and prostrate ground covers. Trees need to be located and identified by type, height, density, and character (canopy or understory) and noted as to their age or condition. There is often a range of grasses, shrubs, and vines that exists on a site prior to our "arrival" which enjoy a healthy relationship with the wildlife and major vegetation. A change in the site's conditions will probably change the quality of ground cover as well. In the process of site analysis, make note of the uniqueness of the wild flowers, grasses, and vines. If lost, many take from two to three years to germinate, even if successfully transplanted. Some may never again be successful. See Chapter 9 for further explanation relative to evaluating existing vegetation and its role in site planning.

Wildlife: The ethical code graphic in Chapter 2 describes a continuum of rights from humankind to other forms of life. Those rights are arranged in a hierarchy, with humanity at the base, and acknowledging the rights of mammals and other animals in the hierarchy. The U.S. Congress believed the nation's laws were inadequate regarding the protection of a limited class of mammals and in 1973 passed the Endangered Species Act (ESA). Of particular importance to a site planner is the expressed purpose of the 1973 ESA to protect species and "the ecosystems upon which they depend." Congress charged the U.S. Fish and Wildlife Service (FWS) and the National Oceanic and Atmospheric Administration (NOAA) with primary responsibilities for administering the list of endangered species and monitoring changes in their status.

In 1978, the ESA was amended to require the mapping of "habitat" zones, except where the publication of such maps might result in harm to the relevant endangered species. Critical habitats, which may be public or private properties, are required to contain "all areas essential to the conservation" of the target species. A significant new policy of the legislation was that federal agencies were now prohibited from authorizing, funding, or carrying out actions that "destroy or adversely modify" critical habitats. It was not uncommon for projects planned and funded with federal dollars to result in losses of endangered species habitats. The conflicting purpose condition was one that was usually lost by the endangered species advocates as environmental impact assessments were commonly biased toward minimizing economic costs.

The years following the passage of the ESA also witnessed occasions in which private landowners would take overt action to reduce or destroy the habitat of an endangered species for fear of losing the development potential of the habitat to public use. To address the issue, the ESA was amended in 1982, in part to provide landowners with incentives to participate in conservation of endangered species. The legislation now requires the preparation of a "habitat conservation plan" (HCP) through which a private landowner may obtain an "incidental take permit." The permits, issued by FWS, can allow otherwise prohibited impacts to endangered, threatened, and other species covered in the permitting documents. A site planner's responsibility then is not concluded with inventory but continues, where appropriate, to prepare an HCP.

Most of the time a site planner simply needs to recognize that the habitat of small birds and animals is directly affected by any major modification of the environment. The variety of species and their migration patterns will be influenced by any change in the site's natural condition. Sensitivity to the various wildlife habitats and their contribution and dependency on the environment are important in good site plans. But on those occasions when it is deemed necessary, an HCP must specify the impacts to species that will occur; the actions that will be taken to minimize and mitigate the incidental damage; the funding available; realistic alternatives, including those not selected; and any other appropriate or necessary actions to support the protection of the target endangered species habitat.

Microclimate: In an examination of any site, consider its latitude, slope, and exposure to the sun on a seasonal basis. Sites sheltered by hills or bluffs may have dramatically different microclimates than those in similar areas with different orientations. The exposure of a building to the sun is one of the most significant issues to be considered in site planning. This impact cannot be underemphasized and is a very powerful and critical part of site analysis. The sun is a source of energy that should be exploited, and understanding its influence and potential contribution is very important.

Air movement is influenced substantially by slope and vegetation. If unspoiled by tree cover, wind speeds at the crest of a hill may be 20% greater than on flat topography. This higher wind speed at the crest generally means that the wind is quieter on the leeward side of a hill as opposed to the windward side. Although the hill crest tends to shelter the leeward side from the more intense winds, it also the case that drifts are deeper off the crest and down the hill (Fig. 7-5).

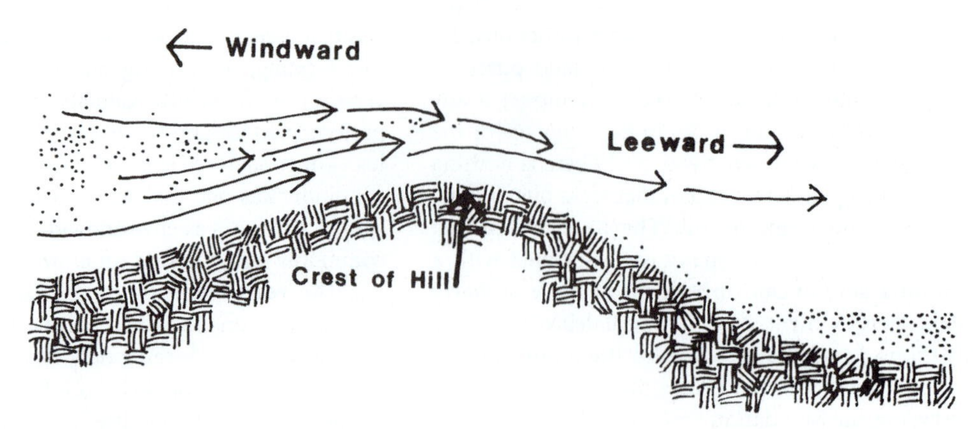

FIGURE 7-5. Wind/snow at the crest of a hill.

Cold air floods are a nocturnal phenomenon on open slopes. A layer of air near the ground is cooled by the ground beneath it, which loses heat by radiation outward to the sky. This film of cold air flows downhill as a shallow sheet, gathering as a stream in an open valley or forming in a pool when blocked by a "dam" of topography, tree cover, or buildings.

Precipitation: Development changes the impact of precipitation on almost every site. The addition of impervious surfaces frequently increases runoff, alters water tables, and has been a principal contributor to increased erosion and aquifer pollution. For these reasons the seasonal patterns, intensities, and major storm periods (snow and rain) need to be considered. The National Weather Service is an excellent source of data associated with precipitation and wind patterns for various regions. This documentation should be integrated into the site planner's analysis (Fig. 7-6). An extensive review of the rationale and methods appropriate to document and interpret microclimate is presented in Chapter 11.

DOCUMENTATION

The most common method of collating and documenting site research is with a series of maps or overlays. The first, or basemap, usually reflects the information identified in parts I and II of the site-planning checklist at the end of this chapter. The basemap is normally limited to elements that lend themselves to graphic presentation. Although zoning might be part of a textual report, boundary lines and constraints should be part of the documentation. The basemap will become the common controlling map for all research. A common method of documentation is to separate specific activities, or site constituents, into their smallest parts. This process is most effective when each environmental constituent (soils, slopes, vegetation, etc.) is translated to maps of the same scale as the base (Fig. 7–7). The value of the method is that it permits the site planner to examine one environmental element, such as surface water, or the collection of numerous elements. This system of overlays can be applied both to a site of a single acre and to a regional planning problem of thousands of acres.

Proper and accurate documentation begins with avoiding secondary and dated sources. Local, state, and federal government agencies are all excellent sources of primary information. It is imperative that the site planner become familiar with the agencies both as a source of information and to become cognizant of their requirements. Since local governments are almost always involved in development, a working understanding of their ordinances, codes, and processes is essential. Such agencies might include the following:

City

- Planning and zoning commissions
- Parks and recreation departments
- Building inspection
- Public works and utilities
- School districts
- Traffic and transportation

County

- County commissioners or judges
- Flood control districts or river commission
- Parks and recreation department
- Health departments
- Highway department

State

- Highway departments
- Water rights commission
- Coastal zone management agency
- Health department
- Department of natural resources or parks and recreation

Federal

- Department of Agriculture
  - Soil Conservation Service
  - Forestry Service
- Department of the Army
  - Board of Engineers for Rivers and Harbors
  - Coastal Research Center
  - Corps of Engineers
  - Mississippi River Commission

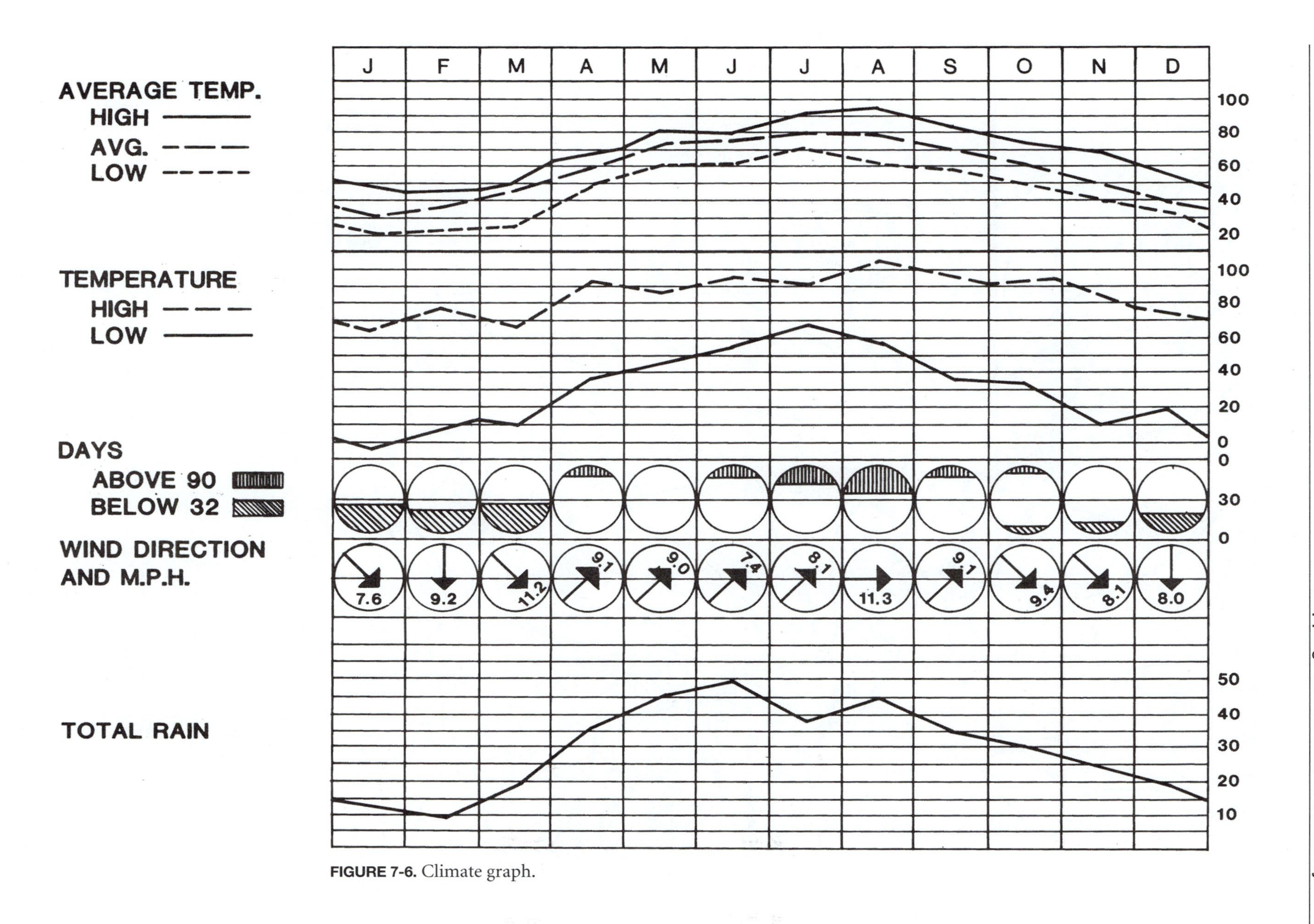

FIGURE 7-6. Climate graph.

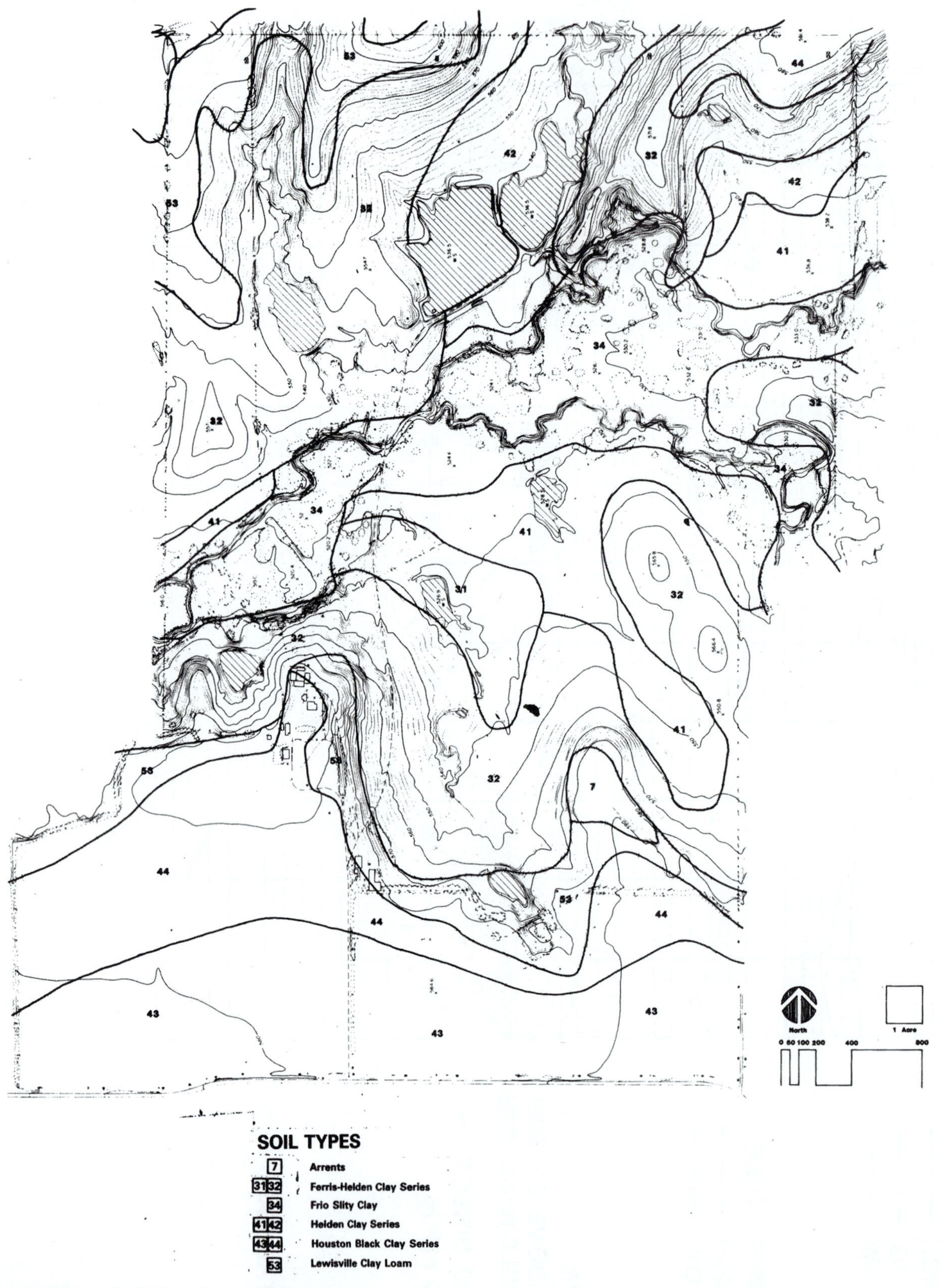

FIGURE 7-7a. Individual inventory map.

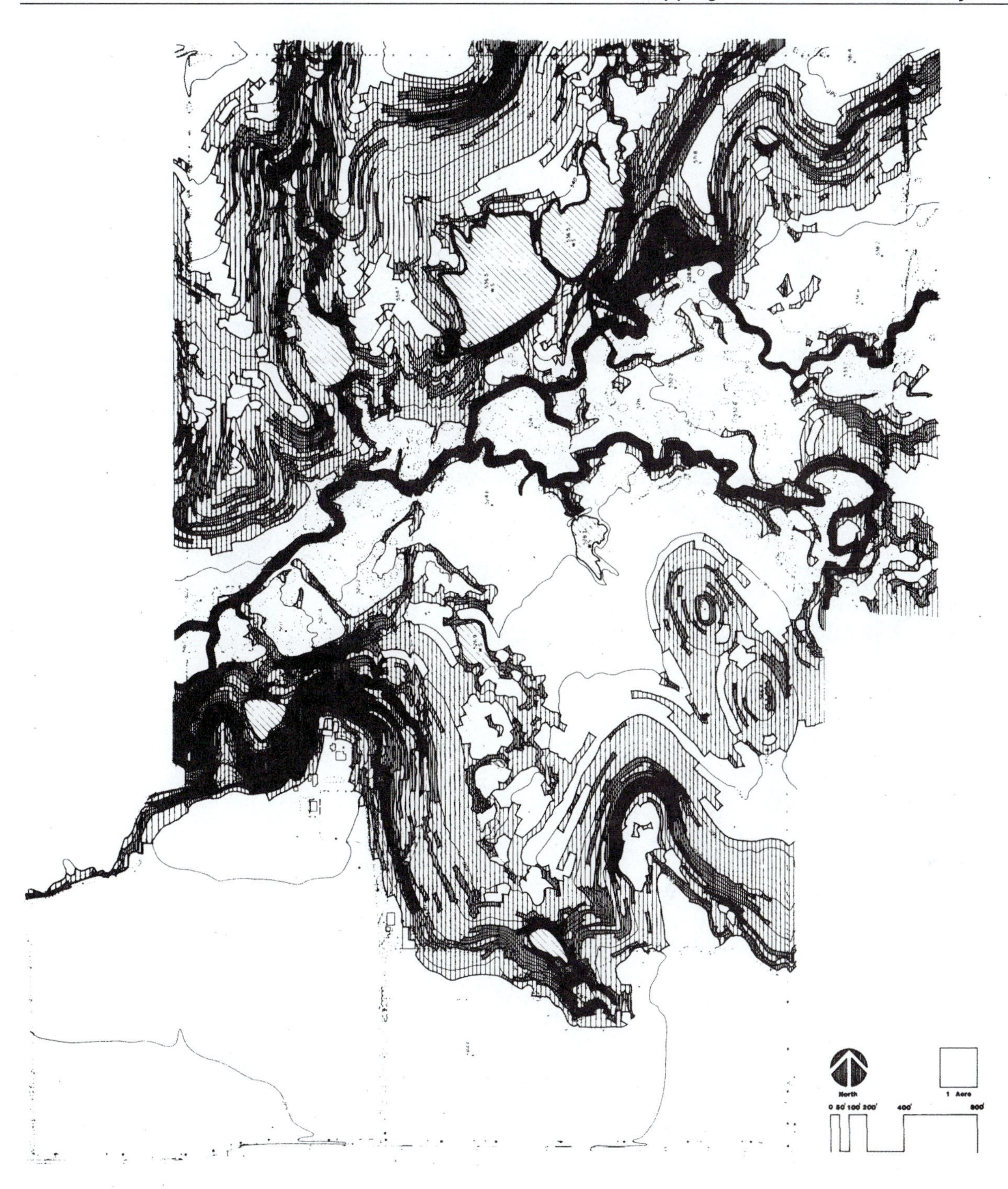

SLOPE ANALYSIS

- 0-5%
- 5-10%
- 10-15%
- 15% & Above

FIGURE 7-7b. *(cont.)*

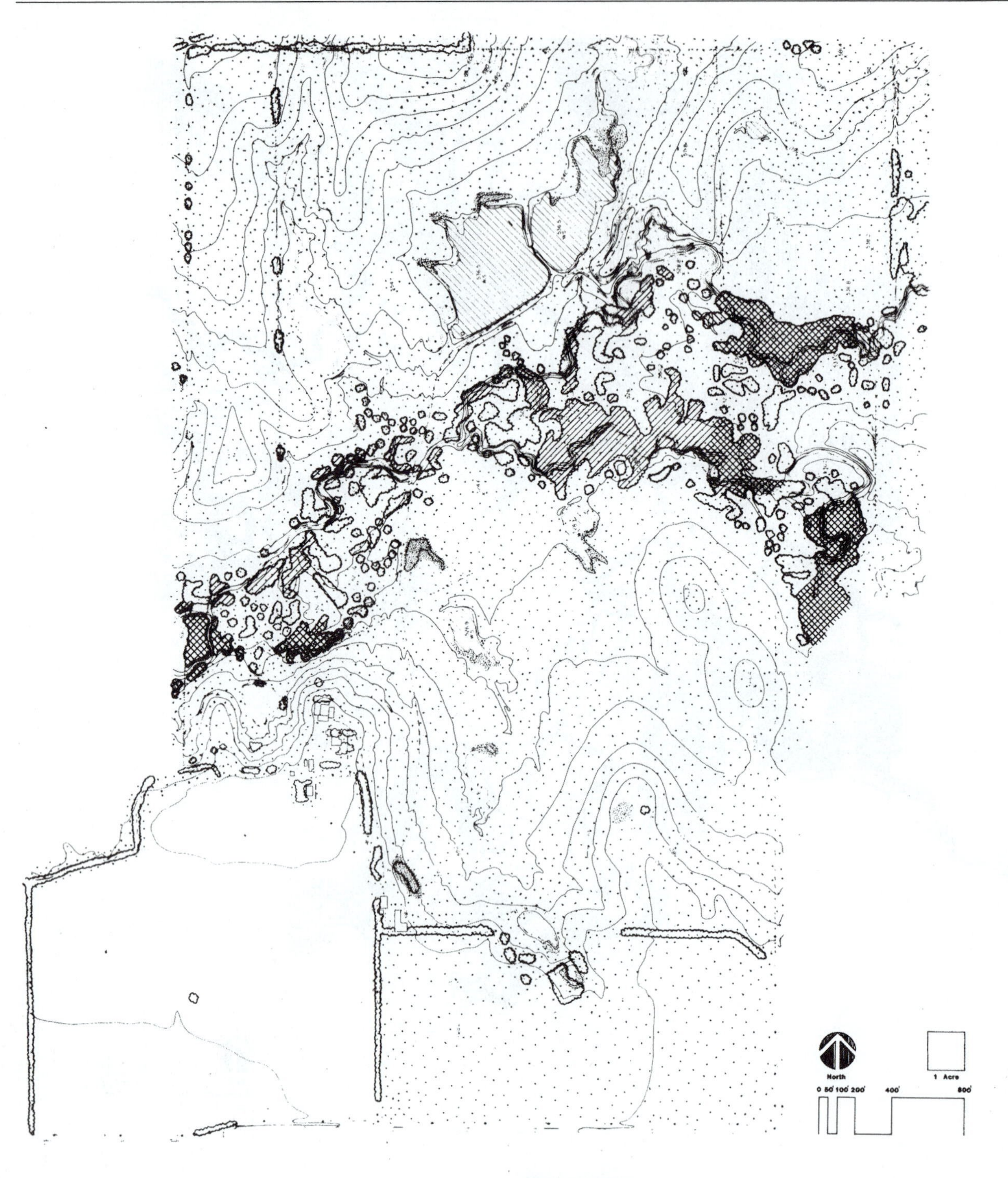

VEGETATION

- Overstory 30' & Above
- Understory Less Than 20'
- Shrubs 18"-8'
- Aquatic Plants
- Ground Covers Less Than 18"
- Cultivated Crops

FIGURE 7-7c. *(cont.)*

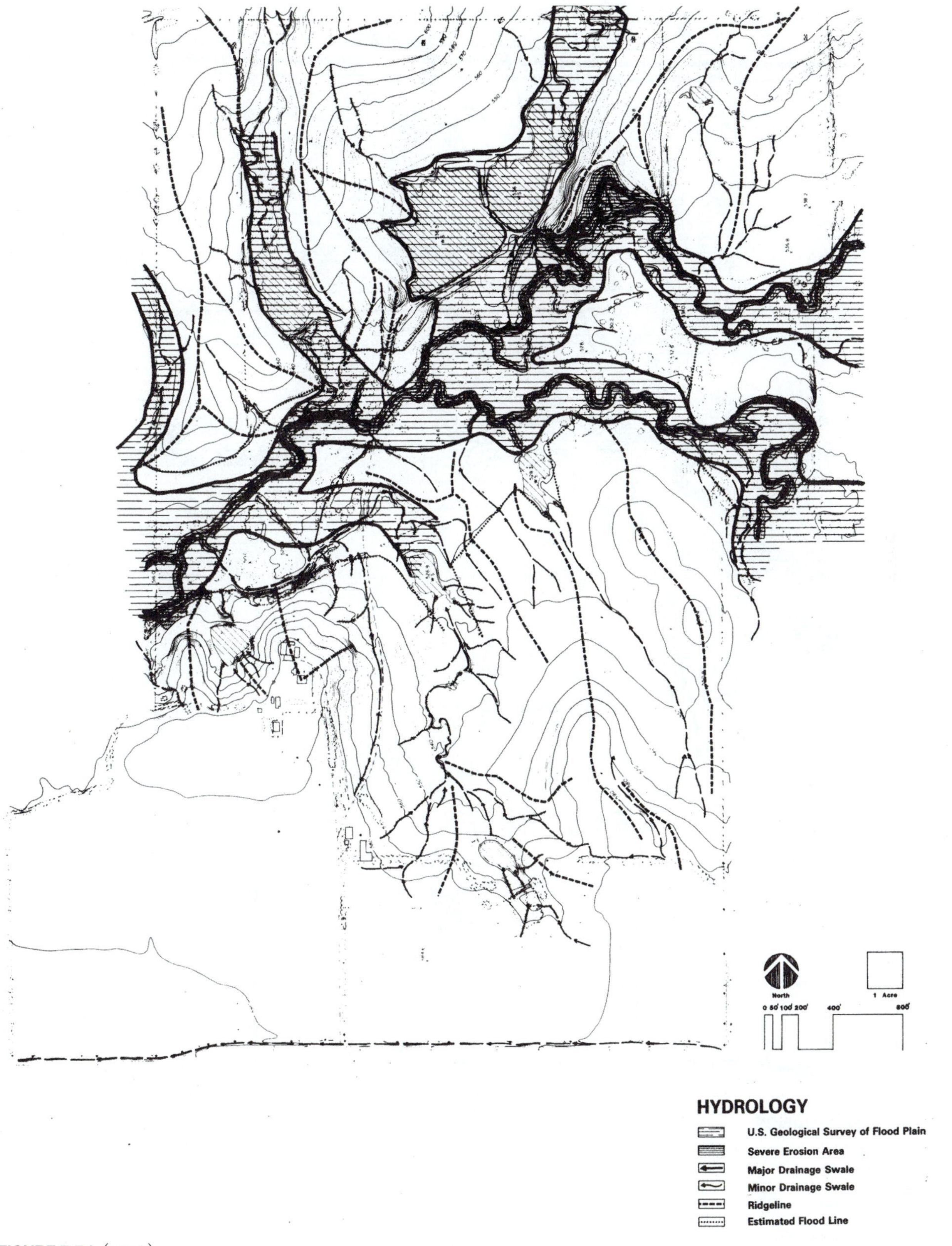
North
1 Acre
0 50' 100' 200' 400' 800'
HYDROLOGY
U.S. Geological Survey of Flood Plain
Severe Erosion Area
Major Drainage Swale
Minor Drainage Swale
Ridgeline
Estimated Flood Line

FIGURE 7-7d. *(cont.)*

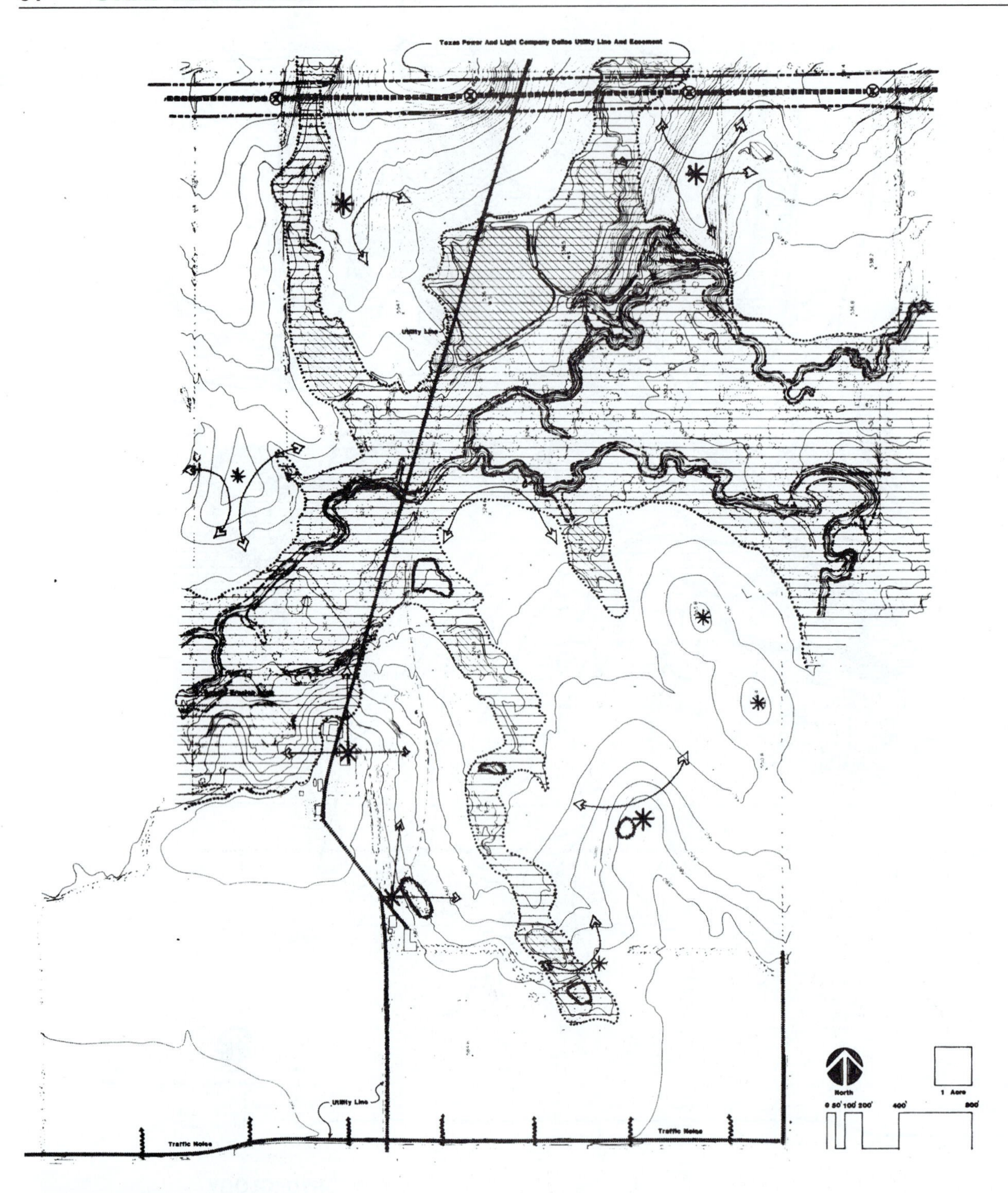

SITE RECONNAISSANCE

- Primary High Point
- Secondary High Point
- Wet Area
- Severe Erosion Area
- Area Unique in Character Due to Topography or Vegetation

FIGURE 7-7e. *(cont.)*

Department of Commerce
- National Oceanic and Atmospheric Administration
- National Ocean Survey

Environmental Protection Agency
Department of Housing and Urban Development
Department of the Interior
- Bureau of Land Management
- Federal Emergency Management Agency
- National Park Service
- U.S. Fish and Wildlife
- U.S. Geological Survey

Three federal agencies collect data on a national scale that can be utilized at the local level. The first is the Soil Conservation Service (SCS). The mission of this agency focuses on three areas: soil and water conservation, natural resource surveys, and community resource protection and development. SCS has a variety of staff nationwide, including engineers, soil scientists, agronomists, biologists, economists, foresters, geologists, archaeologists, landscape architects, cartographers, and environmental specialists.

For the site planner, the soils surveys are extremely valuable, as they provide the base data for land suitability analysis. The research relative to physical properties and chemical characteristics of soils is usually mapped and published on a county-wide basis (Fig. 7-8). The information

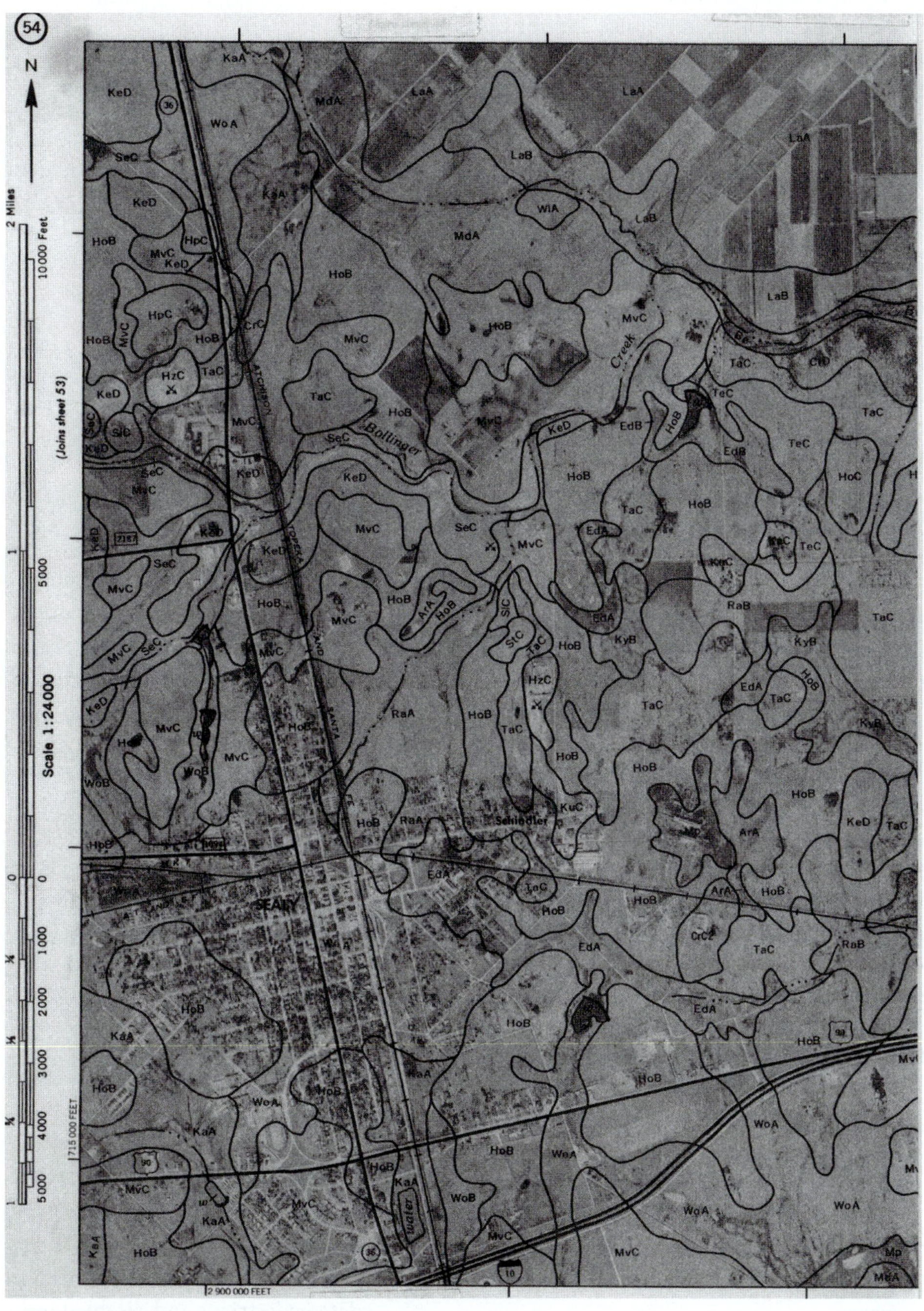

FIGURE 7-8. Soil Conservation Service map.

in the SCS survey can easily be translated and mapped to identify the characteristics of the soils relative to a particular site and their ability to support development. Because of the large land areas and interval of testing, the survey maps should be used as a guide; they are not site specific. As a result, small pockets of soils different from those mapped may be found on a site. Nonetheless, the quality of the dominant soil type should be recognized and considered in the site-planning process. No knowledgeable site planner ignores the importance of understanding the quality of a site's floor: soils.

The second is the U.S. Geological Survey (USGS). The USGS quad sheet at 1 in. = 1,000 ft (1:12,000) is probably the most widely used resource available from USGS. These "topo" maps identify urbanized areas, tree cover, surface water, existing roads, railroads, power lines, buildings, contour lines, and benchmarks (Figs. 7-9 and 7-10). This is a primary resource for situations where more detailed information is unavailable. Although the topo quad sheet is the staple of USGS, the agency has extended its research to include other earth science data. One such area is flood hazard mapping.

Because of the federal government's involvement in flood insurance programs, it charged FEMA with the responsibility of coordinating efforts to minimize flood losses. Part of that oversight role includes mapping those areas of every community that are flood prone. To focus the creation and delineation of data and maps, FEMA created Flood Insurance Rate Maps (FIRM). FIRM are the official maps on which FEMA defines both special hazard areas and risk premium zones of specific communities.

USGS has also mapped land use cover, geological hazards, and surface water. The geological hazards studies focus on earthquake and volcano eruption and provide information on fault systems and active seismic zones. Surface water studies have concentrated on stream discharge and quality analysis and are part of the National Stream Quality Accounting Network (NASQAN) monitoring program. Groundwater is another series of environmental studies under the purview of USGS. Although limited, these studies assess present groundwater supply and identify emerging hydrological problems associated with their continued use.

USGS provides an extensive range of services, from complete cartographic separates to remote sensing. Although the scale of the data is often large, in some cases USGS may be the only source of information.

Site-Planning Checklist and Review

1. *Regulatory, Legal, and Environmental Review*
   - Building envelope study (setbacks and height restrictions)
   - Easement(s) location(s)
   - EPA issues
     - Flood plain documentation
     - Parking requirements
   - Property line survey
   - Special use permit (if required)
   - Utility access
     - Cable TV
     - Electricity
     - Natural gas
     - Sanitary sewer
     - Storm sewer
     - Water
   - Zoning (land use compliance)

2. *Inventory and Analysis*

   Drainage analysis and preliminary grading plan
   - Has a basic understanding of the natural flow of water on and through the site been concluded?
   - Have preliminary decisions regarding the structure's first floor elevation been reached?
   - Have culverts or graded swales been considered as a way to expedite the flow of water?
   - Has a preliminary grading plan that explores changes in elevation for the structure's first floor been developed?

   Microclimate analysis

   Passive solar analysis
   - Has a passive solar analysis mapping the sun's path across the site on December 21 and June 21 been completed?

   Soil analysis

   Slope analysis

   Views and aesthetic analysis

   Tree and "no grading" mapping
   - Have all the trees and understory vegetation with potential been located and mapped?
   - Has a plan describing those areas of the site where no grading is to occur been created?
   - Have paths that provide ingress and egress during construction been defined?

   Conclusions

3. *Planning and Architectural Program Review*

   Impervious surfaces analysis and alternatives
   - Have the areas of all impervious surfaces been reduced to the smallest acceptable or allowable?
   - Are there "open paving" alternatives that could be considered for walkways or parking?

   Parking requirements and alternatives
   - Have the parking requirements been reviewed to determine that they are consistent with the project's needs?
   - If appropriate, is a variance to reduce the parking requirement being considered?

   Pedestrian and bicycle access
   - Have all the considerations necessary to accommodate pedestrians and disabled individuals been reviewed and developed?

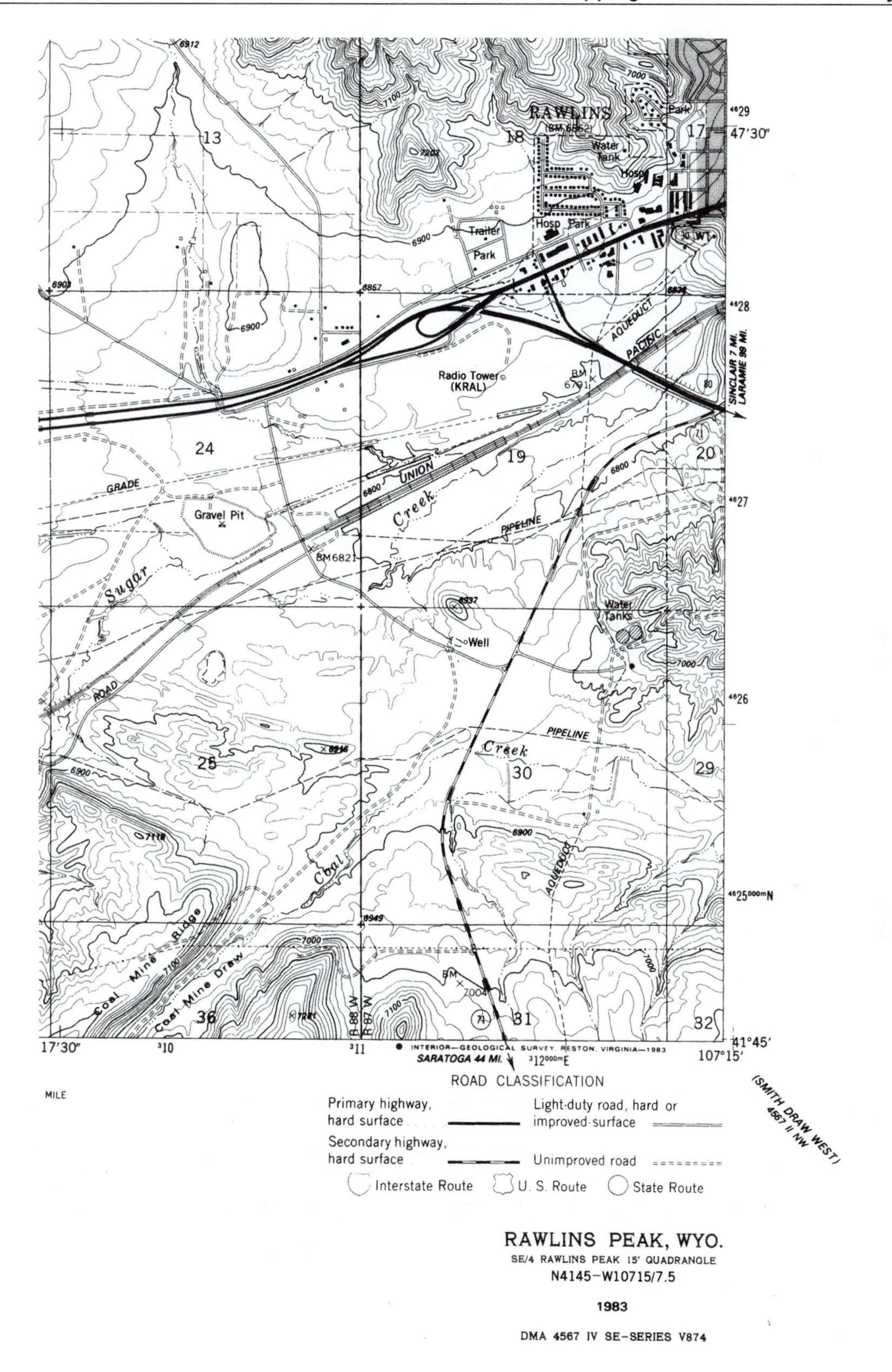

FIGURE 7-9. USGS map.

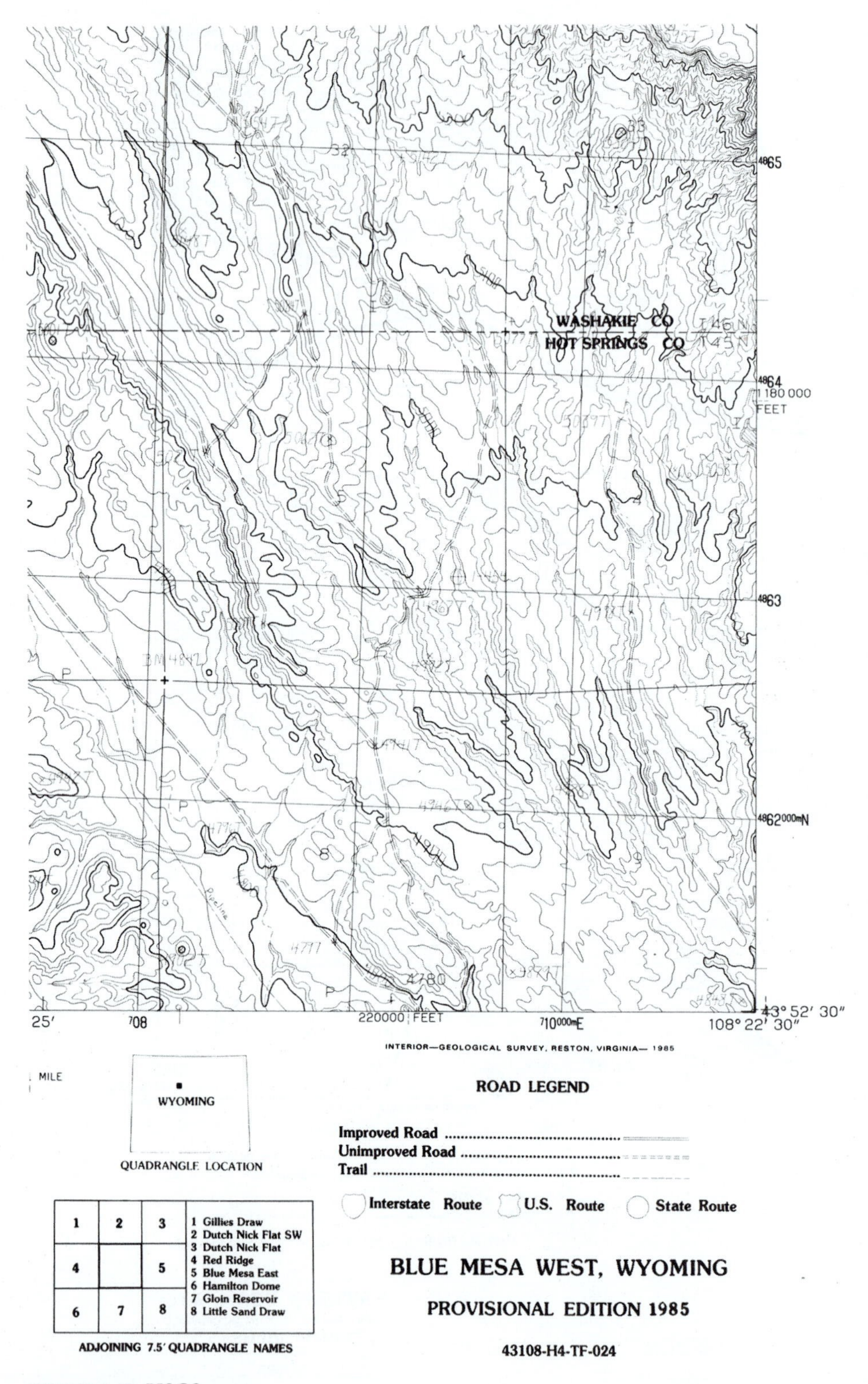

FIGURE 7-10. USGS map.

- Have bike storage racks and storage lockers, where appropriate, been located to accommodate cyclists?

Regulatory review requirements

- Do any aspects of the project require review or permits from federal, state, or local government agencies?

Variances required

Vehicular, service, and emergency access

4. *Design Issues*

Building "footprint"

- Has the architectural program been reviewed to make certain that the first floor plan of the building is the smallest acceptable to the client?

Exterior public and private open space

- Have the site's trees and understory been utilized in the development of public and private exterior open space?

Structure location and orientation

- Has the structure been sited to take advantage of the site's view and inherent natural qualities?

Transition from parking to building entry

- Has the transition from parking to the building entry been designed as an "arrival" to the structure's primary setting?

5. *Digital Resources and Layer Protocols*

Orthophotography

AutoCAD

MicroStation

GIS

The development of the data outlined in this checklist is critical to intelligent site planning, and it is also invaluable in the event of site-related litigation.

CHAPTER EIGHT

GEOLOGY AND SOILS

by Michael E. Ostdick

Geology is the study of the earth. The fundamental principle that underlies most of geology is that the present processes occurring on the earth have occurred throughout geologic time. Subsurface geology and geomorphology provide the site planner with a historical perspective of what has happened to a site as well as what potential problems may be encountered.

There are three basic types of rocks, which are classified according to their origin: igneous, metamorphic, and sedimentary. Igneous rocks are produced from solidifying melt or magma. Metamorphic rocks are rocks that have been changed by high temperature and pressure within the earth's crust. Sedimentary rocks are weathering products of preexisting rocks that are deposited near the earth's surface by wind, water, ice, and biological activity.

SUBSURFACE GEOLOGY

The type of rocks in underlying formations gives a good indication of their ability to accept the load of a building. Different rock types have varying bearing capacities for safe loads on earth foundation beds that need to be investigated when considering extensive building loads (Table 8-1). Equally as important a consideration in known earthquake regions is that buildings built on bedrock will be damaged less than those built on less consolidated, easily deformed material such as natural or artificial fills.

The layer below which rocks and soils are saturated with water is called the water table (Fig. 8-1). This level changes with the seasons. In general, it is a reflection of the surface topography but is more subdued; that is, has less relief than the surface topography. The action of groundwater is controlled by the physical properties of the rocks. The most common reservoir rock—rock porous enough to hold and filter water—is sandstone.

Groundwater is a very valuable economic commodity and is recovered from wells for domestic, industrial, and agricultural use. In areas where groundwater is used extensively, care must be taken that, on the average, no more water is withdrawn in a year than is replaced by natural processes. This can be determined by monitoring precipitation and the water table's fluctuation. The conservation of groundwater is very important because groundwater moves very slowly and many years may be required to replace water hastily removed.

SURFACE GEOLOGY AND NATURAL PROPERTIES

Rocks on the crust of the earth are subject to weathering due to climatic changes. Rainfall and freeze/thaw cycles contribute to the fragmentation of the rock mass. These fragments, together with plant and animal remains, promote soil formation and evolution. Erosion due to rain and snowmelt ultimately creates the three-dimensional forms of swales, valleys, summits, and ridges common in the landscape.

Soils are composed of mineral and organic matter, air, and water. The mineral matter varies in size and texture from fine grains to large rocks. The organic matter, which occurs mainly in the topsoil, accounts for about 1 to 6% of the soil content. It is the topsoil that is critical to all plant life; if completely removed, the newly exposed layer of soil must begin the long process of decay and decomposition to establish a new layer of topsoil.

Every soil type has a distinctive texture. The size and mineral grain content depend on the mineralogical composition, climate, period of weathering, and the method of transportation. Coarse-grained textures include boulders, gravel, and sand, while fine-grained textures include fine

Table 8-1. Safe Loads on Earth Foundation Beds

Material	Load (lb/ft^2)
Hard rock	80,000
Medium rock	30,000
Hardpan	20,000
Soft rock	16,000
Gravel	12,000
Sand, firm and coarse	8,000
Clay, hard and dry	8,000
Sand, fine and dry	6,000
Ordinary firm clay	4,000
Sand and clay, mixed or in layers	4,000
Sand, wet	4,000
Clay, soft	2,000
Alluvial soil	1,000

Source: J. D. Carpenter, ed., *Handbook of Landscape Architectural Construction* (Washington, D.C.: Landscape Architecture Foundation, 1973), p. 239.

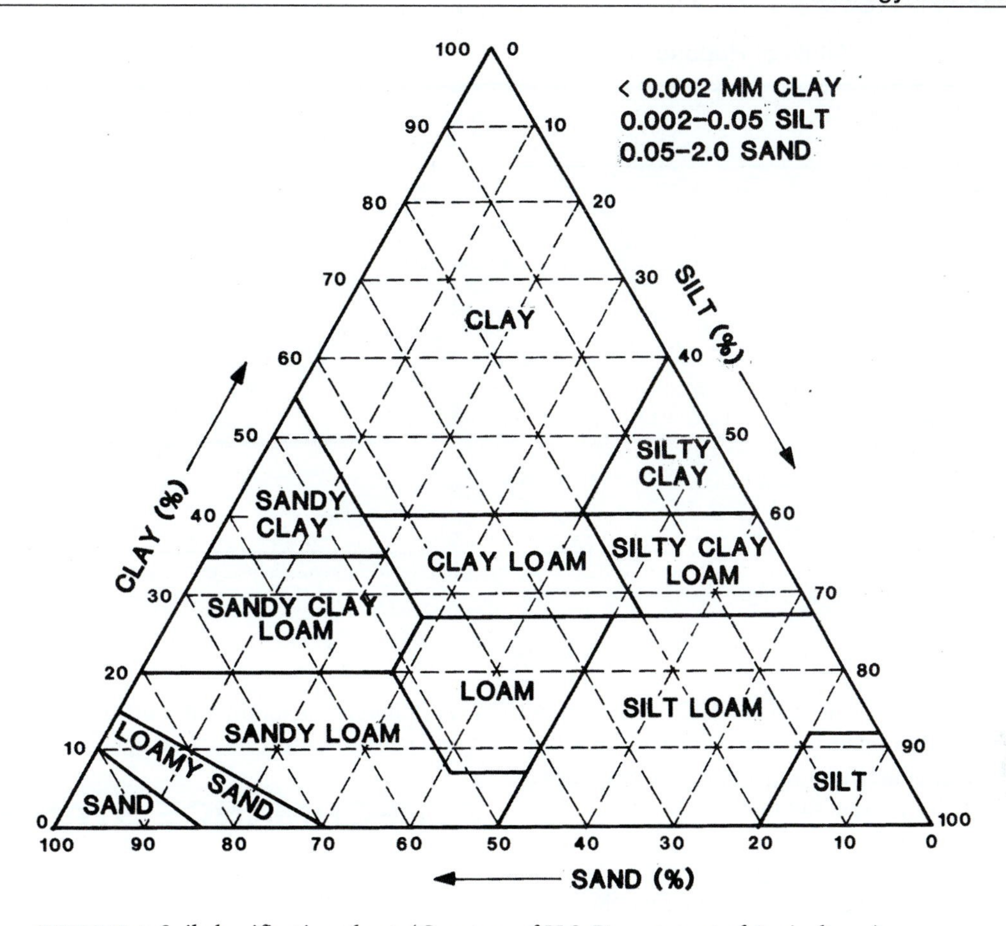

FIGURE 8-1. Soil classification chart. (Courtesy of U.S. Department of Agriculture.)

silts and clays. How these mineralogical components are arranged within the soil mass is referred to as *soil structure.* Soil structure is a critical consideration when designing drainage systems, dams, and embankments.

Soils are classified as being either organic or inorganic, and by grain size. Following are the four classifications for soil by grain size:

1. *Gravel*—particles over 2 mm in diameter.
2. *Sand*—0.05 to 2 mm. The finest grains are just visible to the eye; gritty to the touch.
3. *Silt*—0.002 to 0.05 mm. The grains are invisible but can be felt; smooth to the touch.
4. *Clay*—under 0.002 mm. Smooth and floury or in stiff lumps when dry; plastic and sticky when wet.

STRUCTURAL AND DRAINAGE CHARACTERISTICS

The site planner must be aware of the structural limitations and potentials of a specific soil in order to make the appropriate design and construction decisions. This preview of the engineering characteristics will provide an introduction to the terminology commonly used in a discussion of soils.

Soil elasticity refers to a soil's ability to return to its original shape after temporary deformation as a result of loading. Closely related to elasticity is soil plasticity. This is the trait of soil being deformed under loading and maintaining the deformed shape without cracking or crumbling. Elasticity and plasticity play a significant role in the structural design of building foundations, wall footings, and the preparation of the subgrade or the subbase of paved surfaces.

Shearing strength is the result of friction between the soil particles. The cohesion of these particles, and consequently the shearing strength, vary depending on the soil's water content as well as numerous other factors. Soil stability is directly dependent on the shearing strength of the soil. This characteristic is critical to the soil's ability to support footings and to maintain steep banks and slopes.

The portion of volume change or deformation due to compression through the expulsion of pore water is referred to as soil compressibility. Compressibility is directly related to soil structure and past stress. Compaction is the ratio of volume decrease to air expulsion. For on-grade structures, it is recommended that fill areas for foundations and areas to be paved be compacted to 95% maximum density (ASTM standard). Failure to do so may cause settling and, consequently, foundation slippage and/or fracturing.

Shrinkage and swell result from the buildup and release of capillary tensile stress in pore water and vary with the

Table 8-2. Angle of Repose

Material	Angle of Repose	Slope Ratio
Sand, clean	33°41′	1.5:1
Sand and clay	36°53′	1.33:1
Clay, dry	29°44′	1.75:1
Clay, damp, plastic	18°24′	3.0:1
Gravel, clean	36°53′	1.33:1
Gravel, sand, and clay	36°53′	1.33:1
Soil (average)	33°41′	9.5:1
Soft rotten rock	36°53′	1.33:1
Hard rotten rock	45°	1:1
Cinders	45°	1:1

Source: J. D. Carpenter, ed., *Handbook of Landscape Architectural Construction* (Washington, D.C.: Landscape Architecture Foundation, 1973), p. 227.

moisture demand of clay materials. Soils with severe shrink/swell potential should be avoided where possible, but may be overcome by conscientious planning and responsive design. Some clays are so expansive that elevation changes of 12 to 18 in. are commonplace.

The "angle of repose" is the degree of incline beyond which soil must be reviewed when considering the construction of landforms with new fill. The limiting angles for soils range from 18% for very wet clay and silt to 45% for hard rotten rock (Table 8-2).

Capillarity is the action by which water rises in a channel in any direction above the horizontal plane of the supply of free water. The degree and rate of capillarity are determined by the number and size of channels. Slow in dry soils, more rapid in silt soils, and variable in gravels and coarse sand, capillarity is an important consideration for subgrade pavement design and site construction procedures.

Permeability is the soil's capacity to transport water. It depends on the size and number of continuous soil pores. Permeability varies with the void ratio, grain size, distribution, structure, and degree of compaction. Coarse-grained soils are generally more permeable than are fine-grained soils.

Liquid limit is the minimum moisture content at which the soil will flow under its own weight.

PERFORMANCE STANDARDS

The life-support capacity of soil is based on pH level, organic content, friability, drainage, temperature, and leaching in relationship to the requirements of a particular plant material.

Structural suitability of soils is dependent on soil strength, location, and type. Soil classification systems group soils that will perform in a similar manner when densities, moisture content, water table, climate, and slope are similar. For specific information as to soil performance criteria, refer to the American Association of State Highway and Transportation Officials Classification System and the Unified Soil Classification Systems for proposed use in relation to its structural properties, past performance, and required tests.

Surface and subsurface drainage systems are affected by the structural characteristics of a given soil, moisture content, and location of the water table. The aforementioned considerations should be reviewed when designing subsurface facilities.

CHAPTER NINE

VEGETATION

by Michael E. Ostdick

LANDSCAPE CHARACTER

Regardless of the size of the landscape being viewed, it has an intrinsic character. This character is composed of the basic components of nature: land and rock forms, vegetative patterns, water, and structures. An understanding of these elements is basic to the design of a responsive and creative new landscape.

An examination of the landscape character of the eastern United States reveals that one of its most valuable natural resources is the scenic quality of its rural landscape. The attractiveness can be perceived from a variety of features, with the integral scenic element being vegetative cover and patterns created by that vegetation as a visual resource. In contrast, cropland and grassland are the dominating vegetation of the interior regions of the United States, while the Appalachian region is dominated by mixed hardwoods in the north and pine–hardwood forests in the south. Mixtures of deciduous and coniferous species create considerable contrast, as does the broad variety of shrubs and grasses. Each region possesses well-defined characteristics in terms of terrain and vegetation as well as visual character.

A plant finds its niche based on a number of variables, ranging from soil, orientation, amount of light, and rainfall to the influence of humans and animals. A group of closely growing plants has a natural simplicity. When a landscape has achieved the full climax state of high forest, the dominant trees may be few in number and limited to one species compared with the shrubs and ground-cover layers. They are visually as well as ecologically dominant; thus the qualities of unity and simplicity again relate to this different type of natural landscape. The key issues here are the quality of unity and the degree of simplicity. Contrast in these natural landscapes is usually short-lived, as the bold instances of color vary with the seasons.

There are two ways in which ecology serves as a basis for planting design. When an actual on-site analysis is not possible, it is within the realm of reason to make a conjectural analysis of a site's natural vegetation and to develop a planting plan on that basis. In another approach, the plants are selected in accordance with the soil/climate aspect, appearance, and other relevant environmental determinants. This method is used with the knowledge that if competition from other plants is removed, success is likely. This second approach does not necessarily base plant selection on plant adaptation or aesthetics.

The plants in an ecologically based design should be selected in a way that they will contribute to a biotic community rather than a plant community. In turn, each biotic community should be able to reside congenially in the habitat. In order to have a basis for this type of approach, one must understand the principles of cover and food bondage. *Cover bondage* refers to the link between the dominant plant and the habitat. The dominant plant modifies the habitat, forming microclimates and zones of plant communities.

Food bondage is a three-way link between the flora and fauna and the habitat. One must remember that although wildlife is highly mobile, the animals are restricted by the flora of the habitat. The habitat relies on the flora and fauna to replenish the nutrients of the soil. The goal of any proposed planting design is to develop an ecosystem that provides balance and stability. Care must be taken so that the habitat is not radically changed by human encroachment.

Evaluating Existing Vegetation

The documentation of the existing vegetative cover is an integral step in the site-planning process. Plant materials can be categorized into three general divisions: canopy trees, the understory, and ground covers. Canopy trees are defined as those trees that define or limit the overhead plane (Fig. 9-1). The understory is a combination of small trees, large shrubs, and climbing vines that develop under the canopy trees and can form a physical barrier much like a wall. After the major trees and clusters of understory plant material have been identified and mapped, the last stage is ground cover. This will complete the survey of native plant material, and it provides a picture of what the soils, precipitation, and sunlight will support on a site. It is important to know what works and to match the character of new vegetation with that of what prospers naturally.

Native plants material grows in clusters or communities. If one finds a healthy grove of yaupon holly, other plant material that is drought and heat tolerant and thrives in sandy soil will also likely thrive in the same location. The characteristics that support the growth of one family of vegetation will also support others that mature in the same environment. While introducing other supplemental plants is a part of finalizing a site plan, it is also a necessary step in developing a plan of sustainable diversity. Native wildflowers are an example of diversity that should be supported through careful clearance and grading. Ground covers are grasses, prostrate vines, and wildflowers that define the base plane.

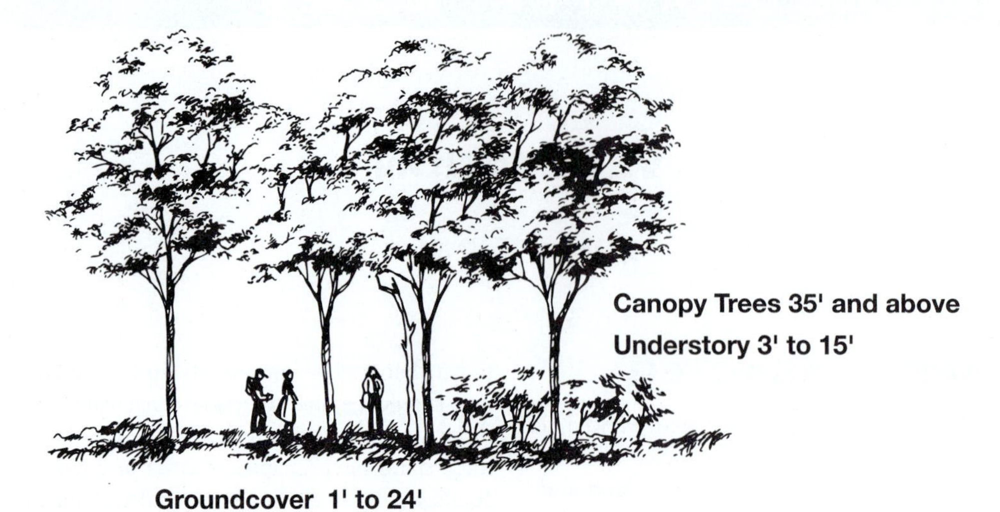

FIGURE 9-1. Size and Scale of Canopy Tree Cover and Understory.

Criteria should be developed for the evaluation of the existing vegetation. The following are some basic standards for evaluation:

1. *Aesthetic value.* Consideration should be given to form, texture, foliage color, flowering habits, autumn foliage, bark and crown characteristics, and type of fruit.
2. *Disease resistance.* Consideration should be given to maintenance costs and the methods for treating the disease.
3. *Life span of the trees.* An obvious aspect of vegetation criteria is its normal expected life.
4. *Wind firmness.* Shallow-rooted trees will blow over easily if they have been growing in a close stand.
5. *Wildlife value.* Plants with nuts and berries provide animals and birds with food during the fall and winter.
6. *Comfort index.* Tree cover can reduce summer temperature by as much as 10°.
7. *Sunlight and heat.* Sudden exposure to direct sunlight and the ability to withstand radiated heat from proposed building and pavement are both necessary considerations.
8. *Space considerations.* The space needed for future growth and the relationship to electric and telephone lines, driveways, walkways, and water and sewer lines must be documented to avoid damage to vegetation and facilities.

Ultimately, the site planner must recognize several basic rules of trees that have grown in the woods:

1. They have shallow roots and should be conserved in a clump.
2. Even mature trees will not survive a violent change in habitat.
3. A change in the groundwater or microclimate will likely be fatal.
4. The area around most trees should never be filled, as filling disturbs the access of air, water, and minerals.

Root systems must have water, sunlight, and food. Wells or retaining walls can be used to hold back dramatic grade changes. When trees are removed from a site, the surface temperature will increase, and erosion often increases unless steps are taken to prevent it. The loss of major trees results in an impact on wildlife and subsurface stability, as well as general aesthetic degradation. Every effort should be made to retain as many as is possible while adopting strict design standards to reduce the amount of ultimate loss due to grading and paving.

THE ROLE OF PLANT MATERIALS

The role of plant materials in site planning and development should not be overlooked or considered a remedial task for the "gardener." The design components available to architects for the design of interior space are pertinent to the design of exterior spaces. The physical materials by which these design concepts are implemented are quite different. The major materials of a building are static elements, whereas the materials of the landscape (such as plants) are constantly changing due to growth and seasons. The base plane or floor becomes grasses or ground cover, the walls become hedges or masses of shrubs, the ceiling becomes the canopy of a tree or a vine on a trellis. Outdoor rooms can be created through a combination of architectural elements. Each element can exist by itself but is enhanced by sensitive blending with others.

Plants play primary functional roles in the following:

1. Wind control
2. Erosion control
3. Energy conservation
4. Wildlife habitat

Wind Control

Three principles can be employed to control the wind: obstruction, filtration, and deflection. Obstruction can be achieved with plant materials by selecting species whose foliage and branching habit are dense. If year-round protection/obstruction is desired, evergreens are a must. Plantings should be spaced such that the mature plants will intertwine/mesh with each other and form a dense obstruction. Where space allows, staggering the plants will help prevent leaks or breaches in the plantings. The specific selection of a plant for this use should take into account soil, climate, and so on. Those plants that are marginally suited for the microclimate of the site should not be considered. Hardy, vigorous plants are a preference rather than plants of questionable hardiness and unreliable growth.

Filtration can be achieved through such simple techniques as selective pruning. For example, say that a site is bordered by a grove of deciduous trees that surrounds an existing lake. It is highly desirable to take advantage of the cool breezes off the lake, yet the trees at the water's edge give a nice sense of space and scale to the property. By selectively pruning and thinning out the lower limbs of the trees, the desired effect is achieved. When selecting trees that will allow air to pass through their structure, consider more upright trees with sparse branching rather than those with hanging limbs and dense branching.

Deflection is redirecting the wind in the hope of either taking advantage of its effect or trying to mitigate its impact. The use of planting masses can guide or direct wind flow to or from an area. See Chapter 11 for more detail on site and climate.

Erosion Control

Plants can play a major role in reducing erosion. Shrubs and grasses that have dense, spreading root systems can bind the soil and prevent unstable slopes from eroding. As discussed in Chapter 16, several products on the market, such as jute mats and plastic webbing, are designed to be used on slopes that are to be planted. The keys to the selection of plants for this use are soil, climate, soil pH, amount of water available, and plant adaptability. The regional offices of the Soil Conservation Service are good sources of information regarding suitable plants for erosion control.

Energy Conservation

The impact of plant materials on energy conservation should not be underestimated. Chapter 11 provides an extensive review of the various ways in which plants influence a microclimate and contribute to energy-efficient site planning.

Wildlife Habitat

Although the process of development reduces the area of the natural habitat for animals, careful and sensitive site planning can limit major incursions. The preservation of existing habitats and their supplementation through the careful selection of plants should be a consideration of the site planner. Another common attribute of the vegetation on every site is the wildlife it attracts. Butterflies are attracted to flowers, and birds use trees as a resource for nesting material, sites, and food. Squirrels and other mammals attracted to vegetation for food can prosper in a healthy wooded setting. All of these smaller creatures contribute another level of richness to a healthy, diverse community of trees and native plants, and their existence on a site should be noted in field research. County agricultural extensions and local environmental clubs are commonly good informational resources regarding wildlife and plants. In addition to the vegetation's value as food and protection for wildlife, the site planner should assess the potential for integrating existing habits with the new use, the character of the new microclimate, ongoing water requirements, and the relative health of the vegetation and its maintenance over time.

The Planting Plan

Developing a planting plan should not be an attempt to reproduce nature. Plants serve a natural and necessary function in the ecological cycle and are thus subject to the hardships and cycles of nature. As such, planting plans should not include rigid principles, the success of which are dependent on the perfection of plants or their maintenance. The development of plants through time and seasonal changes can be an aesthetic contribution to every site. Plants can also be used in the following ways:

- In combination with a building or building complexes for the extension of architectural lines or as a foil to screen or enhance the quality of architecture
- As a silhouette against the sky or reflection in water or as skyline trees
- To define space or a sheltered area, or function as a buffer between activities, or create a sense of openness
- To enhance a view, block a view, frame a view, or even be a view
- As an educational resource

The following are some of the principles that should be considered in the development of a planting plan:

1. Ensure that the plan reflects a predominance of one type of plant or a similar texture, color, or form within any grouping.
2. Exercise some restraint relative to the number of different plants used in the plan.

3. Select plants that have a common soil, climate, and water requirements.
4. Since maintenance of formal (clipped) hedges is a time-consuming operation, consider growth habit and mature size in the selection of plants.
5. In an informal plan, locate plants so as not to rely on the quality of any particular tree.

The plant selection process should be based on at least one of the following criteria:

- A plant's ability to live and flourish in a specific environment
- Knowledge about the existing trees on a site or their healthy growth on adjacent sites
- Plants that have low maintenance and/or low water requirements
- Plants for a specific purpose, such as shade, contrast, color, or size

NATIVE PLANTS AND DROUGHT-TOLERANT PLANTS

Native plants are plants that are indigenous to a site or region. Over the years, certain species have adapted to the climatic conditions of a region and become a significant part of the site's inherent character. The species identified on any given site can be used as an indicator of how well proposed plantings may do and indicate what modifications to existing conditions may be necessary.

The change in our planet's climate, which many thought would never occur, is almost a universally accepted condition. Long-term droughts throughout the United States have left many local governments imposing water consumption limits. As a parallel initiative, other local governments have adopted ordinances requiring new development applicants to employ xeriscape landscape principles in concert with preservation of native plant species. *Xeriscape* is a term that has become part of the vocabulary of those keen on developing planting plans using drought-resistant native plant material. The term *xeriscaping* originated with the Denver, Colorado, water department in 1981. The word is a compound of the Greek word *xeros* (meaning dry) and *scape*, as in *landscape*, and it refers to a theme wherein the selection of plants and design is composed to withstand drought conditions. Among the advocates of xeriscaping are numerous landscape architects who have tried for years to encourage clients to approve xeriscape planting plans designed to remain healthy on the water created by rainfall alone.

For most regions of the United States and most clients, it would be challenging to design a planting plan supported exclusively by natural precipitation. It is not, however, a challenge to design one that minimizes the amount of irrigated water necessary to keep plant life healthy. To many local governments, today more than ever, water conservation and native plant conservation are hand-in-glove concepts that require more than persuasion once accomplished. One should not be surprised to find that a growing number of municipalities across the lower 48 states have adopted landscape planting guidelines with the specific goal of keeping water consumption at a minimum. A variety of municipalities throughout Florida, Texas, Alabama, Georgia, and North Carolina, to mention a few, have adopted native landscape ordinances. The increasing demand for potable water is a particularly serious condition when coupled with fixed safe aquifer withdrawal rates and the shrinking size of recharge areas.

The significance of using native plants should not be underestimated. Native plants are more naturally drought tolerant, as they are more acclimated to the seasonal variations in the microclimate of their environment. This attribute means that their water requirement is often less than that for other ornamental species. In the west and southwestern United States, where water is scarce, drought-tolerant native plants constitute the majority of a designer's plant resources. But regardless of a plant's drought-tolerance capability, there are few plants that can survive without water when first being established, and native plants are no exception. The difference between native plants and ornamental plants is that after the native plants are established, they can be left on their own, without irrigation. Ornamentals, on the other hand, may not last a week, let alone a month or two, without supplemental irrigation.

It is also important to keep a balanced view of native plant use. Common sense suggests that just because a plant is native does not necessarily make it a candidate for any site in every location. Sites vary regarding soil, hydrology, and topography, and the entire site's context should be the first question regarding the application of xeriscape concepts and native plant compatibility. Finally, the site planner needs to be reminded that eliminating harmful, invasive exotic plant species is also a worthy objective. This is a particularly relevant issue in the southernmost states, where the climate sustains a longer growing season for all plants. What appeared to be the introduction of an innocuous species, the kudzu vine (*Pueraria lobata*) (Fig. 9-2) has become a scourge to plant life throughout the southeastern United States. Brought to the Philadelphia World's Fair in 1876 as a fast-growing vine to be used in erosion control, the vine's invasiveness has been so aggressive and difficult to control that it was declared a pest weed by the U.S. Department of Agriculture in 1953. To date the vine covers an area between 7,700 and 12,000 square miles from Virginia to Texas.

Sources for information on native plants and drought-tolerant plants include the regional offices of the Soil Conservation Service and the County Agricultural Agents. These agencies typically have pamphlets available on these topics and have experimental stations that can be visited to see what the plants look like under water-restricted conditions.

a.

b.

c.

FIGURE 9-2. Kudzu vine.

CHAPTER TEN

HYDROLOGY

THE HYDROLOGICAL CYCLE

The influence of water on a site is multifaceted. The most obvious influence is in the form of surface water as it vacillates from amenity to adversary. Understanding how and under what circumstances that change takes place is part of the site planner's obligation. The influence and impact of surface water, for instance, are not always dependent on the capriciousness of nature. Human intervention has created impacts in some circumstances which, if matched by natural phenomena, would have been considered catastrophic.

Compared to the movement of surface water across a site, subsurface hydrology is very obscure. Historically, the site planner's concerns relative to subsurface hydrology were limited to the impact of groundwater on development. More recently, however, the importance of water as a resource and the maintenance of its quality have come to be recognized by all levels of government. But despite general agreement relative to the scope of the problem, water quality, quantity, and equitable distribution continue to elude comprehensive solution in many regions. As with other site components, many local governments play substantive roles in the review of development and water-related issues and thus their function in the process cannot be overemphasized.

Theory and study of the movement of water dates to Anaxogoras of Clazomene (500 to 428 B.C.).[1] His observations of the hydrological process were remarkably accurate considering that they were based on empirical evidence. However, later scholars, such as Plato and Aristotle, disagreed with Anaxogoras's theory. It was not until 1674 that a French research scientist's quantitative analysis of water volumes in the Seine River basin confirmed the concept of precipitation's change into groundwater recharge and runoff, transpiration, and evaporation to form again in the atmosphere as precipitation.[2] Subsequent research continued to expand and confirm the hydrological phenomenon, which, driven by solar radiation, continually recycles the resource (Fig. 10-1).

If solar energy drives the process, surface water must be considered the fuel for the process. Because of the size and extensive heat storage capacity of the oceans, the critical link in the process is the relationship between the atmosphere and the seas. The oceans function as an immense reservoir, which, through evaporation, transfers moisture to the atmosphere, where it is distilled and distributed throughout the globe.

Precipitation is the key conversion of the process that sustains the cycle. Its effect on the site depends on topography, soils, land use, and vegetation. While the topography determines the path and flow of water, the soils attest to a site's ability to absorb precipitation. Land use is a variable that provides the site planner with an insight into an area's runoff potential. Everything in the way of vegetation, from prostrate ground cover and grasses to mature overstory, affects precipitation as it falls to the site.

Rainfall is dispersed by vegetation through transpiration and interception, although there is some problem with differentiating between the two. The leaves of plants intercept water in its free fall to the soil, permitting evaporation from the vegetation. Healthy plants also withdraw water from the soil and transpire moisture through their leaves in the photosynthesis process. Although the relative amounts of water transferred through transpiration and interception are small, both contribute to the entire hydrological process.

The principal storage and fluxes in the hydrological cycle are divided into two general classifications: surface and subsurface water. Examined more closely, precipitation falling on the site is dispersed in various ways. The water may either be absorbed into the soil and become part of the groundwater supply or recharge established aquifers. This absorption process is referred to as infiltration and is the natural downward movement of water through the surface soil. When the zone of aeration (Fig. 10-2), is sufficiently saturated, deep percolation occurs. But until the supply of the water exceeds the holding capacity of the soil, groundwater recharge does not occur. It also follows that if the subsoil in the zone of aeration is poorly permeable, groundwater recharge will occur at a very slow rate regardless of the amount of water in the zone. When this condition occurs and the water moves laterally as well as down, the process is known as *groundwater underflow*.[3]

[1]Harry W. Gehm and Jacob L. Bregman, eds., *Handbook of Water Resources and Pollution Control* (New York: Van Nostrand Reinhold Company, Inc., 1976), p. 16.

[2]Pierre Perrault, *De l'origine fontaines* (Paris: Pierre le Petit, 1674).

[3]Gehm and Bregman, pp. 8–10.

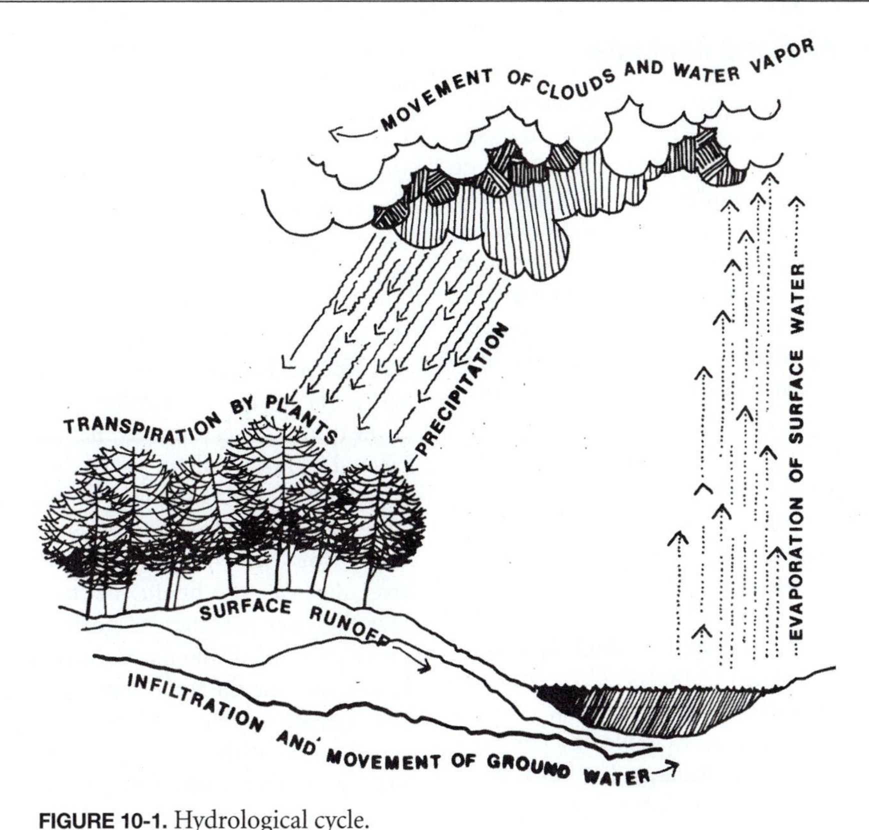

FIGURE 10-1. Hydrological cycle.

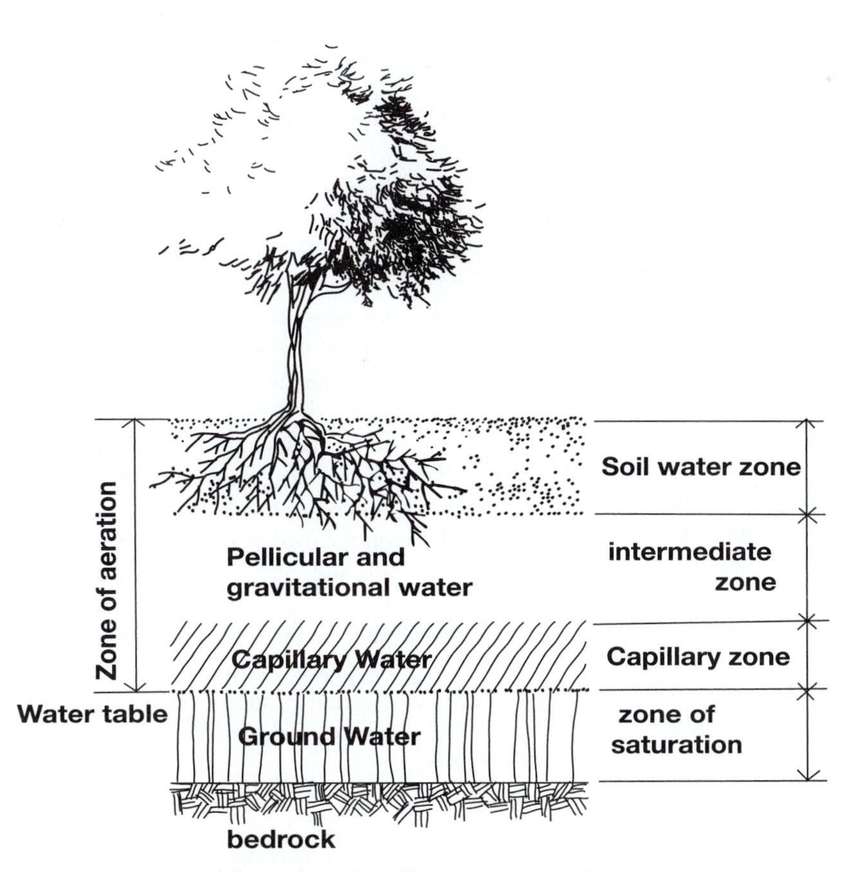

FIGURE 10-2. Section through surface illustrating groundwater.

Resource Distribution and Conflicts

An introduction to subsurface water would be incomplete without an explanation of aquifers. The two general types, confined and unconfined, reflect the geological context of the soils above and below the aquifer. A popular notion is that an aquifer is an underground river; it is more often analogous to a water-saturated container of sand. Although there are unique geological conditions in which impervious soils occur both above and below the more porous aquifer, "confining" its recharge potential to a limited zone, the "unconfined" aquifer stores water in the pore spaces of rock, sand, and gravel.[4]

The capacity for most unconfined aquifers to absorb additional groundwater is limited and many are overflowing. This is evidenced by water that continues to flow in creeks and streams during dry periods. This movement of water from springs and aquifers into tributaries, rivers, and bays by diffused percolation is known as groundwater outflow. Whereas in some cases the supply is modest, in contrast, the Snake River in southern Idaho is fed by a group of springs at the collective rate of 6,000 ft^3 per second.[5]

The large Ogallala Aquifer is an example of a confined aquifer that has a modest annual recharge. The Texas High Plains segment of the aquifer receives little or nothing from adjacent resources and exists primarily on an endowment from a time when there simply was more rain.

Lakes, rivers, and swamps are also contributors to the total surface water area. Depending on the size of the body of water, local climate and hence the hydrological processes can be influenced by lakes. The Great Lakes of the midwestern United States are an example of the effect of the amount of surface water on the climate of an extremely large region. The Great Lakes are the result of unique geological formations; other lakes may simply be wide places in a river. Where the number of circumstances in which lakes were formed by a natural change in the river's edge are limited, the construction of dams (reservoirs) that exploit natural river basins are numerous. Once constructed, reservoirs function in essentially the same way as natural lakes. The thousands of smaller natural lakes are either topographic depressions supplied by groundwater outflow or exist above the water table and provide groundwater recharge through deep percolation.

Swamps, bogs, and wetlands were for years considered nuisances, and their role in a region's ecology was misunderstood. Swamps and wetlands are areas where the water table is essentially level with the land. This physical relationship permits the water to flow unabated from the water table into a topographically shallow elevation that contains the groundwater. Although the lake is usually an easily visible amenity with numerous contributions to make to a site, the swamp was historically something to fill and prepare for development. Since the late 1960s, however, the role of swamps in the food-chain cycle has become common knowledge. Ordinances and legislation controlling the filling of wetlands have been adopted by numerous state and local governments. See page 53 for the salient portion of the Wisconsin Supreme Court's opinion concerning wetlands in that state. The site planner needs to be aware of the environmental implications, as well as the portending disasters of development in swamps, bogs, or marshes.

The special qualities of the river's edge are valuable in the same way that bays, lakes, and the sea contribute to the city. The poetic Chinese reference to the river as "threads of sea" acknowledges the kinship between the two. The deeper truth to this observation is that the river dominates edges beyond its normal flow. Just as the sea responds to climatic changes and affects the land at its edge, so does the river. Although these edge conditions are often the most highly valued for their views and special relationships with the river, they can also be the first to be damaged in the event that the river breaches its normal banks. These areas are probably some of the most critical to the site planner, who knows that the difference between the river's edge and the river's domain can be critical to life and property.

Because of the importance of water in sustaining humanity, conflicts associated with its use have existed since there were two who claimed absolute rights to a single resource. The origin of the word *rival* has reference to those who shared the water of a river or irrigation channel. As a result of such conflicts, laws associated with the quality, quantity, or distribution date to proclamations of Gaius Caesar in 45 B.C. Recognition of the relationship between water and public health came much earlier. Mesopotamian cities following the Mosaic law and Zoroastrian religion had developed sanitary sewer systems by the middle of the third millennium.[6] Contrary, however, to the very early awareness of the import of water quality, later societies were not so adept. As late as the mid-nineteenth century, poor sanitation standards were widespread in English cities and responsible for cholera and typhoid epidemics in London. American cities were no better, as Chicago used Lake Michigan for water supply and sewerage disposal concurrently at the turn of the century.[7]

The law was the process through which conflicts were resolved between civilized people. In the United States, water rights and the legal doctrine were inherited from the English. England's riparian law reflected the environmental context of that water-rich country, which placed little value on the ownership of water. Water was commonly shared by all, and each person owning property adjoining a water source was entitled to use the water. However, each was also obligated to return the water to its course without change in its quality or reduction in its quantity to every downstream riparian property owner. This was a fairly civilized quid-pro-quo relationship, as it held everyone's right to use the

[4] Ibid., p. 37.

[5] Ibid., p. 10.

[6] Charles E. Warren, *Biology and Water Pollution Control* (Philadelphia: W. B. Sanders Company, 1971), pp. 3–4.

[7] Ibid., p. 5.

water to be equal. This concept was easily assimilated into early American law. The eastern seaboard of America was similar to that of England insofar as its water resources were concerned. There was little rationale for changing the concept of water law until the settlement of the arid west, and more particularly, with the discovery of gold. Since the popular mining techniques of the time required the utilization of streams, a reevaluation of the long-established riparian doctrine became necessary. As gold was being mined on public land, custom was that the first to appropriate the water (for whatever use) enjoyed rights superior to those of people downstream. This change in practice resulted in a radical change in the concept of water rights in the west. The antithesis of riparian doctrine emerged as the "doctrine of prior appropriation." This concept, which maintains that the quantity and quality of water can be changed by anyone who appropriates the resource first, is practiced in the 17 states of the lower 48 west of the Mississippi.

Problems associated with water will always focus on quantity (allocation) and quality (pollution control). The arid plainsman dreams of the construction of canals to import water from a better-endowed neighboring state to "make the desert bloom." While hope springs eternal, the financial assistance to support such projects does not, as skeptical voters mount sturdy opposition to dubious water projects in various western states. The reality of the situation is that economic pressures dictate continued use of agricultural techniques that require high water use in the face of reduced water reserves, putting the future of affected farming communities in jeopardy.

If depletion of major aquifers were not enough, continued extraction of groundwater beyond its limits for recharge contributes to yet another water-related malady: subsidence. In most cases the process of pumping oil, gas, and/or water creates little, if any, perceivable reaction on the earth's surface. In some conditions, however, extensive removal of those resources has created collapsed areas in the surface geology. As an example, limited areas of the San Joaquin Valley of California have experienced over 30 ft of surface subsidence since 1925. To aggravate the problem, development eliminates effective groundwater recharge as the resource is being extracted. Although water is not the sole constituent that when removed induces subsidence, it is the only one of gas, oil, and water capable of recharge through natural processes.

Although it is not within the purview of the site planner to solve problems associated with water resource management, it is important to understand the by-products of a society that has both abused and neglected one of its most life-sustaining elements. It is important to understand the concern and rationale for stricter on-site detention policies, storm water management guidelines, and water conservation controls. Each municipality, like individuals, adopts those regulations that reflect its environmental context, local values, and economic health. The problem in the past has been a lack of perspective associated with a value system which failed to recognize that humanity shares a common, critical, and limited provision.

SURFACE WATER: STREAMS, CREEKS, RIVERS, AND WETLANDS

From the beginning of recorded time, one of the most desirable areas of human settlement has been near rivers, lakes, and seashores. The cradle of civilization at the delta of the Tigris and Euphrates valley tells us something of the attraction to the water's edge. The early rationale was easy to understand. The soil of the rivers' bottomland was fertile, rich, and a valuable resource that, with nature's own natural processes, replenished itself. The ebb and flow of the Nile has sustained one of the oldest civilizations on earth beyond the level of its contribution in the food-chain cycle. Similarly, the Yangtze in China and Ganges in India have significance beyond the scope of agricultural resource to their respective countries.

The attraction to the water's edge is natural, since the sea and tributaries provide humans with food as well as a transportation system. The commerce and exchange of ideas and cultures supported by the sea acted as a powerful industry that shaped the establishment and growth characteristics of every coastal nation in the world. The rise and fall of tides have long been well understood and the location of buildings or structures at the water's edge has respected those inevitable changes. In most cases, people have recognized the incredible force in hurricanes and typhoons. Settlements on the water reflect a range of issues, all of which focus on identifying that location which provides the best protection from violent storms and a topographical configuration that supports the unhampered access of deep-draft vessels. Humans had historically accepted the intransigence of nature and knew that decisions which spurned well-established natural patterns jeopardized their safety and well-being.

The industrial revolution provided humanity with the ability to construct machines that could defy (at least momentarily) the forces of nature. Our perceived abilities to control nature appeared to be limited only by our ability to conjure and build the appropriate device. In terms of water and flood control, these devices have taken the form of dams, reservoirs, channel realignment, levees, and seawalls (Figs. 10-3, 10-4, and 10-5). In many communities, popular political strategies

FIGURE 10-3. Reservoir dam.

FIGURE 10-4a. Concrete channel realignment.

FIGURE 10-4b. Concrete bulkhead combined with pedestrian path.

FIGURE 10-5. Gulf coast sea wall.

often support the design of a system that treats symptoms of the problem instead of the problem at its source. Floods, floodplains, and urbanization are interrelated factors that typify the problem–symptom relationship. Floods are not a modern phenomenon and are a problem only when they create property damage or become a threat to human health. Our methods for dealing with the threat of flood often focus on structural techniques, which is natural considering the fact that human-made urbanization contributes to the problem so significantly. Although structural methods have helped curb the destructive power of floods and flooding, they do not solve the problem entirely. The problem is that structures present the illusion of humans conquering the forces of nature through technology and lead many to believe that there is a technical solution to every problem.

Surface Water as a Process

To understand the problem comprehensively involves a recapitulation of the hydrological process, runoff, the drainage basin, and the floodplain. Precipitation falling in a watershed or drainage basin does two things: It is either absorbed into the soil in the process of infiltration, or it moves across the surface of the land, always seeking a lower elevation. This second process is referred to as runoff and occurs when the amount of water applied to a surface occurs at a rate or volume beyond the ground's ability to absorb it. Another condition exists where the surface is completely impervious and will not accept any water. Impervious surfaces are not just human-made (for example, buildings, parking lots, roads) but may also occur in nature.

The runoff from each site in the watershed contributes to the first order of streams. These are the "fingertip tributaries,"[8] the smallest landforms in the system. They have no other contributors, but when merged create a second order of stream. Second-order streams merge to form thirds, and third orders join to create a fourth. The bottom of the drainage basin or floodplain is created by the fourth-order stream[9] (Fig. 10-6), which is formed by long periods of erosion, deposition of nondissolved sediment, and subsurface geology. Although the runoff is incapable of being absorbed, if it traverses any soil, it will transport weathered sediments in three ways:

1. The particles that are carried in chemical solution are called the *dissolved load.*
2. Small particles such as silt and clay are kept in suspension by turbulent water flows. These particles are called the *suspended-sediment load.*
3. "Larger particles of sand and gravel or rock fragments roll, slide, or bounce along the bed of a stream. As the velocity of the stream is reduced, the larger particles of

[8]Arthur L. Bloom, *The Surface of the Earth* (Englewood Cliffs, N.J.: Prentice Hall, Inc., 1969), p. 55.

[9]Thomas Dunne and Luna B. Leopold, *Water in Environmental Planning* (San Francisco: W. H. Freeman and Company, Publishers, 1978), p. 498.

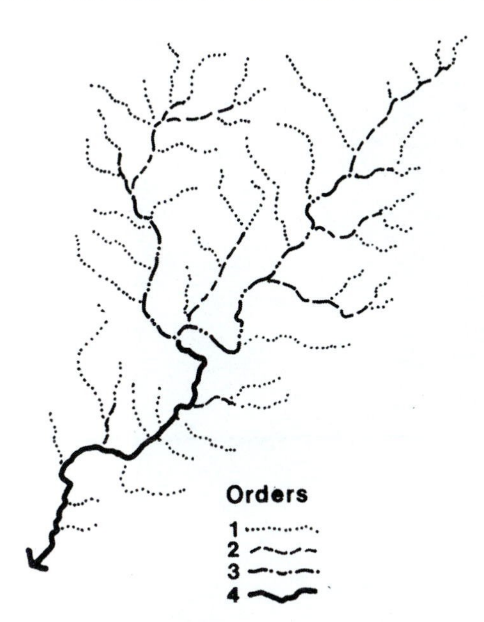

FIGURE 10-6. Four orders of streams.

non-dissolved sediment drop to the bottom."[10] "This sediment is called alluvium. The surface of the alluvium, from the edge of the fourth order stream to the base of the valley walls, is called the flood-plain of the river."[11]

The principal impact of this transportation of particles is erosion. The basic process of stream erosion is the scouring action of the sand and gravel along with the moving water. That process produces three distinctly different types of erosion. The first is an action of "downcutting," which results in lowering the depth of the stream channel. The second and most noticeable is an action called "headward" erosion. This occurs when the velocity and amount of water are such that the water is forced "upslope." Slope retreat is the third major effect of erosion. This sideways erosion causes the valley walls to recede from the channel.[12] All of these result in increased deposition of sediment in the streambed, thus raising the bed and the level of floodwaters.

Erosion not only destroys valuable land, it pollutes rivers, streams, and lakes. A major source of water pollution in many communities is sediment. Sediment in the form of suspended solids in waterways creates special problems to water quality. Increased turbidity reduces the amount of sunlight a stream receives, thus "inhibiting the growth of aquatic plants and habitat. Sediment also settles to the bottom of the stream to form oxygen-demanding sludge."[13] This organic debris increases the amount of decomposing bacteria, resulting in an algae-choked, biologically depleted stream.

[10]Bloom, p. 59.

[11]Ibid., p. 52.

[12]W. K. Hamblin and J. D. Howard, *Exercises in Physical Geology*, 5th ed. (Minneapolis, Minn.: Burgess Publishing Company, 1980), p. 72.

[13]U.S. Dept. of Agriculture, Soil Conservation Service, *America's Soil and Water: Conditions and Trends* (Washington, D.C.: SCS, December 1980), p. 30.

The pollution and sedimentation problems created by urbanization are often more visible than the companion thermal and chemical pollution. Thermal pollution results from a loss of natural vegetation and ground cover to roads, parking lots, and buildings. Although the process must be accepted as going along with urban development, it does raise the temperature of the runoff. The same can be said of the chemical pollution that occurs as precipitation washes petroleum and other chemical residues from parking lots, roads, sanitary landfills, and industrial activities into surface drainage systems and groundwater supply.

Wetlands

The fact that water quality and quantity are related to land and water management was a concept supported by earth scientists, biologists, and citizen activists in the 1960s. Congress responded to the chorus of pleas from across the United States for legislation to change the rules and the role of the U.S. government in the process. In passing the Environmental Protection Act in 1969 and creating the Environmental Protection Agency (EPA) as the watchdog arm of the Congress, lawmakers established a commitment to the management of the nation's water, air, flora, and fauna unequaled in our history.

Soon to follow was the passage of the 1972 Clean Water Act, reinforced by the amendments in the 1977 Water Quality Act. Among the major new components of the 1977 legislation was Section 404, which focuses on wetland protection and the prohibition of wetland filling. If ever there was an environmental condition that was more misunderstood, it is this delicate transition between the land and the sea: the wetland (Fig. 10-7). This zone of water, plants, soil, and habitat for all manner of fish, fowl, and mammals is known by many names—swamp, marsh, wetland, or estuary—depending on the region of its location. Some suggested that the marsh was the last stand of the wilderness. But it was rarely a wilderness to many who saw the marsh as a location for landfills, municipal trash heaps, and industrial waste dumps. Few sought ownership as it required significant filling before it could be considered a plausible building site. For decades it was a genuinely unique occasion when the word *swamp* and the phrase "highest and best use" were used in the same sentence.

As improved as we might think conditions to be, in a 2007 report by National Public Radio on river pollution in cities, it was estimated that approximately half of the rivers, lakes, and bays under EPA oversight were unfit for fishing or swimming. To provide a scale to the level of loss, it has been estimated that the decimation of our wetlands continues at a rate of over 300,000 acres a year. In at least three states—Ohio, Iowa, and California—it is reported that less than 10% of the original wetlands remain today. The federal government's intent to control various aspects of the nation's navigable tributaries and wetlands began with the passage of the Rivers and Harbors Act of 1899. This law prohibited the obstruction or alteration of navigable waters of the United States without a permit from the Army Corps of Engineers. Additional legislation followed in 1912, 1924, and 1948. For decades, legislation that had surface water resources as an

FIGURE 10-7. Typical wetland.

issue of any consequence focused primarily on river and/or Great Lakes navigation. When construction was involved, Congress continued to rely on the Army Corps of Engineers to serve as coordinator or project administrator.

It wasn't until the 1960s that a constituency supported land use and water regulation with a conservation and pollution abatement bias. Research conducted by the EPA documented that the most significant impairment to improving water quality was industrial stormwater discharge into streams, rivers, and lakes. Earth scientists provided the important research and evidence supporting the position that the health of the nation's marshes and wetlands was inextricably linked to the quality of its water. It was clear that the popular practice of locating landfills in marshes and estuaries was the key reason for the pollution and degradation of the water quality, along with the aquatic and bird life unique to lowland habitats. Whereas the Clean Water Act of 1977 established a program for discharge of dredge and fill material in general, subsequent litigation resulted in court decisions with included wetlands and non-navigable waters as aquatic resources to be protected under the law.

The 1977 act covers the following waters of the United States:

- Territorial seas
- Navigable waters, including interstate waters and their tributaries
- Lakes, rivers, streams, and creeks
- Natural ponds, playa lakes, mud flats, sand flats, wet meadows, and wetlands
- Waters that could be used by migratory birds and habitat for threatened or endangered species

At a finer level of detail, the criteria for determining the existence of a wetland on a site are hydrology, hydrophytic vegetation, and hydric soils. These are all common attributes of swamps, marshes, and bogs that are fed or supported by perennial streams or any of the water bodies listed above. A perennial stream is one that has water flowing year-round during a typical year and is located below the level of the water table most of the year. Unlike perennial streams, ephemeral and intermittent streams have water flowing only during precipitation events for a short time and are located above the water table year-round. Intermittent streams exist when groundwater provides sufficient water for stream flow, whereas ephemeral streams have no groundwater support. When executing an inventory of a site's physical resources, notes and photography are essential in documenting the potential for the wetlands. Understand that a wetland does not exist without contributors and an accurate survey that documents the setting as critical to a complete picture.

A part of the picture often overlooked is the potential for habitat of threatened or endangered species. This is an easy oversight as the survey is looking for things that are visible, not those that are attempting to hide from view. Helping us see what is visible is key to site research. One of the few places that fish and wildlife conservation has been supported at an international scale in the western hemisphere is the North American Wetlands Conservation Act of 1989. This agreement is a significant collective effort between the United States, Mexico, and Canada to fund the wetland conservation across the continent; it supports the seasonal nesting for migratory bird populations, which, at this writing, is less than a well-defined geography.

CHAPTER ELEVEN

CLIMATE AND SITE

by George Truett James

The process of planning a site to accommodate buildings and exterior spaces should involve an analysis of climate with a focus on comfort and building energy conservation. In this chapter we focus on the nature of human thermal comfort, the climatic elements that affect the site and buildings, and how the analysis of these elements may inform the site-planning process. Later, in Chapter 20, the focus shifts to how the analysis of external thermal loads (created by the climatic elements) and internal thermal loads (created inside the building by people, lights, and equipment) can provide the site planner with useful information for the preliminary siting, massing, and orientation of buildings. By using these analytic techniques, the site planner will be able to anticipate the impact of climate on exterior spaces and the thermal loading problems that will be encountered by a proposed building, thereby permitting the inclusion of comfort and energy-conscious design criteria during the earliest possible stages of the design process.

COMFORT

A realistic discussion of climate and building energy conservation cannot occur without considering human thermal comfort. It is a premise of this book that through careful analysis, the site planner can:

1. Develop exterior spaces which are sensitive to local climatic norms and extremes, thereby expanding the opportunities for thermally comfortable outdoor experiences
2. Reduce the thermal loads imposed on the site's buildings, thereby promoting thermally comfortable indoor conditions for a minimum expenditure of energy

Although human thermal comfort will not be discussed in great detail here, one should not minimize the importance of comfort as an issue in the design process. It has been estimated that over a normal 40-year life span of a commercial office building, for every dollar spent on a building's construction, operation, and maintenance, a conservative suggested that $25 may be expended on wages, salaries and benefits for the people who occupy the building. Assuming that there is a connection between comfort and productivity, anything that can be done to promote comfort can produce significant financial benefits. Similar benefits may accrue to industrial and retail establishments that provide comfortable conditions for employees and customers. If, through sensitive planning and design, people are provided with comfortable, stimulating places to live, work, and play, and business realizes some economic benefit, the designer partially fulfills his social responsibility to the community and his professional obligation to the client.

What, then, is this commodity that we call "comfort"? Perhaps the surest definition of comfort is "the lack of discomfort." This may seem trite, but it indicates the uncertainty of defining comfort in a generalized sense. Comfort is a function of individual perception, expectation, and need. It depends on our physiological and psychological responses to the environment—a world composed of thermal, luminous, auditory, olfactory, gustatory, and tactile stimuli. All of these elements are important in terms of our sense of comfort, but because we are dealing specifically with climatic issues and building energy use, our discussion of comfort will focus on the thermal environment and, to a limited extent, the luminous environment.

Studies suggest upper and lower limits for humanity's existence, where prolonged exposure results in hypothermia.[1] The ideal combination of environmental conditions can be assumed to occur approximately midway between these two extremes—a set of conditions where thermal stress is minimized and the body's thermal balance is most easily maintained. This suggests that a "zone" can be identified which would represent conditions of thermal comfort in a general sense. A "comfort zone" was postulated in 1963 by Victor Olgyay in *Design with Climate* and was later updated.[2] As indicated in Fig. 11-1, a zone of comfort is identified in what Olgyay calls the "bioclimatic chart." It should be noted that the bioclimatic chart refers to conditions in the exterior environment, and that the comfort zone defines a range of temperature and humidity conditions where a person in normal indoor clothing, doing sedentary or light work in the shade, and with no wind blowing, will presumably be comfortable. The chart indicates the requirements to remedy discomfort when temperature and relative humidity do not fall within the comfort zone. For example, when temperatures are below the comfort zone, the addition of radiation

[1] American Society of Heating, Refrigerating and Air-Conditioning Engineers, Inc., *ASHRAE Handbook—1981 Fundamentals* (Atlanta: ASHRAE, 1981), Chap. 8.

[2] E. Arens, P. McNall, R. Gonzales, L. Berglund, and L. Zeren, "A New Bioclimatic Chart for Passive Solar Design," *Proceedings of the 5th National Solar Conference* (American Section of the International Solar Energy Society, 1980).

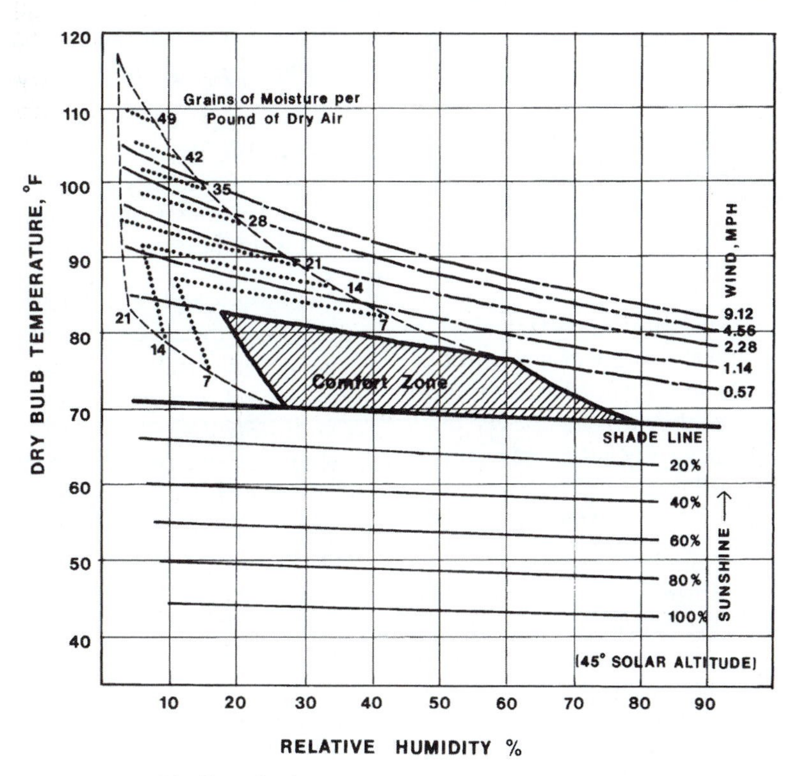

FIGURE 11-1. Bioclimatic chart.

can produce conditions of comfort. With high temperature and high humidity levels, air movement is required to achieve comfort. The chart indicates that the elements which influence thermal comfort are: air temperature, humidity, radiation, and air movement. Clothing and activity levels also have a profound effect on thermal comfort.

The preceding discussion has been a brief introduction to thermal comfort. An enormous amount of detailed information is available on this subject. For a more thorough discussion, the reader is directed to *Design with Climate* by Victor Olgyay; *Man, Climate and Architecture* by Baruch Givoni; *Climatic Design* by Donald Watson and Kenneth Labs; *Thermal Delight in Architecture* by Lisa Heschong; Chapter 1 in *Building Control Systems* by Vaughn Bradshaw; Chapter 8 in the *1981 ASHRAE Handbook of Fundamentals;* and "A New Bioclimatic Chart for Passive Solar Design" by Arens, McNall, Gonzales, Berglund, and Zeren.

THE CLIMATIC ELEMENTS

The four basic climatic elements that influence thermal comfort, building energy use, and the site-planning process are *solar radiation, air temperature, relative humidity,* and *air movement.* To familiarize the site planner with these elements, each will be discussed in terms of:

1. Its physical description
2. The effects that it produces or its behavior
3. Basic strategies for its control or modification
4. The access and interpretation of related data concerned with measuring its magnitude, intensity, or effects

General climatic analysis related to site planning will then be discussed. Attention will also be focused on the difference between normal and extreme climatic conditions. In a general sense, location of exterior spaces, building orientation, massing, and architectural design should be based on normal climatic conditions—the seasonal conditions that can be expected to occur normally or most frequently. Knowledge of the extreme or worst-case conditions that can occur (those conditions with which the mechanical engineer must deal in determining the proper size for mechanical air-handling systems) allows the site planner or architect to compensate for those conditions in order to minimize their impact on exterior spaces and buildings.

Before proceeding with the discussion of climatic elements, it is important to distinguish between the three basic climatic regimes: macroclimate, mesoclimate, and microclimate. Macroclimate refers to the general climatic conditions of a region, while mesoclimate relates to the climate of a particular area or city. An understanding of these regimes is important in any planning process. However, the site planner's influence is exerted most notably on the microclimate—the climate of a specific site.[3] Any time a building is constructed,

[3]C. Thurow, *Improving Street Climate Through Urban Design,* Planning Advisory Service Report No. 376 (Chicago: American Planning Association, 1983), p. 2.

a variety of new microclimates is created in and around the building within the existing microclimate of the site. The ensuing discussion will focus primarily on how climate-responsive site-planning decisions can ameliorate adverse microclimatic conditions, providing thermally pleasant outdoor spaces and reducing climatic loads on buildings.

Solar Radiation

> Radiation is undoubtedly the most important of all meteorologic elements. It is the source of power that drives the atmospheric circulation, the only means of exchange of energy between the earth and the rest of the universe, and the basis for organizing our daily lives.
>
> *Rudolph Geiger*[4]

Description. The sun is the driving force for the earth's climate. A giant fusion reactor, the sun emits electromagnetic energy in three broad ranges of frequencies: the ultraviolet, the visible, and the infrared. The portion of this energy with which we will be concerned is physically perceived as light (the visible frequencies) and heat (the infrared frequencies). Generally, light is referred to as shortwave radiation and heat as longwave radiation.

Solar radiation reaches us in the form of direct, diffuse, and reflected radiation. Direct radiation arrives directly through our atmosphere. Diffuse radiation results from the scattering of light by particulate matter and moisture in the earth's atmosphere. Once solar radiation has reached the earth's surface, it is absorbed, transmitted, or reflected. When shortwave radiation (light) is absorbed, it is converted to longwave radiation (heat). In this manner, the sun heats the earth's surface, driving the climatic cycles of air movement, temperature, evaporation, and precipitation. In addition, the sun's light enables sight and the photosynthetic processes that occur in plants.

The constancy of the earth's rotation and its planetary orbit allow us to predict precisely the location and "apparent" movement of the sun relative to the earth's surface. An understanding of the seasonal variations of solar angles has informed planning decisions for at least two millennia. The ancient Greeks and Romans exploited their knowledge of solar geometry in architecture, site planning, and urban planning:

> Modern excavations of many Classical Greek cities show that solar architecture flourished throughout the area. Individual homes were oriented toward the southern horizon, and entire cities were planned to allow their citizens equal access to the winter sun.[5]

Marcus Vitruvius Pollio, the Roman architect, advised in the first century B.C.:

> If our designs for private houses are to be correct, we must at the outset take note of the countries and climates in which they are built . . . it is obvious that designs for houses ought similarly to conform to the nature of the country and to diversities of climate.[6]

History is replete with examples of cultures that took advantage of the seasonal differences in the sun's position in planning settlements and designing buildings. Although our knowledge of solar geometry may be more precise than that of our predecessors, and our computational devices more sophisticated, the guiding principles are still the same. In the northern hemisphere during the winter months, the sun travels a low path across the southern sky. In the summer, the sun travels a much higher overhead path (Fig. 11-2).[7] The sun's position relative to any terrestrial location can easily be determined and is described by its altitude angle (angle "A") and azimuth (bearing) angle (angle "B") at any given time (Fig. 11-3). Sources and application of solar angles are presented later in the chapter.

Effects. The sun provides light and heat for both our exterior and interior environments. Sunlit exterior spaces are welcomed in the winter. On hot summer days, we seek relief from the sun in cool, shaded exterior spaces. Solar radiation heats the interior spaces of our buildings when it passes through glass and is absorbed and converted to thermal energy by the various surfaces inside the building. This causes a local heating of the air in contact with these surfaces, and the "sealed" nature of the building inhibits the escape of the warmed air. This is commonly referred to as the "greenhouse effect."[8] Solar radiation also heats the opaque surfaces of a building, and additional heat is conducted through these materials into the building.

Passive solar heating of buildings can be beneficial during cold weather, reducing our reliance on mechanical heating systems. However, the solar loads imposed on our buildings during warm periods of the year can greatly increase our dependence on mechanical cooling systems.

The sun provides us with an abundance of natural light, even on overcast days. In addition to the suggested psychological and physiological benefits of natural light,[9] the use of daylighting in our buildings can conserve energy by reducing the amount of energy required for artificial lighting systems, which can be the most energy-intensive

---

[4]R. Geiger, *The Climate Near the Ground* (Cambridge, Mass.: Harvard University Press, 1975), p. 5.

[5]K. Butti and J. Perlin, *A Golden Thread: 2500 Years of Solar Architecture and Technology* (New York: Van Nostrand Reinhold Company, Inc., 1980), p. 3.

[6]Marcus Vitruvius Pollio, *The Ten Books on Architecture*, M. H. Morgan, trans. (Cambridge, Mass.: Harvard University Press; New York: Dover Publications, Inc., 1960), p. 170.

[7]D. Watson and R. Glover, *Solar Control Workbook* (prepared for the U.S. Department of Energy Passive Solar Curriculum Project, *Teaching Passive Design in Architecture*, November 1981), p. 6.

[8]D. Watson and K. Labs, *Climatic Design* (New York: McGraw-Hill Book Company, 1983), p. 20.

[9]"Light as Nutrition," *Progressive Architecture*, April 1981, pp. 176–177.

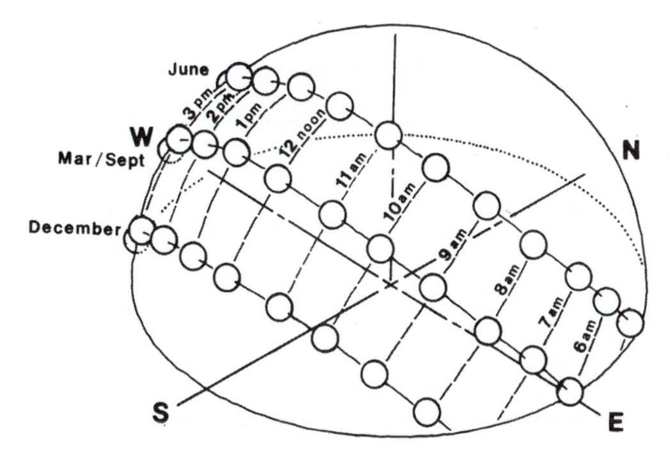

FIGURE 11-2. Seasonal sun paths (northern hemisphere).

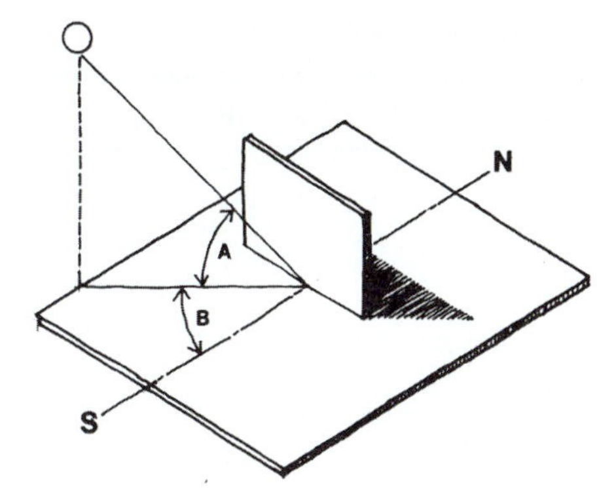

FIGURE 11-3. Sun angles.

systems in certain commercial building types. Parenthetically, lighting systems also generate a significant amount of heat, which can increase a building's air-conditioning requirements. The primary goal in designing for daylighting "is to provide natural illumination while effectively blocking direct solar radiation, since it usually carries uncomfortable levels of heat and light."[10] Generally speaking, diffuse, reflected, and filtered solar radiation are the most useful for daylighting buildings.

Control and Modification. The following discussion focuses on the use of vegetation, architectural elements, and landforms or hardscape to control solar radiation. The four basic control strategies—admission, obstruction, filtration, and reflection—will be discussed as they affect both outdoor spaces and buildings.

Vegetation:

Outdoor spaces. Proper selection and location of vegetation can ameliorate microclimatic conditions by providing exterior spaces with access to solar radiation when temperatures are below the comfort zone, and shade when they are not.

Vegetation produces shade and can intercept and absorb 60 to 90% of incident solar radiation.[11] Dense foliage, multiple foliage layers, or dense canopies obstruct solar radiation, whereas plants with open, loose foliage filter it. The temperatures of surfaces in shade (and consequent air temperatures) can be reduced significantly by shading. During the summer, it is not uncommon for unshaded, dark-colored "hard" surfaces to reach temperatures in excess of 160°F. Oppressive levels of radiant heat generated by such surfaces even after sunset, in addition to their ability to raise the temperature of the proximate air mass, can have a profound impact on perceptions of comfort. The bioclimatic chart (Fig. 11-1) indicates that under calm wind conditions and depending on relative humidity levels, shading is desirable when air temperatures exceed the range 68° to 72°F (the "shade line"). Obviously, shading can be an extremely productive comfort strategy during periods of overheating.

Conversely, exterior spaces that have access to solar radiation can improve or produce conditions of comfort even during very cold conditions. The bioclimatic chart (Fig. 11-1) indicates that a person in normal street clothing exposed to full solar radiation (100%, 45° solar altitude) will feel comfortable when it is calm and the temperature is 44°F. Increasing clothing insulation or activity levels can produce comfort at even lower temperatures with sunshine. It is therefore possible for properly oriented, unshaded exterior spaces that are protected from the wind to be comfortable when ambient air temperatures are well below the comfort zone.

At this point, it is appropriate to examine the distinction between deciduous and evergreen vegetation. Deciduous vegetation normally becomes foliated in the spring and provides shade during the overheated periods in summer and fall. As fall defoliation occurs, solar radiation is no longer obstructed and can be used to warm exterior spaces during winter months. (It should be noted that deciduous trees with dense branching structures can produce a significant amount of shading [greater than 50%] when defoliated.) These selective shading characteristics can be particularly useful in temperate climates with hot summers and cold winters. Evergreen vegetation maintains its foliage throughout the year and can be useful for shading exterior spaces in tropical climates where warm temperatures and sunshine occur throughout the year. Shade trees can be selected with foliation periods that correspond approximately to the times of overheating at a specific site. Table 11-1 lists the foliation periods for a variety of trees. (The observations listed in the table were made in California, Utah, and Oregon and therefore may not be applicable to your site. Local plant guides, experts, or observation can provide information on the foliation characteristics of trees indigenous to

[10]"Through a Glass Brightly," *Progressive Architecture,* November 1981, p. 138.

[11]W. E. Reifsnyder and H. W. Lull, *Radiant Energy in Relation to Forests,* U.S. Department of Agriculture Forest Service Technical Bulletin No. 1344 (Washington, D.C.: U.S. Government Printing Office, 1965) cited by E. G. McPherson, "Planting Design for Solar Control," *Energy Conserving Site Design* (Washington, D.C.: American Society of Landscape Architects, 1984), p. 144.

Table 11-1. Foliation Periods for Various Trees

Common Name	Foliation	Defoliation	Reference
Amur maple	March	October	3
Norway maple	April	November	1
Red maple	April	October	3
Silver maple	May	November	2
Sugar maple	May	October	3
European birch	April	November	1
European hackberry	March	November	1
Common hackberry	April	November	2
English hawthorn	May	December	2
Carrier hawthorn	April	December	3
European beech	May	December	3
European ash	May	November	2
Moraine ash	March	November	1
Green ash	April	October	3
Honey locust	April	October	3
Black walnut	May	October–November	2
Sweetgum	April	November	1
Sweetgum	May	December	3
California sycamore	March	December	1
Cottonwood	April	November	2
Quaking aspen	April	November	2
Pin oak	April	December	1
English oak	May	December	3
Chinese tallow tree	March	November	1
American elm	May	November	2
Siberian elm	April	November	2

Source: Based on information from: E. G. McPherson, "Planting Design for Solar Control," *Energy Conserving Site Design* (Washington, D.C.: American Society of Landscape Architects, 1984), pp. 144–146 citing (1) J. Hammond, J. Zanetto, and C. Adams, *Planning Solar Neighborhoods* (Sacramento: California Energy Commission, 1981); (2) E. G. McPherson, *The Use of Trees for Solar Control in Utah* (Salt Lake City: Utah Energy Office, 1981); and (3) R. L. Ticknor, "Selecting Deciduous Trees for Climatic Modification," *American Nurseryman,* 153(1), 1981, pp. 10–88.

your area.) Determining when overheated periods occur is discussed in the section "General Climatic Analysis."

Vines, used in conjunction with architectural elements (such as arbors), and shrubs are also useful for shading exterior spaces. Ground covers and grasses are useful for shading ground surfaces, and in the summer, grass surfaces can be 10 to 14°F cooler than bare soil[12] and 50°F cooler than pavement.

Buildings. As with exterior spaces, vegetation can be used to reduce solar radiation loads on buildings by shading walls, roofs, and glazed openings. Fig. 11-4 indicates the varying seasonal impact of solar radiation on different building surface orientations. It is apparent that south-facing vertical surfaces receive the most solar radiation during winter months, whereas horizontal, west, and east orientations receive more radiation during summer months. Surface temperatures of walls and roofs exposed to the sun can be more than 40°F warmer than shaded surfaces during the summer.[13] Therefore, residential buildings in temperate climates are often prudently oriented with a long wall facing south to maximize collection of solar radiation in the

[12] V. Olgyay, *Design with Climate* (Princeton, N.J.: Princeton University Press, 1963), p. 51.

[13] E. G. McPherson, "Planting Design for Solar Control," *Energy Conserving Site Design* (Washington, D.C.: American Society of Landscape Architects, 1984), p. 146 citing E. G. McPherson, *The Use of Plant Materials for Solar Control,* unpublished master's thesis, Utah State University, 1980.

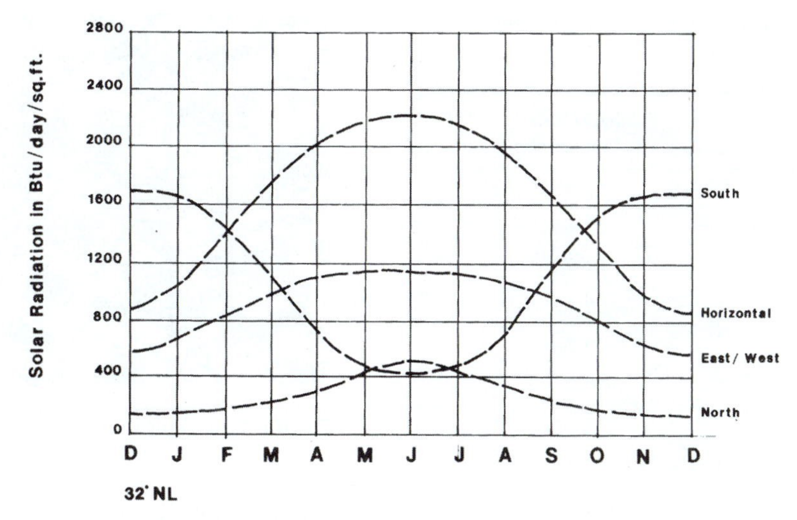

FIGURE 11-4. Seasonal impact of solar radiation. [Data obtained from American Society of Heating, Refrigerating and Air-Conditioning Engineers, Inc., *ASHRAE Handbook—1981 Fundamentals* (Atlanta: ASHRAE, 1981), Table 22A, p. 27.32.]

winter while minimizing summer heat gain. If properly designed, vegetative shading provided for horizontal, west-, and east-facing surfaces can reduce summer solar radiation loads, improve comfort, and reduce air-conditioning requirements. Shrubs or vines used to shade a wall also retard air movement across the wall surface, effectively increasing its thermal resistance and reducing convective heat flow.

Vegetation can have a significant impact on visual comfort in exterior spaces and on the use of daylighting in buildings. Shade trees, which obstruct or filter sunlight, can reduce brightness levels and visual discomfort due to glare, both in exterior and interior spaces. However, reduction of light levels can reduce the amount of daylight in building interiors. Fig. 11-5 indicates that light transmission can be significantly reduced by trees.

Shrubs, ground covers, and grasses are also useful for brightness and glare control. The ability of a surface to reflect shortwave radiation is referred to as its albedo rate. Surfaces with low albedo rates are useful for reducing contrast and controlling glare, whereas those with high albedo rates can help reflect useful daylight deep into building interiors. Table 11-2 lists the albedo rates for a variety of common ground surfaces.

Architectural Elements. Like vegetation, architectural elements can obstruct or filter solar radiation and selectively provide shading for exterior spaces and buildings. Architectural elements are defined in the broadest possible terms and include buildings. Small, free-standing or detached elements, such as gazebos and arbors, have long been used to modify microclimate through shading. Similar shelters are often used at bus stops to provide shade for waiting passengers, as well as protection from wind and rain.

There is also a long history of attached architectural elements that provide seasonal shading of exterior spaces, as well as building surfaces—from the porticoes, arcades, and courts of ancient Egyptian, Greek, Roman, and Islamic architecture, to the outdoor living porches of Southern plantation architecture, the exaggerated prairie-style overhang, and the brise-soleil of more recent times. The similar effect of all of these elements is to provide transitional exterior spaces which can ameliorate microclimatic conditions, increasing the opportunities for comfortable outdoor experiences.

Buildings, as single or collected objects in the landscape, will produce shading in adjacent exterior spaces. Obviously, building shadow patterns will vary both daily and seasonally. Outdoor activities around buildings often tend to migrate with either sun or shade, depending on

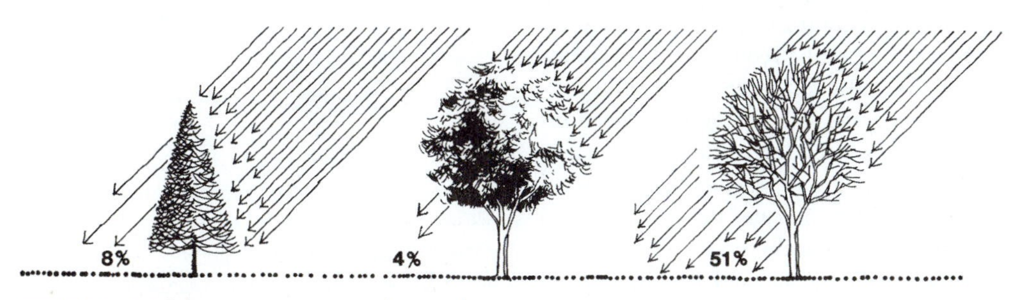

FIGURE 11-5. Light-transmission reduction by trees.

Table 11-2. Albedo Rates for Common Ground Surfaces

Surface	Albedo Rate (% Reflectivity)
Fresh snow cover	75–95
Clean, firm snow	50–65
Old snow cover	40–70
Dirty, firm snow	20–50
Light sand dunes, surf	30–60
Sandy soil	15–40
Meadows, fields	12–30
Dark, cultivated soil	7–10
Water surfaces, sea	3–10
Bare ground	10–25
Rock, gravel	12–15
Dry grass	32
Grass fields, lawns	6
Asphalt	15
Concrete	55

Sources: Based on information from: G. Robinette, ed., *Plants/People/and Environmental Quality* (Washington, D.C.: National Park Service, U.S. Department of the Interior, 1972), p. 70, and *How to Predict Interior Daylight Illumination* (Toledo, Ohio: Libbey-Owens-Ford Company, 1976), p. 11.

individual comfort requirements. These migrations can be observed at locations such as the American town-square courthouse and in many urban parks and plazas. There are numerous examples of settlements in hot, arid climates where buildings are used to provide shade for exterior spaces (Fig. 11-6). However, building shadows produce an interesting set of problems in modern urban high-rise districts, as in the following hypothetical example. Consider the effects produced by a 50-story (600 ft high) reflective glass tower on the surrounding environs. At 32° north latitude (Dallas), the shadow cast by our building, on December 21 one hour after sunrise (8 A.M. solar time), extends over 3,400 ft (0.64 mile). Fig. 11-7 indicates the extent of building shadows for the day of December 21. During winter months, our building casts a shadow across the adjacent park to the north during much of the day, as well as the street-level plaza located on the north side of our building. Shortwave radiation reflected from our building creates glare and visual discomfort in adjacent exterior spaces, but may reduce the heating requirements of buildings in the path of the reflections. Air-conditioning requirements for buildings in the path of summer reflections may be significantly increased by the additional radiation loads imposed by our building, and lawsuits have occurred in similar situations. Fig. 11-8 indicates the areas affected by June 21 reflections. Parenthetically, dense high-rise districts such as those in New York City can have a significant impact on the amount of daylight reaching the street,

FIGURE 11-6. Buildings in hot, arid climates.

which has prompted zoning legislation concerning building form.[14]

In summary, architectural elements and vegetation can be useful devices for controlling or modifying solar radiation. However, their design and location must be carefully considered and must be based on a thorough analysis of microclimatic conditions.

Landforms and Hardscape. In the following discussion, the entire site will be considered a landform, as will smaller landscape elements. "Hardscape" refers to "hard" surfaces constructed in contact with the ground in the exterior environment.

In the site-planning process, the selection of appropriate locations for exterior spaces and buildings related to solar access or shading can be affected by slope direction and gradients on the site. A slope inclined toward the south will receive more winter radiation than will a "flat" area of equal size, and with a gradient of 20%, can expedite the arrival of spring and delay the onset of winter by as much as 3 weeks.[15]

Surface A in Fig. 11-9, normal to the sun's rays, receives maximum radiation. As surfaces rotate away from normal (surfaces B), they receive increasingly less radiation. At 90° of rotation (surface C), the surface receives no direct radiation. Therefore, different slope aspects or orientations produce different microclimatic conditions, as delineated by Gary Robinette in *Energy Efficient Site Design*:[16]

Southeast slope	Most desirable
South slope	Preferred: warm winter, early spring, late fall
East slope	Acceptable: warm winter mornings, cool summer evenings
West slope	Undesirable: hottest summer slope
North slope	Least desirable: coldest in winter

[14]H. Bryan and S. Stuebing, "Natural Light as an Urban Amenity." *Lighting Design and Application*, June 1986, pp. 44–48.

[15]Olgyay, *Design with Climate*, pp. 49–50.

[16]G. Robinette, ed., *Energy Efficient Site Design* (New York: Van Nostrand Reinhold Company, Inc., 1983), p. 13.

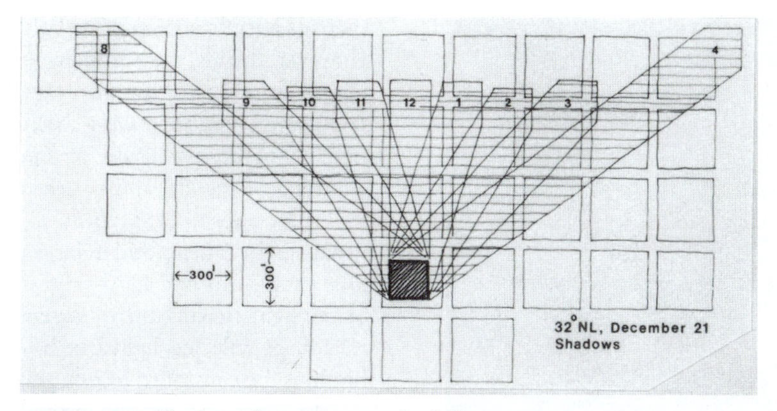

FIGURE 11-7. Shadows from 50-story building.

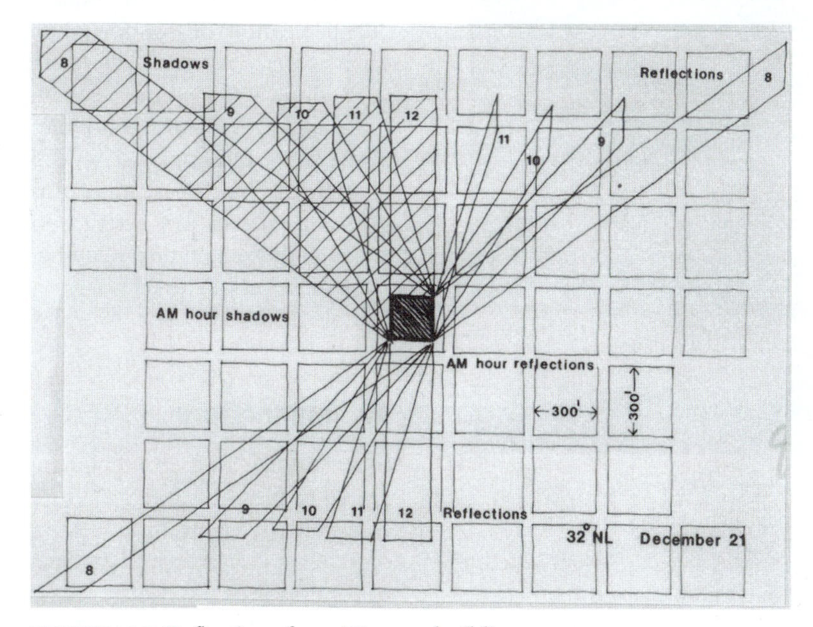

FIGURE 11-8. Reflections from 50-story building.

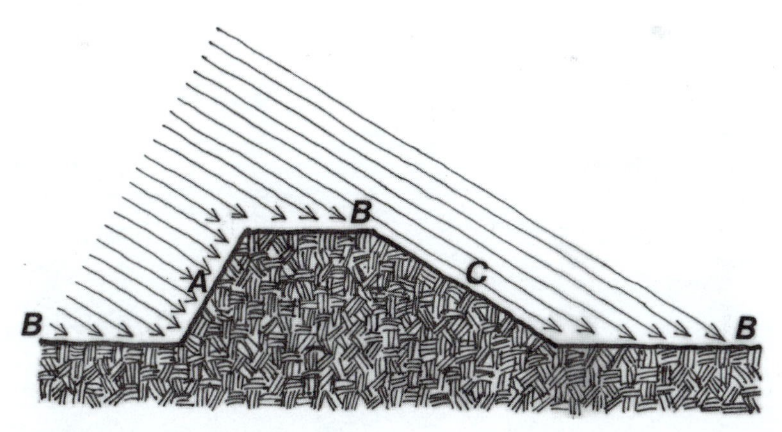

FIGURE 11-9. Solar radiation, slope gradient, and orientation.

When the slope gradient approaches 100%, the slope becomes essentially a vertical "wall." South-facing cliff walls were used to shelter the Native American settlements at Canyon de Chelly in northeast Arizona (Figs. 11-10 and 11-11). As noted earlier (Fig. 11-4), south-facing vertical surfaces maximize winter radiation while minimizing summer radiation. The Native American settlements are recessed into the cliff face, which improves summer shading characteristics. Similar settlements also existed in southern Tunisia and Shensi province in China.[17] Although this is not ordinarily a feasible site-planning option, an understanding of the principles will inform any planning process.

[17] J. M. Fitch and D. P. Branch, "Primitive Architecture and Climate," *Scientific American*, December 1960, p. 140.

FIGURE 11-10. Canyon de Chelly. (Photo courtesy of National Park Service, Department of the Interior.)

Finally, at a smaller scale, landforms and hardscape exposed to solar radiation affect comfort in several ways. Hardscape can be effectively used to collect, store, and radiate thermal energy, ameliorating cold microclimatic conditions. However, during the summer, unshaded hardscape surface temperatures can be significantly higher than those of vegetated landforms, increasing discomfort through the generation of oppressive levels of radiant heat and by further warming of the air. These surfaces continue to radiate heat well after sunset. This effect is commonly experienced in urban environments and in large paved areas such as parking lots. Landforms and hardscape also reflect some percentage of the light striking them, depending on the surface albedo rate. That reflected light can be used to daylight buildings but must be controlled to minimize glare.

In summary, the sun provides us with light and heat, drives the earth's climatic cycles, and significantly affects our perception of comfort. Its position relative to the earth's surface is precisely predictable. If properly designed, vegetation, architectural elements, landforms, and hardscape can be used to modify microclimates by admitting, obstructing, filtering, and reflecting solar radiation. Site-planning decisions have an enormous impact on how solar radiation affects the exterior spaces and buildings we create. A thorough understanding of solar radiation principles and effects is therefore prerequisite to an effective site-planning effort.

Air Temperature

Description. The second climatic element with which we are concerned is air temperature. Dry-bulb temperature, commonly referring to air temperature, is the most common climatic reference when describing weather or feelings of comfort. It is the relative measure of the thermal energy (sensible heat) content of the air, as indicated by an ordinary thermometer.

FIGURE 11-11. Canyon de Chelly. (Photo courtesy of National Park Service, Department of the Interior.)

Temperature differences are the driving force behind the heat-transfer processes of conduction, convection, and radiation which occur between the environment and our bodies and buildings. When a temperature differential exists between two masses (either solid or fluid), thermal energy is transferred *from* the area of higher temperature *to* the area of lower temperature.

Effects. Simply stated, temperature affects our sense of comfort, which in turn affects the amount of energy required to heat and cool our buildings. Disregarding other environmental factors, the relative effect of air temperature can be seen in the bioclimatic chart (Fig. 11-1), where comfort conditions exist in a fairly narrow range of temperatures. It is therefore important for the site planner to understand the dynamic behavior of air temperature. An air temperature profile for March 25, 1979, recorded at the Dallas–Fort Worth International Airport (DFW), is shown in Fig. 11-12. This was a day with cloudless skies, and it will be used to illustrate certain points concerning daily air temperature fluctuations.

The lowest temperature of the day occurs shortly before dawn. After the sun rises, the earth's surface is heated by solar radiation and the surface air layer is warmed by convection. Air temperature rises rapidly and peaks during mid-afternoon. From this high point, the temperature drops at a more gradual rate. As the earth's surface reradiates thermal energy to the clear night sky, the surface air layer slowly cools and temperatures drop until dawn. By contrast, the temperature profile for March 29, 1979, an overcast day, is quite different (Fig. 11-12). The cloud cover inhibits solar radiation from reaching the earth's surface. Some radiation does penetrate, however, which accounts for the slight warming trend in the afternoon. Cloud cover also inhibits night reradiation, which keeps the temperature of the earth's surface and the surface air layer relatively constant. Although general in nature, this overview of daily (diurnal) temperature fluctuations should provide an insight into air temperature dynamics. Methods for analyzing air temperature variations over longer periods of time are discussed in the section "General Climatic Analysis."

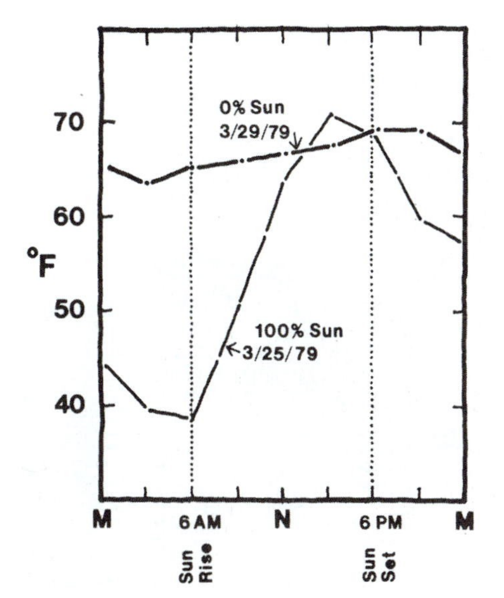

FIGURE 11-12. Temperature profiles at DFW.

Control and Modification. The site planner has few opportunities to control or modify air temperature directly. However, air temperature and comfort can be affected indirectly through control of solar radiation and air movement.

Vegetation. As water changes to water vapor at leaf surfaces (transpiration), an energy exchange occurs. Thermal energy is extracted from the air during the evaporative process, lowering air temperature but raising humidity levels. Studies have suggested that this process can produce a significant amount of cooling in air temperature, and this may be true in the mesoclimatic regime with large, heavily vegetated parcels of land. However, due to the rapid mixing that occurs in the surface air layer caused by wind and temperature differences, the transpirational cooling produced at the microclimatic level by smaller areas of vegetation or individual trees is negligible.[18] The effects of shading, wind shielding, soil-moisture retention, transpiration, and night reradiation from the canopy in stands of vegetation do combine to alter air temperatures substantially, particularly those at ground level (Fig. 11-13). Although temperatures and humidities in the canopy and understory fluctuate markedly, they remain relatively constant near the ground.[19]

Architectural Elements. Fountains and other water features can have a beneficial impact on air temperature in hot, arid climates. The evaporative cooling process lowers air temperature and raises humidity levels, both of which promote comfort in hot, dry conditions. Water features are often located in courtyards protected from the wind, where they generate a "pool" of cool, heavy air that is retained by the court's walls (Fig. 11-14). In cold climates, building surfaces exposed to sun and protected from cold winds can create a "sun pocket" that is capable of raising localized air temperatures as much as 40°F (Fig. 11-15).[20]

Landforms and Hardscape. Other than the effects produced by solar radiation, landforms and hardscape have limited influence on air temperature. However, it is important to note that topographical features on the site, such as valleys, ravines, or natural depressions, have the ability to collect cold air. Cold air flows downhill and will settle in low spots on the site. These areas can be useful for collecting cooler air in predominantly warm climates but pose problems in colder climates.

[18]McPherson, "Planting Design for Solar Control," p. 147, citing B. A. Hutchinson, F. G. Taylor, R. L. Wendt, and The Critical Review Panel, *Use of Vegetation to Ameliorate Building Microclimates: An Assessment of Energy Conservation Potentials* (No. 1913) (Oak Ridge, Tenn.: Oak Ridge National Laboratory, 1982), pp. 17–18.

[19]G. Robinette, ed., *Plants/People/and Environmental Quality* (Washington, D.C.: U.S. Department of the Interior, National Park Service, 1972), pp. 89 and 96.

[20]Robinette, *Energy Efficient Site Design*, p. 42, citing W. Langeweische, "How to Fix Your Private Climate," *House Beautiful*, October 1949, p. 150.

FIGURE 11-13. Vegetation and temperature/humidity changes.

Humidity

Description. The third climatic element to be discussed is humidity. Although it is referred to and measured in a variety of ways, relative humidity is the climatic reference with which most of us are familiar. It is defined as "the ratio of the actual vapor pressure of the air-vapor mixture to the pressure of saturated water vapor at the same dry-bulb temperature."[21]

Effects. Human tolerances for humidity variations are greater than for temperature variations. When air temperatures are elevated, high relative humidities limit the air's ability to evaporate moisture from our bodies, thereby increasing thermal stress and our sense of discomfort. High humidity also produces a significant thermal load on a building's air-conditioning equipment, which expends energy to remove moisture (latent heat) from the air. If high humidities are not sufficiently reduced, mildew problems can occur in buildings. In cold weather, high indoor humidity may result in condensation, usually on interior glass surfaces, which can damage materials and deteriorate finishes.

Low relative humidities can also create problems in buildings. Direct air-heating systems (without humidity control) significantly reduce the percentage humidity of circulating air. In addition to the drying effect on human respiratory passages, low humidities contribute to static electricity problems, which are of particular concern in the computerized work environment. Humidity control in buildings, such as hospitals, can be extremely important. Studies have indicated that the propagation of bacteria is minimized in the relative humidity range 50 to 55%.[22]

Like air temperature, relative humidity is cyclic in nature. In a very general sense, the maximum relative humidity during the course of a day without precipitation coincides with the minimum temperature, and the minimum relative humidity is coincident with the daily maximum temperature. This is certainly not always true but

[21] V. Bradshaw. *Building Control Systems* (New York: John Wiley & Sons, Inc., 1985), p. 26.

[22] M. D. Egan, *Concepts in Thermal Comfort* (Englewood Cliffs, N.J.: Prentice Hall, Inc., 1975), p. 9.

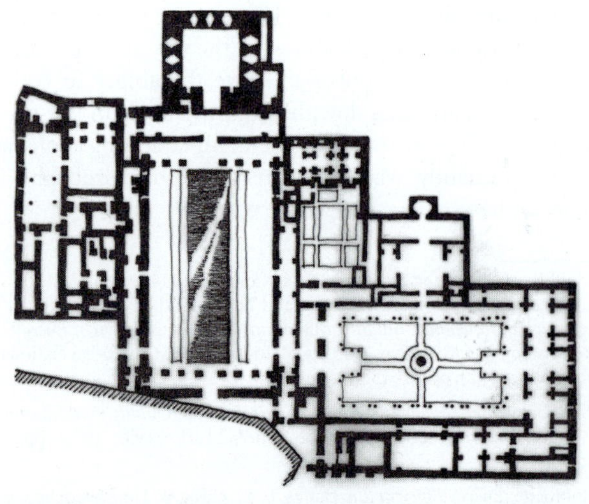

FIGURE 11-14. The Alhambra, Granada.

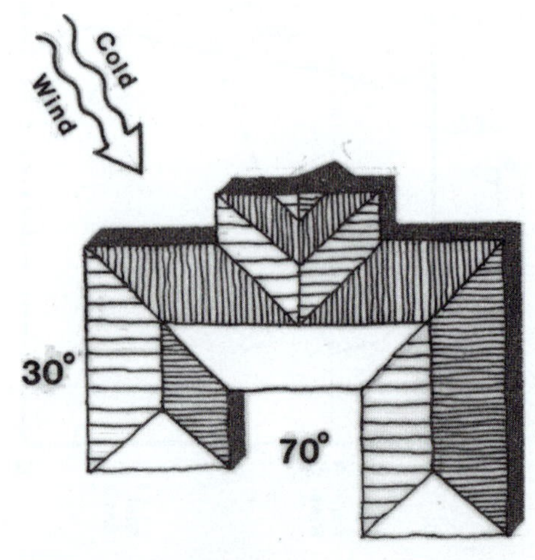

FIGURE 11-15. Sun pocket.

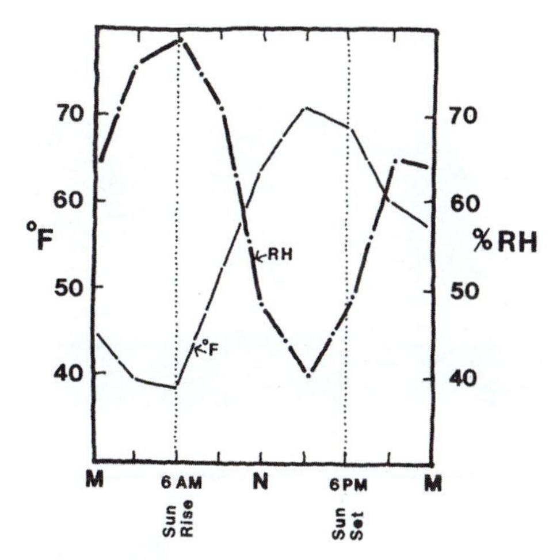

FIGURE 11-16. Air temperature and humidity interrelationship.

serves as a general model of the relationship between relative humidity and air temperature. If the moisture content of the air remains constant over the course of a day, as the air temperature rises, the relative humidity decreases. When the air temperature falls, the relative humidity increases. This type of pattern is illustrated in Fig. 11-16, which shows the temperature and relative humidity profiles at Dallas–Fort Worth International Airport on March 25, 1979. With the exception of 12 midnight on March 26, the correlation holds true.

As indicated earlier in the bioclimatic chart (Fig. 11-1), comfort conditions span a fairly wide range of relative humidity levels (between 18 and 77%). The desirable range exists on the chart between 30 and 65% relative humidity.[23] Analysis of the effects of relative humidity are discussed further in the section "General Climatic Analysis."

Control and Modification. As with air temperature, the site planner's options for significant modification of relative humidity are limited. The primary concern with humidity is when air temperatures occur above the comfort zone. In hot, dry conditions, moisture can be added to the air (evaporative cooling or evapotranspiration from plants) to induce comfort (Fig. 11-1, upper-left portion of bioclimatic chart). In hot, humid conditions, air movement is the only means to alleviate discomfort (Fig. 11-1, upper-right portion of the bioclimatic chart).

Vegetation. As indicated previously (Fig. 11-13), relative humidities often exceed ambient levels and air temperatures are diminished in dense stands of vegetation close to the ground. This situation could produce beneficial effects in hot, arid climates, but it is often difficult to maintain dense vegetation in such climates. In hot, humid climates, dense vegetation may be detrimental to comfort by increasing relative humidity and inhibiting air movement. Therefore, planning solutions using vegetation should be designed to provide effective shading which does not significantly retard air movement. Thinning underbrush and removing lower limbs of trees to a height of 8 to 10 ft promotes comforting airflow.

Plants are natural humidifiers and can be used to provide supplementary humidity in buildings. This is an appropriate strategy when air-heating systems reduce humidity levels in buildings during cold periods.

Architectural Elements. When properly designed and protected, water features can raise humidity levels and lower temperature. This is essentially an adiabatic process where heat is not gained or lost but transformed from sensible to latent heat. Consider a hypothetical perfectly protected courtyard with a fountain in a hot, arid climate, where the temperature is 90°F and the relative humidity is 20%, a point above Olgyay's comfort zone (Fig. 11-1). If enough moisture is evaporated from the fountain to raise the relative humidity of the air mass in the courtyard to 40%, the air temperature theoretically will drop to approximately 79°F, a point almost in the comfort zone. At 50% relative humidity, the temperature would drop to 75°F. (These temperature–humidity combinations occur along a line of constant enthalpy on the psychrometric chart. Refer to the discussion of psychrometric processes in *Mechanical and Electrical Equipment for Buildings,* 6th ed., Chap. 5, by Benjamin Stein and John Reynolds for a detailed discussion.)

Because of the number of variables involved and the dynamic nature of the climatic elements, the effects of elevating humidity levels in exterior spaces are extremely difficult to quantify. For the purposes of this discussion, it is sufficient to understand that evaporative cooling can be useful in hot, arid climates. Increasing humidity levels in hot, humid conditions will increase discomfort, although active water features may offset this by providing "psychological cooling." Finally, it should be noted that areas of the site with poor drainage or low areas with standing water can elevate localized humidity levels, adversely affecting comfort in hot, humid climates.

Air Movement/Wind

Description. The final climatic element requiring discussion is air movement or wind. The site planner should be familiar with how specific site characteristics affect air motion and how local wind patterns vary seasonally. Knowledge of the speed and direction of the air mass close to the ground and its associated temperature can provide valuable information for siting buildings and planning exterior spaces, so as to protect against winds that are undesirable while taking advantage of those which are useful.

The distribution and characteristics of air movement over a region are affected by both global and local factors:[24]

1. The seasonal global distribution of air pressure
2. The earth's rotation about its axis

[23]Olgyay, *Design with Climate,* p. 18.

[24]B. Givoni, *Man, Climate and Architecture,* 2nd ed. (New York: Van Nostrand Reinhold Company, Inc., 1981), p. 8.

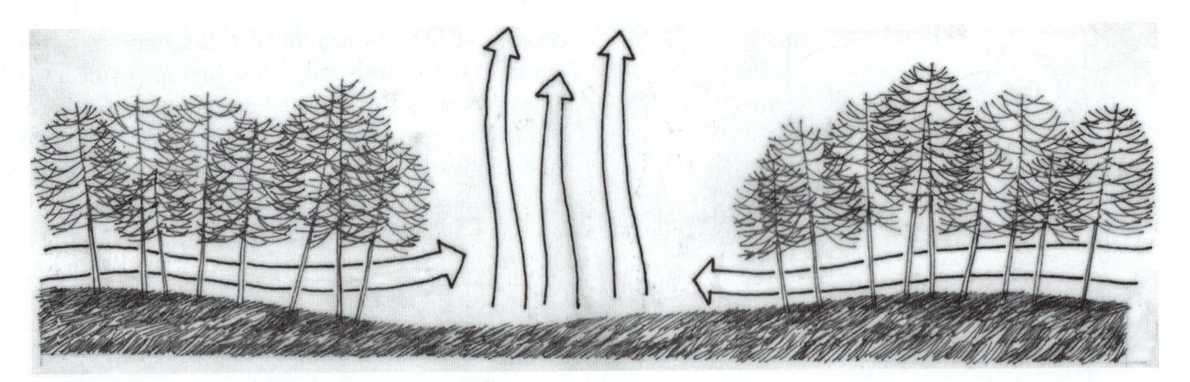

FIGURE 11-17. Flow of air from forest to plain.

3. The daily variations in the heating and cooling of the land and sea
4. The topography of the region and its surroundings

Temperature and pressure differences affect local air movement and the microclimate of the site. Air flows from high-pressure areas to low-pressure areas. G. Z. Brown[25] describes an example of a meadow surrounded by a forest. As solar radiation heats the meadow air, its density and pressure decrease, causing it to rise. This air is replaced by cooler air flowing from the higher pressure zone of the forest (Fig. 11-17). Similarly, sun-warmed air in a valley flows uphill during the day (Fig. 11-18). The airflow is reversed at night as cold, dense air falls down the valley slopes (Fig. 11-19).[26]

Large bodies of water also affect daily air movement patterns, because they are normally cooler during the day and warmer at night than the adjacent land surface. Consequently, during the day as the warmed air over land rises, it is replaced by breezes of cooler air off the water (Fig. 11-20). The land surface and adjacent air mass cool rapidly at night, and the airflow reverses (Fig. 11-21).[27] Parenthetically, large bodies of water are cooler in summer and warmer in winter than inland areas, and therefore moderate seasonal temperature extremes of coastal sites.[28]

Moving air behaves like a fluid and will take the path of least resistance as it flows over, under, around, and through objects in the landscape. Air flows smoothly around a streamlined object (Fig. 11-22). As illustrated in Figs. 11-23 and 11-24, when moving air encounters a nonstreamlined

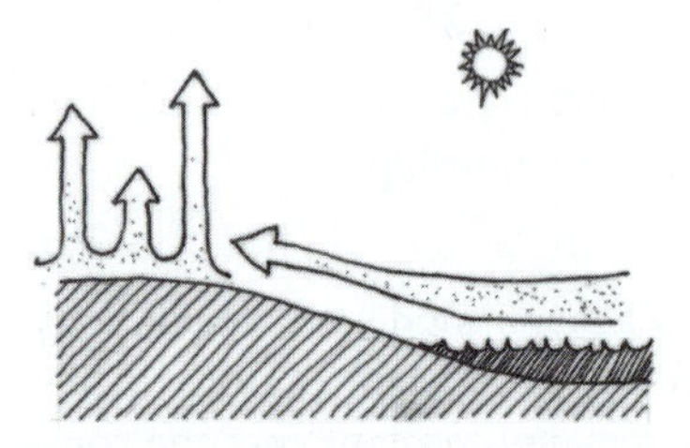

FIGURE 11-20. Air movement shoreline—day.

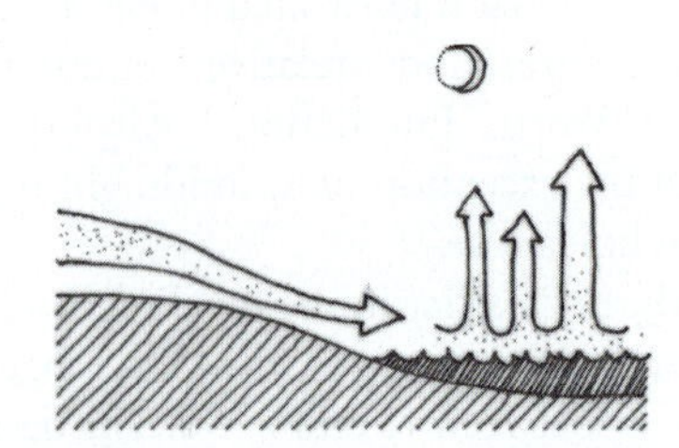

FIGURE 11-21. Air movement shoreline—night.

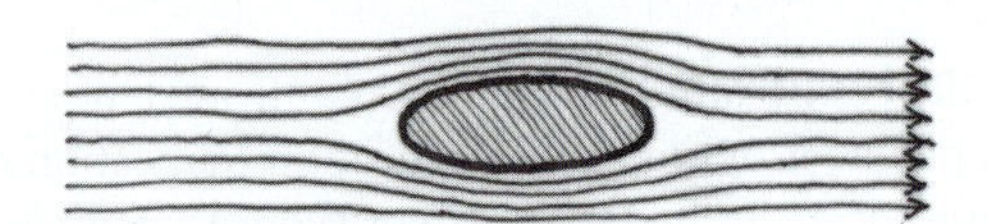

FIGURE 11-22. Air movement around streamlined object.

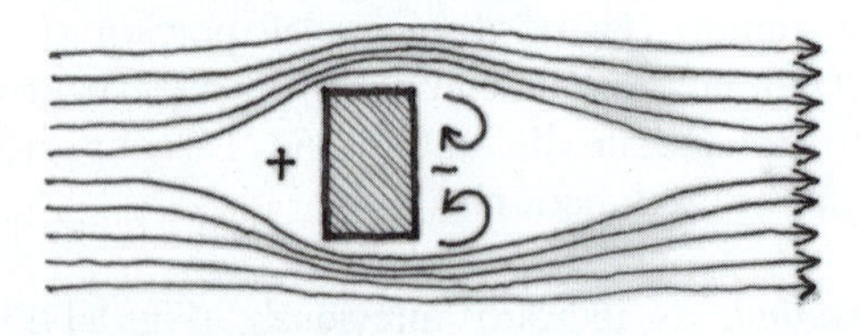

FIGURE 11-23. Air movement around rectangular object.

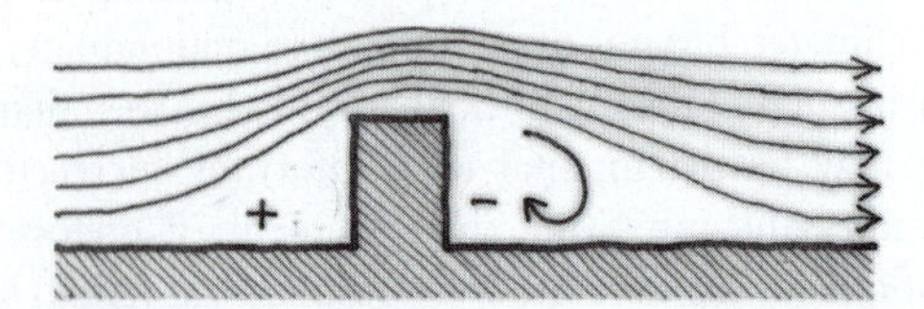

FIGURE 11-24. Air movement around rectangular object.

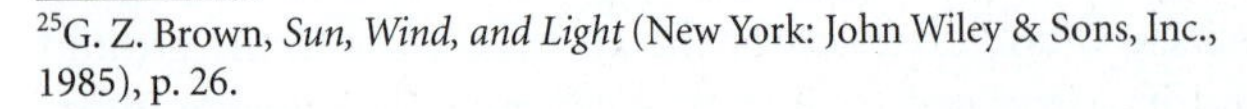

25 G. Z. Brown, *Sun, Wind, and Light* (New York: John Wiley & Sons, Inc., 1985), p. 26.

26 Ibid., p. 27.

27 Olgyay, *Design with Climate*, p. 51.

28 Ibid.

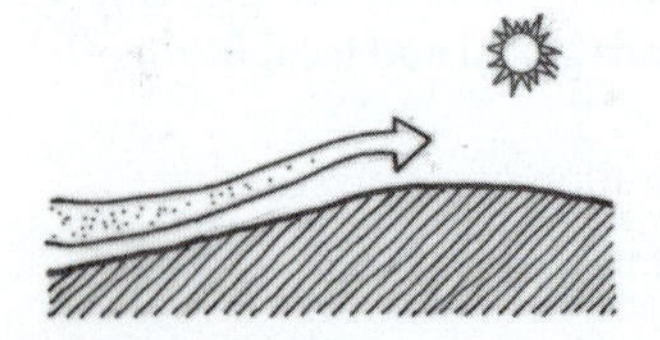

FIGURE 11-18. Air movement uphill—day.

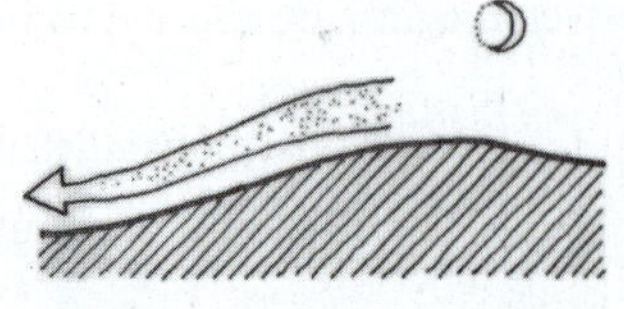

FIGURE 11-19. Air movement downhill—night.

or "bluff" object, a high-pressure area and pressure eddies are created on the windward side of the object. As the air moves over or around the object, it compresses and its speed increases. A low-pressure area is created on the leeward side of the object, producing suction eddies and a turbulent wake. Having passed the object, the air mass begins to resume its original pattern of flow. The area of protection between the object and the point where the wind resumes its normal flow is referred to as the "wind shadow." The highest wind speeds occur at the windward corners and along the sides of the object parallel to the flow, while a relatively calm area is produced at the leeward side. Any modification or addition to the object will alter the pattern of airflow around it. The object under discussion could be dense vegetation, a building, landform, or hardscape. The general principles of air motion apply to all.

Effects

> Cold winds are disagreeable, hot winds enervating, moist winds unhealthy.
>
> *Vitruvius,* The Ten Books on Architecture[29]

Air movement affects both our perception of thermal comfort and the amount of energy required to maintain thermally comfortable conditions in buildings. Cold winds increase our sense of discomfort by increasing the rate of heat loss from our bodies. This effect is commonly described by the wind chill factor, which combines the effect of wind and air temperature to produce an "apparent" air temperature lower than the actual air temperature. For example, at 30°F, a wind velocity of 25 mph produces a wind chill temperature of 0°F.[30] Cold winds can also affect comfort-energy requirements in buildings by increasing heat loss rates through the building "skin" and by increasing infiltration or air-leakage rates.

Temperate winds can be useful in promoting thermal comfort in both outdoor and indoor settings. In the outdoor environment, air movement promotes the evaporative cooling of our bodies and is essential for reducing discomfort when high temperatures are associated with high relative humidities. As indicated by the bioclimatic chart (Fig. 11-1), a wind speed of approximately 2.5 miles per hour can produce comfort when the temperature is 80°F with 80% relative humidity. It is also apparent that air movement is important even when relative humidities are not high (Fig. 11-1, e.g., 90°F, 30% relative humidity). Temperate winds are also useful for ventilating buildings and cooling occupants, thereby reducing mechanical cooling requirements.

Control and Modification. When discussing the control or modification of air movement, it is important to understand the nature of air motion and how it can be affected by design decisions and various site features. Seasonal wind patterns are fairly predictable, and effective site planning for wind control can divert discomforting winds while admitting or channeling useful winds. The identification of seasonal wind patterns is discussed in the section "General Climatic Analysis."

Vegetation. Vegetation can control or modify wind by obstruction, diversion, guidance, and filtration. When vegetation is used to protect exterior spaces and buildings from cold winter winds, it is referred to as a *shelterbelt* or *windbreak* (Figs. 11-25 and 11-26). The direction of seasonal winds and those winds associated with cold weather events, such as winter cold fronts, are predictable with some certainty, allowing the site planner to determine the most effective location for shelterbelts.

Airflow is affected by the geometry and density of the shelterbelt. Fig. 11-27 indicates the general effects on air flowing perpendicular to a moderately dense shelterbelt 30 ft tall. Airflow can be affected for a distance 5 to 10 times the barrier height on the windward side, and up to 30 times the barrier height on the leeward side.[31] When airflow is oblique to the barrier, the area of protection will be reduced.

[29]Vitruvius, *Ten Books on Architecture,* p. 24.

[30]ASHRAE, p. 8.17.

[31]Robinette, *Plants/People/and Environmental Quality,* p. 78.

FIGURE 11-25. Vegetation as windbreak. (Photo by Kenneth Roemer.)

FIGURE 11-26. Vegetation as windbreak. (Photo by Kenneth Roemer.)

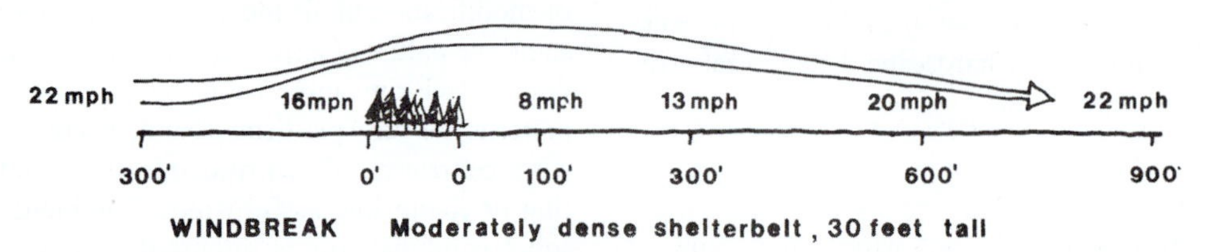

FIGURE 11-27. Windbreak: effects on airflow.

The relative density or penetrability of the shelterbelt affects leeward wind speeds and the size of the wind shadow. The suction eddies on the leeward side of a barrier pull the airflow back toward the ground. Dense barriers produce more suction than penetrable barriers. Dense vegetation provides a greater reduction of wind speeds close to the shelterbelt but a shorter wind shadow. The effects of more penetrable vegetation extend farther beyond the shelterbelt because airflow through the barrier reduces the suction pressures on the leeward side. The graph in Fig. 11-28 indicates the effects of shelterbelt density on airflow.[32]

Evergreen vegetation is effective for year-round wind control, while deciduous vegetation is effective when foliated. It should also be noted that a shelterbelt that is open near the ground, such as a screen of high-branching trees, measurably increases wind velocity immediately beneath and just beyond the screen.[33] A detailed discussion of shelterbelts can be found in *Plants, People, and Environmental Quality* by G. O. Robinette.

In addition to diverting discomforting winds, properly designed windbreaks can direct desirable winds through exterior spaces. The windbreak illustrated in Fig. 11-29 is designed for a site where cold winds arrive primarily from the north and northwest, and temperate winds are from the south. The windbreak protects the building and yard space

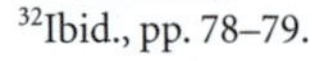

[32]Ibid., pp. 78–79.

[33]Ibid., p. 77, citing R. F. White, *Effects of Landscape Development on the Natural Ventilation of Buildings and Their Adjacent Areas,* Texas Engineering Experiment Station Research Report 45 (College Station, Tex.: Texas Engineering Experiment Station, 1945), p. 25.

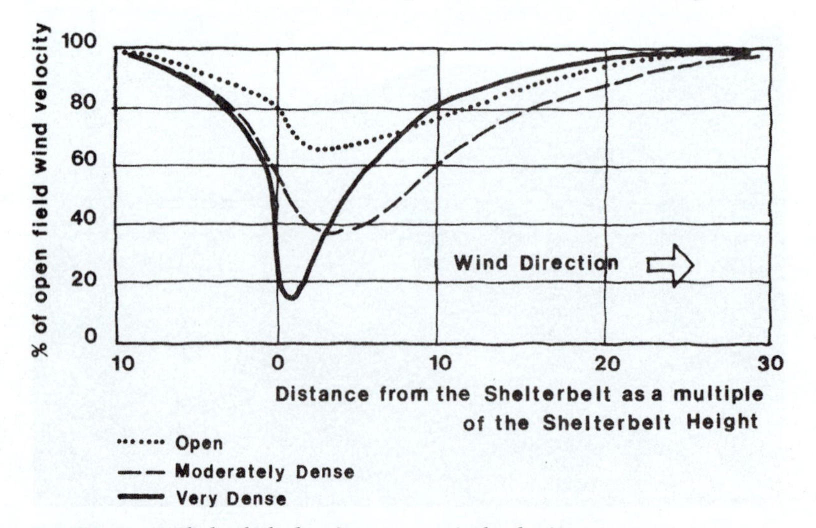

FIGURE 11-28. Shelterbelt density versus wind velocity.

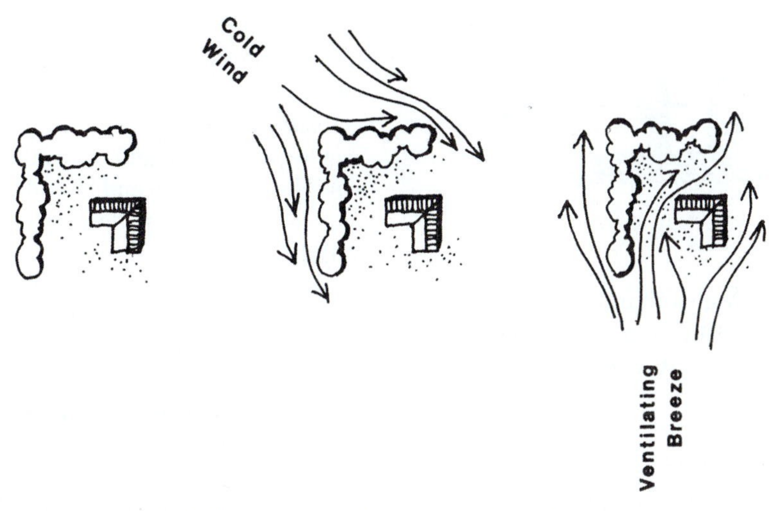

FIGURE 11-29. Windbreak designed for seasonal wind patterns.

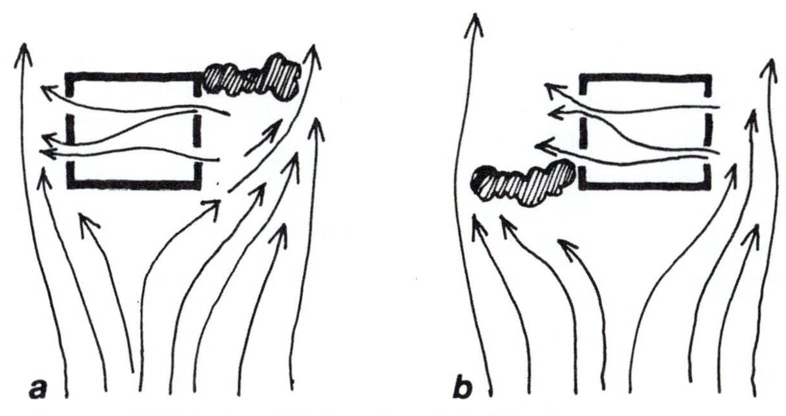

FIGURE 11-30. Diversion of airflow through building.

from cold winter winds while helping to channel southerly winds through the side and rear yards.

Smaller screens of vegetation can also be used to direct airflow through exterior spaces and buildings. Fig. 11-30 illustrates how two different dense hedge placements produce essentially the same effects on airflow through a small building.[34]

Architectural Elements. The primary architectural elements that influence airflow are buildings, walls, and fences. Like vegetation, buildings, or groups of buildings, can be used for wind control. In cold climates and hot, arid climates, buildings and exterior spaces often huddle together, avoiding discomforting winds (Fig. 11-31). In hot, humid climates, buildings can be spaced apart to admit comforting winds (Fig. 11-32). Small elements like gazebos are usually designed to provide access to comforting winds as well as shade.

Large urban structures can produce extremely discomforting, even dangerous winds at street level. The urban pedestrian is often faced with wind gusts that may

[34]Olgyay, *Design with Climate,* p. 102.

FIGURE 11-31. Settlement in hot, arid climate.

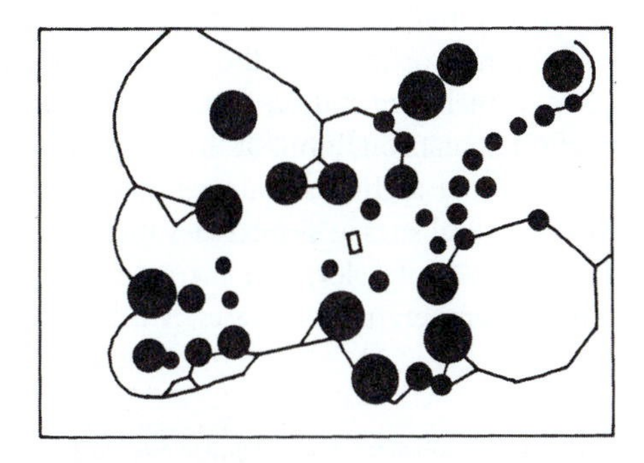

FIGURE 11-32. Settlement in hot, humid climate.

FIGURE 11-33. Berm deflecting airflow.

make walking difficult, and with airborne dust and debris. An excellent discussion of urban wind and its control can be found in *Improving Street Climate Through Urban Design* by Charles Thurow.

Walls and fences, as objects in the airstream, can be used to divert or channel winds and produce effects similar to the shelterbelts discussed previously. Again, the effects on airflow are dependent on barrier geometry (height and length) and penetrability.[35]

Landforms. As wind flows over and around hills, the highest wind velocities occur at the crest and on the windward side of the hill. Protected areas occur on the leeward side. Landforms such as berms also affect airflow and can be used to divert or channel winds (Fig. 11-33).

INFORMATION SOURCES AND INTERPRETATION

Information on the climatic elements is readily available, and an analysis of local climate can be used to inform decision making in the site-planning process. Initial site-planning decisions which consider the impact of climate can eliminate some difficult compromises that might otherwise occur in later stages of the design process. The acquisition of climatic information is discussed in this section. Discussion of the analysis of this information follows, in the section "General Climatic Analysis."

Ideally, instrumentation would be set up on the site to monitor the microclimate for an extended period of time prior to the site-planning process. The expense and time requirements make this an impractical approach, and the site planner must rely on other sources of information. The climatic information the site planner often ends up using has been recorded somewhere quite remote from the site being analyzed. Often referred to as "microclimatic data" and representing the general climatic characteristics of a region, it is recorded in or close to many major urban areas. It is in reality a record of the microclimate at the weather recording station. Data may therefore require some adjustment to more closely reflect actual conditions at the building site.

Every aspect of the earth's climate reflects differences between urban and rural sites at similar elevations and latitudes on the globe. As site elevation increases above the weather station, air temperature decreases. A rule of thumb for estimating these effects is that temperature will drop 1°F for each 330 ft of elevation increase in the summer and each 400 ft in the winter.[36]

Coastal sites and sites located only a few miles inland may experience significant differences in temperature, wind flow, radiation, and humidity levels.[37] Due to the unpredictability of these differences, local experts should be consulted.

Wind instruments at weather stations are usually located well above ground level (22 ft at Dallas–Fort Worth International Airport, as indicated in *Local Climatological Data*). Since wind velocity generally increases with elevation, the wind velocity close to the ground at a nearby building site is generally less than weather station data would indicate.

The primary source of climatic information used in this text is the National Climatic Data Center (NCDC) under the National Oceanic and Atmospheric Administration (NOAA) in Asheville, North Carolina. Data should be obtained for a recording station close to the building site, or a station with similar climatic characteristics. A list of weather-recording stations for which local climatological data can be obtained from National Oceanic and Atmospheric Administration in Washington, D.C.

Solar Geometry

The sun's position relative to any point or object on the earth's surface is described by its altitude angle (angle A) and azimuth (or bearing) angle (angle B) (Fig. 11-3). The altitude angle indicates the vertical position of the sun, while the azimuth angle describes the horizontal position of the sun relative to *true south.* (It should be noted that magnetic orientation and solar orientation are different. Magnetic north is different from true north.[38] All ensuing references to orientation will relate to solar orientation.) Solar angles can be obtained from the *sun angle calculator,* a "tool" that provides a quick graphic method for determining altitude and azimuth angles for various latitudes. It also provides other useful information on profile and incidence angles. Copies of the former Libbey–Owens–Ford sun angle calculator can be obtained from the Society of Building Science Educators (SBSE), at www.SBSE.org. Because of the graphic format, the sun angle calculator provides an "image" of solar geometry and seasonal sunpaths that is not readily apparent from solar angle tables. This device is extremely useful for determining shadowing and designing sun-shading devices for buildings.

The altitude and azimuth angles can be used to project plan shadows to determine the location and extent of daily and seasonal shadows on the building site. Fig. 11-34 illustrates the determination of shadowing on a flat site from a 30-ft-tall cone-shaped tree and a square building of the same height, when the azimuth angle is 30° west of true

[35]For more information, see Robinette, *Energy Efficient Site Design,* pp. 68–70.

[36]Olgyay, *Design with Climate,* p. 44.

[37]Ibid., p. 45.

[38]*Sun Angle Calculator* (Toledo, Ohio: Libbey–Owens–Ford Company).

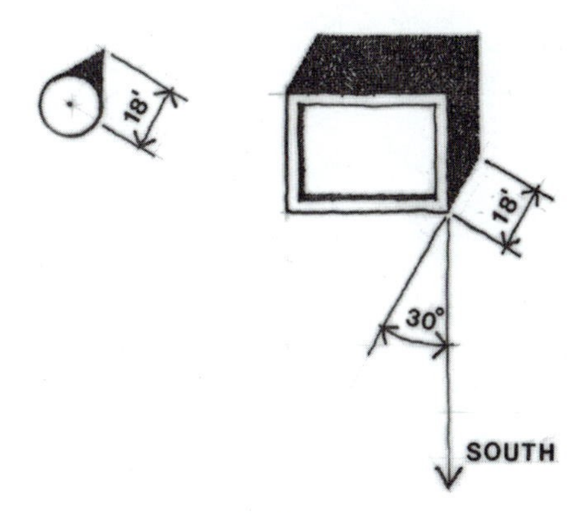

FIGURE 11-34. Shadow projections.

south and the altitude angle is 59° (32° north latitude, September 11, 1:00 P.M.). The length of the shadow is determined by the equation

$$SL = \frac{H}{\tan(\text{Alt})}$$

where

SL = shadow length
H = height of the object or edge casting the shadow
Alt = sun's altitude angle

$$SL = \frac{30}{\tan(59°)} = \frac{30}{1.6643} = 18.02 \text{ ft}$$

The shadow length of 18 ft is then scaled along the azimuth lines projected from the objects (Fig. 11-34).

This process is normally complicated by topographical changes and shadows falling across other objects. When this occurs, plan/section/elevation projections are often required to determine the precise location of shadows. It is critical for the site planner to understand solar geometry and its implications for site design, and seasonal shading comparisons for a site can be developed using the techniques that have been described. However, even relatively simple sites require a tedious series of calculations and drawings to produce seasonal shading comparisons. A simplified method for determining seasonal site shading is available—the *sun peg* (Fig. 11-35). This tool, when used in conjunction with a site model, quickly indicates site shading for each hour of the twenty-first day of each month of the year.

Sun pegs are simple to construct and use. Photocopy the sun peg chart from Appendix B that most closely corresponds to your latitude. Attach it to a stiff piece of cardboard or foam core and "erect" a gnomon peg to the height indicated on the chart. Place the sun peg on your site model, with the north arrow aligned to true north on the site, making sure that the sun peg lies "flat." By tilting the site model and sun peg in the sunlight, the gnomon's shadow can be manipulated to indicate the time of day and month of the year, and seasonal shading comparisons can quickly be developed.

One final tool will be mentioned. An inexpensive, home-made *solar transit* was developed by Charles Benton, James Akridge, and Scott Wright (Georgia Institute of Technology) (Fig. 11-36). This is a very useful instructional device that provides a good sense of seasonal solar geometry, and can be used to establish solar access and shading for specific points on the site.[39]

[39]Instructions for constructing and operating the solar transit can be found in the following document, which was distributed to schools of architecture as part of a curriculum development package: C. Benton and J. Akridge, "Solar Site Analysis," *Laboratory Exercises* (prepared for the U.S. Department of Energy Passive Solar Curriculum Project, *Teaching Passive Design in Architecture,* June 1981).

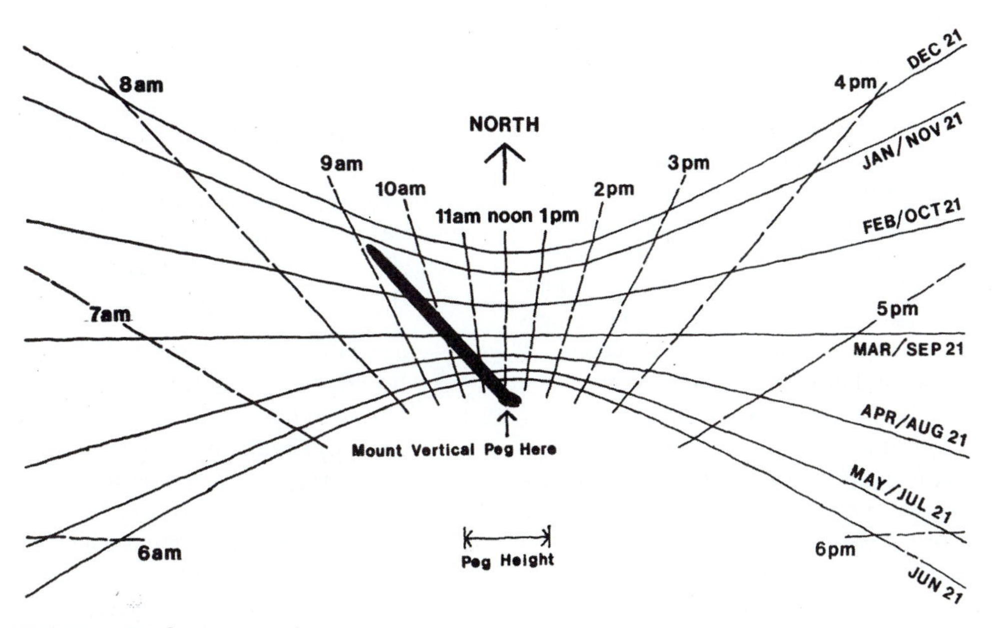

FIGURE 11-35. Sun peg.

(a)

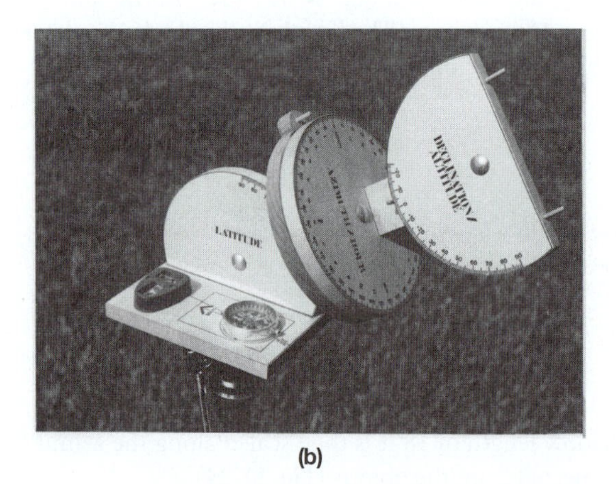

(b)

FIGURE 11-36. Solar transit.

Climatological Data

There are a number of publications that can be used for climatic analysis, available from the National Environmental Satellite, Data, and Information Service at the National Climatic Data Center (NCDC), Asheville, NC 28801-2696. They can also be found on the web at www.nesdis.noaa.gov:

Local Climatological Data—Annual Summary with Comparative Data

Information on normal climatic conditions is available for individual locations listed in Appendix A. In addition to the one-page "Narrative Climatological Summary" included in the *Annual Summary,* "Normals, Means, and Extremes" for the weather recording station are identified (Table 11-3) and are the source of information dealing with normal ranges of temperature, and humidity, and sky conditions.

As shown in Table 11-3, normal temperature information is presented under "temperatures (°F)" in the second, third, and fourth columns. For the month indicated, the second column represents the normal daily maximum temperature, the third column the normal daily minimum temperature, and the fourth column the normal daily average temperature. Under "relative humidity (%)" are listed normal relative humidity values for midnight, 6 A.M., 12 noon, and 6 P.M. (listed in military time). Finally, in the first three columns, under "mean number of days" ("sunrise to sunset"), information on sky conditions is presented (the mean number of "clear," "partly cloudy," and "cloudy" days for the month indicated). The application of this information is discussed in the section "General Climatic Analysis."

Comparative Climatic Data for the United States

For our purposes, this publication provides essentially the same temperature and sky condition information as that in *Local Climatological Data,* with less humidity information. Information is listed for more than 250 locations around the country. A typical format is shown in Table 11-4.

Airport Climatological Summary—(CLIM90)

In addition to temperature, relative humidity, and sky condition information, Table 11.A in this publication provides information on wind direction and coincident wind speed for each month of the year as well as totals for the year. Table 11-5 indicates the format of the information. The percentage of observations occurring for a given direction and individual wind speed group is given. The total percentage of observations for a given wind direction is listed

Table 11–3. Normals, Means, and Extremes[a, b]

Month	Temperatures (°F) Normal Daily Maximum	Normal Daily Minimum	Normal Monthly	Extremes Record Highest	Year	Record Lowest	Year	Normal Degree Days Base 65 °F Heating	Cooling	Precipitation (in.) Water Equivalent Normal	Maximum Monthly	Year	Minimum Monthly	Year	Maximum in 24 hr	Year	Snow, Ice Pellets Maximum Monthly	Year	Maximum in 24 hr	Year
e				12		12					22		22		22		22		22	
J	55.7	33.9	44.8	88	1969	4	1964	626	0	1.80	3.60	1968	0.19	1971	2.39	1973	12.1	1964	12.1	1964
F	59.8	37.6	48.7	87	1969	12	1971	456	0	2.36	6.20	1965	0.15	1963	4.06	1965	3.7	1975	3.7	1975
M	66.6	43.3	55.0	96	1974	19	1965	335	25	2.54	6.39	1968	0.10	1972	2.99	1958	2.5	1962	2.5	1962
A	76.3	54.1	65.2	95	1972	30	1973	88	94	4.30	12.19	1957	0.92	1959	4.55	1957	0.0		0.0	
M	82.8	62.1	72.5	96	1967	42	1971	0	236	4.47	12.64	1957	1.06	1961	4.86	1965	0.0		0.0	
J	90.8	70.3	80.6	105	1972	51	1964	0	468	3.05	6.94	1962	0.40	1964	3.11	1966	0.0		0.0	
J	95.5	74.0	84.8	106	1974	59	1972	0	614	1.84	11.13	1973	0.09	1965	3.76	1975	0.0		0.0	
A	96.1	73.7	84.9	108	1964	56	1967	0	617	2.26	6.85	1970	0.01	1973	3.30	1970	0.0		0.0	
S	88.5	66.8	77.7	102	1963	46	1971	0	381	3.15	9.52	1964	0.23	1956	4.76	1965	0.0		0.0	
O	79.2	56.0	67.6	96	1963	37	1966	60	141	2.68	9.22	1959	g	1975	5.91	1959	0.0		0.0	
N	67.5	44.1	55.8	88	1965	22	1975	287	11	2.03	6.23	1964	0.20	1970	2.83	1964	g	1974	g	1974
D	58.7	37.0	47.9	84	1966	10	1963	530	0	1.82	6.99	1971	0.21	1955	3.10	1971	2.6	1963	2.5	1963
YR	76.5	54.4	65.5	108	Aug. 1964	4	Jan. 1964	2382	2587	32.30	12.64	May 1957	g	Oct. 1975	5.91	Oct. 1959	12.1	Jan. 1964	12.1	Jan. 1964

Month	Relative Humidity (%) Hour 00 (Local Time)	Hour 06	Hour 12	Hour 18	Wind[c] Mean Speed (mph)	Prevailing Direction	Fastest Mile Speed (mph)	Direction	Year	Pct of Possible Sunshine	Mean Sky Cover, Tenths, Sunrise to Sunset	Mean Number of Days Sunrise to Sunset Clear	Partly Cloudy	Cloudy	Precipitation .01 in. or More	Snow, Ice Pellets 1.0 in. or More	Thunderstorms	Heavy Fog, Visibility 1/4 Mile or Less	Temperatures (°F) Max. 90° and above[d]	Max. 32° and below	Min. 32° and below	Min. 0° and below	Average Station Pressure (Ml.) Elevation 525 ft m.s.l.
e	12	12	12	12	22	10	22	22			22	22	22	22	22	22	22	22	12	12	12	12	3
J	75	82	61	59	11.5	S	46	36	1957		6.1	10	6	15	7	1	1	3	0	1	14	0	998.4
F	71	80	57	52	12.3	S	51	36	1962		5.6	10	6	12	7	1	2	2	0	1	10	0	997.9
M	72	80	57	51	13.3	S	55	29	1954		5.8	10	7	14	7	1	4	1	1	0	4	0	992.2
A	75	85	59	54	13.1	S	55	32	1970		6.1	9	7	14	8	0	7	1	1	0	1	0	993.9
M	80	88	62	59	11.4	S	55	14	1955		5.8	8	11	12	8	0	7	f	3	0	0	0	991.5
J	74	86	57	52	11.0	S	52	32	1955		4.8	11	11	8	6	0	6	f	19	0	0	0	992.8
J	68	81	50	45	9.7	S	65	36	1961		4.3	15	9	7	5	0	5	0	27	0	0	0	994.6
A	70	83	53	48	9.3	S	73	36	1959		4.3	15	10	6	5	0	5	f	25	0	0	0	994.8
S	79	88	60	58	9.7	S	53	11	1961		4.8	13	8	9	7	0	4	f	11	0	0	0	995.3
O	76	85	56	57	9.9	S	44	27	1957		4.5	14	7	10	5	0	3	1	2	0	0	0	997.6
N	75	82	56	58	10.9	S	50	34	1957		5.0	13	5	12	6	0	2	1	0	0	3	0	997.4
D	76	81	60	60	11.2	S	53	32	1968		5.6	12	6	13	7	f	1	3	0	f	9	0	997.8
YR	74	83	57	55	11.1	S	73	36	Aug. 1959		5.2	140	93	132	79	1	45	11	88	2	39	0	995.4

[a]*Means and extremes are from existing and comparable exposures. Annual extremes have been exceeded at other sites in the locality as follows: Highest temperature 112 in August 1936; lowest temperature –8 in February 1899; maximum monthly precipitation 17.64 in April 1922; minimum monthly precipitation 0.00 in November 1903; maximum precipitation in 24 hours 9.57 in September 1932.*

[b]*Normals, based on record for the 1941–1970 period; date of an extreme, the most recent in cases of multiple occurrence.*

[c]*Prevailing wind direction, record through 1963; wind direction, numerals indicate tens of degrees clockwise from true north, 00 indicates calm; fastest mile wind, speed is fastest observed 1-minute value when the direction is in tens of degrees.*

[d]*70°F and above at Alaskan stations.*

[e]*Length of record, years, through the current year unless otherwise noted, based on January data.*

[f]*Less than one-half.*

[g]*Trace.*

Source: National Climatic Data Center, *Local Climatological Data—Annual Summary with Comparative Data* (Asheville, N.C.: National Oceanic and Atmospheric Administration).

Table 11–4. Normal Daily Maximum Temperature (°F)

Normals 1951–1980	Years	Jan.	Feb.	Mar.	Apr.	May	June	July	Aug.	Sept.	Oct.	Nov.	Dec.	Annual
Birmingham C.O., Ala.	30	51.7	56.6	64.8	75.0	81.7	88.2	90.6	89.8	84.1	73.5	62.3	54.5	72.7
Birmingham AP, Ala.	30	52.7	57.3	65.2	75.2	81.6	87.9	90.3	89.7	84.6	74.8	63.7	55.9	73.2
Huntsville, Ala.	30	49.4	53.9	61.9	73.0	79.9	86.8	89.4	89.2	83.5	73.4	61.6	53.0	71.2
Mobile, Ala.	30	60.6	63.9	70.3	78.3	84.9	90.2	91.2	90.7	87.0	79.4	69.3	63.1	77.4
Montgomery, Ala.	30	57.0	60.9	68.1	77.0	83.6	89.8	91.5	91.2	86.9	77.5	67.0	59.8	75.9
Anchorage, Alaska	30	20.0	25.5	31.7	42.6	54.2	61.8	65.1	63.2	55.2	40.8	27.9	20.4	42.4
Annette, Alaska	30	37.3	41.3	43.3	49.1	55.7	60.5	64.1	64.6	60.0	51.8	44.2	39.9	51.0
Barrow, Alaska	30	−8.0	−13.8	−9.7	5.4	23.6	37.4	44.6	42.4	33.8	18.9	4.6	−7.0	14.4
Barter Is., Alaska	30	−8.5	−14.3	−9.5	7.0	26.1	38.1	45.3	43.7	35.0	20.4	5.7	−6.7	15.2
Bethel, Alaska	30	11.8	13.1	19.3	31.7	48.7	58.9	62.1	59.4	52.0	35.6	23.7	11.3	35.6
Bettles, Alaska	30	−6.5	−0.6	12.1	31.6	52.5	67.2	69.0	62.7	48.8	25.5	6.2	−4.7	30.3
Big Delta, Alaska	30	0.8	11.0	23.0	40.5	56.5	66.7	69.4	65.0	52.7	31.7	14.6	2.3	36.2
Cold Bay, Alaska	30	32.8	32.2	33.6	37.7	44.3	50.0	54.9	55.4	52.0	44.2	38.8	33.9	42.5
Fairbanks, Alaska	30	−3.9	7.3	21.7	40.8	59.2	70.1	71.8	66.5	54.4	32.6	12.4	−1.7	35.9
Gulkana, Alaska	30	0.2	13.8	27.2	41.8	54.6	64.9	68.4	64.8	54.0	35.7	14.7	1.8	36.8
Homer, Alaska	30	27.0	31.2	34.4	42.1	49.8	56.3	60.5	60.3	54.8	44.0	34.9	27.7	43.6
Juneau, Alaska	30	27.4	33.7	37.4	46.8	54.7	61.1	64.0	62.6	55.9	47.0	37.5	31.5	46.6
King Salmon, Alaska	30	20.0	22.9	27.6	38.7	50.8	58.8	62.9	61.3	54.7	40.5	30.0	19.7	40.7
Kodiak, Alaska	30	36.6	34.8	38.3	43.3	48.4	56.1	59.3	61.0	55.9	47.1	39.9	35.2	46.3
Kotzebue, Alaska	30	3.7	1.3	8.0	21.5	38.6	49.8	58.7	56.9	46.9	27.8	13.6	2.2	27.4
McGrath, Alaska	30	−1.1	9.1	21.5	37.5	54.9	65.5	68.0	63.5	52.7	31.8	13.4	−0.8	34.7
Nome, Alaska	30	13.4	11.8	15.7	25.8	42.1	52.0	56.6	55.7	48.5	33.8	22.7	12.0	32.5
St. Paul Island, Alaska	30	30.4	26.4	28.2	32.0	38.6	45.1	49.4	50.8	48.4	41.7	37.2	32.2	38.4
Talkeetna, Alaska	30	17.9	25.3	32.0	43.7	55.9	64.8	67.8	64.5	55.6	39.8	26.3	17.7	42.6
Unalakleet, Alaska	30	10.3	12.0	18.0	30.1	45.5	55.3	60.5	58.3	50.5	33.3	19.3	8.2	33.4
Valdez, Alaska	30	29.3	30.0	36.2	43.9	51.1	57.5	62.1	61.5	54.1	43.2	33.8	25.1	44.0
Yakutat, Alaska	30	30.1	34.8	37.2	43.2	49.7	55.6	59.5	59.7	55.3	47.2	38.4	32.7	45.3
Flagstaff, Ariz.	30	41.7	44.5	48.6	57.1	66.7	77.6	81.9	78.9	74.1	63.7	51.0	43.6	60.8
Phoenix, Ariz.	30	65.2	69.7	74.5	83.1	92.4	102.3	105.0	102.3	98.2	87.7	74.3	66.4	85.1
Tucson, Ariz.	30	64.1	67.4	71.8	80.1	88.8	98.5	98.5	95.9	93.5	84 1	72.2	65.0	81.7
Winslow, Ariz.	30	45.0	53.2	60.7	70.0	79.9	91.0	94.5	91.1	85.2	73.1	57.9	46.0	70.6
Yuma, Ariz.	30	68.6	73.9	78.5	85.7	93.6	102.9	106.8	105.3	101.4	90.9	77.4	69.1	87.8
Fort Smith, Ark.	30	48.4	53.8	62.5	73.7	81.0	88.5	93.6	92.9	85.7	75.9	61.9	52.1	72.5
Little Rock, Ark.	30	49.8	54.5	63.2	73.8	81.7	89.5	92.7	92.3	85.6	75.8	62.4	53.2	72.9
North Little Rock, Ark.	30	48.6	53.6	62.0	73.1	80.3	87.5	91.6	90.7	84.0	74.5	61.2	51.9	71.6
Bakersfield, Calif.	30	57.4	63.7	68.6	75.1	83.9	92.2	98.8	96.4	90.8	81.0	67.4	57.6	77.7
Bishop, Calif.	30	52.9	58.2	63.4	70.9	80.2	90.4	97.5	95.2	88.2	77.1	63.5	55.1	74.4
Blue Canyon, Calif.	30	43.5	44.9	45.3	51.3	60.3	69.2	77.7	76.5	72.4	62.8	51.4	46.3	58.5
Eureka, Calif.	30	53.4	54.6	54.0	54.7	57.0	59.1	60.3	61.3	62.2	60.3	57.5	54.5	57.4
Fresno, Calif.	30	54.2	61.2	66.5	73.7	82.7	91.1	97.9	95.5	90.3	79.9	65.2	54.4	76.1
Long Beach, Calif.	30	66.0	67.3	68.0	70.9	73.4	77.4	83.0	83.8	82.5	78.4	72.7	67.4	74.2
Los Angeles AP, Calif.	30	64.6	65.5	65.1	66.7	69.1	72.0	75.3	76.5	76.4	74.0	70.3	66.1	70.1
Los Angeles C.O., Calif.	30	66.6	68.5	68.7	70.9	73.2	77.9	83.8	84.1	83.0	78.5	72.7	68.1	74.7
Mount Shasta, Calif.	30	42.1	47.3	50.9	57.9	67.0	75.4	85.1	83.3	77.5	65.4	50.9	43.9	62.2
Red Bluff, Calif.	30	53.9	60.0	64.0	71.2	81.2	90.3	98.2	95.7	90.4	78.8	63.6	55.2	75.2
Sacramento, Calif.	30	52.6	59.4	64.1	71.0	79.7	87.4	93.3	91.7	87.6	77.7	63.2	53.2	73.4
San Diego, Calif.	30	65.2	66.4	65.9	67.8	68.6	71.3	75.6	77.6	76.8	74.6	69.9	66.1	70.5
San Francisco AP, Calif.	30	55.5	59.0	60.6	63.0	66.3	69.6	71.0	71.8	73.4	70.0	62.7	56.3	64.9
San Francisco C.O., Calif.	30	56.1	59.4	60.0	61.1	62.5	64.3	64.0	65.0	68.9	68.3	62.9	56.9	62.5
Santa Barbara, Calif.	30	63.3	64.4	65.0	66.6	68.4	71.1	73.6	74.8	75.0	72.5	69.0	64.8	69.0

Source: National Climatic Data Center, *Comparative Climatic Data for the United States* (Asheville, N.C.: National Oceanic and Atmospheric Administration).

Table 11–5. Wind Direction Versus Wind Speed (Percent Frequency of Observations) All Weather[a]

Wind Direction	Wind Speed (knots)										
	0–3	4–6	7–10	11–16	17–21	22–27	28–33	34–40	Over 40	Total	Average Speed
N	0.4	2.7	2.5	1.9	0.2					7.6	8.3
NNE	0.3	1.4	1.7	1.1	0.2					4.7	8.6
NE	0.4	1.1	1.6	0.7						3.8	7.7
ENE	0.5	0.8	0.9	0.1						2.3	6.2
E	0.5	2.7	1.8	0.5						5.5	6.7
ESE	0.3	3.2	1.6	0.6						5.6	6.6
SE	0.4	2.9	3.8	1.5	0.0					8.7	7.7
SSE	0.3	2.7	5.7	4.2	0.7	0.0				13.7	9.6
S	0.6	3.3	7.0	9.2	2.1	0.3				22.4	11.0
SSW	0.1	1.8	2.9	3.1	0.6	0.0				8.6	10.2
SW	0.4	1.0	0.8	0.5	0.2					2.9	7.6
WSW	0.2	0.4	0.5	0.2						1.2	6.9
W	0.2	0.8	0.4	0.1	0.0					1.5	6.3
WNW	0.1	0.4	0.3	0.3	0.1					1.1	8.9
NW	0.1	0.8	0.6	0.5						2.0	8.0
NNW	0.2	1.7	1.4	1.0						4.4	7.9
CALM	4.2									4.2	
Total	9.0	27.6	33.4	25.5	4.1	0.4				100.0	8.5

[a]*All weather: all wind observations.*

Source: National Climatic Data Center, *Airport Climatological Summary—(CLIM90)* (Asheville, N.C.: National Oceanic and Atmospheric Administration).

under "total." This information will be used to construct wind roses for climatic analysis.

SMOS—Summary of Meteorological Observations, Surface: Part E—Psychrometric Summaries
This publication provides data on wind direction and associated air temperature, as illustrated in Table 11-6, in addition to a great deal of other psychrometric information. Data is recorded at present and former military installations.

The first three publications can be considered interchangeable, and provide all the temperature, humidity, and sky condition information required for general climatic analysis. One of these inexpensive publications should be ordered from NCDC. The third publication provides necessary wind information and should be ordered for the weather station nearest your site. The last publication listed is used in this book to clarify certain issues regarding the bivariate frequency analysis of wind direction and temperature and is not essential for climatic analysis as described herein.

The preceding discussion has focused on a range of climatic information sources. During analysis, it is important to remain aware of the differences between the building site and the weather recording station. Following is a discussion of climatic data analysis as it relates to the site-planning process.

GENERAL CLIMATIC ANALYSIS

In this section we focus on planning for thermal comfort in exterior spaces. In this regard, the purpose of climatic analysis is to identify those areas on the site which are most appropriate for the location of exterior spaces and buildings. During the initial site microclimate analysis, it is important to remember that at least six new submicroclimates will be created in, on, and around a building when it is placed on the site. Therefore, the initial analysis should always be reevaluated when locating buildings on the site.

The initial site microclimate inventory should establish the general climatic conditions of the building site. The first step is to identify normal temperature and humidity conditions and plot them on the bioclimatic chart (Fig. 11-1). This provides the site planner with an "image" of local temperature and humidity ranges and roughly identifies those times of the year when conditions of thermal comfort are likely to exist. In addition, appropriate remedial measures are suggested by this analysis for those conditions occurring outside the comfort zone.

As discussed previously in the section "Humidity," the lowest daily temperature and highest daily humidity can be considered to occur at the same time of day. Similarly, it can

Table 11–6. Percentage Frequency of Air Temperature Versus Wind Direction

93901	Dallas, Texas	Jan. 1973–Dec. 1977	All	All
Station	Station Name	Years	Month	Hours

	Wind Direction										
Temp.	NNW & N	NNE & NE	ENE & E	ESE & SE	SSE & S	SSW & SW	WSW & W	WNW & NW	Calm	Total Freq.	Percent of Total
122 +											
117–121											
112–116											
107–111			50.0						50.0	2	.0
102–106			23.5	29.4	17.6	5.9	5.9		17.6	17	.1
97–101	4.8	1.6	11.1	19.0	46.0	7.1	4.0	.8	5.6	126	.9
92–96	1.3	3.9	5.8	16.4	57.0	8.4	2.0	1.9	3.4	537	3.7
87–91	3.7	4.7	9.5	19.4	48.0	6.4	2.8	1.7	3.8	811	5.6
82–86	3.7	4.6	8.9	17.2	49.6	8.6	2.3	1.4	3.7	1247	8.5
77–81	4.8	4.3	7.6	17.3	45.5	9.5	2.4	2.0	6.7	1739	11.9
72–76	7.9	4.6	7.5	15.9	41.8	10.3	2.0	2.2	7.9	1833	12.5
67–71	9.4	7.3	7.3	12.8	38.0	7.9	3.8	5.3	8.2	1426	9.8
62–66	12.3	4.9	7.6	12.8	33.3	8.1	4.2	7.0	9.7	1338	9.2
57–61	17.2	6.5	5.9	9.8	27.4	8.9	6.5	7.9	9.7	1133	7.8
52–56	15.9	6.3	6.8	9.5	24.7	10.1	5.1	11.5	10.1	1087	7.4
47–51	18.9	6.4	7.3	7.3	17.2	9.5	7.2	14.4	11.8	977	6.7
42–46	23.6	6.6	5.8	9.2	15.2	8.5	5.4	14.1	11.6	850	5.8
37–41	22.7	6.7	6.1	4.9	12.3	10.4	6.1	16.0	14.8	656	4.5
32–36	32.3	8.1	1.5	2.3	6.4	13.5	3.3	15.8	16.8	393	2.7
27–31	33.9	7.0	3.5	1.6	7.8	12.8	3.9	17.1	12.5	257	1.8
22–26	47.7	5.5	.9		6.4	10.1	.9	22.9	5.5	109	.7
17–21	56.3	6.3	2.1		2.1	10.4		8.3	14.6	48	.3
12–16	26.3					26.3		47.4		19	.1
7–11									100.0	1	.0
2–6					100.0					1	.0
−3–1											
−8–4											
−13–9											
−18–14											
−23–19											
−28–24											
−33–29											
−38–34											
−43–39											
−48–44											
−53–49											
−58–54											
−59 and lower											
Totals	12.4	5.6	7.1	12.6	33.8	9.2	3.9	6.9	8.6	14607	100.0

Source: National Climatic Data Center, *SMOS—Summary of Meteorological Observations, Surface: Part E—Psychrometric Summaries* (Asheville, N.C.: National Oceanic and Atmospheric Administration).

be assumed that the maximum daily temperature coincides with the minimum relative humidity. (These assumptions apply to a day without precipitation.) Using the temperature and relative humidity data obtained from either *Local Climatological Data* or *Comparative Climatic Data* (see "Information Sources and Interpretation"), plot the point representing normal daily maximum temperature and the lowest relative humidity for January on the bioclimatic chart. Then plot the normal daily minimum temperature and the highest relative humidity value. A "line" representing the temperature–humidity conditions for a normal day of the month in question is produced by connecting these two points. Repeat this process for each month, or every other month, to generate a picture of the normal temperature–humidity conditions over the course of a year. Fig. 11-37 represents the normal conditions at Dallas–Fort Worth International Airport (DFW). The temperature–humidity data used to plot the month of January were obtained from *Local Climatological Data*—"Normals, Means, and Extremes" (Table 11-3). For example, the points representing January are:

Point 1: 55.7°F 59% relative humidity

Point 2: 33.9°F 82% relative humidity

Whichever data source is being used, associate the lowest relative humidity value listed with the daily maximum temperature, and the highest humidity with the lowest temperature. (Four daily humidity values are listed in *Local Climatological Data*; only two are listed in *Comparative Climatic Data*.)

The DFW example (Fig. 11-37) suggests a temperate climate with periods of overheating on summer and fall days, where wind and shade are necessary to improve thermal comfort. The underheating that occurs on winter and spring days can be remedied with solar radiation. Since each normal day "line" represents 24 hours, the line can be subdivided to indicate time of day in order to understand when solar radiation will be available. It is assumed that the maximum temperature on the January day occurs in the afternoon at 4:00 P.M. and that the minimum temperature occurs 12 hours before, at 4:00 A.M. These two times can be assigned to opposite ends of the "line" as illustrated in Fig. 11-38. Here the line is represented as a double "bar" and daylight hours are differentiated from night by the cross-hatching. Sunrise and sunset times can be estimated from the sun angle calculator for any month. This permits the visualization of time along the normal day line.

The results of this analysis are interpreted in Table 11-7, which indicates the desirable conditions for exterior spaces in the Dallas–Fort Worth area under normal temperature and humidity conditions. By grouping "like" conditions, two primary categories emerge: times when sunny, calm conditions are desirable; and times when shade and wind are desirable. From a practical standpoint, it is important to remember that this analysis is based on "normal" conditions, and that daily conditions can vary considerably. (The record high temperature for January in the Dallas–Fort Worth area is 88°F.) Therefore, site-planning solutions should be able to accommodate varying microclimatic conditions, such as the provision for shady areas in January in Dallas–Fort Worth.

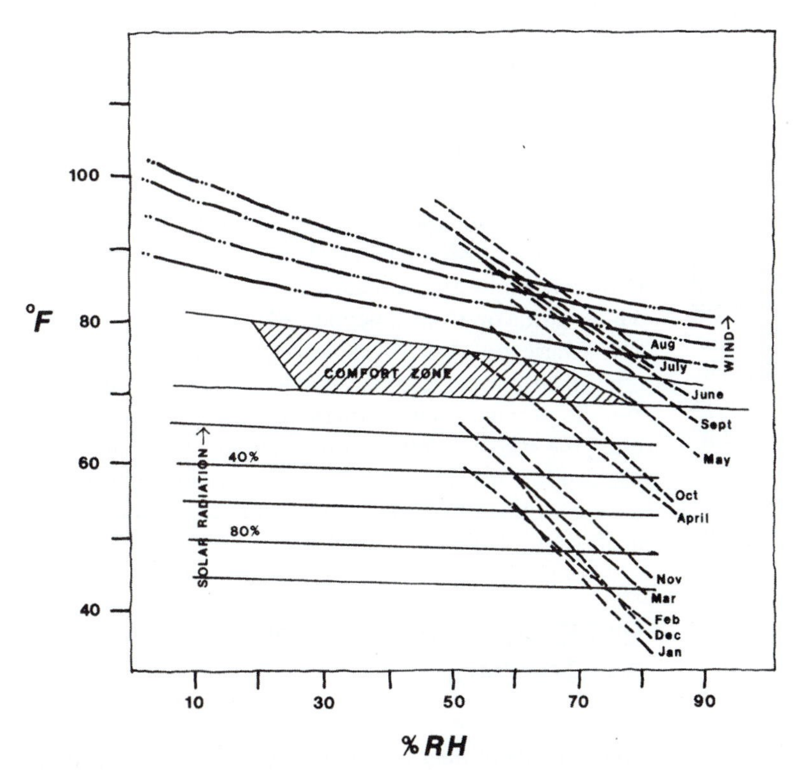

FIGURE 11-37. Normal conditions at Dallas–Fort Worth.

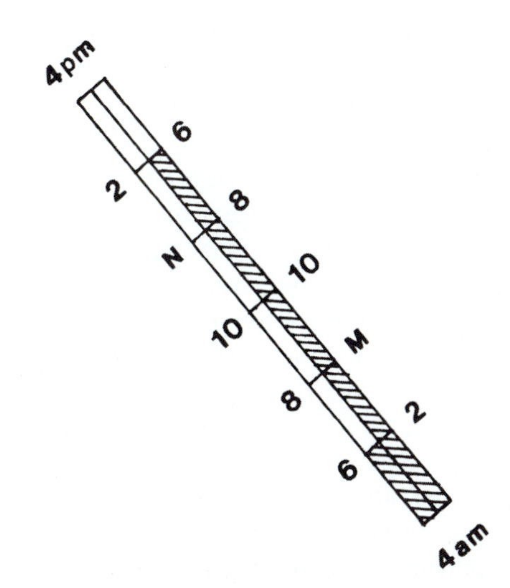

FIGURE 11-38. January temperature as "bar."

Table 11-7. Bioclimatic Analysis Results for Dallas–Fort Worth

Month	Sun	Shade	Wind	Calm
May		×	×	
June		×	×	
July		×	×	
August		×	×	
September		×	×	
October	×	×	×	×
November	×			×
December	×			×
January	×			×
February	×			×
March	×			×
April	×	×	×	×

A knowledge of solar geometry, a sun peg, and a site model allow precise determination of those areas of the site which will have access to the sun when it is desirable. Areas with desirable shading characteristics can also be identified. Knowledge of local plant materials makes it possible to select deciduous trees with foliation periods that correspond roughly to the microclimatic needs which were identified in the analysis (foliation in April, defoliation in October would be ideal in the Dallas–Fort Worth area).

The bioclimatic analysis indicated that comforting winds should be admitted during periods of overheating, and that winds should be blocked during underheated periods. The primary source directions of both comforting and discomforting winds should be determined in order to develop responsive design solutions for exterior spaces.

The information found in Table 11.A of the *Airport Climatological Summary* (see Table 11-5) can be used to construct what is commonly referred to as "wind rose" diagrams. These diagrams graphically depict the directional frequency of winds for each month of the year. The percent frequency of observations listed in the "total" column (Table 11-5) is graphed for the appropriate direction. The monthly wind roses for Dallas–Fort Worth are shown in Fig. 11-39. (The percentage of observations that were "calm" is shown in the center of the wind rose.)

The January wind rose indicates that the majority of winds arrive from the south and north. Moving toward the summer months, the northerly components steadily decrease. The southerly components increase to a peak in July. In August, the hottest Dallas–Fort Worth month, the winds undergo a transition toward the east. Beginning in September, the northerly components begin to resume some relative prominence, until the northerly and southerly components again approach a "balance" in December and January.

As seasonal temperatures increase (from winter to spring and summer), the percent frequency of winds from southerly directions increases. It can be inferred, with some certainty, that the source of warm, temperate winds is from these southerly directions. As seasonal temperatures decrease (from summer to fall and winter), the percent frequency of winds from northerly directions increases, even though the prevailing wind direction is still from the south. It can be surmised, perhaps with less certainty, that the northerly directions may be the source of cold, discomforting winds. Information from the "Narrative Climatological Summary" contained in *Local Climatological Data,* personal knowledge of area climatic patterns, or information provided by local experts or residents can usually verify the primary source directions of cold and temperate winds.

A more precise determination is possible using *Summary of Meteorological Observations: Part E—Psychrometric Summaries,* which contains percent frequency observations relating wind direction to air temperature (see Table 11-6). Wind roses for discrete air temperature "ramps" or "bins" (Fig. 11-40) clearly indicate that the coldest winds (2 to 31°F) arrive primarily from northerly directions. (Note that only 3% of all observations occur in the 2 to 31° temperature bin.) Comfort and ventilating winds (62 to 86°F bin, 51.9% of observations) arrive primarily from southerly directions. The trend from warm-south to cold-north is clearly identifiable in this sequence of diagrams.

The bioclimatic analysis (Fig. 11-37 and Table 11-7) indicated that wind is desirable from May through September. The prevailing direction for winds during these months is primarily out of the south, south-southwest, and south-southeast. These winds can be useful in improving thermal comfort and should be admitted to exterior spaces. The cold, northerly winds that occur during winter months should be blocked. It should be recognized that most but not all discomforting winds (2 to 31°F bin) come from the northerly directions, and that most but not all temperate winds arrive from the southerly directions.

The site planner's focus in the Dallas–Fort Worth area should be the provision of sunny exterior spaces protected against cold, northerly winds during underheated periods (November into April), shady exterior spaces which admit winds during overheated periods (May through September), and a variety of in-between spaces that accommodate the other-than-normal climatic conditions which will occur and the varying comfort criteria of the people who will be using the spaces. In other words, exterior spaces should be planned that will accommodate a variety of climatic conditions and provide a range of opportunities to locate thermally comfortable places to inhabit.

SITE-MICROCLIMATE MAPPING

Having completed an analysis of normal climatic conditions, the site planner is equipped with the necessary information to analyze the microclimates of a specific site. This

FIGURE 11-39. Wind roses.

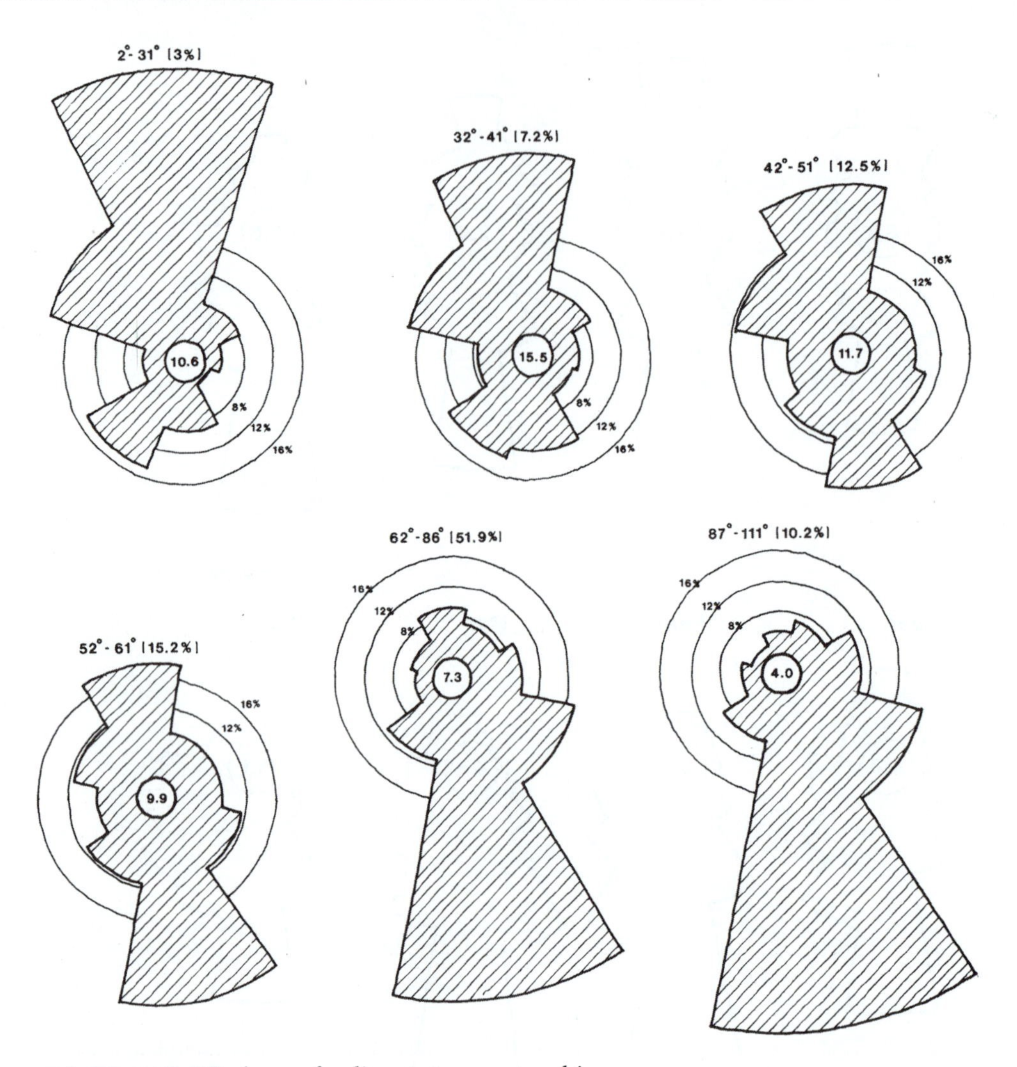

FIGURE 11-40. Wind roses for discrete temperature bins.

process, which will be referred to as site-microclimate mapping, was demonstrated in *InsideOut* by G. Z. Brown, John Reynolds, and Susan Ubbelohde and again by G. Z. Brown in *Sun, Wind and Light.* It will indicate which areas of the site are most suitable for the location of exterior spaces.

The bioclimatic analysis indicates seasonal requirements for sun and wind. As defined by *InsideOut* and illustrated in Fig. 11-41, four general conditions can exist on a site at any given time:

A: areas with sun and wind

B: areas with sun protected from the wind

C: areas with shade and wind

D: areas with shade protected from the wind

Category B was identified as the desirable combination during underheated periods, while category C was desirable during overheated periods. Using a site plan gridded to some convenient scale (15- to 30-ft cells or greater, depending on drawing scale or the site), this identification system can be used to "map" site microclimates. The initial mapping should identify the predominant conditions existing in each cell for winter and summer extremes (January and August for Dallas–Fort Worth conditions). A B condition (sun with wind protection) is desirable for exterior spaces in the winter, while a C condition (shade and wind) is desirable during

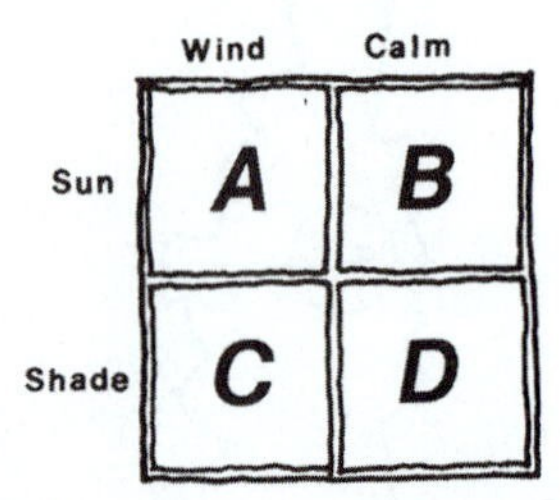

FIGURE 11-41. Bioclimatic analysis matrix. [From G.Z. Brown, J.S. Reynolds, and M.S. Ubbelohde, *InsideOut: Design Procedures for Passive Environmental Technologies* (New York: John Wiley & Sons, Inc., 1982) pp. 3–12.]

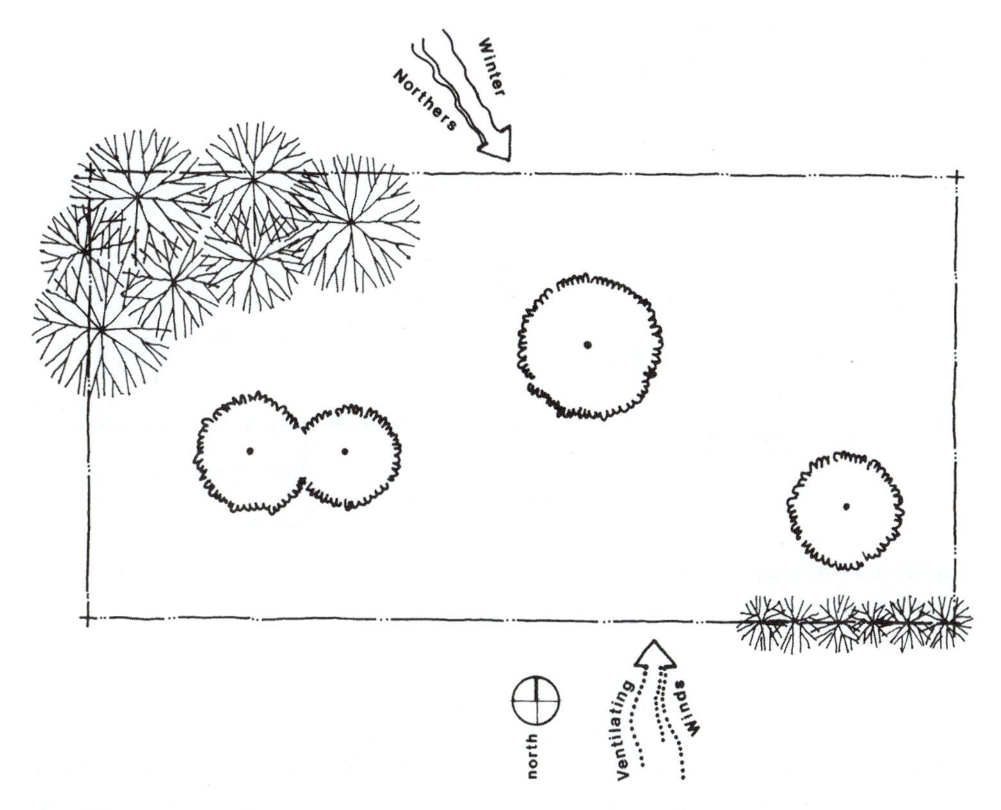

FIGURE 11-42. Hypothetical site.

summer. A cell on the gridded site plan that is mapped as B/C (winter/summer) would most likely have characteristics promoting comfort year-round. A C/B cell would be least suitable for exterior spaces. Keep in mind that the grid lines do not represent absolute cell boundaries. The grid is simply a device used to identify general areas on the site.

Fig. 11-42 illustrates a small, hypothetical "flat" site in the Dallas–Fort Worth area. There are dense screens of evergreen vegetation at the northwest and southeast corners of the site which will be effective windbreaks. There are four mature deciduous trees (branching begins 8 ft above the ground) providing shade in summer and admitting sunshine in winter. All trees are 30 ft tall. Cold winds come from the north and northwest. Ventilating winds come from the south.

Fig. 11-43 illustrates the microclimate mapping of the site based on the climatic analysis discussed previously.

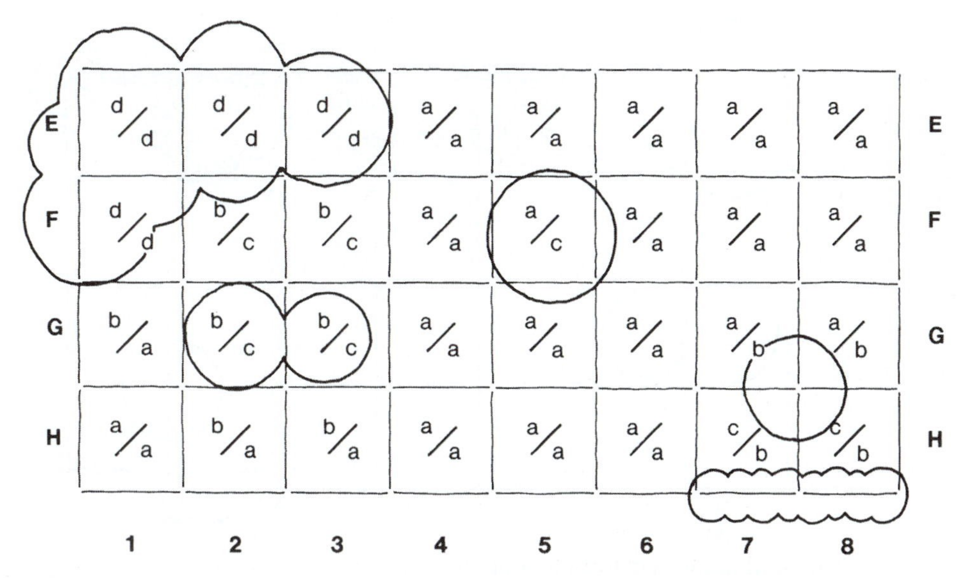

FIGURE 11-43. Microclimatic mapping of Fig. 11-42.

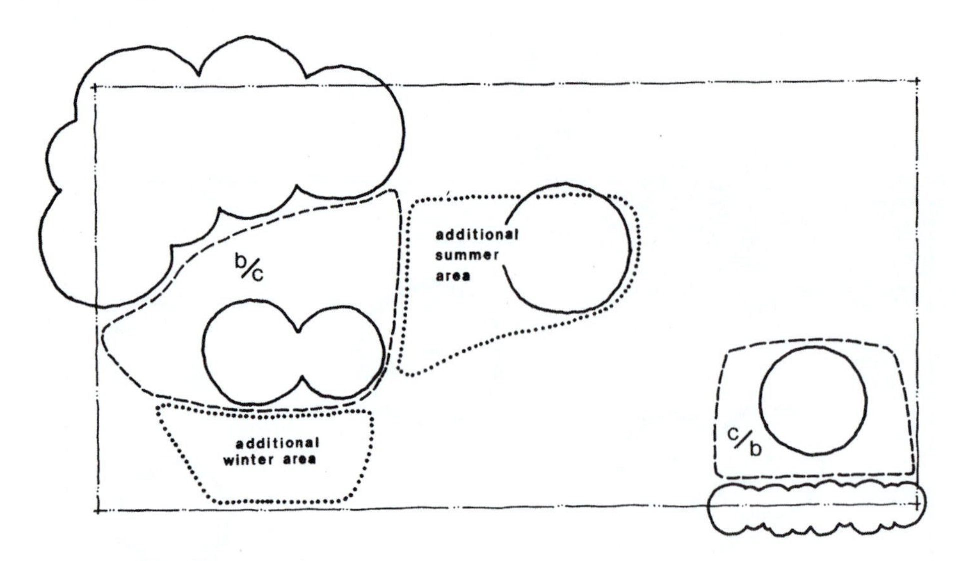

FIGURE 11-44. Area suitability of hypothetical site.

Some of the cell conditions may be uncertain or imprecise, and qualified judgments are often necessary. Cells F-2, F-3, G-2, and G-3 (B/C cells) define the area most suitable for developing exterior space, while cells H-7, and H-8 (C/B cells) are least suitable. Fig. 11-44 shows these areas as well as other areas which are seasonably suitable. A site model with a sun peg can clarify which areas are predominantly sunny or shady.

With intervention, any or all of these conditions of course can change. A conditions can be changed to desirable B or C conditions using architectural elements and/or vegetation. For example, the placement of a wall or fence along the north property line in concert with deciduous trees can convert cells E-4 through E-8 from an A/A to a B/C designation (Fig. 11-45). This wall and the vegetation can be designed to provide not just a continuous B/C set of conditions, but a range of microclimatic conditions, thereby increasing the opportunities for comfort with "other-than-normal" climatic conditions. Fig. 11-46 illustrates an "environmental option wall" that is designed to produce a range of microclimatic conditions.

The site-microclimate mapping process identifies general areas that may be appropriate for the location of exterior spaces, and indicates what must be done to alter the microclimate of other areas on the site to increase comfort opportunities. Finally, when buildings are being located on the site, the microclimates should again be mapped to reflect changes created by the buildings. The process of determining the appropriate location for a building on the site is similar, but in addition to comfort the focus must

FIGURE 11-45. Modifications to hypothetical site.

FIGURE 11-46. Environmental options wall.

include building energy conservation. A building responds not only to the climatic elements (external thermal loads) but also to heat generated inside the building (internal thermal loads). Methods for anticipating the thermal loading problems of buildings are discussed in Chapter 17. These methods provide the site planner with information that can be used for preliminary siting, massing, and orientation decisions.

PART **FOUR**

THE PLAN FOR CHANGE

- **CHAPTER 12:** Earthworks: Shaping the Site
- **CHAPTER 13:** Grading
- **CHAPTER 14:** Retaining Walls
- **CHAPTER 15:** Topography and Surveys
- **CHAPTER 16:** Storm Water Management and Erosion Control

CHAPTER TWELVE

EARTHWORKS: SHAPING THE SITE

Adapting a site's form to accommodate structures is a principal objective of most site plans. Where the quality of the work will vary, the primary intent of the plan is to direct the changes in the site's character to integrate human-made uses. Although this integration is accomplished primarily through grading, the English term "earthworks" appears to be more appropriate, as it includes drainage and erosion control as well. Grading, erosion control, and drainage are all part of the same process to find that compromise which accepts the change with a minimal amount of environmental damage. Although grading is often considered as a perfunctory exercise of site planning, an ill-conceived grading plan often reflects basic flaws in reconciling the scope of issues associated with earthworks.

The purposes of grading fall into four basic categories. First, it involves the effort to model the topography to accept a structure and its immediate land area. Buildings naturally require that some areas must be made level, but only those areas to receive buildings should be level, as other parts of the site without slope would not drain properly.

Second, grading for drainage involves the manipulation of the topography into a coordinated system that directs water away from those areas that need to be kept dry. Erosion control is a concomitant element of this task.

The third most common problem in earthworks is the grading associated with the access and location of those facilities to accommodate the automobile. This would include roads as well as parking, since location and configuration of these facilities are some of the most problematical for site planners. Grading associated with pedestrian movement, recreational activities, or special design issues make up the fourth category of earthworks problems. This would include, but not be limited to, sidewalks, ramps, stairs, retaining walls, berms, parking, drives, and field games.

The notion of architecture without architects generates an image and quality of buildings and communities worth a second look in terms of site planning and earthworks. There are numerous examples of vernacular architecture where the building and site appear as an integrated, unpretentious achievement. These are often the result of cultural precepts associated with the site, or simply the fact that the technology to change the site radically was unavailable. Unlike today, where Gulliveresque equipment is everywhere, the ancients were not only apprehensive of radical change to the earth, but the methods or techniques by which the land could be reshaped were limited. Indeed, more careful and sensitive site planning would probably occur tomorrow if every cubic yard of soil were to be moved by human hands. This is not to say that technological advancement has been counterproductive to the development of a humane environment; but didn't we recently rediscover the value of the "low-tech" operable window and the ceiling fan?

CONCEPTS AND VOCABULARY

The Contour Line

To understand grading, one begins with an understanding of contours and the contour line. Although the principles associated with the contour line are simple, their manipulation in the grading process is one that continues

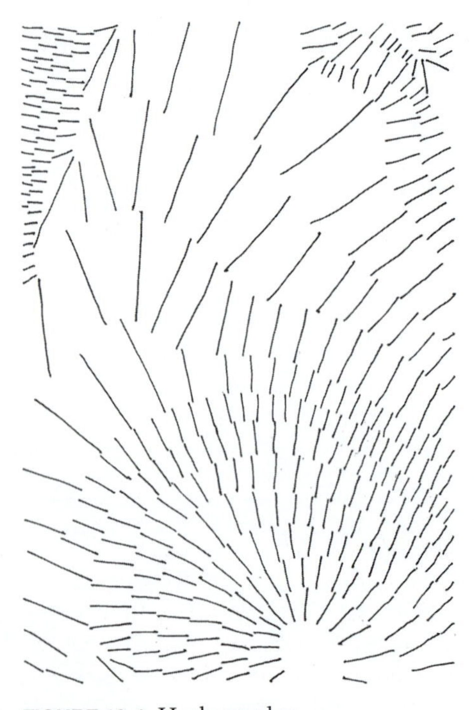

FIGURE 12-1. Hachure plan.

FIGURE 12-2. Water's edge is visible contour line.

to be a source of consternation to the novice. This is due, in part, to the nature of contour lines, as they are a two-dimensional device used to represent three dimensions. Nothing in architectural drawing is similar to contour lines, and although the contour line as a concept may be easy to grasp, it seems that the ability to visualize the vertical distance (contour interval) on a map lies at the base of the enigma.

Although the technique was abandoned in favor of contour lines, hachures still provide the site planner with a valuable graphic technique. Hachures are drawn perpendicular to a contour line, and the closer that two contour lines are to each other, the more frequent the number of hachure lines (Fig. 12-1). Hachures are a valuable tool, as they provide the planner with an image of the site as well as an understanding of the drainage pattern. Since the hachure line is the shortest distance between two contour lines, it also represents the flow of water across the site. The darker area depicts the steeper slopes and rapid flow of water, the lighter areas, the shallow slopes and gradual runoff.

The contour line was apparently first used as a method to record the edge of a river. The various levels of water in the river were probably obvious to the observer and hence provided the basis for the notion that a line at the water's edge at a given elevation also existed in dryland forms (Fig. 12-2). Compared to the use of hachures, the contour line was much less time consuming to delineate. Furthermore, the necessity to locate buildings, roads, orchards, and vegetation on a map with hachures was not always the clearest of drawings (Fig. 12-3). The role and function of the contour line are such that they must be understood completely to comprehend site planning and the grading process properly.

Since, by definition, the contour line is a line on a plane that connects all points of the *same* elevation, the first rule is

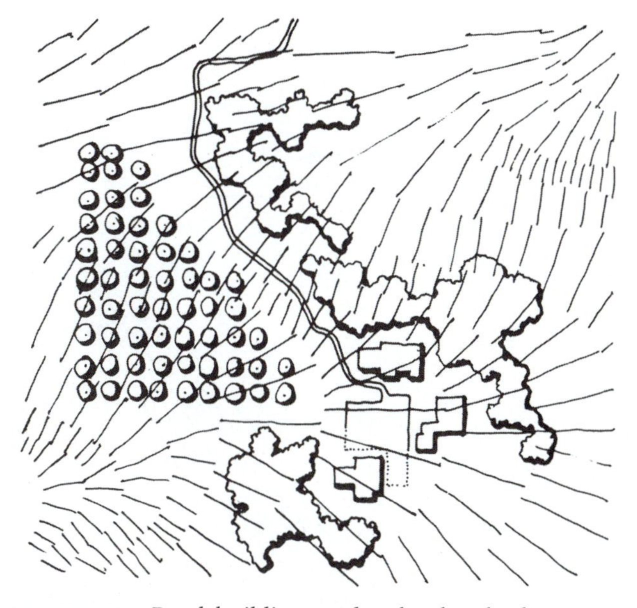

FIGURE 12-3. Road, buildings, and orchard on hachure map.

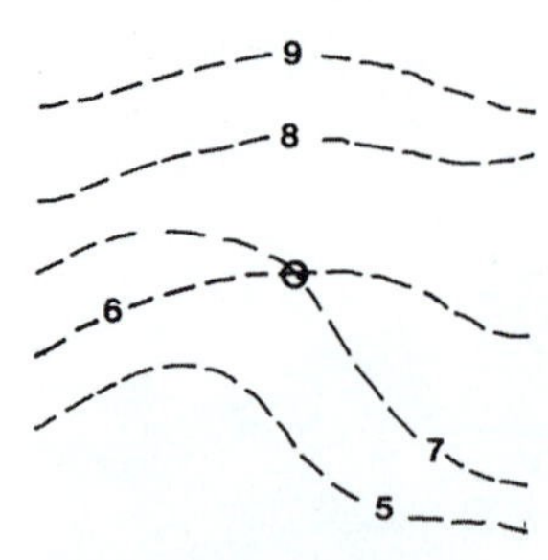

FIGURE 12-4. Contour lines crossing.

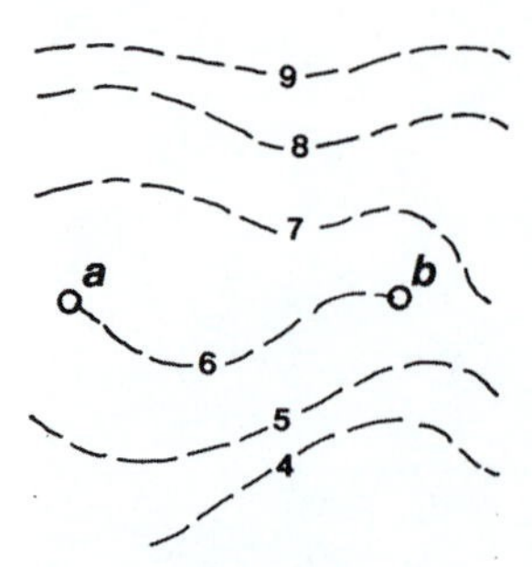

FIGURE 12-5. Contour lines beginning and ending *a* to *b*.

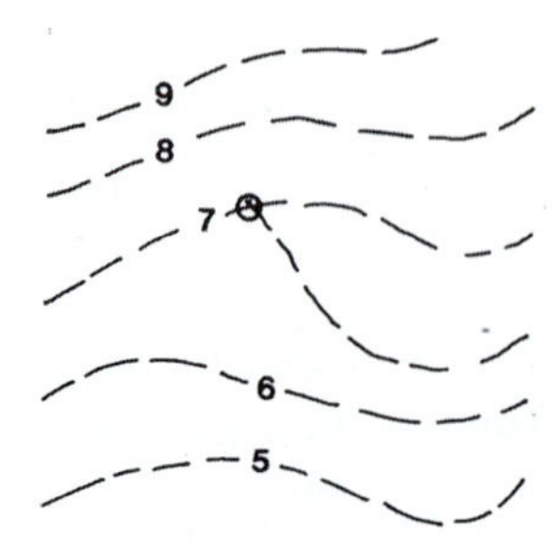

FIGURE 12-6. Contour lines forking.

that contour lines never cross (Fig. 12-4). The only exception to that rule is a natural topographical overhang or natural bridge. The second rule is that a contour line will either close on the site plan or run to the edge of the site. They never exist as a line that begins at a point *a* and ends at a point *b* on the plan (Fig. 12-5). The third rule is that they never fork or merge, as in Fig. 12-6. There are, however, common occasions in which contour lines of the same elevation may be represented side by side (Fig. 12-7). The fourth holds that the interval between contour lines is constant. This is to say that a topographical map of a specific scale will represent the contour lines at a constant interval, for example: 1 ft, 2 ft, 5 ft, and so on. This constant vertical interval is necessary to avoid circumstances in which varying intervals would develop a distorted picture and an inaccurate interpretation of a site's physical form. Only in unusual circumstances in which contour lines are omitted because they are too close to be drawn and visualized is this rule compromised (Fig. 12-7).

An additional note is necessary here concerning very flat sites. On those occasions in which the topographical change is very modest, consider interpolating between the 1-ft interval to locate an intermediate 6-in. contour. This additional interval provides a much sharper picture of the site whose physiography is subtle and visual topographical relief is absent. During the regrading process, this additional interval will become invaluable as change to the site and the design of a positive drainage pattern becomes a challenge.

Topography

On the topographical map, all existing contour lines are dashed and every fifth contour is drawn darker. The darker lines provide an additional interval level and assist in the visual interpretation of the site. During the site-planning

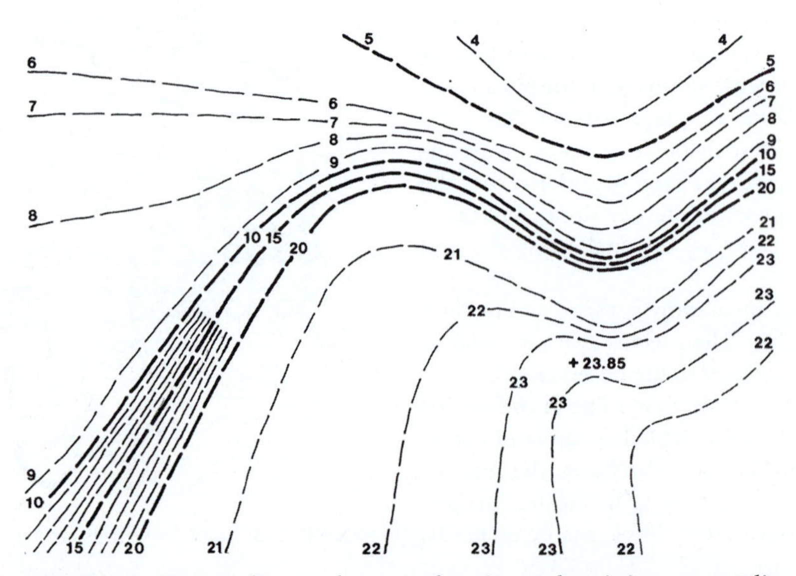

FIGURE 12-7. Contour lines at the same elevation and omitting contour lines.

process the contour lines that need to be changed (proposed contour lines) are drawn solid. This change between existing and proposed contours becomes easy to read and identifies areas of topographical "cut" or "fill."

Contour lines provide a picture of the site's topographical relief. Contour lines delineate areas that appear similar, such as:

1. Ridge (Fig. 12-8) or valley (Fig. 12-9)
2. Depression or hill (Fig. 12-10)
3. Convex or concave slopes (Figs. 12-11 and 12-12)

Sections through these landforms reveal not only the dramatic differences but also the essential importance of the section. Just as no architect or engineer would attempt to design a structure with plans or elevations alone, the section is equally as important to the site planner. The section is of immense value in reconciling the perceived vertical distance illustrated on the plan to the real change that occurs on the site.

Although the vocabulary of landforms and topographical features varies, all are variations on basic forms, and the best way to understand their function is to draw them. When possible, build a model. That "a picture is worth a thousand words" and "a model is worth a thousand pictures" are never more true than when dealing with topography. Although the experienced site planner has developed a sense of understanding topography, laypersons have serious problems

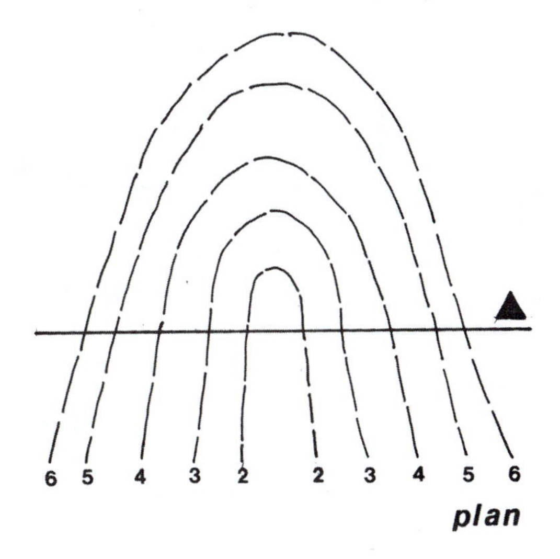

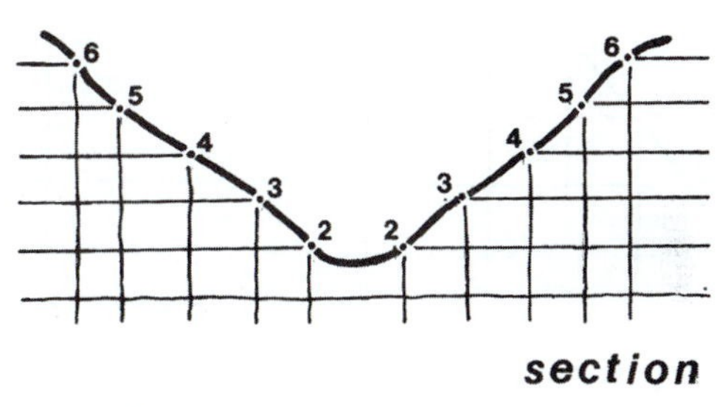

FIGURE 12-9. Valley.

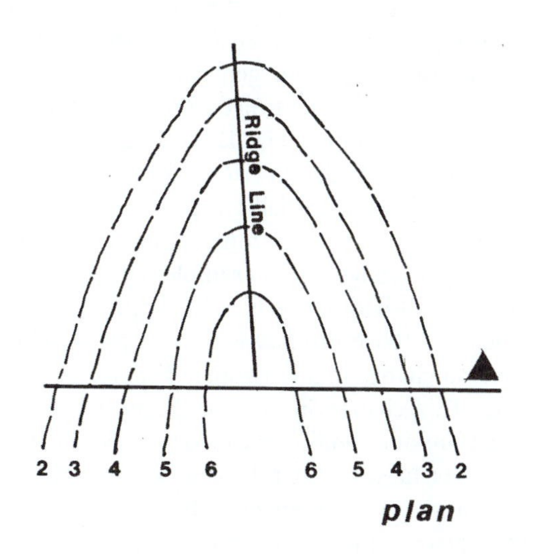

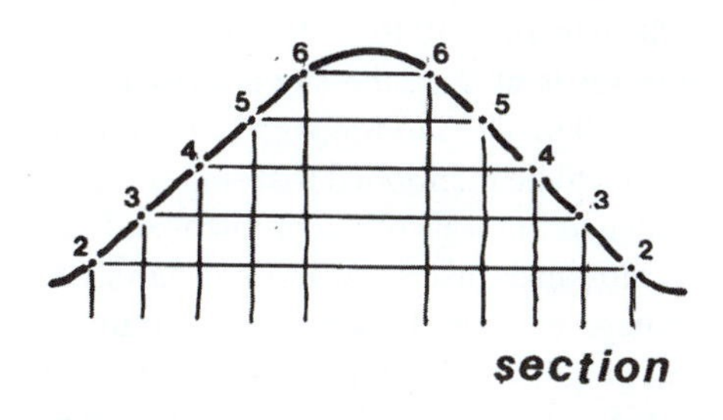

FIGURE 12-8. Ridge.

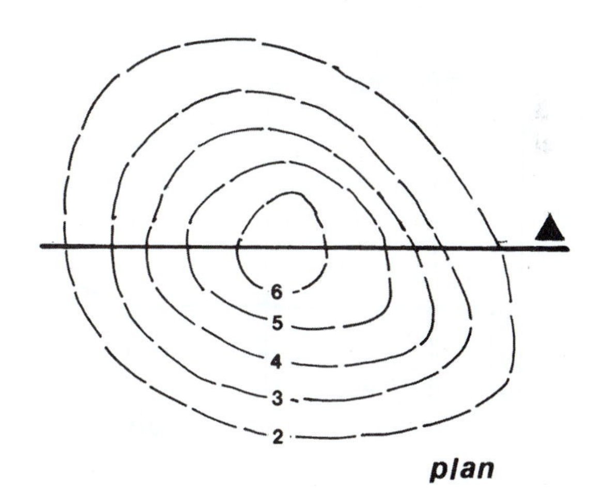

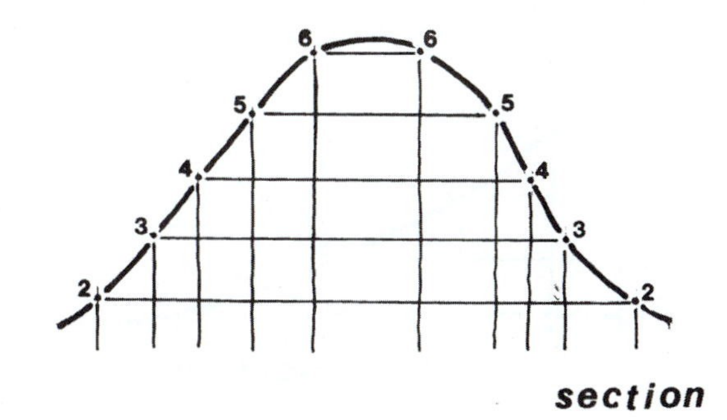

FIGURE 12-10. Hill.

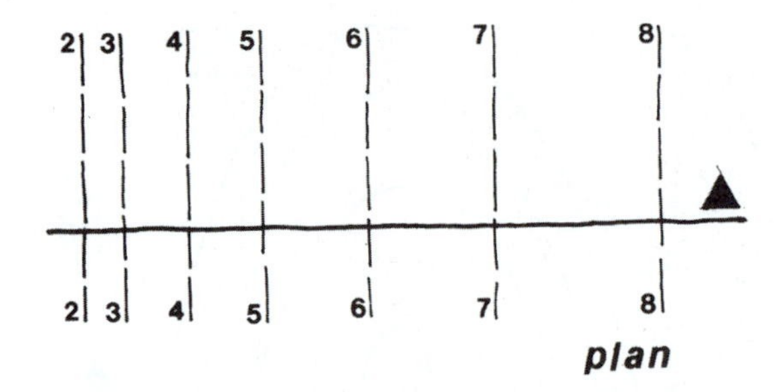

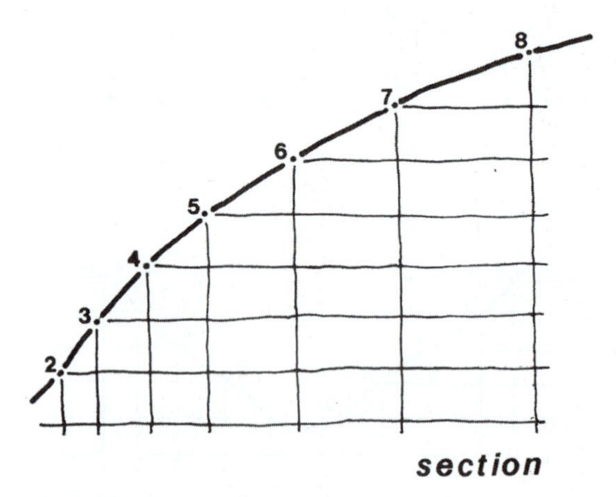

FIGURE 12-11. Convex slope.

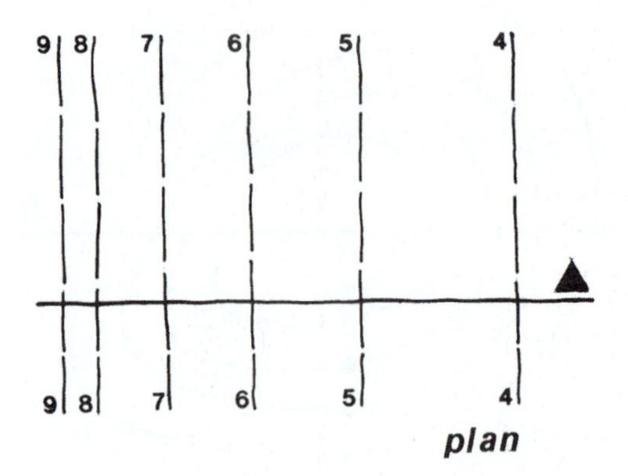

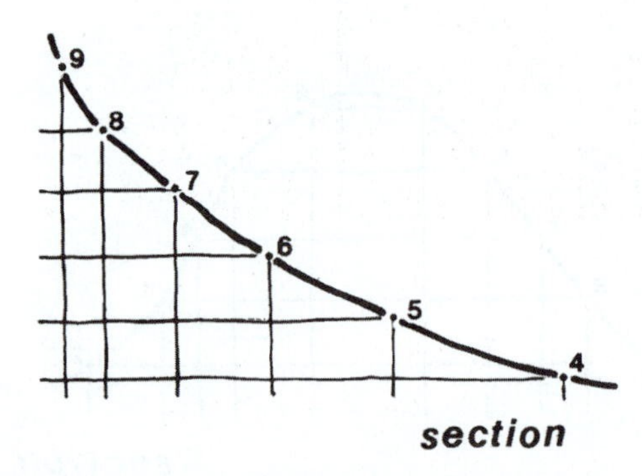

FIGURE 12-12. Concave slope.

visualizing a site's character. For that reason, models are highly recommended as a method to ensure better communication between planner and client (Fig. 12-13).

Summits and depressions are spot elevations of the landform and identified with an × on the topographical map (Fig. 12-14). The hachure lines on the low side of the contour line graphically define the depression. Spot elevations are also used to identify human-made elements on a site such as the top of a curb, a slab, a wall, or a drainage structure. A *benchmark* is a primary spot elevation or reference point. Benchmarks are located on U.S. Geological Survey maps as well as on topographical surveys made for site-specific problems (Fig. 12-15, on page 124). Spot elevations are also used to define critical elevations for drainage patterns. As an example, an 8-in. swale is often very important to the development of a system and can only be identified with spot elevations. Natural features such as trees and rock outcroppings are also often identified with spot elevations, as they influence design and grading decisions.

Slope Analysis

Quantifying the character of any design problem is an appropriate step in the process of finding a solution. Each site has a number of qualities and characteristics that can be quantified, and among those is its relative slope. Since describing an area as steep or flat is both insufficient and simplistic, slope analysis is a method that permits the graphic identification and illustration of various slopes on a site.

The term "slope" will be used in two ways in this section on earthworks; the first will be as a function of slope analysis. The process of slope analysis is necessary to make certain that there is some general agreement between the character of the human-made improvements and the quality of the topography. Fig. 12-16 (see page 125) illustrates some acceptable design criteria for various site activities. Slope analysis essentially expresses the change in vertical elevation over a horizontal distance of 100 ft expressed as a percentage. As an example, a vertical change of 2 ft across a 100-ft distance is a 2% slope. A 4-ft vertical change across a 100-ft distance is a 4% slope. It follows that a 4-ft vertical change in a horizontal distance of 50 ft is an 8% slope. This change in vertical distance is commonly measured from one contour line directly to the next.

It appears, however, that there are no uniform percentages employed for slope analysis. If the only topography available is at a 5-ft interval, it's easy to see that the ratios for slope analysis will be simpler to identify in 5% increments, that is 0 to 5%, 5 to 10%, 10 to 15%, and so on. If the topographical interval is at 2 ft, the slope analysis is easier to break down into 0 to 4%, 4 to 8%, 8 to 12%, and so on.

It also needs to be mentioned that slopes are sometimes analyzed relative to the scale of the problem and topographical interval combined. Instead of a slope analysis developed as the percentage of slope between two contour lines, the slope analysis might be developed for some constant distance. As an example, on a 200-acre site with topography at

(a)

(b)

FIGURE 12-13. Topographical model.

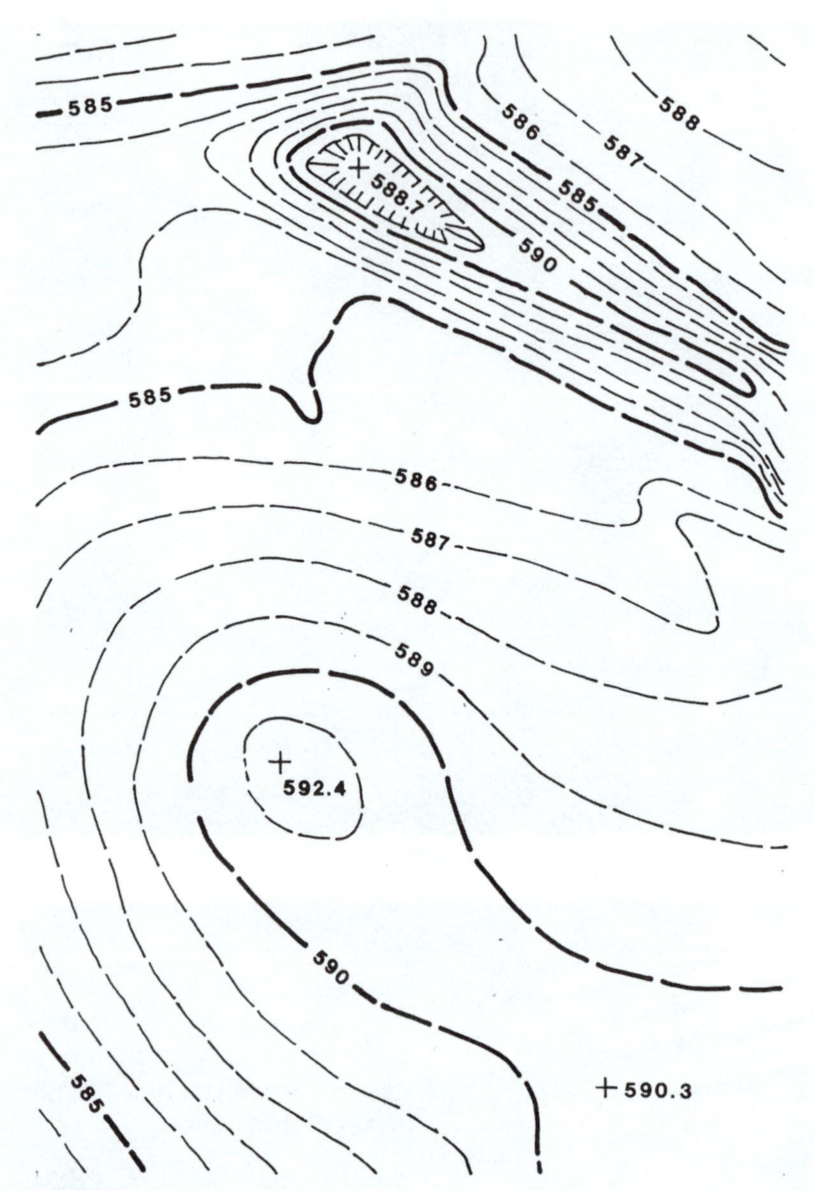

FIGURE 12-14. Summit and depression.

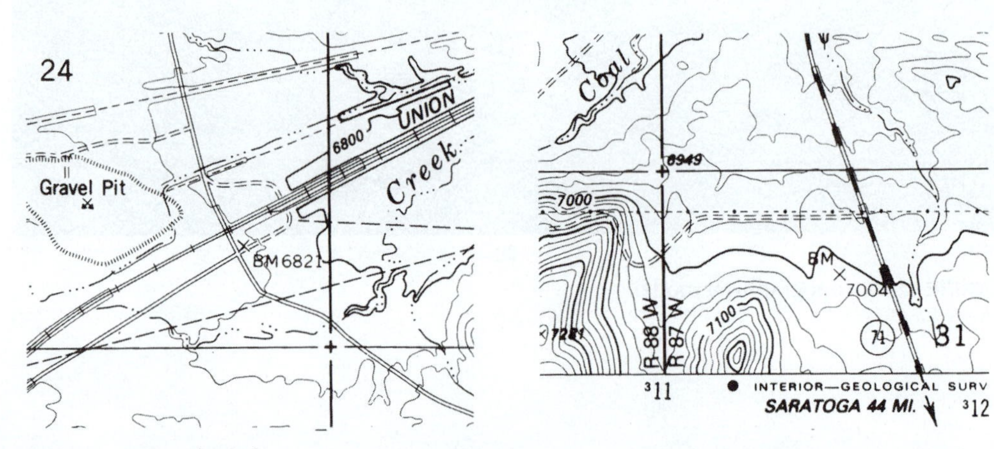

FIGURE 12-15. Benchmark.

AREA	FUNCTION	SLOPE IN PERCENT	
		MAX.	MIN.
Streets & Drives		5%	1% [a]
		8%	.05% [b]
Ramps		10%	1% [a]
		15%	NA [b]
Walkways Approaches and Entrances		4%	1% [b]
		5%	0.5% [a]
Service Areas and Collector Walks		8%	0.5% [a]
		10%	0.5% [b]
Terrace and Sitting Areas		2%	1%
		2%	0.5% [b]
Lawn Area and Playgrounds		3%	2% [a]
		4%	0.5% [b]
Swales		10%	1% [a]
Grassed Banks	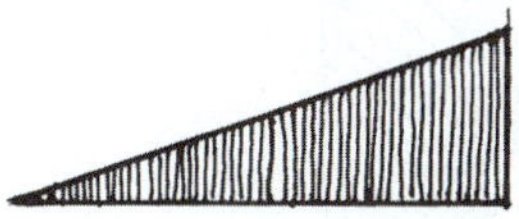	33% (3:1) [a]	NA
		25% (4:1) [b]	NA
Planted Banks (unmowed vines or ground cover)	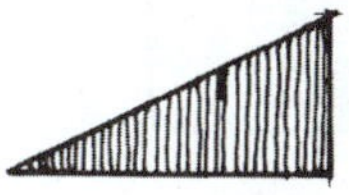	50% [a]	NA
		2:1 [b]	NA

[a]Richard Untermann, *Grade Easy* (Washington, D.C.: Landscape Architecture Foundation, 1973), p. 73.
[b]Public Housing Design, National Housing Agency (Washington, D.C.: Federal Public Housing Agency, 1946).

FIGURE 12-16. Slope criteria.

2-ft contour intervals and at a scale of 1 in. = 100 ft with high topographical relief, the site planner may choose to execute the slope analysis as a change across 50 ft, not from one contour line to the next. This method has some validity, as it permits the slope analysis to reflect the scale of the problem as well as the existing resources.

On a smaller site the opposite may be more appropriate. Fig. 12-17 is just such a case. This is a 20.8-acre site where the topographical interval is 2 ft, the map scale was 1 in. = 50 ft, and the slope analysis was executed from one contour line to the next. The level of detail is much finer, and for most site-planning problems, this is the preferred methodology.

Following the determination of the various slope categories, the development of a map that depicts a site's slopes becomes the next step. Since the basis for determining the slope is a function of contour interval and the distance between two contour lines, that relationship can be expressed using the following formula:[1]

$$\text{Distance between contours} = \frac{\text{Contour interval}}{\text{Percent of slope}} \times 100$$

[1]Harvey Rubenstein, *A Guide to Site and Environmental Planning*, 2nd ed. (New York: John Wiley & Sons, Inc., 1980), p. 17.

As an example, if the slope category is 4 to 8% on a map with a 2-ft contour interval, the distance between two contour lines for these respective slopes would be:

$$\text{Distance} = \frac{\text{Contour interval}}{\text{Percent of slope}} \times 100$$

$$D = \frac{2\text{ ft}}{4\,\%} \times 100 \qquad D = \frac{2\text{ ft}}{8\,\%} \times 100$$

$$= 50\text{ ft} \qquad = 25\text{ ft}$$

Therefore, for an area to be in a 4 to 8% slope category, the distance between two contour lines (with a 2-ft interval) could not be *more* than 50 ft to be a minimum of 4% slope, nor *less* than 25 ft to be a maximum of 8% slope.

Slope analysis is one of the most basic components of site analysis and no research is complete without it. This information is a crucial step in determining the appropriate location for all site improvements and thus is an essential part of intelligent decision making.

The second way in which "slope" is used as a term is to describe vertical change in small areas, particularly as a design criterion for cut and fill. In areas where the site planner must maintain a specific angle of repose, slope as a percentage

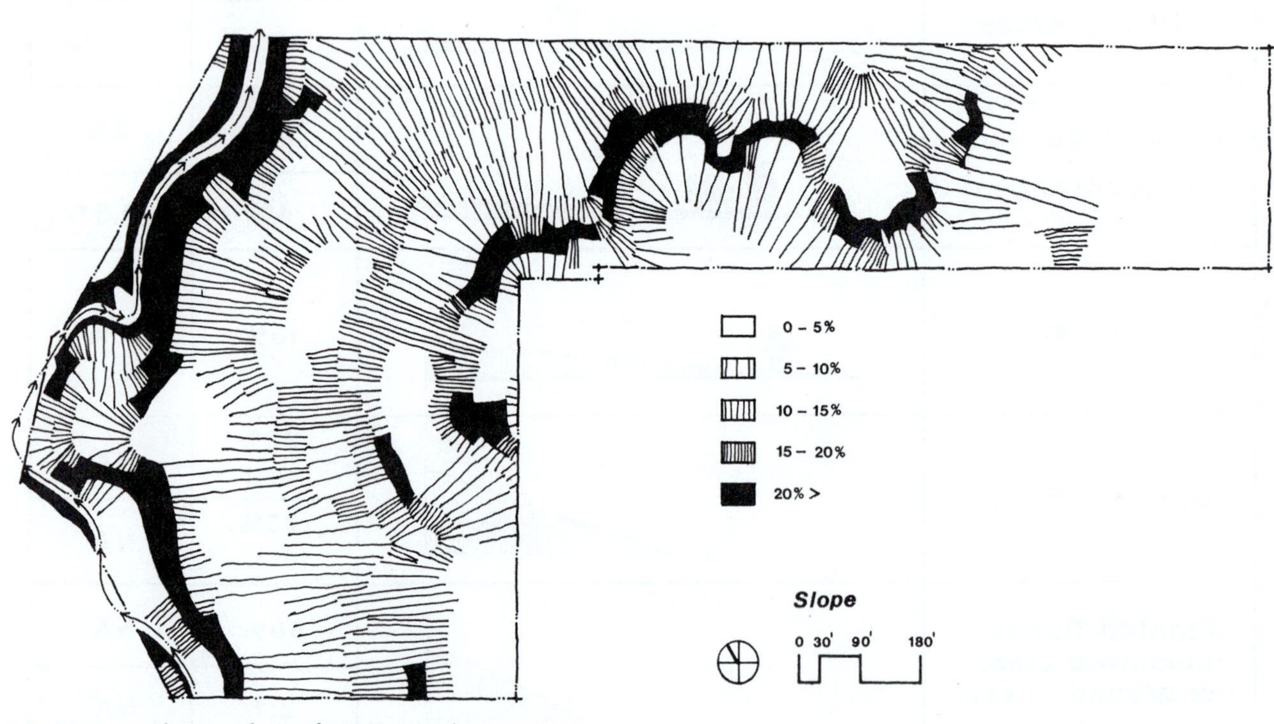

FIGURE 12-17. Slope analysis of a 20.8-acre site.

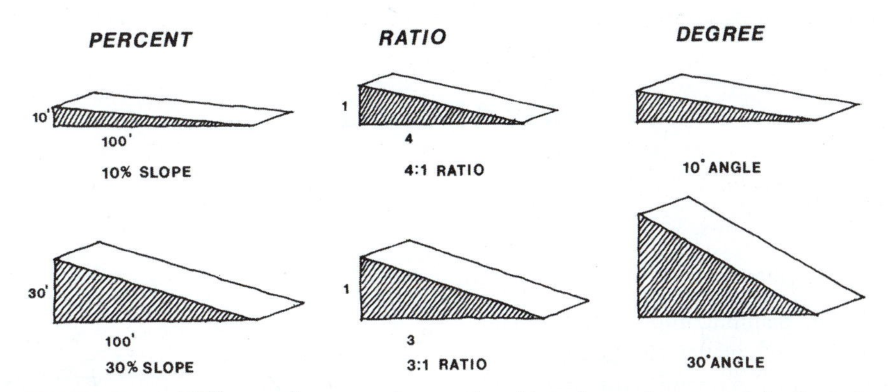

FIGURE 12-18. Difference between slope ratio of 3:1, slope of 30%, and incline of 30°.

of change over 100 ft is inappropriate. Another common expression refers to the slope as a ratio between a horizontal dimension and 1. In Chapter 13 reference is made to an unstabilized "cut" of 3:1. This translates to mean that the length of the slope (perpendicular to the contours) will be 3 horizontal feet for every foot of vertical change. Fig. 12-18 illustrates the difference between the slope ratio of 3:1, a slope of 30%, and an incline of 30°.

Since substantial quantities of grading outside a building's perimeter will occur in less than 100 ft, the value of the slope as a percentage has little applicability. The key to remembering the differences between the two types of slope is to understand that each serves an entirely different function: one in site analysis, the second in grading. Second, because of those differences, each is expressed in two entirely different mathematical forms.

CHAPTER THIRTEEN

GRADING

By its very nature, site planning implies change. The question comes down to how that change can be accomplished with minimal economic costs and environmental damage. As mentioned earlier, the grading plan is discovered or emerges as a result of good research, careful documentation, detailed analysis, creative synthesis, and patience. Grading does not lend itself to the quick-fix, big idea.

An evaluation of a site's more buildable areas must be based on an understanding of its topography, and the quantitative character of that is slope analysis. Since grading involves the modification of existing contours, the location of a building site or parking lot should reflect the information developed in the slope analysis. Before beginning any grading, review the slope analysis discussion in Chapter 12. If during the grading process, it appears that the results will create major environmental damage or loss of vegetation, consider relocating the structure. With the slope analysis, a locational decision concerning roads, parking, or building site can be made within a more enlightened framework. Grading may now be considered within the context of all site constraints.

Keeping water from draining into the building is the obvious, primary mission of the grading plan. Since grading and drainage must be thought of in a symbiotic way, each must be considered as a function of the other. In a pursuit to shape the site to accommodate an activity, a positive drainage pattern must be an integral aspect of the plan. The development of a site for a building implies that at least parts of the area must be level, and any topographical change must reflect earthworks that will induce water to run away from the immediate building area. To reiterate a point made earlier, the two characteristics of water that need to be remembered are that it always runs downhill and perpendicular to the contour line (Fig. 13-1).

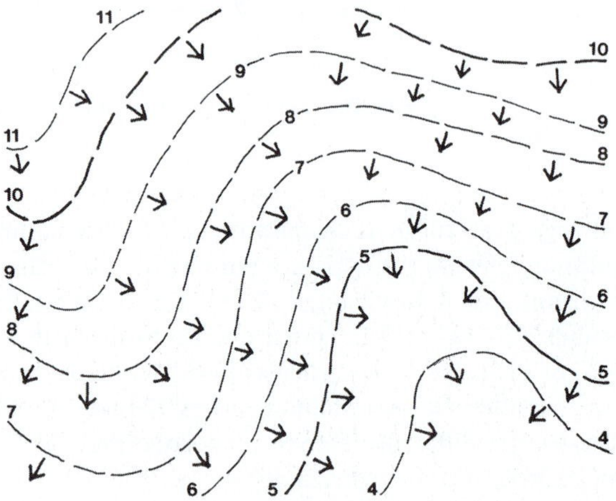

FIGURE 13-1. Contour lines and flow of water.

Richard Untermann in *Grade Easy* suggests three basic topographical forms that might be employed in grading.[1] The sloping plane is the first and most common landform used because of its infinite varieties (Fig. 13-2). The sloping plane may also emerge as a convex or concave slope. Although it is possible that a site can be graded using the sloping plane alone, one must realize that the water will eventually accumulate at the bottom of the "plane" and cannot, in all cases, simply become someone else's problem.

When the case is such that the water must be restricted to a site, a partial answer may be the use of a valley as a method to control the runoff (Fig. 13-3). This is the second basic form and is used in situations where the valley is as shallow as a swale (Fig. 13-4) or as pronounced as a creek (Fig. 13-5). Although the swale is a common method of carrying water around a building, it must be designed so that the water has a positive flow. Critical to the success of the

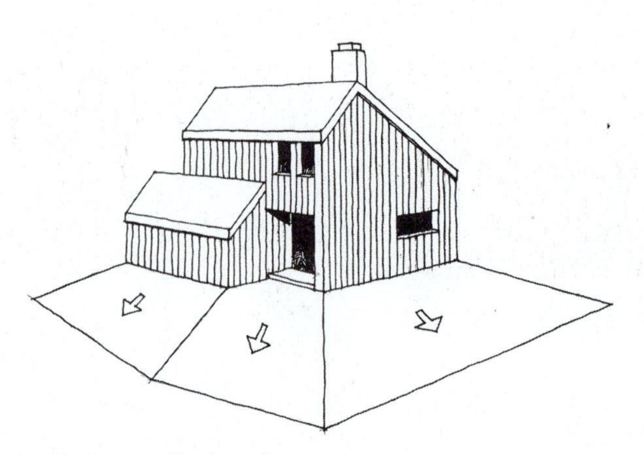

FIGURE 13-2. Sloping plane.

[1]Richard Untermann, *Grade Easy* (Washington, D.C.: Landscape Architecture Foundation, 1973), pp. 52–53.

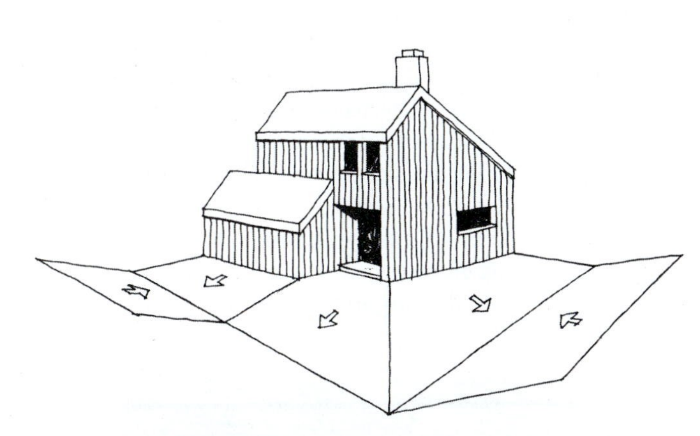

FIGURE 13-3. Valley.

FIGURE 13-4. Swale.

FIGURE 13-5. Creek.

swale is remembering that the minimum recommended slope to achieve positive drainage in turf is 1%.

The funnel (Fig. 13-6) is the third form normally considered as a primary grading configuration. This shape is an aggregation of the two discussed earlier, but it has one characteristic that the others do not have. A graded funnel form will, in most cases, require the inclusion of a drainage structure at the base of the funnel. Although the addition of an area drain is not a monumental endeavor, it must be noted that its omission could create a very undesirable condition. Later we address in more detail the various drainage structures ordinarily used in the development of a grading plan.

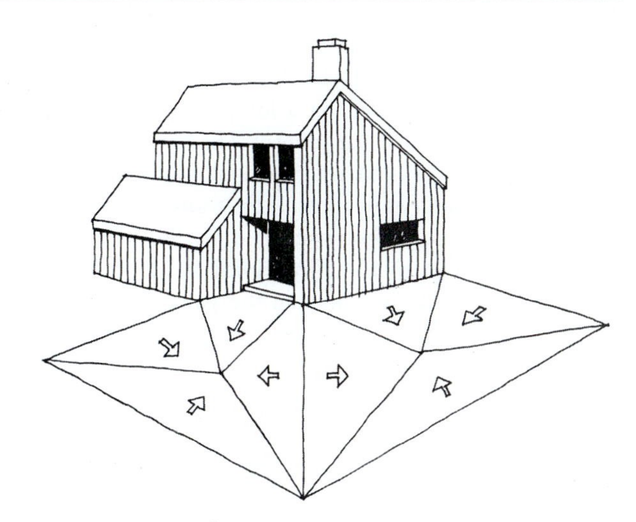

FIGURE 13-6. Funnel.

The grading process requires a careful change of the contours so that they support the integration of building and site. Smooth grades and convex earth forms are methods used to develop a site that reflects a natural aesthetic. Geometric lines, harsh angles, and steep grades sometimes create difficult transitions between site and building. Urban environments often require an architectural response, especially when the development program demands a highly articulated hardscape (Fig. 13-7). The message here is that the site planner must be cognizant of the context wherein each of those methods would be appropriate and intelligent.

In a review of earthworks, the site planner needs a working knowledge of grading to save a tree. Grading in this

FIGURE 13-7. Articulated hardscape.

sense means: Do not fill (add soil) or cut soil or roots back within the drip line (Fig. 13-8). Respecting this first rule and using common sense will go a long way toward keeping most vegetation out of harm's way. This criterion is based on the knowledge that a tree's health is, in large part, determined by the health of its roots. If the roots are damaged or impaired, the leaf and limb systems will reflect that damage (Fig. 13-9 and Fig. 13-10). If regrading inside a tree's drip line is unavoidable, the limbs on the same side should be cut back proportionally to reduce the possibility of disease attacking any withering branches. If changing the grade within the drip line appears imminent, use stacked stone walls at the edge of the drip line to ensure a tree's access to moisture (Fig. 13-11). The technique is not only a functional alternative to save vegetation, but it also adds to the aesthetic quality of the earthworks as well. Before the process gets to the problem stage, check to see if the building's location can be adjusted and, with it, the grading plan. If the area is being graded, to accommodate pedestrian traffic, try to use wood decking and avoid grading (Fig. 13-12).

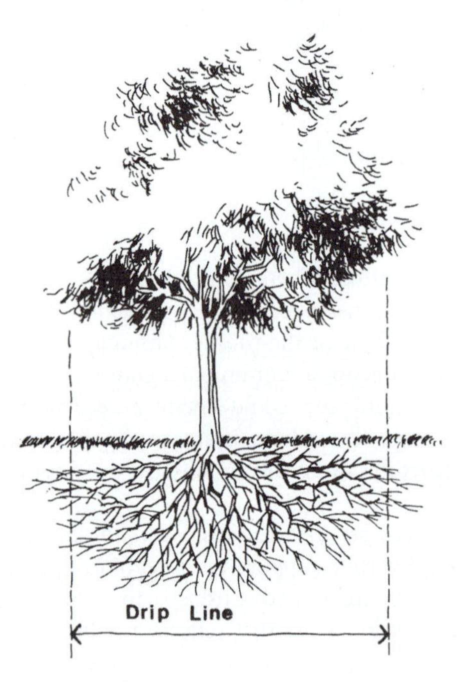

FIGURE 13-8. Grading near tree.

FIGURE 13-10. Trees lost due to paving inside the drip line.

FIGURE 13-9. Trees lost due to paving inside the drip line.

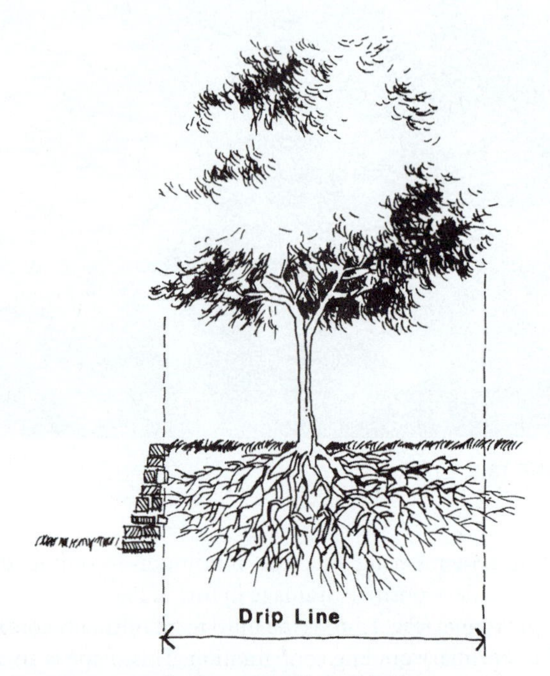

FIGURE 13-11. Stacked stone wall at drip line.

FIGURE 13-12. Trees through wood decking.

A more extensive examination of trees in the urban environment is covered in Chapter 8 and should be consulted in the event that more complex problems with trees arise.

The importance of topsoil to a site also needs to be emphasized. Topsoil is rich with the nutrients that foster and maintain healthy plant growth. Although some parts of the continent are still rich with the resource, other areas have had topsoil stripped by erosion or the development processes. When the topsoil depth is too shallow to support revegetation, sandy loam or an equivalent substitute is introduced to the site. Needless to say, when an area must be regraded, the topsoil stripped from that area needs to be retained on the site and utilized in the regrading process. The distribution of topsoil across the site is obviously crucial to the site's ability to support new vegetation.

Every site will have its own set of restrictions, whether physical or legal. Physical restrictions must be acknowledged. When possible, rather than changing the site, consider modifying other design parameters to resolve conflicts between building and site. As an example, consider reducing the area of the site to be regraded. Sometimes a client's program is sufficiently flexible to permit reducing a building's "footprint" by increasing its height. A design solution that is two stories instead of one can, in some cases, favorably accommodate a site's natural attributes. Reducing the area to be graded also implies an examination of the paving area committed to the automobile. Auto access and parking can create an enormous demand for paving. A careful examination of the design criteria for auto circulation should be made and every effort expended to keep the amount of paving to a minimum without jeopardizing safety.

THE GRADING PROCESS

The grading process begins with establishing the parameters of the grading problem. In addition to the known, existing spot elevations that control and/or influence every grading plan (roads, curbs, sewers, easements), each site has a property line beyond which *no* regrading may be done. These "givens" should identify significant vegetation, rock outcroppings, or problems such as depressions or drainage pockets. The site-planning checklist in Chapter 6 will provide some guidance to this research. Secure with a knowledge of contours and topography, the implications of change are next. Some site modifications become part of a project when the program implies change. Normally, that change means that the contour configuration will be modified by a "cut," a "fill," or a combination of the two. Just as the nomenclature suggests:

1. A cut is the removal of a portion of a site's natural topography (Fig. 13-13).
2. Fill is the process of blending additional earth into a site's existing topography (Fig. 13-14).
3. A combination of the two is shown in Fig. 13-15.

The preferred method of solving grading problems is to balance cut and fill. The emphasis is on *preferred* because, in some cases, the combination is not appropriate. There are some distinctive advantages and disadvantages of cut and fill. One positive aspect of cutting is that the bank from which the cut was made was relatively stable. Undisturbed earth is generally stable, as it has had years to compact and, as a result, often permits more economical grading solutions.

Balancing the cut and fill also resolves the question of acquisition of new fill or the disposal of any residual. Since either acquisition or disposal has a dollar cost associated with it, the financial benefits of balancing the cut and fill are obvious. The balancing concept by its nature implies that any fill will use soils from the same site; as such, the bonding action is often more successful. There will, however, be circumstances in which a site plan may need to be developed using only a cut or fill, and attempts at balancing may only create problems. Certainly, any time that topographical

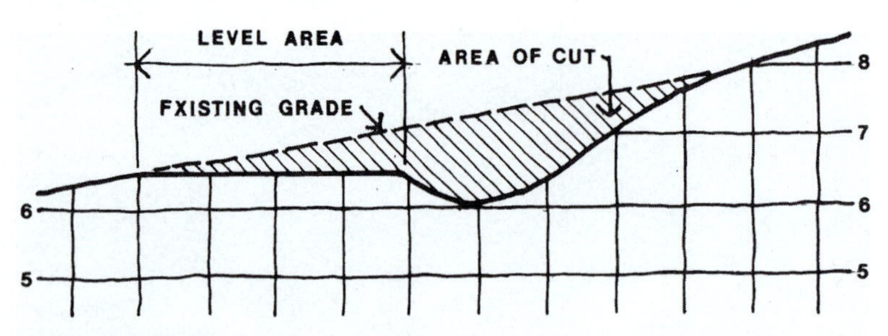

FIGURE 13-13. Cut (section).

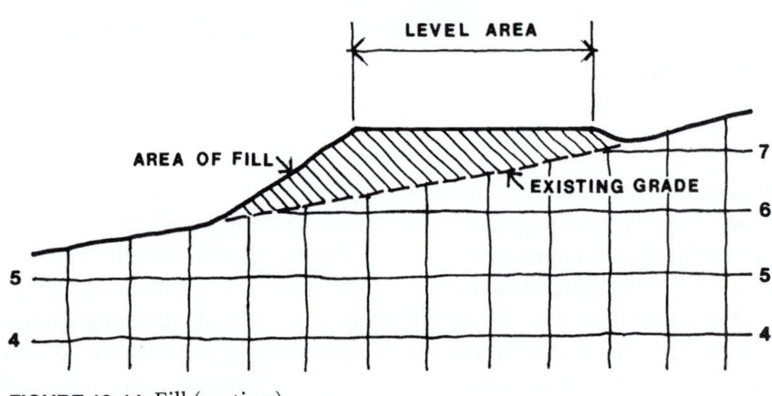

FIGURE 13-14. Fill (section).

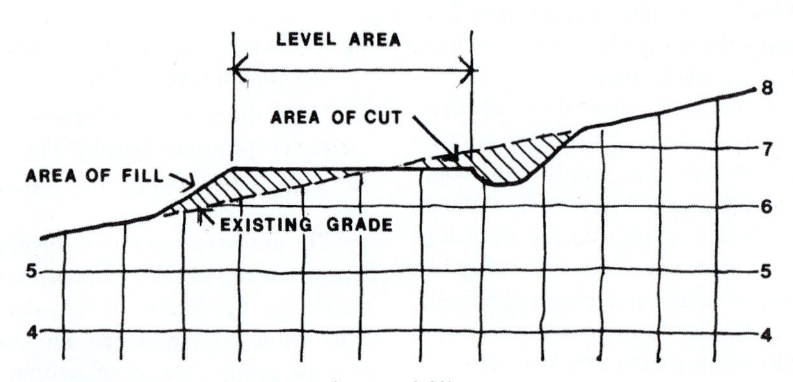

FIGURE 13-15. Combination of cut and fill.

modifications are excessive, destroy major trees, or create drainage problems, the grading plan should be changed.

In those situations in which the topography is too steep for a fill, a cut may be the only legitimate alternative. Under circumstances where a cut is required on a steep slope, the grading plan may also necessitate a retaining wall to avoid erosion or spalling. A disadvantage to the cut lies in the need to dispose of the residual. Although in some communities, fill may be sold, in other municipalities, finding a place to dump fill has become a serious problem. Some local and state governments have adopted very stringent ordinances concerning fill, and for that reason, consultation with local authorities is highly recommended. Small sites can be a problem in balancing cut and fill, and generally, lend themselves to regrading by cutting alone. Opportunities for filling on small sites are obviously limited, as indiscriminate fill can create serious problems for existing trees. Often, grading by fill is done simply to make an area usable. Low or poorly drained areas are commonly filled to develop positive drainage patterns for a site, but there are a few guidelines that should be recognized before regrading by fill. First, some caution must be exercised in filling on a slope. Because of the lack of long-term bonding between the existing grade and the new fill, the potential for erosion is very high. Second, any construction on fill must be designed to respect the quality of the fill.

Following the decisions associated with the location of the structure and the documentation of all known fixed or controlling elevations, the next step involving the grading of a site begins with identification of the finish floor elevation. Make note of the fact that the process began by establishing a spot grade on the slab from which other decisions were made. A site is planned and graded based on spot grades. Contours are then changed to reflect the spot-grading decisions. The following chart can be used to determine the various decimal intervals of 12 in.:

Decimal Equivalents (Inches to Feet)

1 in. = 0.083 ft	5 in. = 0.416 ft	9 in. = 0.75 ft
2 in. = 0.166 ft	6 in. = 0.5 ft	10 in. = 0.833 ft
3 in. = 0.250 ft	7 in. = 0.583 ft	11 in. = 0.916 ft
4 in. = 0.333 ft	8 in. = 0.667 ft	12 in. = 1.0 ft

If the site is to be graded using a cut alone, specify an elevation for the finish floor 0.5 ft above the nearest contour line on the low side of the building (Fig. 13-16). This requires that all contours above the finish floor elevation of the building be "wrapped" around the building. Where each of the new contour lines crosses an existing contour line of higher elevation, we know that a "cut" occurs. Fig. 13-16 illustrates the cut necessary as the new contour line 6 crosses the existing contour 7, and 7 moves up toward 8. Note that the proposed contours are carefully curved and shaped back into the existing contour lines. This is referred to as "meeting the grade." While an effort must be made to merge the proposed contour with the existing contour as soon as possible to minimize the extent of the cut, some care must be taken that merging does not result in an abrupt or harsh topographical change.

The contour change and its relationship to the slab must also conform to some accepted slope criteria. Remember that "slope" is the relationship between horizontal distance and vertical rise and is also expressed as a ratio (i.e., 3:1 or 4:1). This concept is critical to the grading process, for without established criteria, the relationship between contour lines becomes arbitrary and will vary between unnecessarily shallow and inappropriately steep. Accepted criteria for grading an unstabilized cut are a maximum 3:1 slope and 4:1 fill. Since the fill is not as secure in its bonding to the existing soil, the slope must necessarily be more gradual.

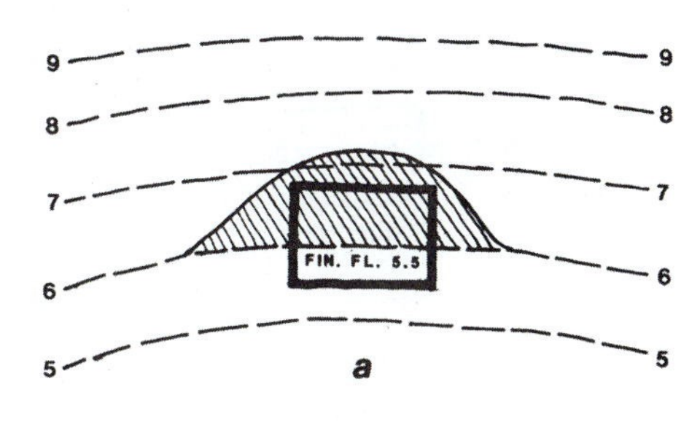

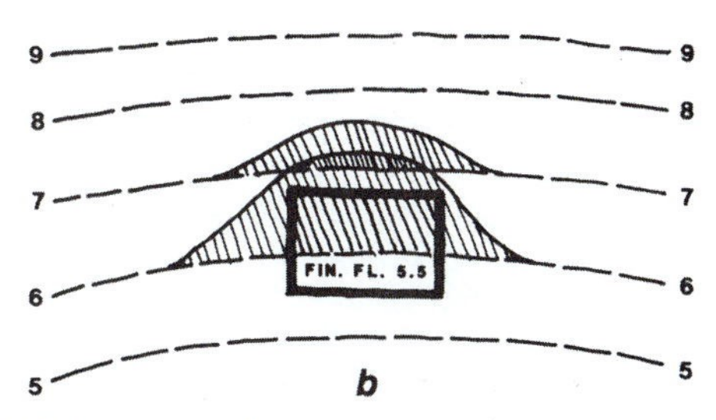

FIGURE 13-16. Cut-in plan.

Since stabilized cut and fill slopes are a function of the techniques employed, they are addressed more extensively later (Chapter 16). For now, however, stabilized cuts can be as steep as 1:1 and stabilized fill 2:1. These guidelines will permit the site planner to develop a grading plan knowing that there are alternatives and base criteria for these options.

Following the established criteria, subsequent contours can be located beginning with the lowest and, in turn, progressing to the higher contour elevations. When complete, the plan should identify each existing and proposed contour line, as well as all spot elevations critical to the grading effort. The necessity of identifying the spot elevations surrounding the structure in sufficient detail to know that the grade slopes away from the building is a common oversight. Remember, leave nothing to chance.

To grade an area level using fill only, the process is simply the reverse of the cut. A finish floor elevation is identified which is 0.5 ft above the highest contour line that passes through the area to be graded level. Having established that elevation, the highest contour line in the area to be graded is "wrapped" around the slab over the lower contour lines at a distance that reflects the slope criteria mentioned earlier (Fig. 13-17). Similar considerations concerning meeting the grade are still applicable and, in some cases, more obvious when executed poorly on a fill. Buildings that appear to be rising out of the topography, demanding dramatic alterations in a site's character, detract not only from the total aesthetic quality of a building and site, but also become maintenance problems when soils are given to spalling under heavy moisture.

Make note of the fact that in the examples of both cut and fill, the long sides of the rectangular slabs were oriented parallel to the contours. This is a basic corollary associated

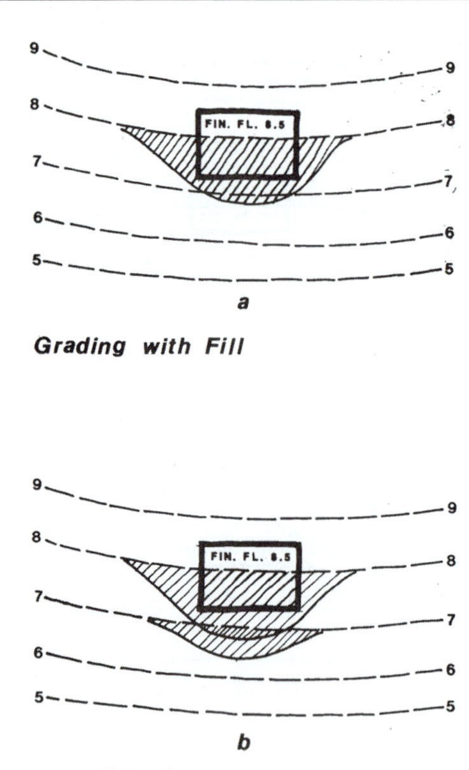

FIGURE 13-17. Fill-in plan.

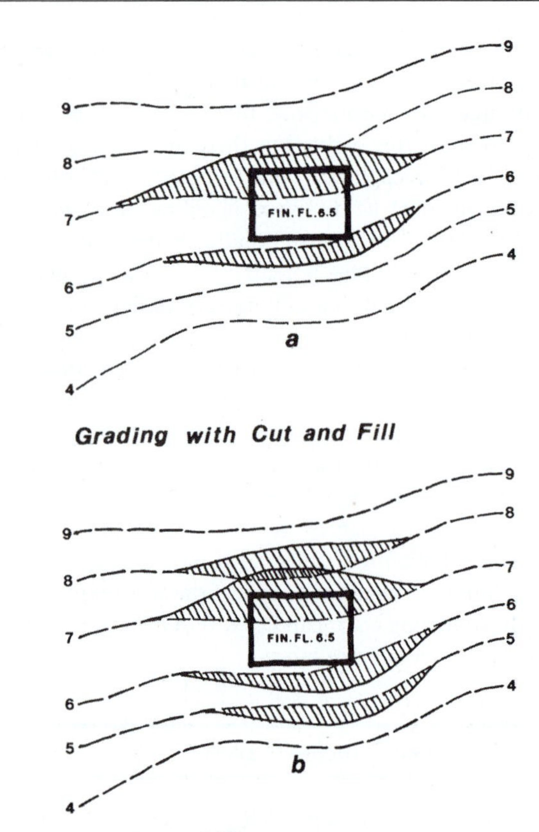

FIGURE 13-18. Cut and fill.

with a structure's location on a slope and should be considered as a primary criterion in the process. The location of any structure on a site should take into account its orientation relative to the direction of the contours, as the implications of cut and fill can be significant.

Having reviewed the methods of cut and fill separately, the third alternative, which combines the two techniques, needs to be examined. Pursuing the concept that the best method is to balance the cut and fill, the finish floor level is established at 6.5 ft above the contour line that appears to be midway between the highest and lowest elevations covered by the slab (Fig. 13-18). This median elevation implies that half of the area to be graded will be accomplished by a cut and the other half by a fill. After determining the desired elevation of the slab and those of the surrounding grade, the contour lines are then "wrapped" above (cut) or below (fill) the slab to reflect the new elevations. Fig. 13-18 illustrates the extent of the cut and fill. The same criteria concerning slopes and meeting the grade are still in effect. Additional earthworks techniques associated with retaining walls, stairs, ramps, and terracing will all be explored but will also rely on the basic concepts of cut and fill.

Two important aspects of the process need to be mentioned now. First, it has been suggested that good grading is discovered, not imposed. If, in the grading process, it appears that the results cause major environmental damage or radical loss of vegetation, consider relocating the structure. Another part of the site may be more accommodating. Solutions to site-planning problems that require extensive grafting or extensive removal might need to be reevaluated. Second, if holding or maintaining one elevation for the slab in every direction creates large areas of cut or fill, consider breaking or separating the slab vertically (Fig. 13-19). Keep in mind that this must be done in conjunction with the spatial planning of the building. Properly executed, the change in vertical dimensions can emphasize or define the limits or functions of a particular area of the architectural plan and greatly add to its aesthetic quality. This concept applies to building as well as site scale. This technique could be applied to tennis courts, parking lots, or any situation where the change in elevation would reduce the amount of site grading. The bifurcated road is an example of a two-way thoroughfare that has been separated in sections to reduce the amount of cut and fill (Fig. 13-20). The value of this strategy is not only one in which large parking areas may be broken and their nuisance quotient reduced, but also one in which there is the potential to enhance the character of an architectural space (Fig. 13-21, on page 136).

To summarize cut and fill, remember that when proposed contour lines cross existing contour lines that are numerically higher, the change is a cut (Fig. 13-16). When the proposed contour lines cross existing contour lines that are numerically lower, the change is a fill (Fig. 13-17). Both techniques, individually and compositely, must become a working part of the site planner's lexicon of skills.

For some reason, many think of grading as the movement of contours. Invariably, when presented with a grading problem, students begin drawing contour lines instead of

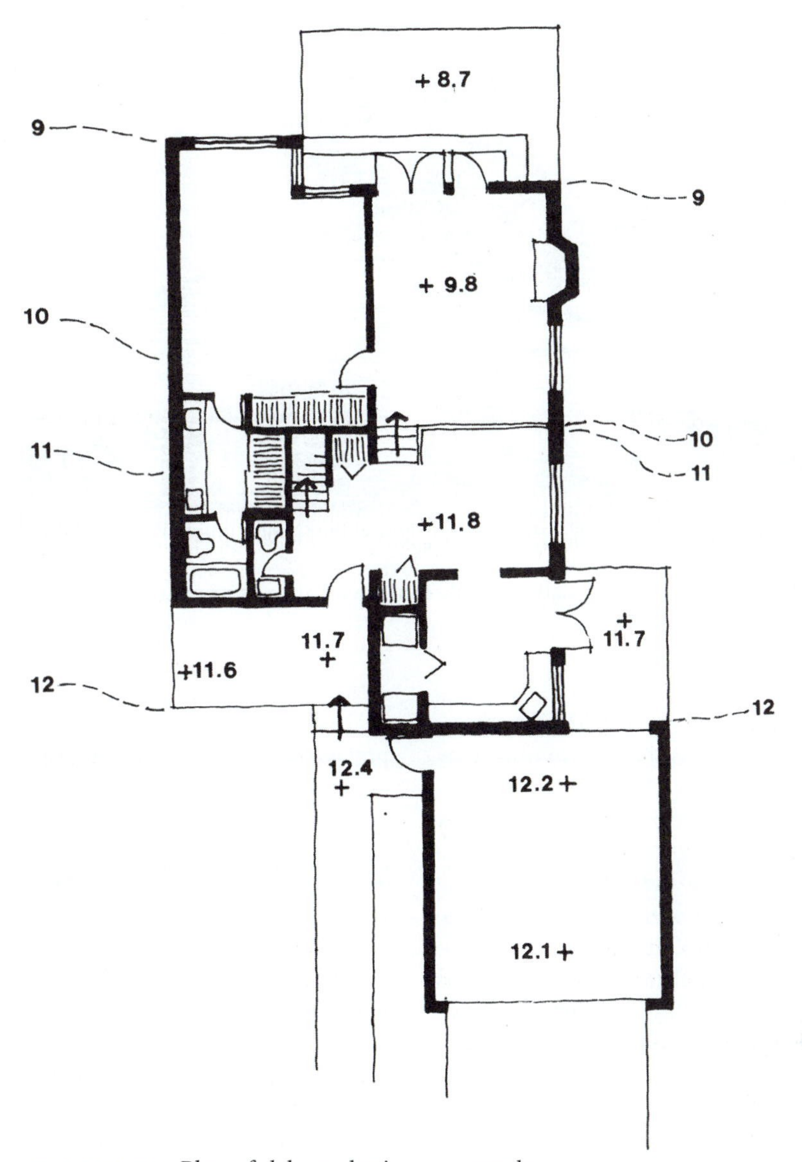

FIGURE 13-19. Plan of slab on sloping topography.

FIGURE 13-20. Bifurcated road.

FIGURE 13-21. Section illustrating slab vertically separated to accommodate sloping topography.

developing the site with spot elevations first. This was mentioned earlier but bears repeating, as the urge to draw may overpower the necessity to think.

To facilitate the grading process, Table 13-1 is a matrix of distance and slope. The left margin is distance; the horizontal line is slope. To determine the change in grading necessary for a particular problem, identify the slope at the top of the chart; read down the column to locate the distance that corresponds to the problem. Use combinations of even numbers: 50 plus 4 for 54 ft. The chart can eliminate a lot of time-consuming math while respecting good grading practice.

The final step in the process requires a review of those circumstances in which no grading should be done. Richard Untermann in *Grade Easy* has suggested that the following conditions be avoided or reevaluated:

1. Grading that results in radical loss of vegetation and/or topsoil
2. Grading that interrupts the natural drainage
3. Grading that results in aesthetic degradation
4. Grading on difficult slopes (excess of 25%), in floodplains, estuaries, or bogs, or in other environmentally unique conditions
5. Grading in areas susceptible to natural disasters, such as mud slides or along earthquake fault lines[2]

Each of these circumstances is obviously a harbinger of future problems to either client, user, or community. Although the values of some communities are such that development would not be permitted in any of these conditions, others perceive development in any form as a means to expand the tax base. What we often fail to realize that if communities and professionals abdicate the responsibility to define the limits to development, private self-interest never will.

[2]Ibid., p. 87.

GRADING AND DRAINAGE

Grading Principles

Two of the principle goals in the development of a grading plan are:

1. Keep unwanted water from entering a building
2. Keep surface runoff from creating damage to property or people during periods of heavy rainfall and subsequent runoff

For these good and sound purposes, the site planner must begin by considering drainage a function of grading, just as the architect considers structure an integral part of architecture. The scope of problems to resolve in every grading plan will range widely, as much depends on a site's ecology and context. Regardless of whether the site's context is an urban area, where 80 to 90% of a site may be impervious surfaces, or a suburban area, where those percentages can drop from 90% to 40%, decisions relative to on-site runoff cannot be dismissed casually.

To avoid the problems associated with drainage and runoff, the site planner needs to recognize that nothing can be left to chance. While each set of site circumstances is different, each has a common set of criteria and concepts. The first criterion is that water always flows downhill, from a higher to a lower elevation, and perpendicular to the contours. Although this may be an elaboration of the obvious, faulty drainage diagrams and ridge-line identification are not limited to student site-planning exercises; mistakes associated with analysis errors have also been built.

Table 13-1. Percent of Slope

DISTANCE	SLOPE (%)												
(ft)	0.5	0.75	1.0	1.25	1.5	1.75	2.0	2.5	3.0	3.5	4.0	4.5	5.0
1	0.005	0.08	0.01	0.013	0.015	0.018	0.02	0.025	0.03	0.035	0.04	0.045	0.05
2	0.01	0.015	0.02	0.025	0.03	0.035	0.04	0.05	0.06	0.07	0.08	0.09	0.10
3	0.015	0.023	0.03	0.38	0.045	0.05	0.06	0.075	0.09	0.105	0.12	0.135	0.15
4	0.02	0.03	0.04	0.05	0.06	0.07	0.08	0.10	0.12	0.14	0.16	0.18	0.20
5	0.025	0.04	0.05	0.063	0.075	0.09	0.10	0.125	0.15	0.175	0.20	0.225	0.25
6	0.03	0.05	0.06	0.075	0.09	0.105	0.12	0.15	0.18	0.21	0.24	0.27	0.30
7	0.035	0.05	0.07	0.088	0.10	0.123	0.14	0.175	0.21	0.245	0.28	0.329	0.35
8	0.04	0.06	0.08	0.10	0.12	0.14	0.16	0.20	0.24	0.28	0.32	0.36	0.40
9	0.045	0.064	0.09	0.113	0.14	0.158	0.18	0.225	0.27	0.315	0.36	0.405	0.45
10	0.05	0.75	0.10	0.125	0.15	0.175	0.20	0.25	0.30	0.35	0.40	0.45	0.50
15	0.075	0.11	0.15	0.188	0.23	0.265	0.30	0.375	0.45	0.53	0.60	0.655	0.75
20	0.10	0.15	0.20	0.25	0.30	0.35	0.40	0.50	0.60	0.70	0.80	0.90	1.0
25	0.125	0.19	0.25	0.31	0.38	0.44	0.50	0.63	0.75	0.88	1.0	1.12	1.25
30	0.150	0.23	0.30	0.376	0.45	0.53	0.60	0.75	0.90	1.05	1.2	1.35	1.50
35	0.175	0.26	0.35	0.409	0.53	0.62	0.70	0.88	1.05	1.24	1.4	1.58	1.75
40	0.20	0.30	0.40	0.50	0.60	0.70	0.80	1.0	1.2	1.4	1.6	1.71	2.0
45	0.225	0.34	0.45	0.56	0.68	0.79	0.90	1.13	1.35	1.58	1.8	2.03	2.25
50	0.250	0.38	0.50	0.62	0.75	0.88	1.0	1.25	1.5	1.76	2.0	2.24	2.50
55	0.275	0.41	0.55	0.68	0.83	0.97	1.1	1.38	1.65	1.94	2.2	2.47	2.75
60	0.30	0.45	0.60	0.75	0.90	1.06	1.2	1.5	1.8	2.12	2.4	2.7	3.0
65	0.325	0.48	0.65	0.81	0.98	1.15	1.3	1.63	1.95	2.30	2.6	2.93	3.25
70	0.350	0.53	0.70	0.82	1.0	1.24	1.4	1.76	2.10	2.48	2.8	3.16	3.50
75	0.357	0.57	0.75	0.94	1.1	1.33	1.5	1.89	2.25	2.66	3.0	3.39	3.75
80	0.40	0.60	0.80	1.0	1.2	1.4	1.6	2.0	2.4	2.8	3.2	3.6	4.0
85	0.425	0.64	0.85	1.06	1.3	1.49	1.7	2.13	2.55	2.98	3.4	3.83	4.25
90	0.45	0.676	0.90	1.12	1.35	1.58	1.8	2.26	2.7	3.16	3.6	4.06	4.5
100	0.50	0.175	1.0	1.25	1.50	1.75	2.0	2.5	3.0	3.5	4.0	4.5	5.0

Source: Richard Untermann, *Grade Easy* (Washington, D.C.: Landscape Architecture Foundation, 1973), p. 75.

Although the natural methods by which rainfall might be removed from a site might also appear obvious, it is important that they be identified, as they combine and become part of the hydrological cycle:

1. Surface runoff is the first and most visible method of removing water from a site. Just as the name implies, the precipitation that is not absorbed into the soil accumulates across the surface of the site and is collected in swales or into subsurface storm sewer systems. Ultimately, all of the runoff in a watershed is combined into freshwater tributaries which find their way to the sea. The development process almost invariably creates additional runoff. As such, some consideration must be given to the direction of the flow as well as the quantity of the impending runoff.
2. Subsurface runoff is the second method by which water is removed from a site. This is the process in which water permeates the surface, is absorbed by the soil, and becomes part of the groundwater supply or an aquifer. Since subsurface runoff is a very important contributor to the quantity of water in both groundwater and aquifers, it is particularly important for the site planner to consider ways in which the runoff may be transferred into subsurface systems. As mentioned in Chapter 10, subsidence, or the settling of the earth's crust, follows the extraction of groundwater at a rate faster than replenishment of the resource. This particular environmental malady is the result of public policies that permit water removal as a priority, without guidelines as to groundwater or aquifer replenishment. Obviously, any effort that is cognizant of the potential environmental problem and facilitates the natural processes should be pursued.
3. Evaporation of water from plants or bodies of surface water (lakes, creeks, etc.) is the third method by which water is removed from a site. The notion of utilizing plant life to retain water molecules on a site

may appear as a rather modest means of removing negligible amounts of water. When the alternatives to be considered are vegetation vis-à-vis an impervious surface, there is no question which contributes more to the elimination of surface water and reinforces the natural systems.

4. Transpiration from vegetation is the fourth method of removing rainfall from a site. Plants and vegetation absorb water in photosynthesis, contributing to the hydrological cycle. As with evaporation, although the amount of water absorbed through transpiration may be negligible, the process is still preferred to an impervious surface. Although the primary method that will be employed to remove rainfall will generally be surface runoff, evaporation and subsurface infiltration should be considered as potential contributors to that process.

In the design of a grading plan for the site, the principal objective of the plan is to collect, transfer, and dispose of surface water.[3] Within that framework, there are two basic methods: open and closed systems. Surface water can be accommodated in low density/minimal coverage situations with "open" surface systems which control the water through the shape and form of the topography. Although more expensive, "closed" systems (subsurface piping) are sometimes necessary due to a site's location, off-site drainage, or land use. In many cases, the most cost-effective method will be one that employs a combination of open and closed systems. Just as in grading, where cut and fill are often part of the same plan, the same applies to drainage. There are sometimes conditions where one method or another will be the only appropriate solution. There is also a set of circumstances in which closed systems would need to be used in spite of the site's appearance. In some places where the topographical relief is too slight or shallow, closed systems are almost mandatory. Recall that there are minimum slopes to accommodate positive drainage. If the natural gradient is less than the criterion, water will pond, creating a nuisance and hazard to health.

Drainage: Considerations and Plan

The grading and drainage process begins with three primary considerations. Determine, first, where the water is coming from; second, where it needs to go; and third, how it traverses the site. Just as the site planner knows that the context of every site mandates an examination of the land beyond the property line, that requirement is never more obligatory than in the case of drainage. An analysis of the site and its context relative to the development of a grading plan should document, as a minimum:

1. The topographical characteristics, identifying high and low points, drainage swales, ridge lines, and the extent of the immediate water shed
2. Any unusual soil types, such as gravelly or sandy soils, which percolate satisfactorily, as well as clay or silty soil, which do not
3. Fixed elevations or points on the site, such as roads, trees, adjacent buildings, or rock outcroppings
4. All areas that need to be kept dry and their corresponding elevations
5. The location and extent of the existing sewer systems

Finally, develop a drawing that illustrates the existing drainage pattern from the areas adjacent to the site, across the site, and beyond. It is imperative to look beyond today. Recognize that undeveloped sites will change and could, in that process, create problems for those who did not anticipate the potential for change. The factors that need to be considered in addressing drainage problems on-site are (1) topography, (2) soil types, (3) vegetation, (4) size of property, (5) land use, and (6) precipitation rates. Land use should give some insight as to the amount of impervious surfaces or open space that would probably occur on a site. A knowledge of the site's soil and vegetation can provide an index of absorption attainable by the natural systems. The size of the area and amount of rainfall are statistical quantities that provide the site planner with the final two parameters necessary for an understanding of drainage and surface runoff. See Chapter 16 for an explanation of storm water management methods and runoff analysis.

No development question can be resolved without an understanding of its economic implications. The financial considerations of a grading/drainage plan need to be understood, as they are a part of the entire building budget. As an example, surface drainage systems which permit water to flow across sloping planes of planted or paved areas are often more cost-effective than a closed system of catch basins and subsurface pipes. On occasion, the rate of runoff can be ameliorated with on-site detention ponds (Fig. 13-22). On-site retention often results in a reduction in the amount and size of the pipes, which can be translated into lower financial costs. Although more environmental objectives can be attained through the use of retention ponds, the site planner must also be aware of the economic implications.

Since erosion control is an integral part of the grading/drainage plan, there are a few necessary precautions that must be taken concerning the plan in the construction process. Water can be a genuine problem and create substantial erosion damage. For that reason, avoid stripping an entire site of its natural vegetation. Some thought should be given to regarding the site in stages, where erosion can be controlled in a more effective way. Finally, all slopes should be replanted immediately following regrading.

Listed here are a few widely accepted guidelines to follow in the development of the grading plan and design of the drainage system:[4]

[3] Ibid., p. 47.

[4] Ibid., p. 48.

FIGURE 13-22. On-site detention pond.

1. New runoff must never be purposefully redirected from its natural course on one property so as to become a nuisance on another property. It is acceptable in most communities to continue a process or system that occurs naturally, but do not increase the flow artificially.
2. Always consider some method to retard the velocity of the water so that it might be absorbed into the soil. Riprap and stone lining are often employed as techniques to dissipate the force of the water in swales or creeks. In reducing the rate of flow, the technique dramatically reduces the scouring action of water and increases the rate of absorption (Fig. 13-23).
3. Design the grading and drainage plan so as to respect, reinforce, and duplicate the existing natural systems.
4. While on occasions there are few alternatives, a drainage plan with more than one outlet course is considered good planning.
5. Avoid draining large paved areas across pedestrian paths. Catch basins and trench drains can be used to collect the substantial quantities of runoff created by parking lots or pedestrian plazas.
6. Identify any areas that appear to be appropriate for drainage structures. Sinks, depressions, or long channels are primary candidates for catch basins or drains.
7. In the design of subsurface systems, begin at the higher elevations of the site and work toward the lower. This method should result in a drainage system that resembles a tree form, with its smaller branches (pipe size) at the ends and higher elevations progressing to larger sizes as it approaches major limbs (mains), or the trunk, at the bottom of the water shed or drainage area. The pipe size should increase as the system collects and transfers the runoff to the lower elevations.

Surface drainage systems are generally preferred to underground systems for two reasons: cost and ecology. Since surface systems are more preferred and gravity is the vehicle, some minimal grades must be respected. Water will move downhill only if there is enough change in elevation to permit that movement. Fig. 12-16 identifies some of those minimum and maximum grades.

Avoid the design of a system that necessitates the location of a drainage line through a foundation or under a slab. The problems associated with a drainage line that ruptures or leaks are the ingredients of an architect's bad dream. One should avoid cutting a hole in a grade beam for a pipe; for similar reasons, one avoids cutting holes in floor beams to accommodate air-conditioning ducts. The process suggests

FIGURE 13-23. Riprap or stone at edge of creek.

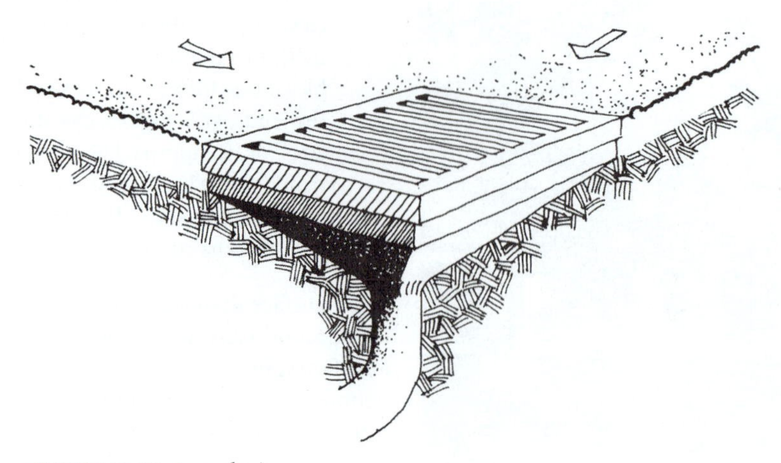

FIGURE 13-24. Area drain.

that some criteria be established which will facilitate objective decisions. As a minimum, those criteria might include:

1. The minimum grades to drain, taking into account the natural and human-made character
2. The maximum grades for cut or fill without stabilization
3. The maximum grades for a cut or fill with stabilization
4. The stabilization methods

Basic Subsurface Systems and Structures. Having established those conditions under which the plan will be developed, some mention of drainage structures and their function is in order. Subsurface systems are designed to resolve the same objectives as those of surface systems: collect, transfer, and dispose. Surface runoff is collected in area drains (Fig. 13-24), catch basins (Fig. 13-25), and trench drains (Fig. 13-26). An area drain should be located at the lowest point in a drainage area. It is conceptually like a big shower drain through which all of the water falling in a specified area passes. "Trench drain" is a term that has been given to any linear drain. This structure is often used at the bottom of a slope where water needs to be collected to protect an adjacent area.

The catch basin is also an area drain, but it has been designed to "catch" debris in its base below the pipe that transfers the water to a point of disposal. When the lid of the catch basin is lifted, the debris and sediment that would otherwise have clogged the drainage system can be removed. Curbs and gutters supplement the area drains and catch basins as methods of collecting surface runoff.

Subsurface collection is often accomplished with perforated pipe set in ditches filled with gravel. The holes are on one side of the pipe and are pointed "up" to catch the water that seeps down through gravel ditches (Fig. 13-27a). A minimum percent slope on the pipe is necessary to provide

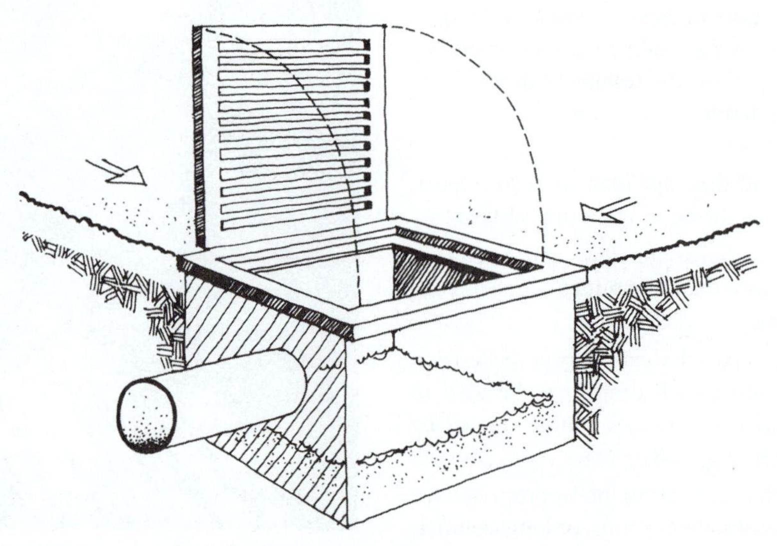

FIGURE 13-25. Catch basin.

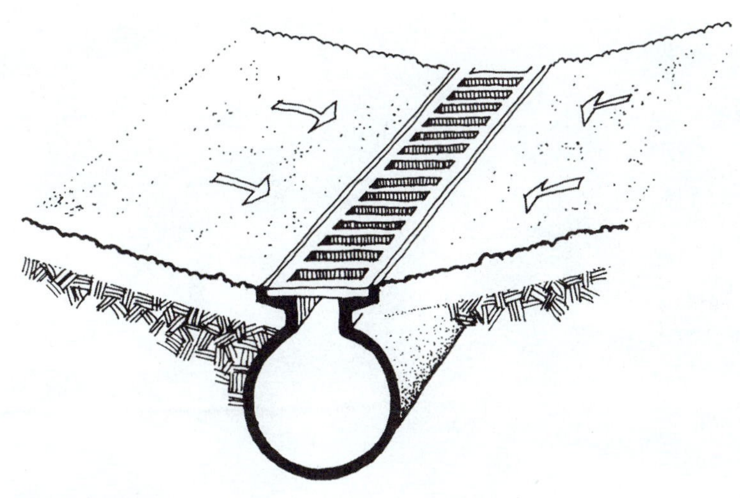

FIGURE 13-26. Trench drain.

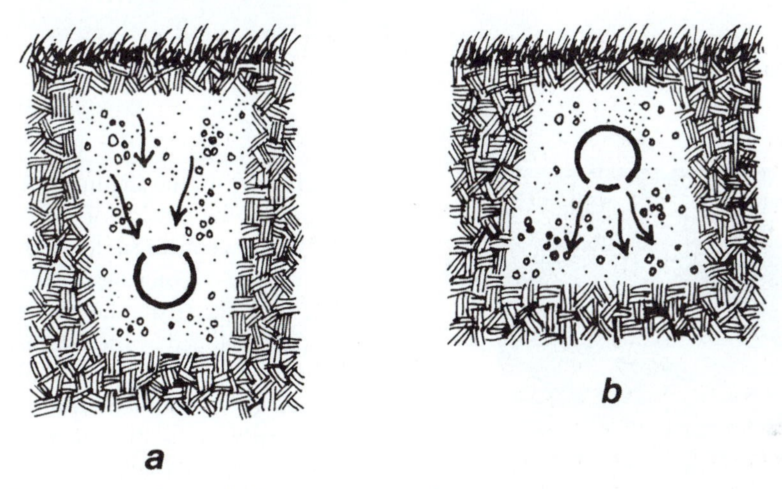

FIGURE 13-27. "French" drain—perforated pipe.

positive drainage. By reversing the orientation of the perforations, the runoff collected can be transferred in the same system and disposed of on-site (Fig. 13-27b). Water runs through the system, draining through the pipe and gravel, recharging the groundwater. The system is environmentally sound and keeps runoff and erosion at a minimum, but it is an expensive way to control on-site runoff. All alternatives have both economic and environmental costs. The problem for the site planner is to develop a grading and drainage plan that respects a site's unique ecology while minimizing the obligations of capital costs.

GRADING FOR STREETS AND ROADS

Grading for a road or any circulation system relies on the basic concepts discussed in Chapter 12. The road appears to be a more complex grading problem because it has a constant slope and is a continuous system. The function of roads is such that they serve first as an auto path and second as an adjunct to a drainage system. A road must be designed in conformance with strict design standards, maintaining appropriate grades, curvatures, and sight distances. Since most communities have established street and road design criteria, these important variables will be discussed only conceptually.

As the character of natural topography is such that it is almost never level, for safety purposes, roads should be designed to minimize the fluctuations across variable terrain. Keep in mind that neither the road nor the pedestrian path is really level: both are designed to permit movement on constant slope or grade (Figs. 13-28 and 13-29). Both systems simply require design considerations that will permit travel horizontally and vertically. One common and sensible objective for any system is the design of a route that is as short and direct as possible. Accepting that as a worthy objective must be balanced against the implications of that decision. Although the shortest route may have certain desirable advantages, sometimes the change in the natural topography is so great that a road alignment of the "shortest distance"

FIGURE 13-28. Road on slope.

FIGURE 13-29. Pedestrian path on slope.

approach could result in extensive cutting (Fig. 13-30). A road designed in an alignment that is at or nearly perpendicular to the topographical fall is described as "bucking the grade." The road, like any other site "improvement," must be designed to minimize the cut and fill necessary, while responding to prescribed design criteria. For that reason, unless the natural topographical slope is very shallow and requires only minimal grading, a common design guide is to align the road's centerline with or parallel to the contour lines (Fig. 13-31). This method can keep large cut or fill areas to a minimum while the alignment changes vertically. In circumstances in which the technique may demand "switchbacks" in the alignment, care must be taken to make certain that the outside of the curve is elevated to accommodate the centrifugal force of the vehicle through the curve (Fig. 13-32, on page 144). The common crown section would only exacerbate the problem created by the centrifugal force developed by the vehicle moving through the turn.

Because of a road's section and length, the grading problem for roads often creates genuine consternation for the beginner. Some thought must be given to considering the design of the road, first, as the function of the road's centerline. If the road is initially thought of as a centerline, the decisions associated with its slope, curvature, or connections may be simpler to achieve. By reducing the road to a line, the question of cut or fill can also be evaluated in a very straightforward way. If the objective is to balance the cut and fill, the point at which the centerline of the road and the contour line cross simply becomes the center of the road (Fig. 13-33, on page 144). The "high" side to the centerline is a cut, and the "low" side is a fill.

If the topography is too steep for a successful fill, the point at which the outer edge of the road crosses the contour is used as the control point on the low side (Fig. 13-34, on page 144). The contour line is then adjusted to reflect the crown of the road and the extent of the cut. If the road must be graded on fill, the point at which the high side of the road crosses the contour line becomes the control point (Fig. 13-35, on page 145). The contour line is adjusted accordingly to meet the grade of the existing topography.

Most often, "What do I do now" problems occur when the road alignment or configuration does not match the examples in the text. Three problems are rather common and need to be mentioned: (1) the road crosses a series of contours; (2) the road crosses a drainage swale; and (3) the road crosses a steep slope.

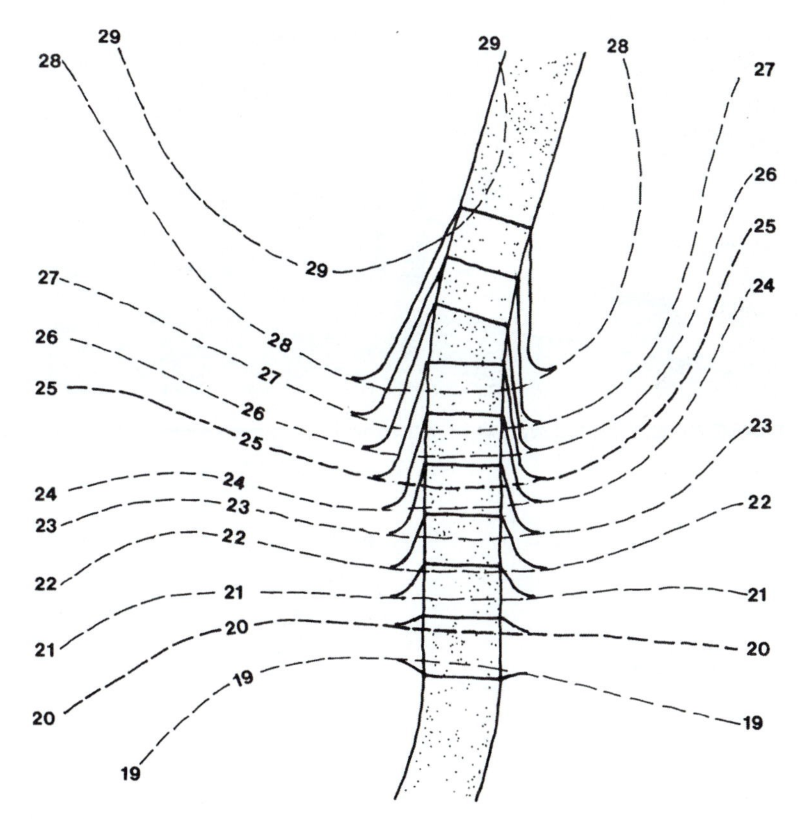

FIGURE 13-30. Plan of road "bucking the grade."

In the design of the road across a series of contours, the site planner needs to keep in mind the criteria established for the road. One of those criteria reflects the need for a constant slope (Fig. 13-36, on page 145). This is critical to its design, as a road that might continuously rise or fall with the natural contour could create problems with sight distances and intersections. So, to respect the uniform slope design criteria, the decisions concerning the road's alignment

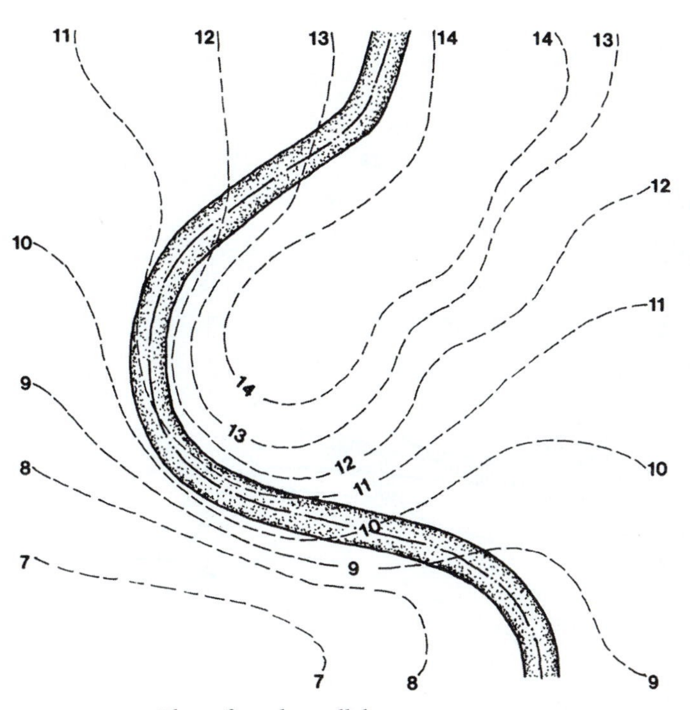

FIGURE 13-31. Plan of road parallel to contours.

FIGURE 13-32. Elevated section of switchback.

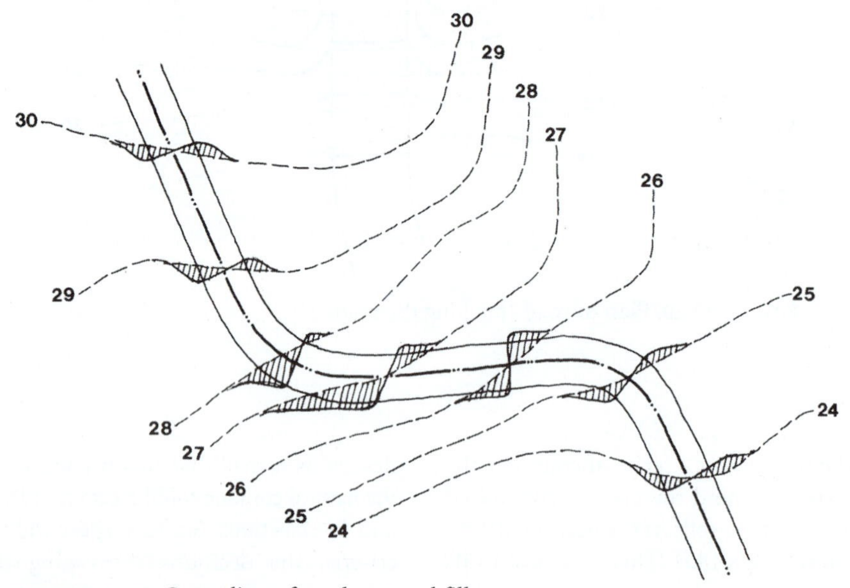

FIGURE 13-33. Centerline of road: cut and fill.

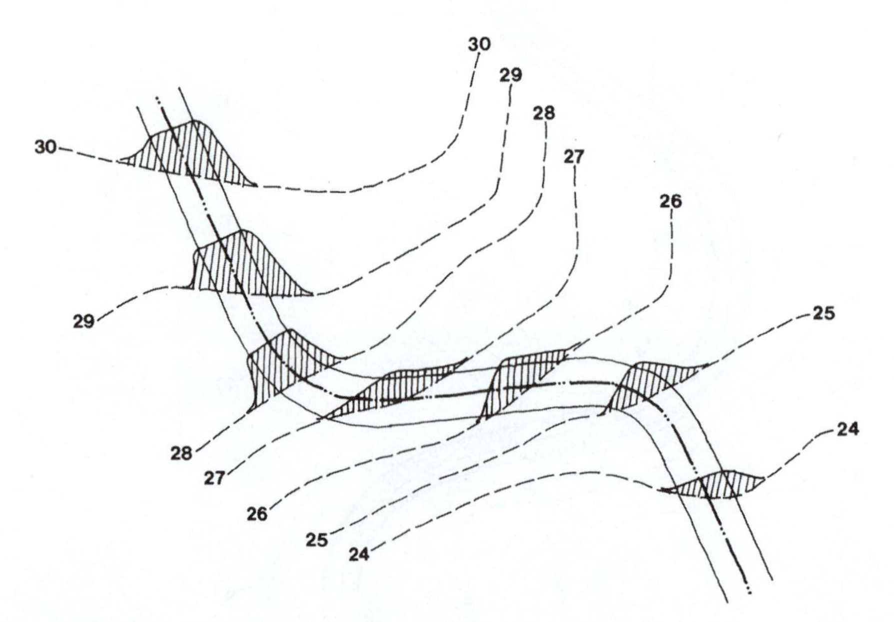

FIGURE 13-34. Centerline of road: cut only.

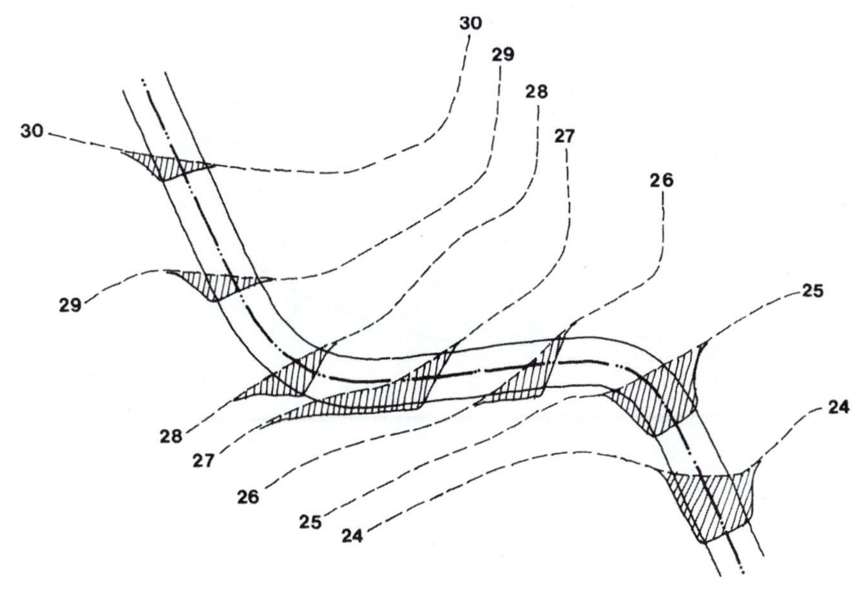

FIGURE 13-35. Centerline of road: fill only.

would begin as they did earlier—with the road's centerline, the slope, and the first contour. The criteria determine the specific elevations of the road's centerline and, as such, the locations of each new contour line in the series (Fig. 13-37).

The second problem mentioned considers those circumstances in which the road crosses a drainage swale and the question concerning the intersection between the road and the topographical depression. Depending on the length and depth of the swale, the road may necessitate either a culvert or a bridge. Modest topographical changes that accommodate minor amounts of water can be resolved with a culvert (Fig. 13-38). If substantial amounts of water are visible on the site in a natural water course, a bridge may be necessary. Additional riprap to reduce the erosion probability in a water course could also be an important adjunct to the design. These are considerations to support an enlightened response to the role of both creek and road and to separate their functions (Fig. 13-39, on page 147).

The third problem involves the encounter between the road and radical topographical slope. Even though the road's centerline is "running" with the contour lines, the road's section across the slope can sometimes create problems.

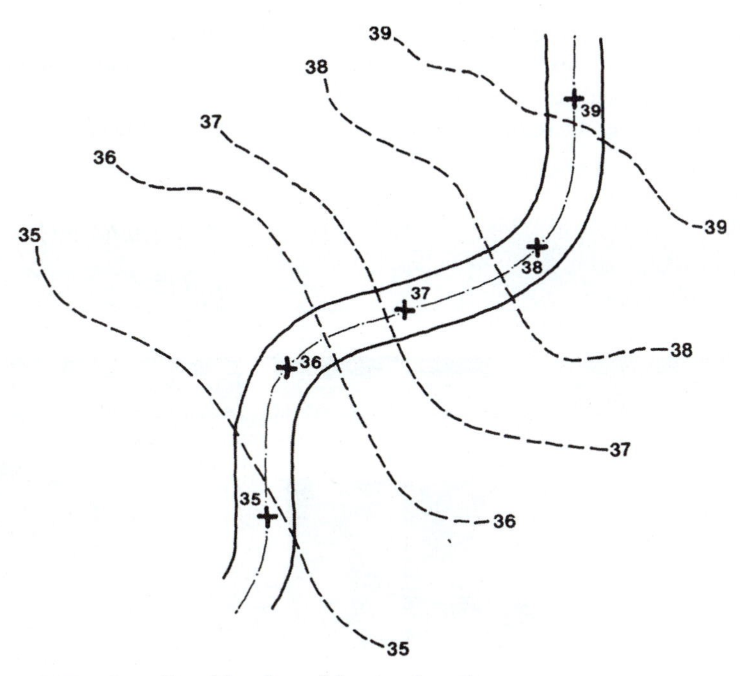

FIGURE 13-36. Road in plan with spot elevations.

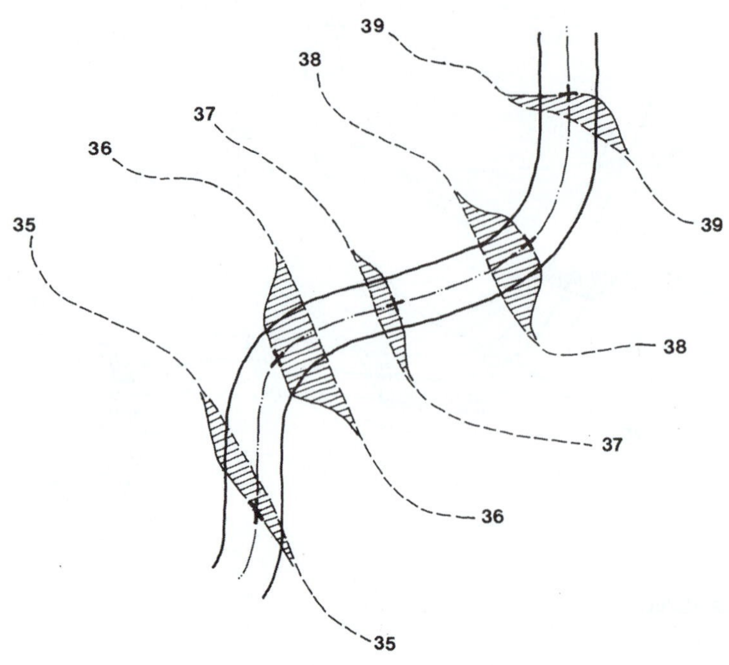

FIGURE 13-37. Road in plan with new contour lines.

One alternative is a bifurcated section (Fig. 13-20). The technique can be employed where the number of moving lanes is four or more (two each way). It would be inappropriate to separate a two-lane street and create a situation where a stalled vehicle in either lane would render a section of the road impassable. The other alternative is to consider the road's section in typical cut and/or fill terms. Although a cut is more stable than a fill, caution should be exercised, as heavy rainfall can induce mud slides and create disastrous problems. As mentioned earlier, fill is more unstable than cut, and soils with any history of bonding problems should be avoided. Often in these cases, the only way to resolve the situation is with a retaining wall on the high side of the slope (Fig. 13-40). Although it is not an inexpensive device, the retaining wall is a very positive method of resolving a situation that is potentially dangerous. In resolving those situations, there is an old canard in offices that bears repeating: "Why is it that we always find the time and money to do it right the second time?"

The road also serves as a drainage system and there are some basic configurations with which the site planner needs to be familiar. First, the typical cross section (Fig. 13-41) has a "crown" in the center, with the curb and gutter at the edges. The high crown forces the water to the edges and keeps standing water out and away from the path of moving vehicles. This is a primary criterion for the design of any street or thoroughfare. If water is trapped and left to stand on a road, it can induce hydroplaning in auto/truck tires, which reduces the basic adhesion necessary for control of a vehicle. Although a crown of 6 in. is common, always check with the appropriate governmental authority, as their design standards will vary to reflect local climate and environmental conditions.

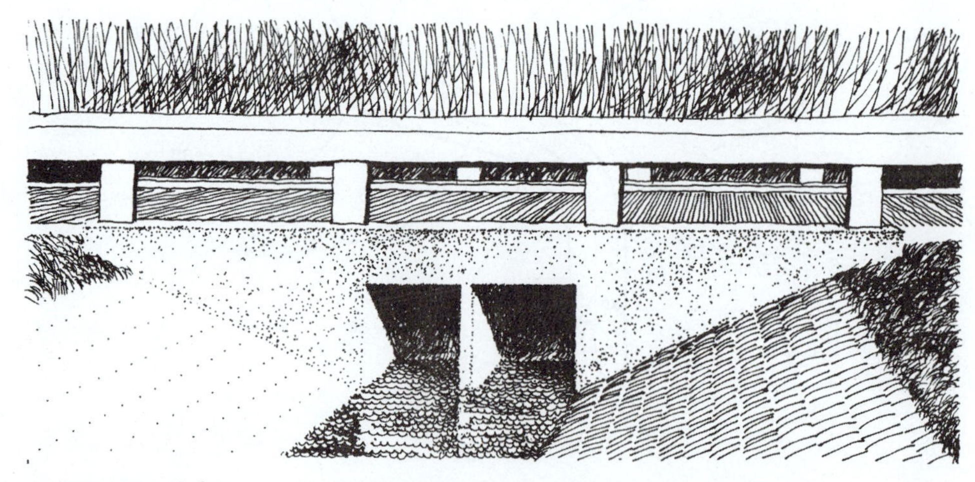

FIGURE 13-38. Culvert.

FIGURE 13-39. Small vehicular bridge over creek.

FIGURE 13-40. Retaining wall on high side of bifurcated road.

On occasions, the road may be sloped from side to side (Fig. 13-42). This is sometimes necessary to accommodate a natural topographical slope where use of the typical crown section cannot produce the necessary positive drainage. Under some circumstances, the road may be designed in a concave shape, eliminating all together the need for a curb. The example in Fig. 13-43 is an illustration of a private road in a small (60-dwelling-unit) subdivision. There are two important variables associated with this design. The first is the shape and the absence of the traditional curb. The

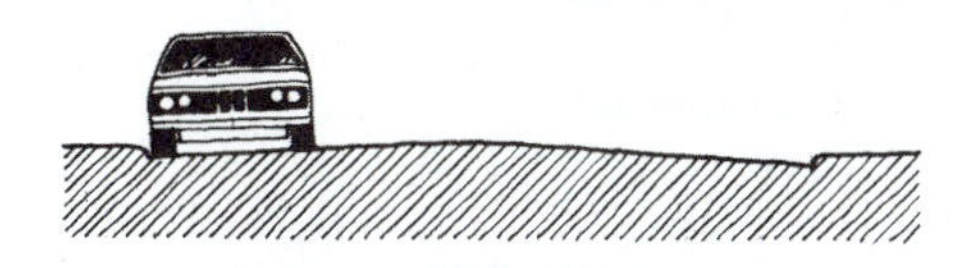

Crown Section

FIGURE 13-41. Section through road with crown.

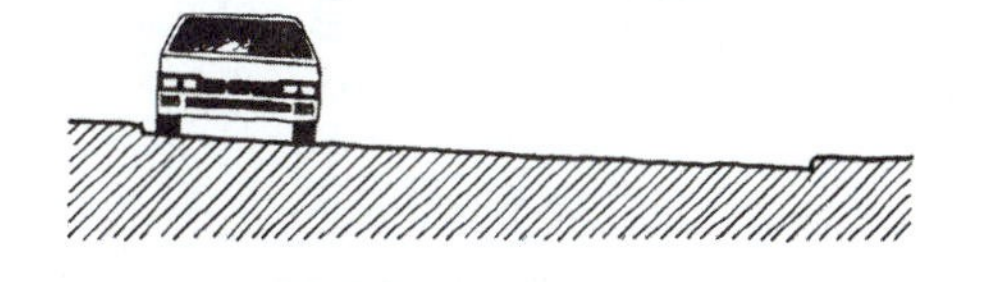

Sloping Section

FIGURE 13-42. Road sloped from side to side.

FIGURE 13-43. Road with concave section.

second is the paving material. This material has an "open" joint and, as such, accepts the runoff from adjoining lots and paving. The total runoff from the development and the street to the sewer is significantly smaller than that of concrete or asphalt.

It must be recognized that this particular solution is uniquely appropriate to the conditions associated with this site. Although the concave section contains and controls surface runoff for this community, it is successful because the entire site, including roads and paving material, was considered as an integrated system.

The negative aspects of this example required that the sewer be located under the paving section. Under most conditions today, utilities are located in either the public right-of-way or easements at the edge of the paving section. Fig. 13-44 depicts a utility location which recognizes the fact that utility systems often need both repair and expansion. As such, a separate right-of-way is a more cost-effective alternative to the traditional thinking, which required removing the paving to gain access to the utilities.

ESTIMATING CUT AND FILL

It is helpful to have a general understanding of the cumulative volume of cut and fill required for a finished grading plan. An imbalance will require either the disposal or acquisition of fill, and both of those actions have cost implications. While a resulting imbalance is sometimes the only way a grading plan can accomplish other design goals, knowing the costs keeps the site planner ahead of the curve.

While there are computer programs that can calculate the final cut and fill requirements, the quick sketch methods will provide a picture sufficient for comparison between alternative grading plans. One of the most common methods is the grid method. This strategy is very simple when the site has a constant moderate slope and the site itself is a relatively simple polygon.

Begin by sketching an orthogonal grid overlay of the site (Fig. 13-45). There are two criteria to consider in establishing the interval between grid lines. First, the grid interval should reflect the size of the site. For example, if the site

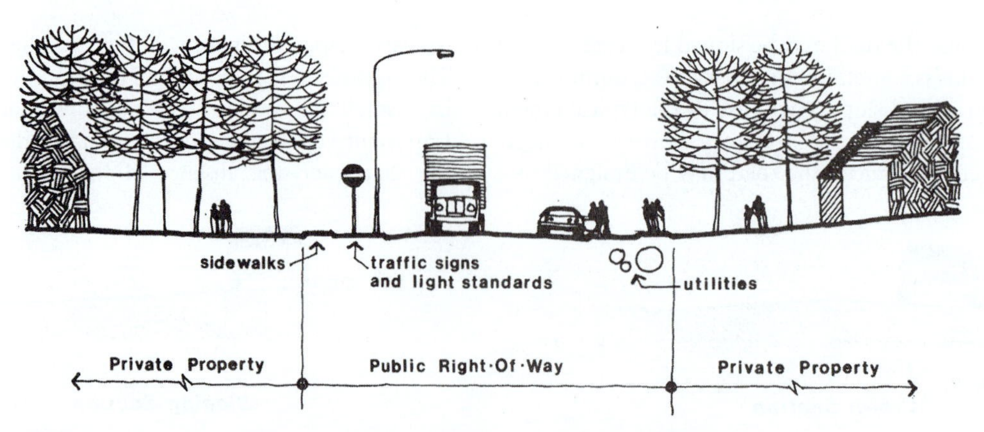

FIGURE 13-44. Section through road, illustrating location of utilities.

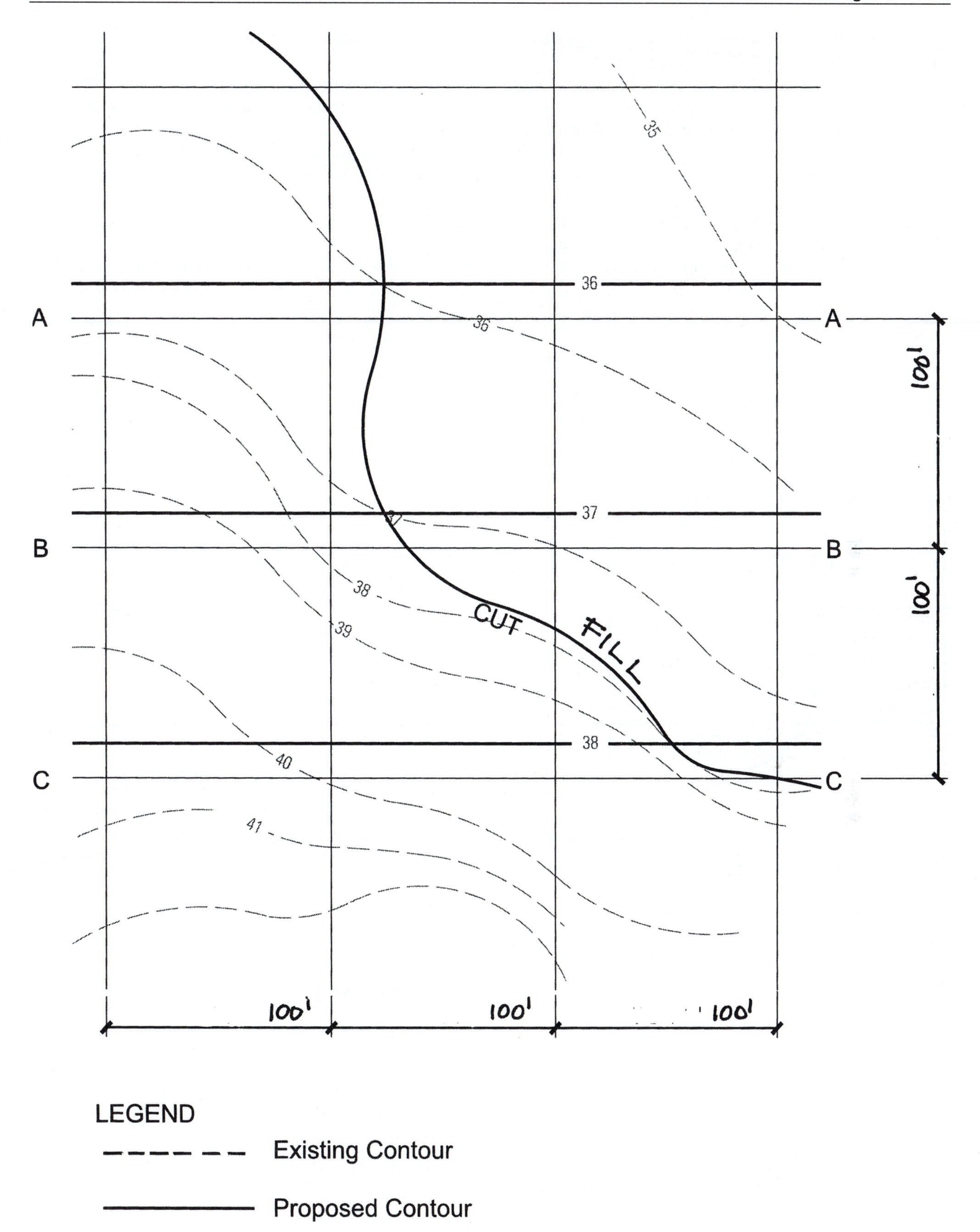

FIGURE 13-45. Project site showing 100' grid.

is a 60 × 150-ft residential housing lot, a 20-ft-square grid would be appropriate. If the site is 500 ft × 800 ft (10 acres +/-), a 100-ft-square grid would be more appropriate. Next, if the site has high topographical relief, the distance between grid lines should be shorter. Conversely, if the site has only moderate relief, the grid interval, or cell size, can be larger. Some trial and error will need to take place before you arrive at what you consider the proper grid to be.

Second, place the grid overlay created over the proposed grading plan and estimate the final spot elevations of each x/y coordinate on the grid (Figs. 13-46 and 13-47).

Third, determine the differences between the spot elevations of the existing grades and the final spot elevations at each x/y coordinate on the grid.

Fourth, calculate the amount of either cut or fill required for each cell in the grid. At this step of the process, look to see if your original assumptions regarding the results of the grading plan coincide with the analysis. It is not necessary to complete the process if it is evident that changes need to be made in the proposed grading plan (Fig. 13-48, on page 152).

After all of the interim adjustments have been made to the final grading plan elevations, calculate the total amount of cut or fill necessary to achieve the desired plan. Understand that this technique is used as a method to preliminarily determine the magnitude of the resulting cut and fill imbalances of the final grading plan. There are other methods for calculating cut and fill, but none are less tedious or more simple than the grid method.

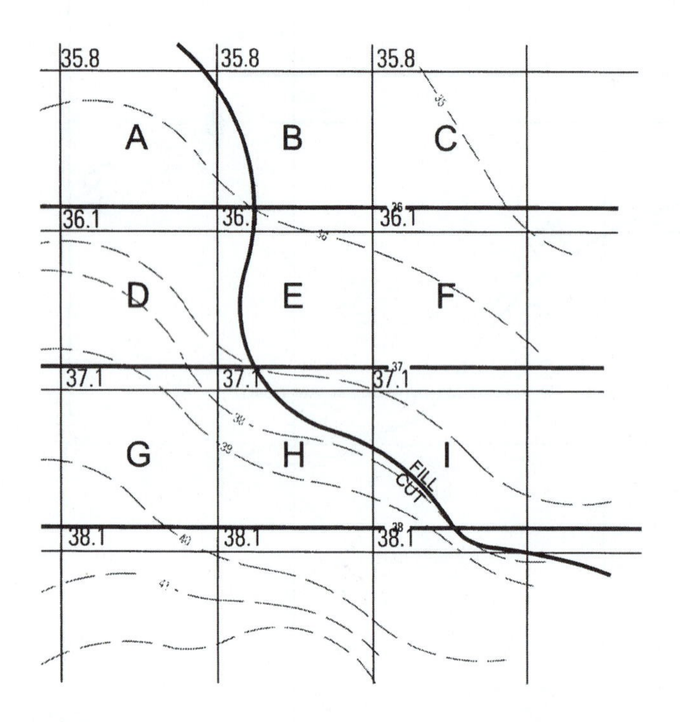

Proposed Elevations at Grid Points

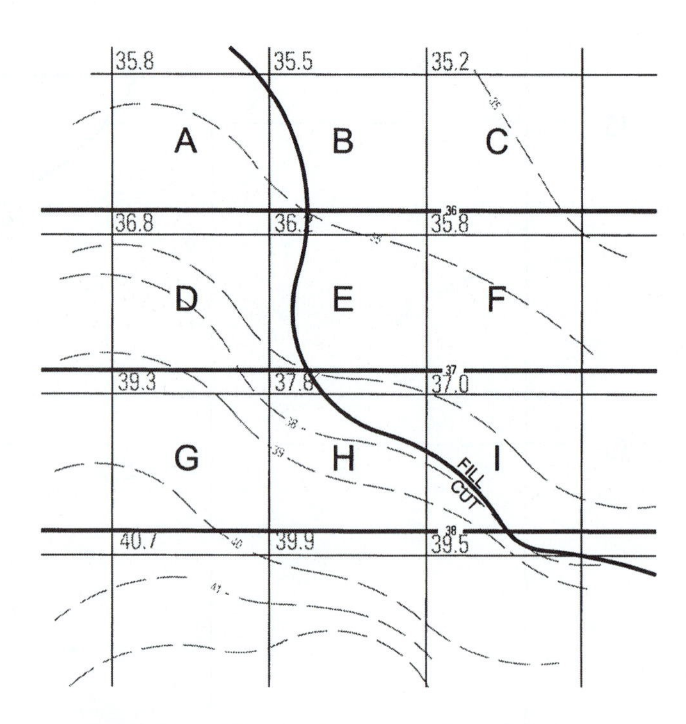

Existing Elevations at Grid Points

LEGEND

Existing Contour

Proposed Contour

38.1 Grid Line with Elevation

FIGURE 13-46. Elevations at grid points.

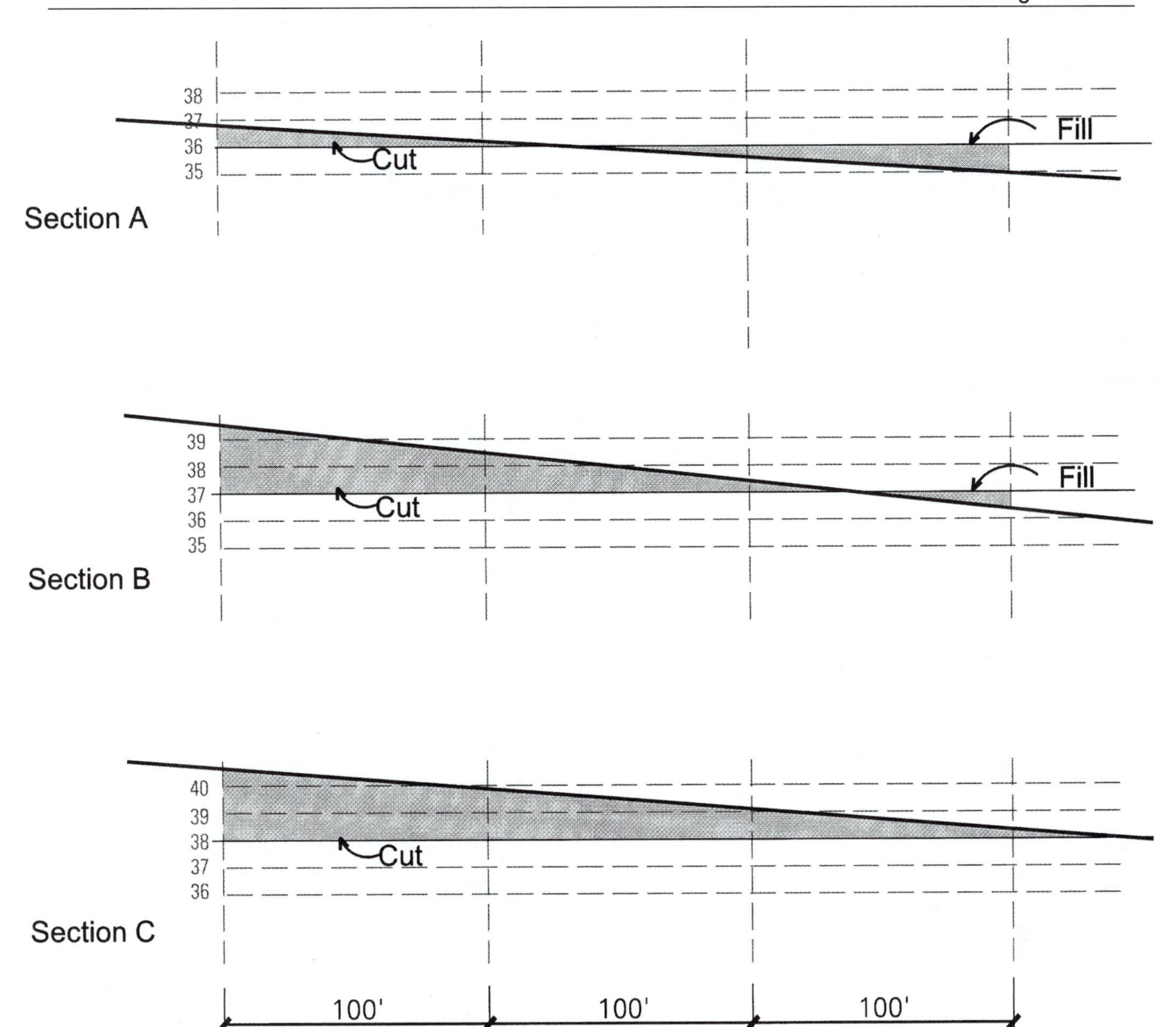

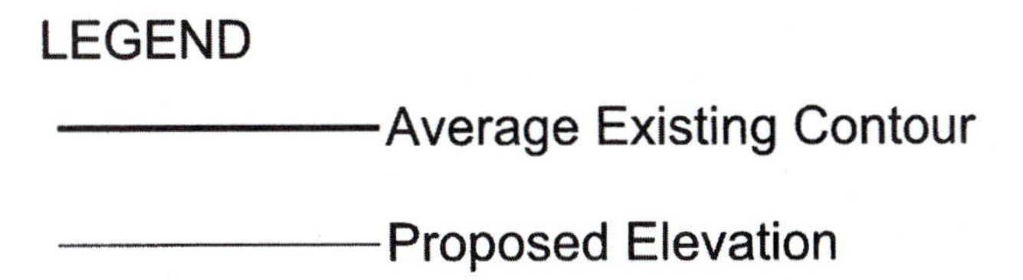

FIGURE 13-47. Project cross sections.

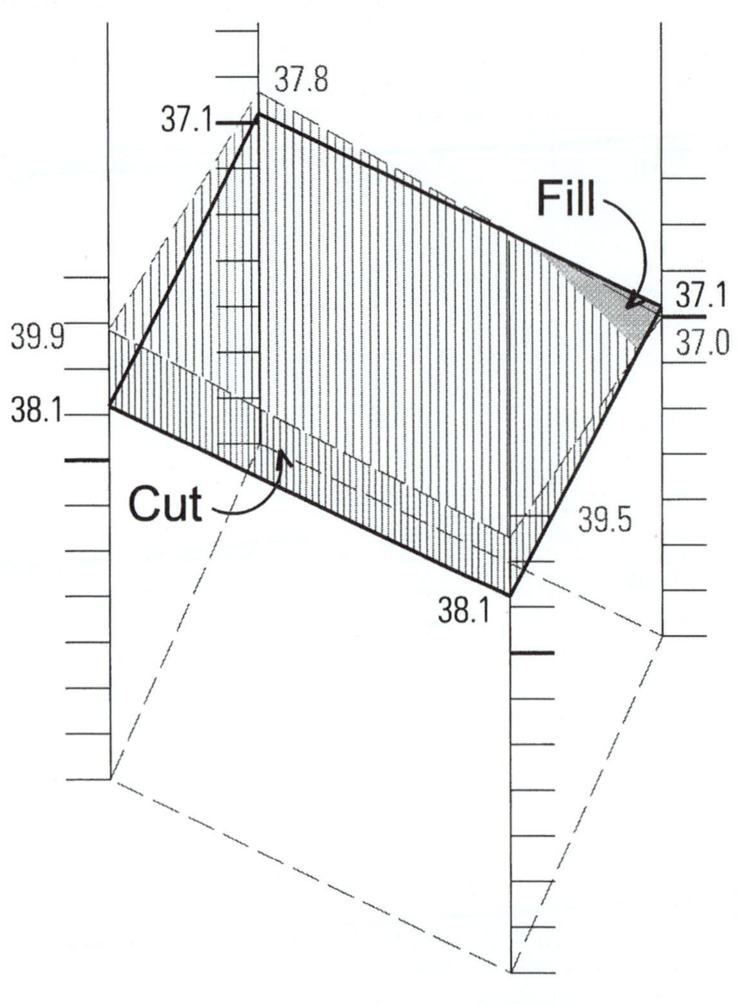

Grid 'H' Diagram

Grid 'H' is 100' x 100'

Average existing Elevation is
(37.8 + 37.0 +39.5 + 39.9) / 4 = 38.5

Average Proposed Elevation is
(37.1 + 37.1 + 38.1 + 38.1) / 4 = 37.6

38.5' - 37.6' = 0.9' Average elevation Change

333.3 cy Cut

LEGEND

37.0 Existing Contour

38.1 Proposed Contour

Grid Line

FIGURE 13-48. Cut and fill diagram.

CHAPTER FOURTEEN

RETAINING WALLS

by Ernest Buckley

STRUCTURAL DESIGN REVIEW

Significant differences in finished surface elevations are frequently a part of designs, sometimes as an unavoidable consequence of the topography and sometimes as a means of dramatizing features for purely aesthetic reasons. In all cases, retaining walls must be properly designed to resist the lateral loads imposed by soil over the long term. All too often, too little attention is given to the structural requirements and the soil's characteristics, and retaining wall failures are the result.

Properly designed and constructed retaining walls are relatively expensive; therefore, their use should be avoided whenever possible. Differences in elevation can be accommodated more inexpensively by graded slopes. Slopes are, however, contingent on a soil's stability and thus can become subject to erosion. The expense of retaining walls is frequently justified to accommodate abrupt changes of grade and reduce the potential for erosion.

Where a change of grade must be made vertically, the retaining wall is, in effect, a dam. Although soil normally appears to be a hard, stiff, and stable material, it is important to recognize that under some conditions of saturation or near saturation the soil mass will slide, slip, or flow. The soil can act as a viscous fluid, and the retaining wall must be able to prevent the sliding or flowing action under all foreseeable conditions.

The retaining wall design is, in some respects, analogous to a building foundation (see Fig. 14-1). Basement walls restrain the earth outside by lateral reactions produced by the footing and by the structural elements at the ground level. Free-standing walls are designed to be stiff enough to transfer lateral loads to the reaction at the footing. In either case, the design of the wall and footing must take into account active earth pressure and, where it

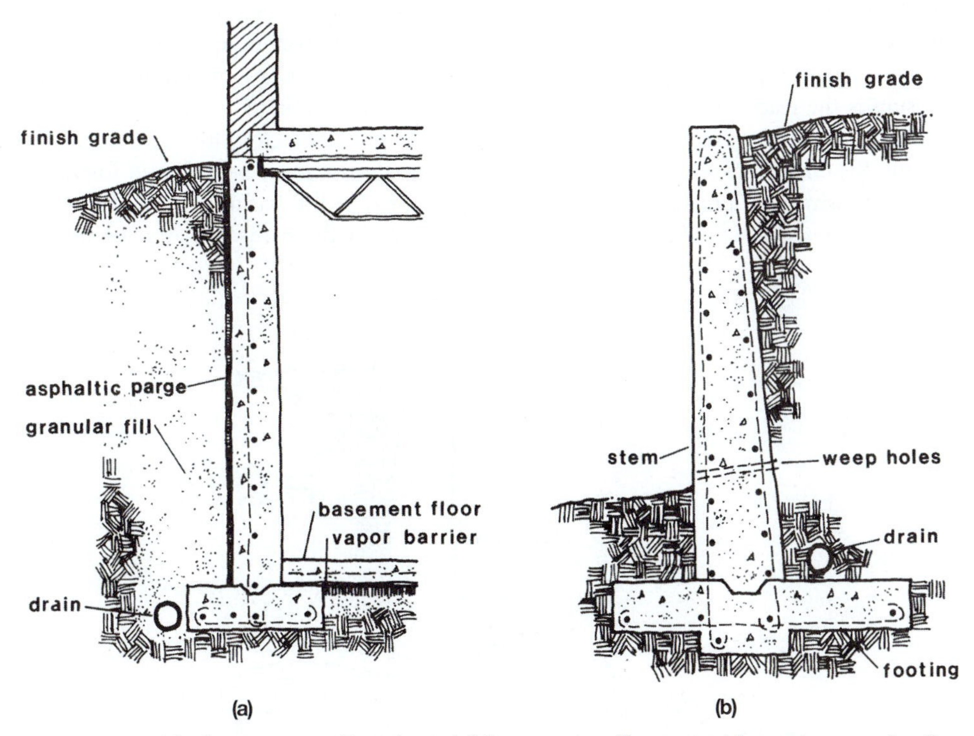

FIGURE 14-1. The basement wall analogy. (a) Basement walls must withstand external soil pressure by reaction at floor and footing levels. (b) Free-standing walls react by cantilever action about the footing.

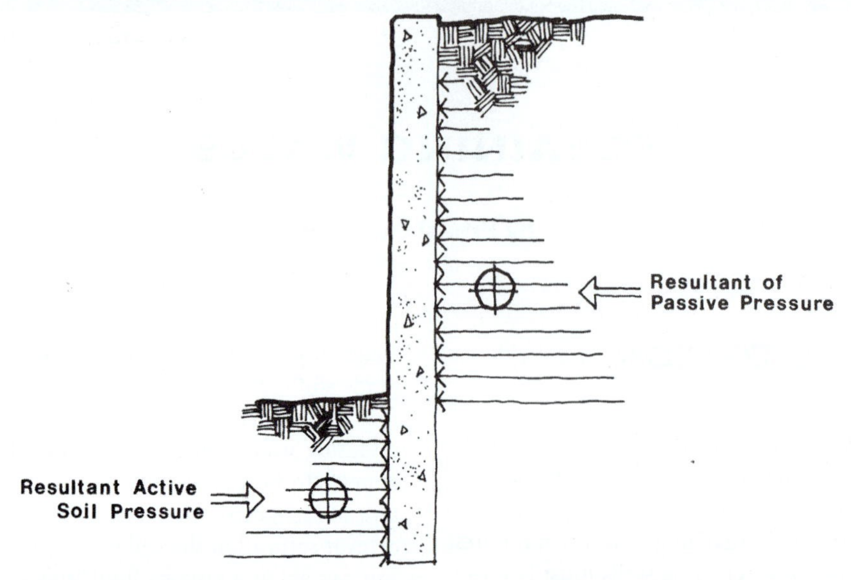

FIGURE 14-2. Passive and active earth pressures. (a) Passive soil pressure is the resistance of earth to displacement and can be taken advantage of in retaining wall design. (b) Active soil pressure is assumed to be analogous to fluid pressure. Equivalent fluid pressure depends on soil density, depth, and internal friction.

exists, passive earth pressure. Fig. 14-2 provides a further explanation.

Active earth pressure exerts horizontal force against any structural element that prevents the soil from sliding or creeping (flowing) into a state of natural equilibrium. The pressure varies directly and linearly with the vertical difference in elevation across the retaining wall. The type of soil and the soil moisture (degree of saturation) also affect the magnitude of the active soil pressure.

Passive soil pressure is the reaction of the soil itself to applied lateral forces. This reaction is related to the compressive strength of the soil. Retaining walls can be designed to take advantage of the passive soil pressure reaction by extending deep into the soil. Resistance to sliding is increased by the deeper wall, but usually, the effect should be neglected in the calculation of the overturning moments produced by the active loads. In other words, the footing of the retaining wall must be wide enough and strong enough to resist the tendency to overturn. The footing should be proportional, so as to provide a factor of safety against overturning of 2.0 and against sliding of 1.5 (see Fig. 14-3).

Retaining wall design begins with consideration of the *in situ* soil conditions, the limits of excavation required, and the type of backfill to be used. In many cases, the soils at the site are not satisfactory for backfill and, instead, select granular materials must be acquired for this purpose. It is important that the backfill material be permeable to permit drainage. A drain tile line may be installed to ensure that

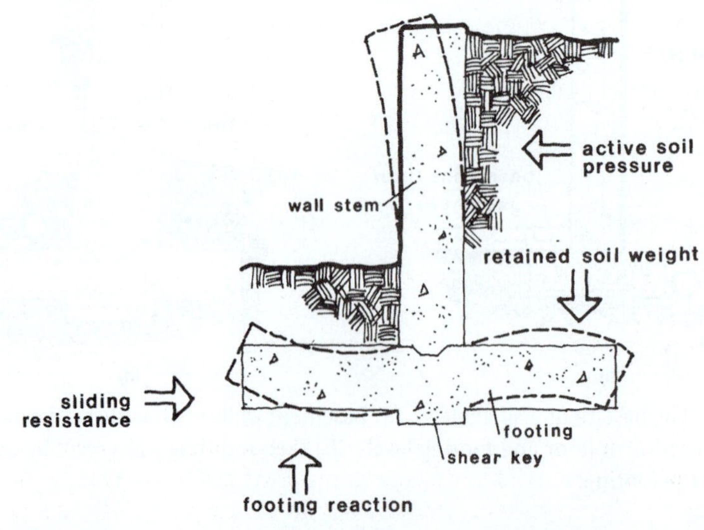

FIGURE 14-3. Loads, reactions, and deformations of a retaining wall.

water is not permitted to build up to saturate the backfill. As an alternative or in addition to the tile drain, weep holes may be installed to penetrate the wall at 4 to 8 ft on centers.

The best retaining walls and footings are constructed of reinforced concrete. For the purpose of design, a segment of the wall 1 ft in length is usually isolated for the analysis of loads, shears, and moments, first on the vertical wall stem and then on the footing segment. For cantilevered wall, obviously, since shear and moment approach zero at the surface, a thicker wall and more steel is required at the top of the footing than at the top of the wall. Tall walls are, therefore, often built with a batter (slope) to the surface of the wall against the backfill (see Fig. 14-4). To increase the reactive capacity of the footing to lateral sliding, the footing for tall walls (8 ft or more) often incorporates a shear key. Low walls (less than 8 ft) are usually of constant thickness, and shear key on the bottom of the footing is usually omitted.

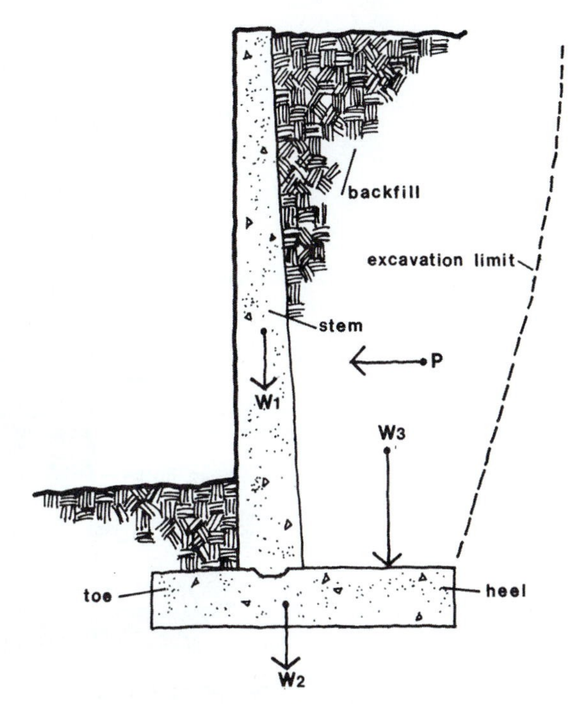

FIGURE 14-4. Typical reinforced concrete retaining wall. Loads on the wall include its own weight (W1), the weight of the footing (W2), the weight of the backfill on the heel of the footing (W3), and the linearly varying active soil pressure (P). For stability, the resultant of the reaction must be in the middle third of the footing.

Materials other than poured-in-place reinforced concrete can be used to construct retaining walls that will perform satisfactorily. Brick, concrete masonry units (block), or stone masonry can be used. Treated timber or railroad ties can also be considered. Whatever material system is selected, the structural response must be of the same capacity as the equivalent reinforced concrete retaining wall.

As construction costs escalate, it is not surprising that various alternatives are considered. Sometimes, aesthetic factors dictate the selection of alternative materials. As a general rule, however, it is usually the case that reinforced concrete will be the least costly. The inherent material system characteristics of the alternatives, if properly accounted for in the design, will ordinarily result in higher cost.

All forms of masonry retaining walls must be reinforced. Brick and stone masonry must be a minimum of two wythes. All masonry wall stems must be designed to be supported by and to act integrally with an adequate footing. Brick masonry design should be in accordance with Brick Industry Association (BIA) Technical Notes 17E and F. Fig. 14-5 illustrates the principal considerations relative to stone masonry construction.

Concrete block walls, or concrete masonry units (CMUs), can be designed for single-wythe construction if careful

Masonry Retaining Walls

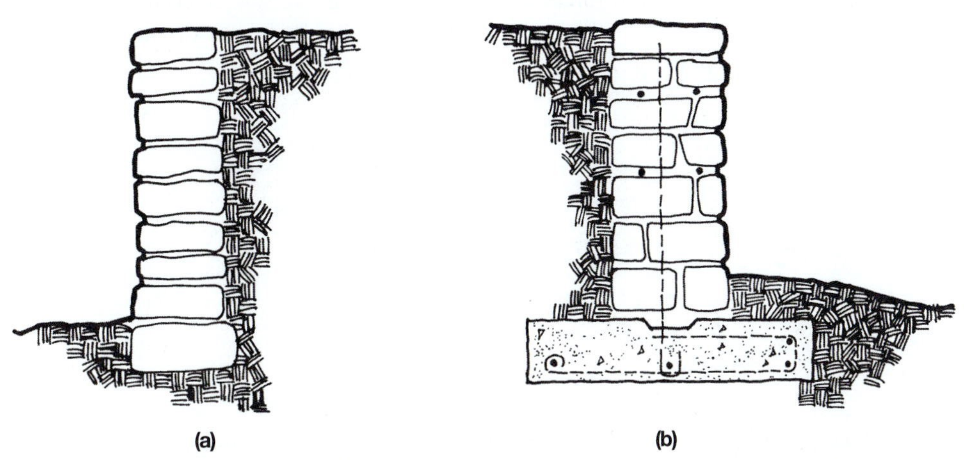

FIGURE 14-5. Stone masonry wall construction. (a) Unreinforced, single wythe of stone masonry is obviously *unstable*. An adequate footing is critical to adequate performance. (b) Double wythe of stone masonry, reinforced, and on a properly designed footing, will perform adequately.

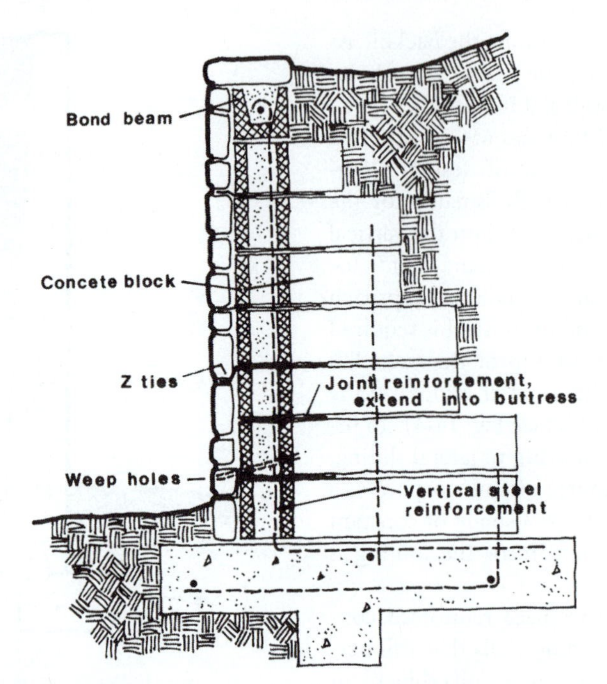

FIGURE 14-6. Concrete block retaining wall.

detailing is provided. An example is illustrated in Fig. 14-6. Concrete block is best used as a backup for a stone or brick veneer, as shown. Horizontal joint reinforcement and vertical reinforcement in grouted cores are essential. Buttresses, extending back into the retained soil, will contribute to wall stem stability. A properly designed reinforced concrete footing is required.

Alignment of a wall is also important. Curvature of a masonry wall, if concave outward, can add appreciably to stability. Lateral loads produced by soil induce compression stresses that can readily be accommodated by the masonry materials. However, a curved wall, concave inward, subjects the masonry material to tension and the likelihood of failure is enhanced (Fig. 14-7). Similarly, masonry tree wells are effective, since the wall is placed on compression by the external active pressure. Tree pedestals must be heavily reinforced to resist the tension forces developed (see Fig. 14-8).

Although the need for economy is recognized and appreciated, one cannot help but be dismayed at the apparent boundless optimism of those who install some retaining wall configurations. Examples are seen of reinforced masonry with no footing at all. Stone masonry of single-wythe construction, on roughly constructed footings, are all too common.

In recent years, the quest for economy, supported by ill-considered design, has led to widespread use of salvaged railroad ties as a retaining wall material system. For a rustic effect and for very low applications, this innovation may be justified. In most cases, however, the railroad ties represent the ultimate in inadequacy. Serious failures occur more often than not (Fig. 14-9).

The most common use of railroad ties, as shown by Fig. 14-10, is as a surface treatment for a slope. The wood members are simply stacked against the slope, spiked together, and expected to prevent erosion. There is no struc-

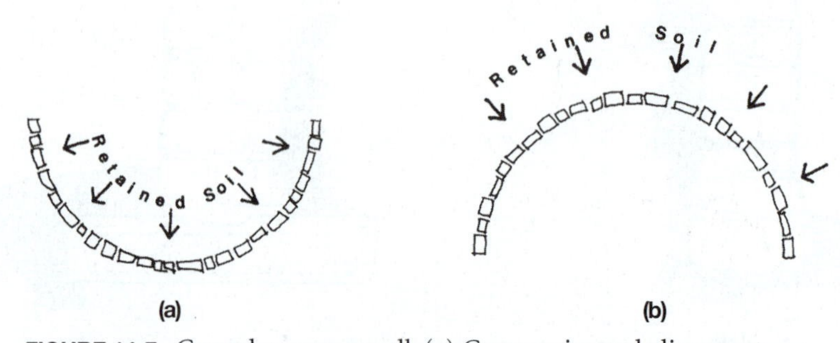

FIGURE 14-7. Curved masonry wall. (a) Concave inward alignment produces internal tensile stress. Reinforcement is required. (b) Arch effect adds stability to masonry wall when concaved outward, since internal stress is compression.

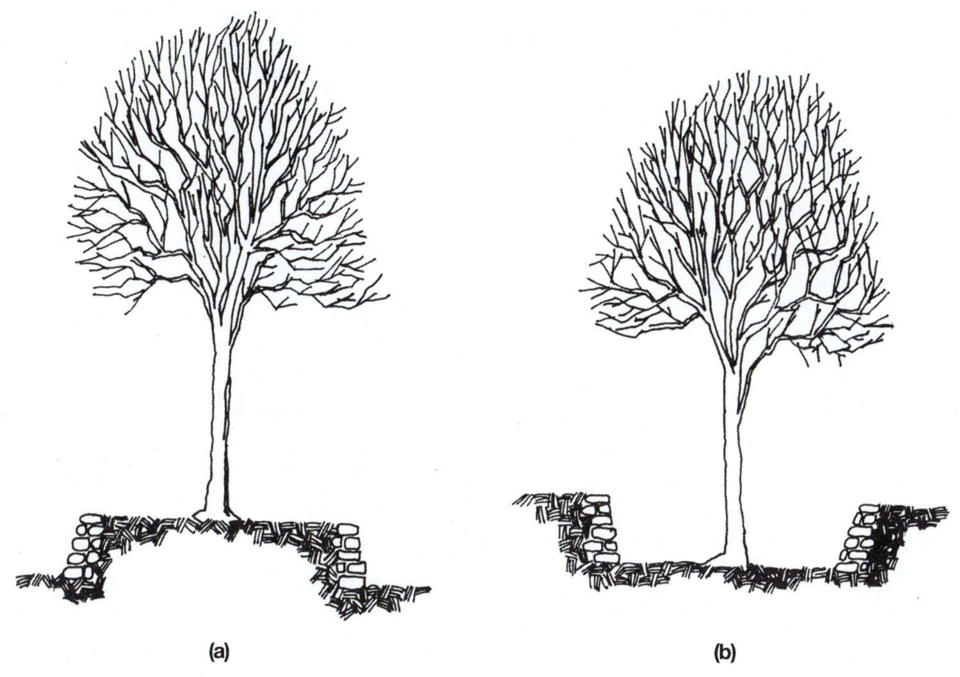

FIGURE 14-8. Tree pedestal and tree wall. (a) Tension stress will result in failure unless resisted by steel reinforcement. (b) Compressive stress developed contributes to stability.

tural integrity as far as resisting active lateral loads is concerned. The system is similar in concept to stone riprap, and it is relatively ineffective simply because of the relatively low density of the wood material as compared to stone. Lateral sliding or fluid movement is uninhibited. Erosion behind the retaining wall members also occurs, even in the case of granular stable soils. Vertical walls formed of railroad ties can be used for very low walls. Even in these cases, it is difficult to maintain alignment, and loss of soil by erosion can be expected. Stacking of more than three units in height should be avoided.

High walls of wooden ties are often seen with an occasional tie installed in a "deadman" orientation. The deadman extends back into the fill material behind the wall. Apparently, the designer believes that stability can thus be gained. Actually, the deadman (anchor) is almost completely ineffective. The bond stress between the soil and the surface of the anchor tie (deadman) is very low, and relatively small forces can pull out the anchor. The shear surface on which the retained soil slides or flows is, more often than not, well beyond the length of the anchor. The configuration shown

FIGURE 14-9. Railroad-tie wall failing.

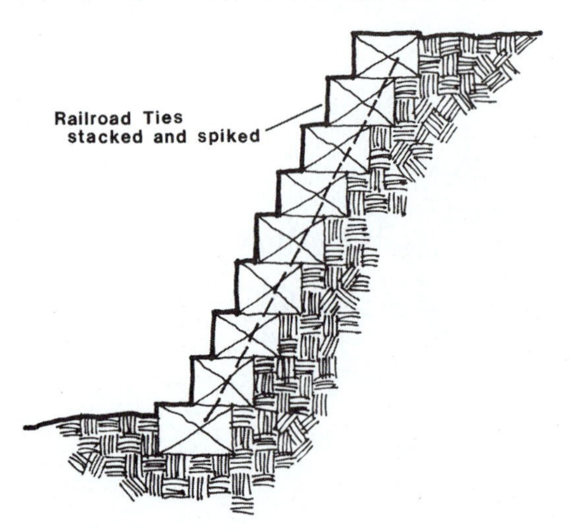

FIGURE 14-10. Treated-timber retaining wall. Actually, this typical configuration is not a retaining wall at all. It is an attempt to stabilize a slope by surface treatment, analogous to using stone riprap.

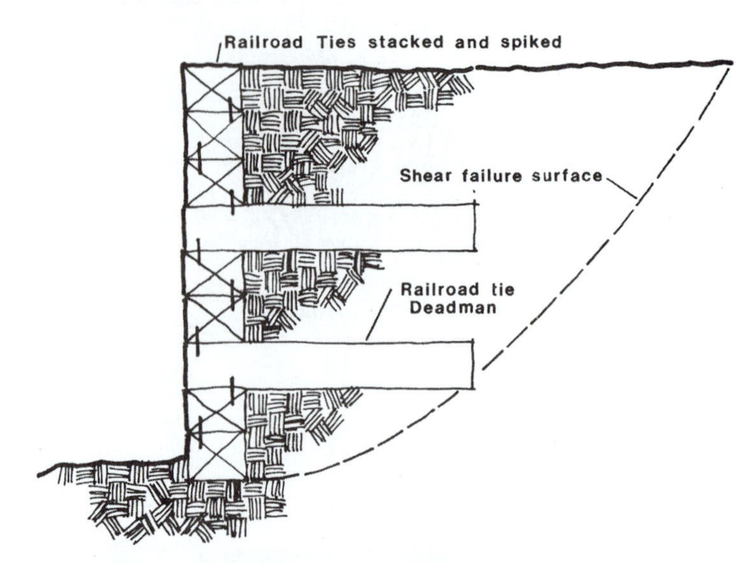

FIGURE 14-11. Deadman concept. Stability is expected from the tier installed as "deadmen," extending back into the fill behind the wall.

by Fig. 14-11 is inherently *unstable.* A more appropriate solution for the tall, heavy timber retaining wall in Fig. 14-12 would have been a mechanically stabilized earth wall or reinforced concrete.

Retaining walls are a useful device for effecting the desired site surface configuration. Their use can result in good aesthetic quality and overall economy. Attempts to economize in the design and construction of the retaining walls themselves, however, are almost sure to degrade quality.

FIGURE 14-12. Timber retaining wall.

PRECAST CONCRETE RETAINING WALL SYSTEMS

The techniques, methods, and configurations employed in heavy timber and stone retaining wall construction have been matched by those in precast concrete systems. The innovative strategy, although commonly more expensive in initial construction, has numerous structural and aesthetic benefits over heavy timber. A variety of companies are using the construction technique, which borrows a little from every traditional method. While each precast concrete manufacturer's specifications and installation methods will vary, the following provides the site planner with a general overview as to the process and how it might differ from other retaining wall construction.

To begin, the site planner needs to have reached some preplanning decisions. Among those decisions will be whether the wall will be constructed as a vertical wall or a battered wall. Fig. 14-13 illustrates a section through a *battered* wall; a term used to describe a wall designed with a predetermined slope. The value of the battered wall is that its slope reduces the dead load of the earth to be retained. Consult the manufacturer's guidelines as to the angle of the slope and what, if any, devices are used to maintain a constant slope in the construction of the wall. First, regardless of manufacturer, a foundation must be designed with sufficient structural integrity to support the lateral dead load and the weight of the precast concrete wall. The foundation base should also be at a depth sufficiently low to keep, at a minimum, the first

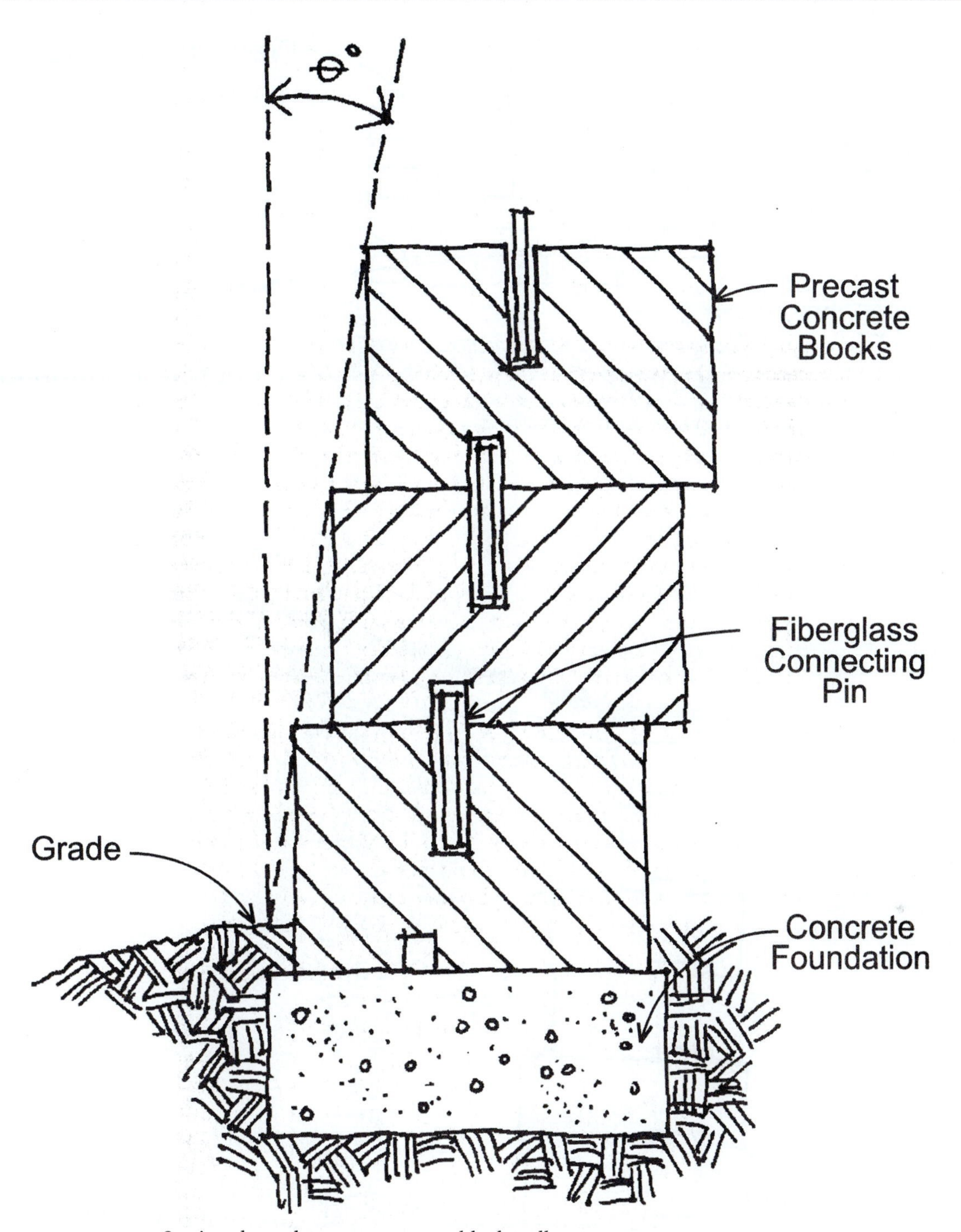

FIGURE 14-13. Section through precast concrete block wall.

course of precast units below the desired finish grade (Fig. 14-14). The site planner should make certain that the grading plan prepared for the site reflects all of the spot elevations that are critical to the construction of the retaining wall. Additionally, organic soils, or those with a high clay content, should be avoided in the construction of the retaining wall as they hold water and do not compact properly.

One precast technology designed to resist the lateral dead load of the earth and provide for vertical structural integrity is an interlocking ridge-and-recess system, illustrated in Fig. 14-15. This particular alternative system also uses units of the largest depth at the base of the wall and tapers to more shallow units as the lateral loads become smaller. Like almost all other retaining walls, this system recognizes the necessity to disperse the accumulation of water and uses crushed stone backfill from top to base, along with drain tile along the base.

Another precast concrete wall system uses fiberglass pins to tie the blocks vertically. Individual blocks are aligned

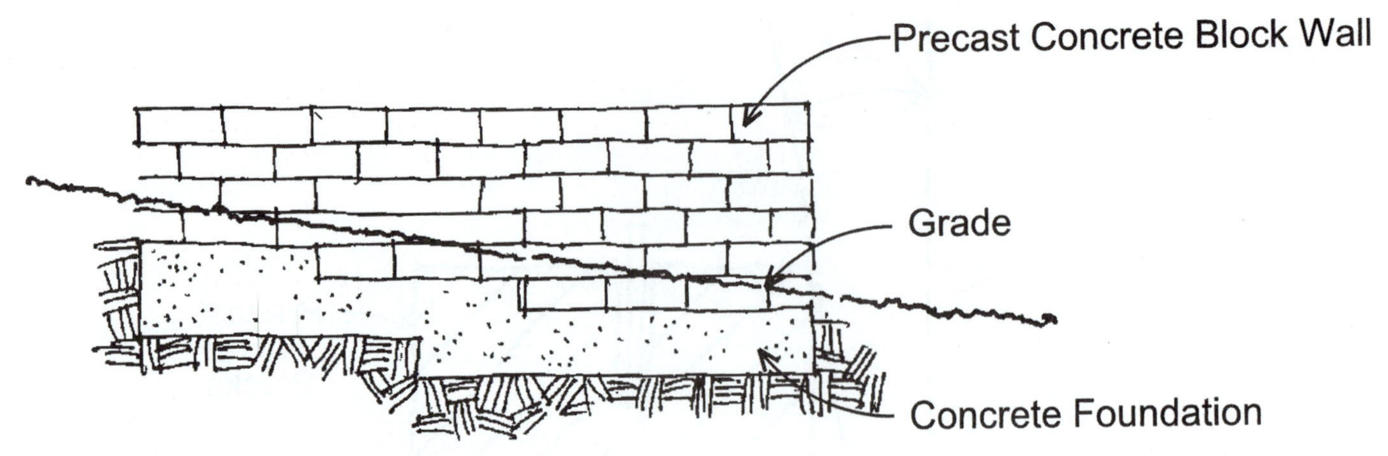

FIGURE 14-14. Section/elevation at precast concrete block wall.

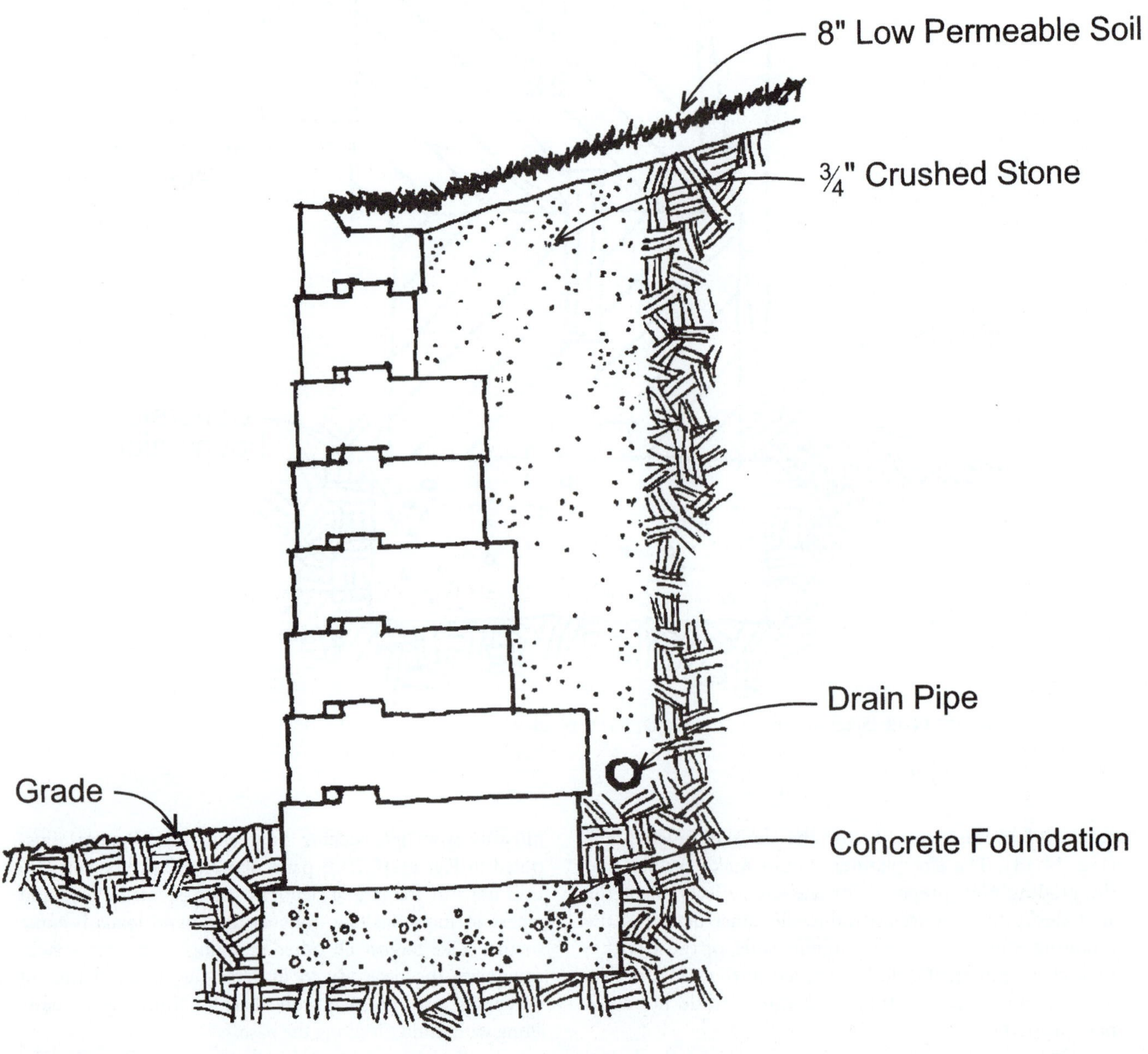

FIGURE 14-15. Section through precast concrete interlocking block wall.

vertically, using fiberglass pins set in the voids of the units of each block for the length of each row of blocks (Fig. 14-13). After the first courses of precast have been set and the backfill has been compacted, a geogrid reinforcement layer is anchored to the precast units and stretched over the compacted backfill. The purpose of the geogrid is to provide the lateral reinforcing tie for the wall. The process of laying a row of blocks, compacting the backfill, and anchoring the blocks with a geogrid at regular intervals is common to a variety of systems. The use of a geogrid reinforcement is common to many manufacturers of precast wall systems. Once at the top, a cap row of precast units can be set using a bonding material, such as a flexible epoxy, to prevent the cap blocks' removal. The variety of precast concrete systems and structural techniques is expanding. If conditions warrant the use of a concrete retaining wall, one should look beyond the traditional poured-in-place method for both structural and aesthetic reasons.

MECHANICALLY STABILIZED EARTH (MSE) WALLS

In circumstances in which a poured-in-place concrete retaining wall will be required, mechanically stabilized earth (MSE) walls are also an alternative. In those circumstances where heavy loads created by road embankments or problem soils create design issues, MSE walls are an alternative course of action. One of the companies that has an extensive amount of experience with MSE walls is Reinforced Earth (Terre Armée). It needs to be emphasized that MSE is a sophisticated process used to reconcile complex retaining wall applications. While the MSE process has enjoyed wide success in transportation and environmental problems, it is not an appropriate application for low walls with moderate design loads.

The MSE technology was developed by the French architect/engineer Henri Vidal in the late 1950s. It has been reported that Vidal noticed that the load-bearing capacity of sand was increased dramatically when pine needles were added to alternative layers of sand. Five years of testing passed before Vidal's theories and invention were ready for real-world application: a 15-m (59-ft)-high retaining wall in the French Alps. Following the success of his first commission and an additional eight years of work, Vidal named his invention *reinforced earth* and filed for patent rights in Europe. He later established a company under the name Reinforced Earth and began marketing the MSE technology in the United States in 1972. Approvals from U.S. agencies and their applications using Reinforced Earth were as early as 1973 from the Federal Highway Administration (FHWA) and 1975 from the U.S. Army Corps of Engineers.

The Mechanically Stabilized Earth Process

The manufactured parts used in the MSE application include interlocking precast concrete wall panels and deformed, high-adherence galvanized steel reinforcing straps. The construction process begins by cutting back a sufficient distance from the foot of the wall to create a foundation footing for the first row of precast panels. The first row of panels is set, anchored to the footing, and braced to remain stable during the filling process (Fig. 14-16). The next step is to create the first layer of a mechanically stabilized earth mass. Unlike the traditional poured-in-place concrete wall, which is designed to resist movement based on the width and depth of the steel reinforced concrete wall and its lateral reinforcing, the MSE wall's strength is based on its capacity to resist movement as the compacted fill is also reinforced

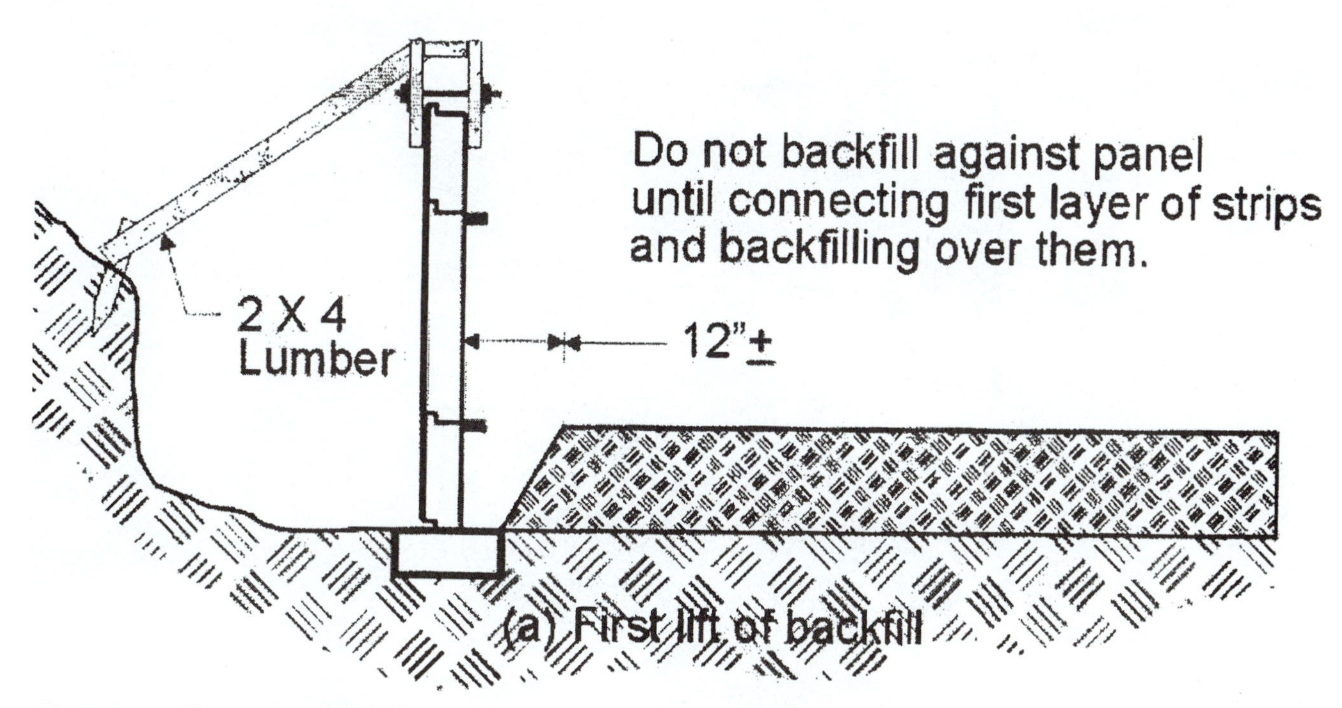

FIGURE 14-16. First course of backfill.

FIGURE 14-17. Installing reinforcing strips. (Courtesy of Reinforced Earth Co.)

with deformed, high-adherence, galvanized steel reinforcing strips laid horizontally across the fill (Fig. 14-17).

To create a precast concrete wall facing that has both continuity and integrity, metal plates are set along the edges of the panel forms prior to pouring the precast concrete panels. When the precast panels have been properly set, the plates in each panel are welded together, thus creating an integral wall and edge to the steel reinforced compacted earth fill (Fig. 14-18). Similar to the process in setting the first row of precast panels, the second row is an

FIGURE 14-18. Setting bottom full panel. (Courtesy of Reinforced Earth Co.)

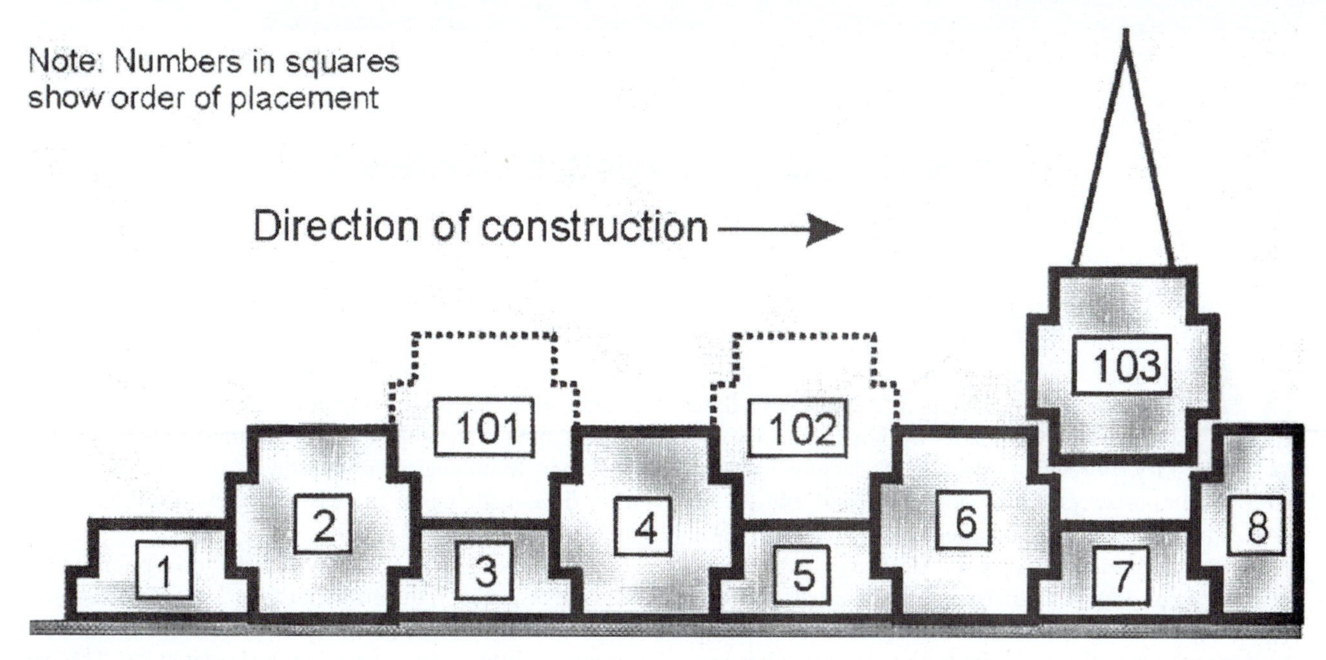

FIGURE 14-19. Installing the next row of panels.

interlocking panel set in place usually using cranes (Fig. 14-19). As with the first row, the steel plates anchored along the edges of each panel are welded together. The second layer of ties is subsequently covered by a layer of compacted earth fill, and the process is repeated until the desired height of the wall is achieved (Fig. 14-20). The result is a wall of steel-reinforced earth fill of substantial depth edged by interlocking precast concrete panels that are welded together.

MSE has demonstrated success to a broad range of design problems and applications ranging from major transportation to environmental issues around the developed and developing worlds. The range of applications vary from a 160-ft-high wall at the Tacoma, Washington, airport to avalanche barriers in Iceland. For further reading on the research, development, and design theories of MSE, download *Mechanically Stabilized Earth Walls and Reinforced Soil Slopes Design and Construction Guidelines* from www.fhwa.dot.gov/engineering/geotech/retaining/100317.cfm.

FIGURE 14-20. Completed structure. (Courtesy of Reinforced Earth Co.)

CHAPTER FIFTEEN

TOPOGRAPHY AND SURVEYS

Most of the time your request to a client to acquire a topographical survey at a 1- or 2-ft interval will be respected without quarrel. Although an explanation is necessary, more often than not, the client will understand the importance of this document and engage the services of a professional engineer or registered surveyor. However, on occasions when a site has been altered after a survey or your empirical analysis of the site tells you that something is not right, you need to know how to execute a small topographical survey on your own. Although it is possible and appropriate under some circumstances to undertake a *limited* topographical check, unless you have the professional qualifications, *do not* engage or contract to do work that involves checking or *describing any property line, easement, or road right-of-way.* The legal implications associated with defining a property line's location, bearing, and distance are significant and in most states the purview of the licensed civil engineer or land surveyor.

In situations in which an area needs to be checked or the topographical interval is too large to permit detailed grading, instead of interpolating, survey the site. The method is fairly simple. Before going to the site, lay out a square grid on a print of the most current topographical information available. If possible, use one of the longer property lines of the site as a baseline. The baseline can be on the property line or in some other way "fixed" relative to the boundary line survey (Fig. 15-1). This is absolutely crucial to the process, for if the baseline is not correct, the grid and all points on the grid will be incorrectly located. Ultimately, spot elevations at the intersection of each grid line (A/1, B/2, C/9, etc.) provide a matrix of spot elevations (Fig. 15-2). This matrix of spot elevations provides a base of known points that must be related to a benchmark, or reference point, to be of any value. Just as knowing the exact location of the grid baseline is imperative to knowing the location of all subsequent points on the grid, the benchmark is critical to knowing the relative elevation of all points on the grid.

Where the grid interval is such that there are large areas of unknown elevation, intermediate grid lines may be introduced. In Fig. 15-3, horizontal grid line 6 + 85 is just such a line. Its nomenclature indicates that it is line 6 plus 85 ft. Other spot elevations not on the survey grid can be located and identified by using a transit, a compass, and a tape.

The equipment necessary to execute the survey properly includes a transit and rod (Fig. 15-3), a steel tape (minimum 100 ft), pins, and a hammer (Fig. 15-4). Solicit instructions relative to setting up the transit before going to the site. Although some of the older instruments must be leveled in two directions, technological advancements in surveyors' equipment have made newer models much less time-consuming to set up.

After setting the transit up over a known point on the baseline (Fig. 15-5), the grid coordinates are located on the site with the pins. The grid is orthogonal, and the surveyor simply needs to make certain that the grid is at 90°. This can be checked with the compass on the instrument. This should be a fairly straightforward process of using the transit to establish a true orthogonal pattern and the tape to establish the proper grid coordinates.

Following the establishment of the grid, the first reading to be made and documented is that of a benchmark. A benchmark is a known, predetermined spot elevation on the site (Fig. 15-6). Some research is necessary to determine the location of a benchmark on the site prior to the survey. The local public works departments can be helpful in establishing these important spot elevations. Determining the location and documenting the elevation of a benchmark is critical, because without some known elevation on the site, all subsequent rod readings will simply be a series of "relative" elevations. As an example, if the rod reading of a known benchmark of 564.3 ft is 7.22 ft, all subsequent readings and interpolations will be based on that known relationship. When it is necessary to move the transit, it is imperative to take a new rod reading of the benchmark. Failure to do this *every time* the transit is moved will result in major errors in the interpolation process. Although secondary benchmarks may be necessary, their evaluations must be read, documented, and interpolated relative to the original benchmark.

It must also be mentioned that a bona fide benchmark may not be on or near the site. Conditions may be rural, or it may for some other reason be impossible to locate a benchmark. Under those circumstances, simply use any permanently fixed point. In urban settings, curbs or water meters might serve the purpose. In rural conditions, a small concrete pile might need to be set. Either of these conditions recognizes that while the survey will not reflect the accurate elevation above sea level, properly executed it will be accurate relative to that particular site and its "benchmark."

Rod readings must be taken at each coordinate point on the grid to identify the elevation of that specific point. Viewing

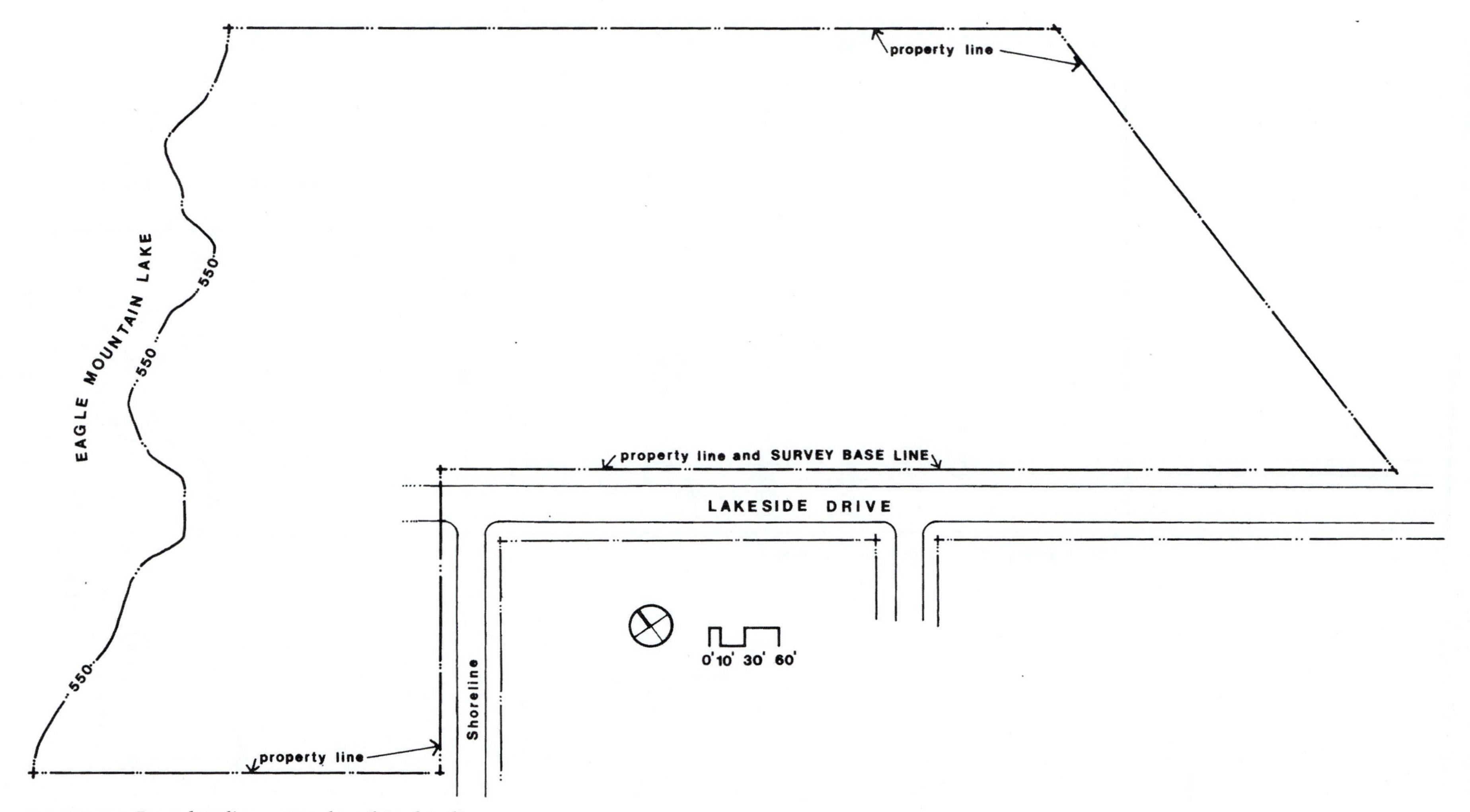

FIGURE 15-1. Boundary line survey describing baseline.

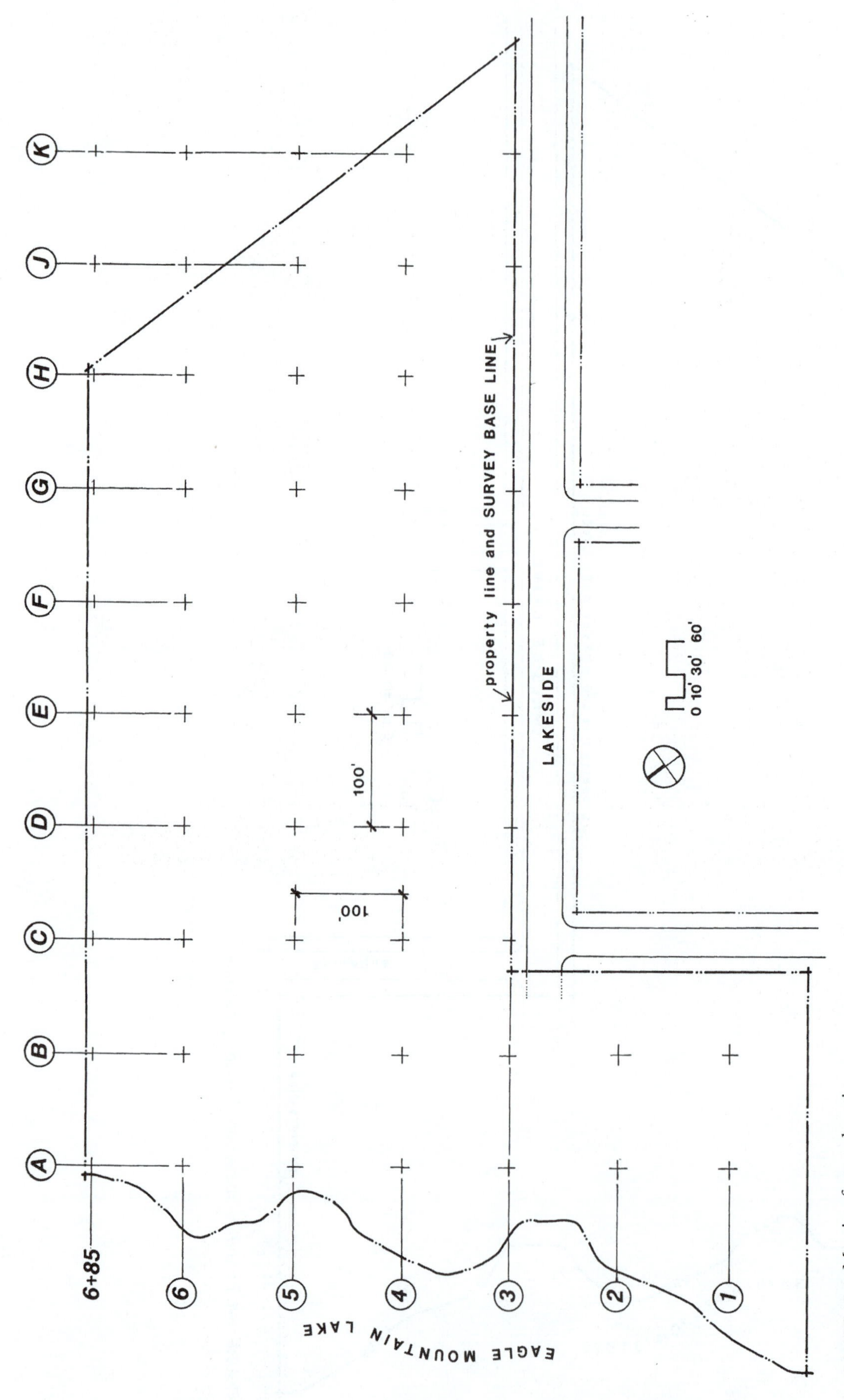

FIGURE 15-2. Matrix of spot elevations.

FIGURE 15-3. Transit and rod.

FIGURE 15-4. Steel tape, pins, and hammer.

the rod through the transit, the elevation is taken at the center cross hair in the scope of the transit. To permit a detailed fix on any point, the rod has been separated into feet, tenths, and hundredths of a foot. As an example, the reading of the rod in Fig. 15-7 is 4.12 and in Fig. 15-8 is 6.97. The space between the black hundredths divisions is sometimes unclear, as the rod is often smaller in the transit scope than shown in the examples.

FIGURE 15-5. Transit set up over known point.

FIGURE 15-6. Benchmark close-up.

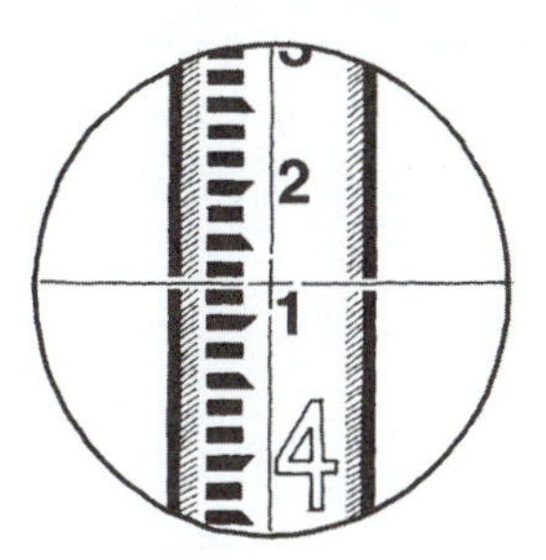

FIGURE 15-7. Rod reading.

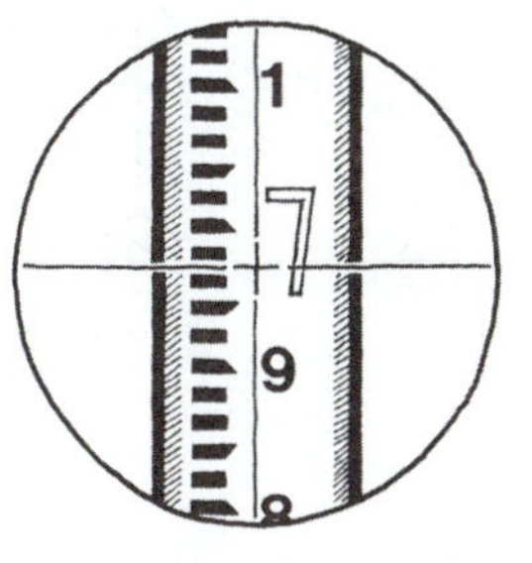

FIGURE 15-8. Rod reading.

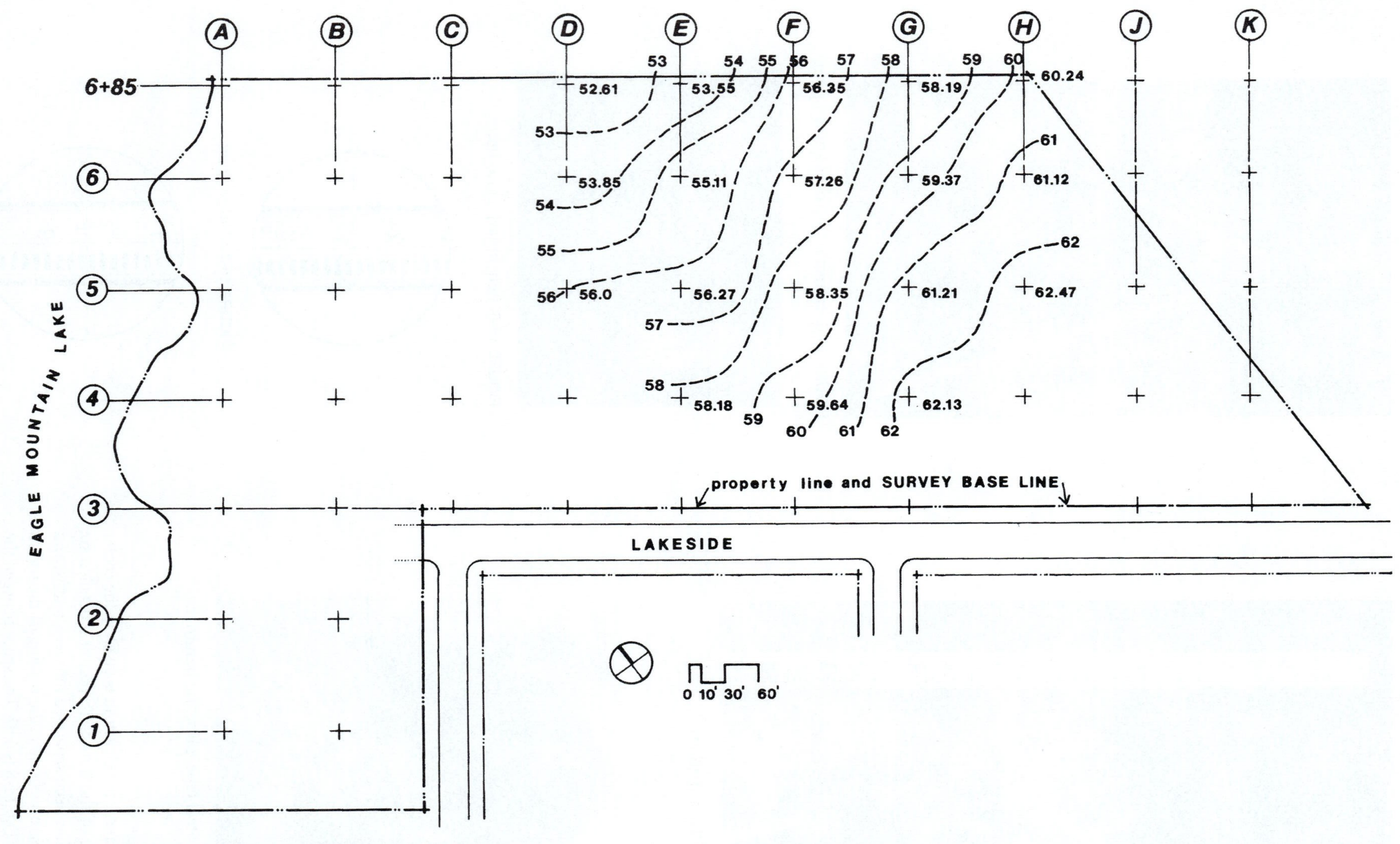

FIGURE 15-9. Grid of spot elevations and contour lines.

To identify the exact location of the whole-number contour line between any two spot elevations, the following formula is applied:

$$Y = \frac{B - X}{B - A}(L)$$

where

B = highest elevation (spot grade)
A = lowest elevation (spot grade)
X = whole-number contour sought
L = distance between the two spot elevations
Y = distance from the highest elevation down to the whole-number contour

In cases where there may be two or more whole-number contours between two spot elevations, treat each contour line in the formula separately, as though there were no other contour lines. This technique is quite accurate and produces consistently good results.

Example:

	Spot Elev. A	Spot Elev. B
	+	+
B = 94.1	93.6	94.1

A = 93.6

L = 50 ft

X = 94

$$Y = \frac{94.1 - 94.0}{94.1 - 93.6}(50)$$

$$= \frac{0.1}{0.5}(50)$$

Y = 10; therefore, the contour line 94 ft crosses 10 ft below 94.1, on a line between 94.1 and 93.6.

As mentioned earlier, the method is rather straightforward and is intended only to provide a check on a site that appears to have been modified, or as an elaboration on *known* topography where interpolation would be inappropriate. The spot elevations taken at the intersections of the grid now permit the identification and location of each contour line on the site. Fig. 15-9 is an example of a contour map and base spot grade matrix.

This will, hopefully, reinforce the idea that the grading process is pursued through the establishment of spot grades first. Contour lines are then changed to accommodate the spot grades. The most common mistake in grading is one in which contour lines are changed without due consideration of spot elevations. The results are often radical and unnecessary changes because the site planner failed to remember that the contour line at its *smallest* increment represents a 1-ft change. The spot elevation can become any number and thus more accommodating grading is plausible.

CHAPTER SIXTEEN

STORM WATER MANAGEMENT AND EROSION CONTROL

This final segment in the topic of earthworks focuses on the concept of storm water management and erosion control. While the topic is an integral aspect of grading and drainage, there are some very specific standards and procedures that one must follow in the preparation of a storm water pollution prevention plan (SWPPP). SWPPP is a term that the site planner must understand as the Environmental Protection Agency (EPA) and many state agencies have established guidelines and processes that require approval and compliance. Recall from Chapter 9 that the EPA adopted specific criteria and standards to reduce pollution from sediment to the nation's surface water system and protect wetlands. The requirements in submitting an SWPPP are a part of and an extension of that law.

Since regulations are always an evolving design parameter, no attempt will be made here to discuss anything other than general concepts related to the preparation of an SWPPP. It must be emphasized that federal and state environmental agencies have established roles and reviews in the SWPPP process. Therefore, before initiating anything other than the data collection that due diligence would necessitate, it is not recommended that the site planner begin any design work that might later be considered premature. Due diligence demands that you make no assumptions; contact should be made with the appropriate agency, and the names, dates, and times of your conversations with agency staff should be recorded as part of your research. Do not assume that because you prepared an SWPPP last year, you have the most current data and information necessary to prepare a SWPPP. A request should be made of the agency for all current applicable regulations. At a minimum, the Web site of the agency and e-mail address of the staff person should become part of the project file. Of course, develop an assessment of the resources and a preliminary estimate of the potential runoff, as well as where the storm water needs to go before you meet with agency representatives. This chapter will focus on concepts and techniques that facilitate resolving the problems created by precipitation and surface runoff. To begin with, storm water management systems generally have two purposes: (1) the control of water to prevent or minimize damage to property or physical injury, and (2) the control of water to minimize or eliminate an inconvenience in the use of the site during less significant storms.

The basic concepts associated with storm water management are few and, for the most part, follow what most would consider to be good common sense.

First, when plausible, water that falls on a site should be retained on that site to the extent that the rate of discharge is no more than that prior to the development.[1] This is often achievable in suburban sites and in environments in which the impervious surfaces are less than 90%. Where appropriate, storm water systems should be designed to permit aquifer recharge. Conversely, storm water management systems should avoid recharge in those soil conditions where groundwater effects might be harmful. Aside from the environmental advantages, on-site retention often reduces the community's investment in storm sewers, expanded floodplains, or channelization programs. Permanent or temporary ponding is a common method of controlling on-site water but must be considered as a part of the entire development process.

Second, the optimum design of any storm water collection and storage system is a balance between capital costs, environmental quality, risk, and operational costs.[2] Each design alternative must be examined through a process that considers all of the foregoing issues, for ignoring any one invites a decision that could result in substantial economic or environmental loss.

Next, emphasis should be placed on natural engineering techniques, that is, those consistent with the natural processes and resources. This obviously begins with recognition that precipitation and storm water are a part of the hydrological cycle. These concepts also require an understanding of the site's natural ecology and the methods by which the existing natural systems dispose of precipitation. An understanding of the character and physiography of the site permits the recognition of valuable opportunities, as well as imminent disaster.

Further, the responsibility for genuine storm water management rests with the municipality and the developer. Do not be misled by polemics that portray the developer as the villain, complete with black hat and muddy boots. The fact of the matter is that no developer can do anything less than that which the local or state governments permit through their ordinances and regulations. The developer is not a nonprofit or philanthropic institution and thus should not be chastised when his or her proposals do not coincide with your perceptions or wants. The author's recommendations to local governments and citizens are to say what you want, be specific about it in terms of quality and quantity, and finally, assume

[1] *Residential Storm Water Management* (Washington, D.C.: Urban Land Institute, American Society of Civil Engineers, and the National Association of Home Builders, 1975), p. 11.

[2] Ibid.

that you will be disappointed. If a local government employs strict regulations and has rigorous controls, the results will more likely reflect environmental quality that has not been left to chance or the developer's CPA's ledger.

Finally, recognize that each site involves a different set of circumstances, land use, occupancy, and environment and will require different solutions. The critical aspect of each site-planning problem lies in the design criteria established for that particular site. Criteria must be clear and definitive, for without some accepted standard or criterion, every solution is acceptable.

FLOODS: DESCRIPTION AND REGULATION

Floods are some of the most insidious natural disasters, as many begin as heavy rain storms that linger too long. We must be reminded that the physical issues and financial losses that follow a flood are not erased when the water recedes. They are only beginning. One of the reasons that disasters created by floods are so fragmented and ill coordinated is that rivers, streams, creeks, and marshes traverse various municipal and state boundaries and rarely reflect the geographic limits of watersheds. In addition, counties and townships are often hampered with minimal data, diminishing financial resources, and marginal intra-regional cooperation in their effort to develop programs and policies that reflected local environmental development conditions.

In response to congressional directives to develop coordinated unified efforts to minimize flood losses, the federal government charged the Federal Emergency Management Agency (FEMA) with this responsibility, regardless of the meteorological event that creates flooding. FEMA has a key oversight role in mitigating the disaster of flooding, and it is also the primary source for data to assist in flood prevention by determining the flood-prone potential of a community's streams, creeks, and rivers. To facilitate the dissemination of data and maps of flood-prone areas, FEMA has developed the Flood Insurance Rate Maps (FIRM). These are the official maps on which FEMA has delineated both the special hazard areas and the risk premium zones of specific communities. Community officials use these maps in preparing master plans for a community, administering floodplain management regulations, and mitigating flood damage. Design professionals, insurance brokers, and developers also use FIRM to identify structures and properties in insurance risk zones. Recent advances in technology have permitted FEMA to create more accurate floodplain updates more quickly and help communities to address the changes created by urbanization. FEMA's Web site, www.fema.gov, is the source of FIRM for the site planner charged with preparing an SWPPP.

For years, communities used a patchwork of regulations and approved development in floodplains that only exacerbated flooding. Development in floodplains and in coastal zones frequently inundated by hurricane storm surge was all too common, as it was viewed by many as an increase to the local tax base. Few, however, asked about the cost of this development to the community. And little appeared to be learned as the cycle of flooding, property loss, reconstruction, and flooding was routine. For example, aside from those cases where the public has outright ownership, there are few cases where any public policy has prohibited barrier island development. The public will to establish unpopular policies has been shallow and spotty. The first and most popular flood control strategy for many communities has been structures or projects such as:

1. *Dams and reservoirs:* These are designed to hold or retard the flow of water in a watershed and release the water at a flow rate that will minimize downstream damage.
2. *Levees:* These are designed to confine water in a river, creek, or stream to a channel course.
3. *Channel improvements:* These usually involve methods to increase a channel's capacity to improve its ability to discharge water.

These projects are popular as they express the public's intent to control or minimize the disaster of the impending water. In those cases where they are the most successful, they succeed when used in concert with smaller, less visible projects.

The second strategy of flood control is regulation, which can be separated into two subdivisions: site and nonsite. Site-specific regulations control the land use within the stream corridor and in the buffer zones immediately adjacent to the corridor. In addition to protecting life, minimizing public expenditures, and preventing and/or reducing flood damage to property, land use control strategies strive to define and allocate land to its most appropriate use.[3] Some of the instruments of regulatory approach are subdivision regulations, floodplain zoning, local building codes, and parks and open-space acquisition programs. Other municipal policies that might restrict the extension of public facilities (water, sewer, roads, etc.) would obviously retard the future development of flood-prone areas. Generally, the most effective regulatory strategy couples land use controls and public policy.

Nonsite regulations complement the site-specific regulations and focus on land use or development regulation in the entire watershed. It must be emphasized that neither site-specific nor nonsite regulation alone is usually a sufficient approach. Nonsite regulations reflect the more holistic perspective of flood control mechanisms and are devised to reduce the impact of future growth within the watershed in flood-prone areas. Conceptually, the nonsite regulations seek to lessen the negative environmental impacts with runoff control ordinances and/or erosion and sedimentation control ordinances.

Faced with shrinking budgets and rising demands for services, many municipalities can no longer find the financial resources or public support for additional storm sewers. Numerous local governments have preferred to adopt

[3]Jon A. Kusler and Thomas M. Lee, "Regulations for Flood Plains," American Society of Planning Officials, Report No. 277, February 1972, p. 7.

mechanisms that limit the amount of runoff from any specific development to the hydraulic capacity of the existing, in-place system. There are variations on these policies which may, for example, require on-site retention of a specific storm with a companion release rate. Although the criteria may vary from one watershed to another within a community, normally the objectives are the same. In some communities, the developer may be given the flexibility to use a variety of methods, provided that the impact on the existing systems does not exceed prescribed limits. In many cases, this on-site storm water detention not only prevents additional runoff from flooding downstream areas, but is frequently more economical than standard storm sewer construction. It also forces the site planner and client to evaluate the existing environment carefully as part of the planning process.

Although site-specific policies are critical to address the problem at its source, it is equally important for communities to understand the importance of adopting land-use control regulations for properties in flood-prone zones. Without both programs in place, communities are often reminded that half-hearted efforts are rewarded with equal results.

Floodplains

In spite of the public's growing awareness of the hazards of floodplain development, construction within the 50-year floodplain remains an immensely popular enterprise. The questions concerning an ordinance's legal status begin with the data used in the documentation of the floodplain. This is the most critical, and often the most expensive, phase of a regulatory flood program. The development of base maps and flood data is a prerequisite to the preparation of a floodplain management program. Although various federal agencies will have information affecting the delineation of a regulatory flood and discharge, the first source agencies will be:

1. The Federal Emergency Management Administration (FEMA)
2. Flood Insurance Rates Maps (FIRM)
3. The Army Corps of Engineers
4. The U.S. Geological Survey
5. The Soil Conservation Service

Since the vocabulary of floodplain management can be confusing, an overview of a few concepts is essential to an understanding of the process of delineating the various intervals of the flood. First, for every flood discharge or flood frequency, there is a different floodplain, flood fringe, and floodway. A flood discharge is essentially the translation of a flood into that quantity of water which flows past a specific point in 1 second and is expressed in cubic feet per second. A flood frequency is the translation of a storm's intensity (inches of rainfall/hour) into a probability factor. This probability is based on the documented rainfall for the region, as well as the hydrological characteristics of the water course. A rain of *n* inches per hour in one region may occur so infrequently that it may happen only 5% of the time. This 5% occurrence can be translated into a frequency term and in doing so is then expressed as a 20-year storm (Fig. 16-1).

Two things are important to remember. First, these intensity levels will vary widely from one region to the next. Second, it is possible for any storm (20-year, 25-year, 50-year,

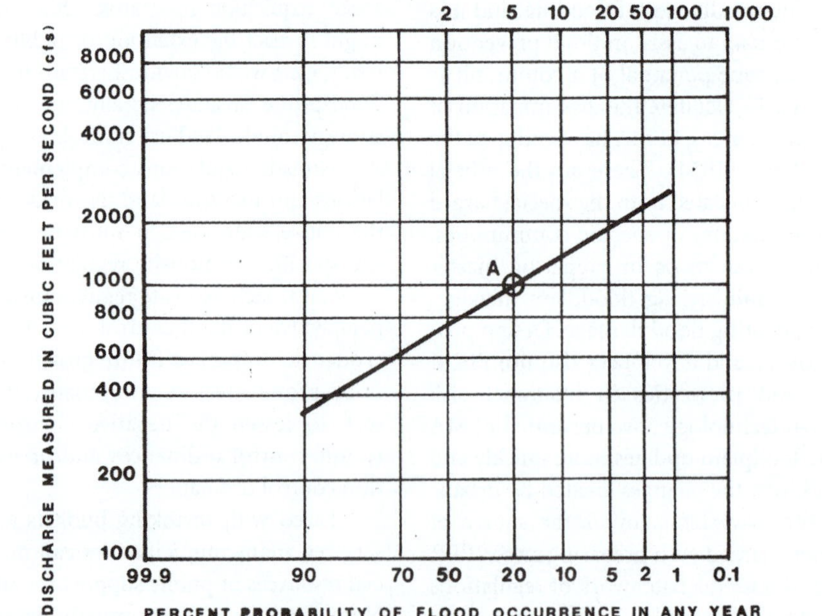

FIGURE 16-1. Flood discharge versus frequency curve.

etc.) to occur more often than once every 20, 25, or 50 years. Recall that this is a probability index and is based on existing data. To recapitulate, a storm of such intensity that it would occur 50% of the time is considered a 2-year storm. Fig. 16-2 illustrates the conditions that might occur under a variety of storms within the same natural channel. One final aspect of the frequency/discharge ratio is that the intensity (inches/hour) is not a geometric relationship. The 10-year storm is not double the intensity of the 5-year storm. Knowing the precipitation rates and intervals, then, provides the

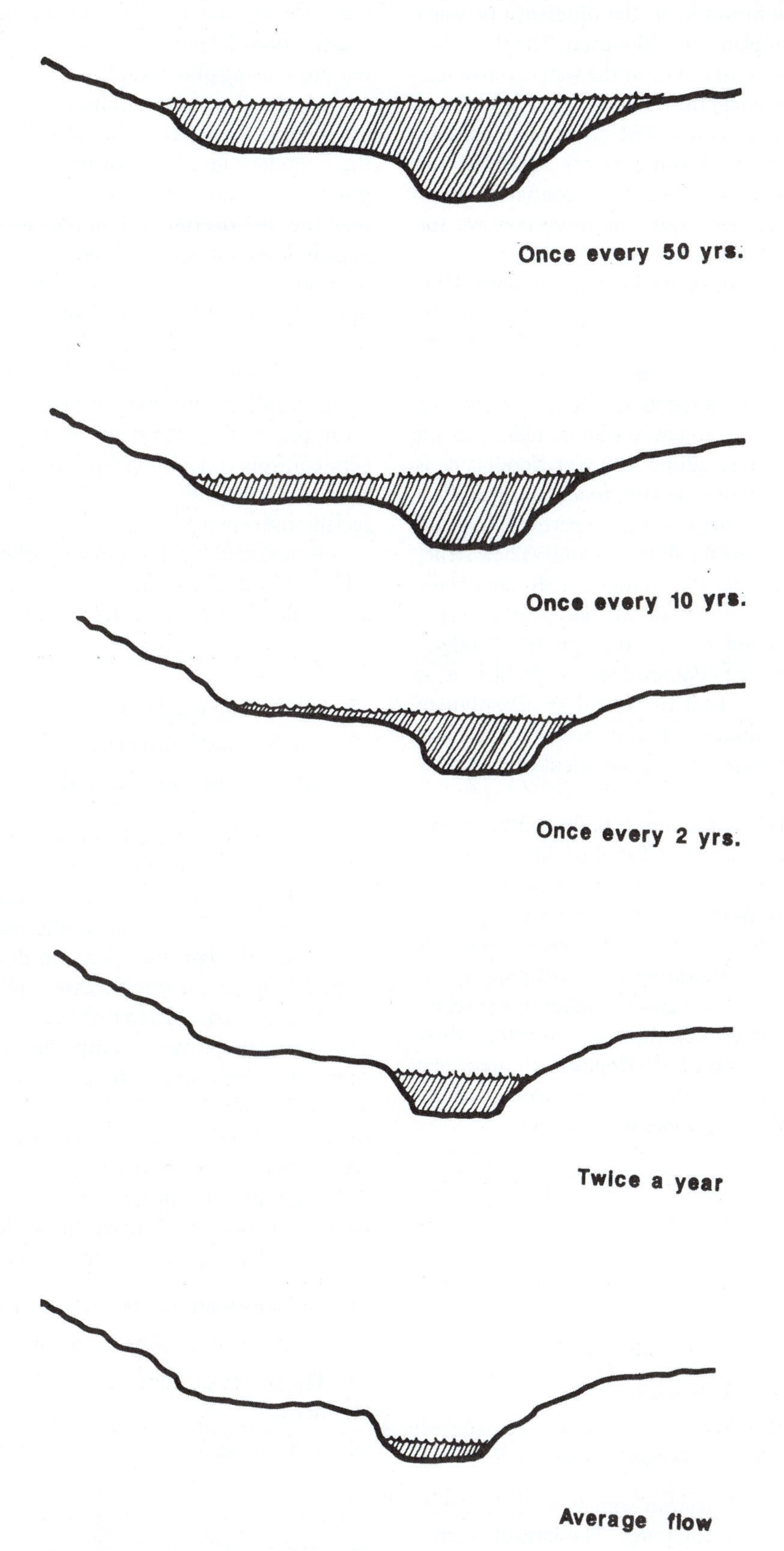

FIGURE 16-2. Diagram of floodplains.

community with the first level of information necessary to document its various flood-prone areas. Fig. 16-3 is a map published by the U.S. Department of Agriculture that portrays regional variances of rainfall for a 10-year storm of 30-minute duration.

There is often confusion as to the difference between the definition of a floodplain and a floodway. The *floodplain* is that area beyond the natural flow of the water course that is necessary to accommodate the flow of water for a known frequency (10-year, 20-year, etc.). The limits of this floodplain line are a function of two primary variables: the amount of water in the floodplain and the configuration or form of the floodplain. Obviously, the more concave the section in the floodplain, the larger its holding capacity.

The *floodway* is that zone of the floodplain that would be required to pass an amount of water equal to a flood of the same period at an elevation 1 ft above the line of the floodplain (Fig. 16-4, on page 176). This interrelationship between the floodplain and floodway prescribes, for example, that the flow capacity of the 100-year floodway, with its additional 1 ft above the elevation of the natural 100-year floodplain, is equal in capacity to that of the 100-year floodplain. The effect of this in various communities has been to permit filling and construction in the 100-year floodplain as long as new structures maintained finish floor elevations 1 ft above the 100-year floodplain. This policy would conceptually permit the normal 100-year flood to pass through the floodway without physically entering buildings. The only problem is, as we know, that floodplains tend to expand as development continues. Unless a community has an active program to maintain the runoff at current levels, the extent of the floodplain is likely to increase.

Table 16-1 (page 177) is a summary of flood data classifications, criteria, hydraulic factors, and applicability. This information is particularly useful in evaluating the accuracy of a floodplain. Since classes B and C are based on extrapolated data, the delineation of the floodplain with either method may invite a legal challenge. Common sense should prevail. The experienced site planner approaches with caution any development near a known floodplain. In those circumstances where the floodplain is undocumented, development in areas that have characteristics similar to the following should be reviewed extensively for flooding before site planning:

1. Relatively low, gentle-sloping terrain confined by a valley wall
2. Soils of high water table and poor drainage
3. Healthy vegetation of the species that flourish in wet, flooded soil conditions
4. Land areas adjacent to water courses with:

   (a) Outer edges of steep slopes
   (b) Traces of "oxbow" lakes
   (c) Natural terraced benches against valley walls

Local and regional history can sometimes help determine the extent of an inundated area. The lack of intense agriculture in a flood-prone area may be the result of several generations of farmers. Remote sensing involves infrared satellite photography that can be used to identify areas of standing water. Although the scale of this photography is very large, and detailed information will require enlargement, in the absence of other documentation, remote sensing may be the only source of hard facts. The Earth Observation Satellite Company (EOSAT) is the source of remote-sensing photography.

Most urban communities have identified and delineated both 50- and 100-year floodplains. However, knowing the boundary lines has not necessarily retarded the sale or purchase and ultimate development of properties within either the 50- or 100-year floodplain. Some municipalities permit development in a floodplain provided that the developer increase, by dredging or other earthworks, the capacity of the floodplain in an equal amount. This concept is based on the assumption that floodplain levels are fixed and unchanging. Depending on the community and its land development policies, this may or not have been a well-founded assumption. Although many local governments have stringent controls concerning "fill" in a 50-year floodplain, others are silent and consider floodplain fill to be a land reclamation enterprise.

To reiterate the point made earlier, the prevalent methods used by local governments to control land uses and hazards in flood-prone areas are the following:

1. Zoning
2. Subdivision regulations
3. Building and housing codes
4. Sanitary and water well codes[4]

These methods are independently successful, and collectively they provide local government with a comprehensive package of implementation procedures. But regardless of what methods are adopted, the regulations should work to protect adjacent upstream and downstream landowners from direct and indirect increases in flood damage.

Zoning is considered first because it addresses the problem of land use, coverage (impervious surfaces), and density comprehensively and at the site source. Zoning is employed to restrict from the floodplain development uses which, when inundated by water, could create health or safety hazards either by chemical decomposition or floating debris. Although the constitutionality of some floodplain ordinances has been challenged, those that are upheld usually have the following qualities in common:

1. Similar situations are treated equally.
2. The regulation is based on solid data.
3. The threat of flood damage is balanced with land use needs.
4. Some realistic private economic land use is permitted.

[4]Ibid., p. 8.

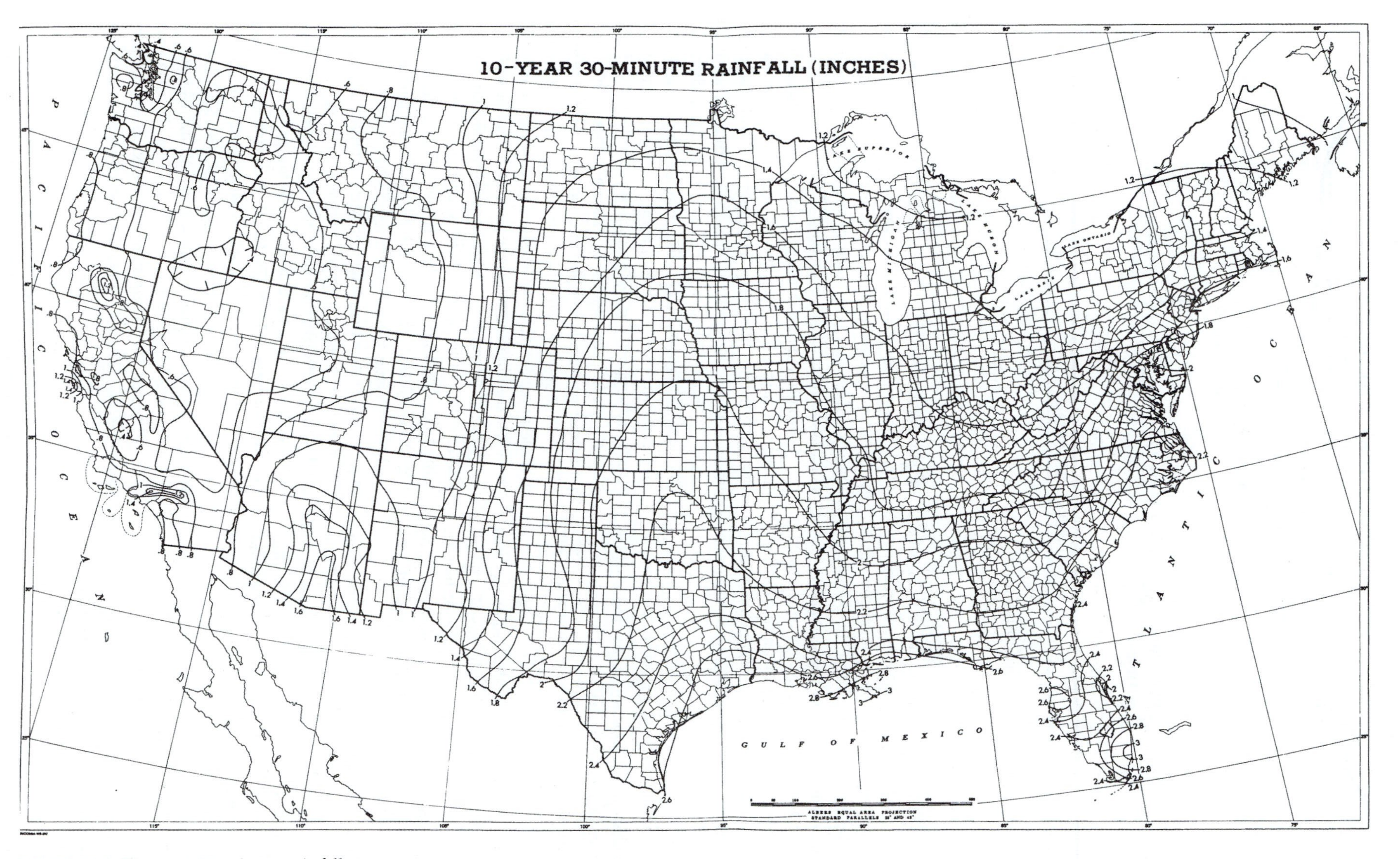

FIGURE 16-3. Ten-year 30-minute rainfall.

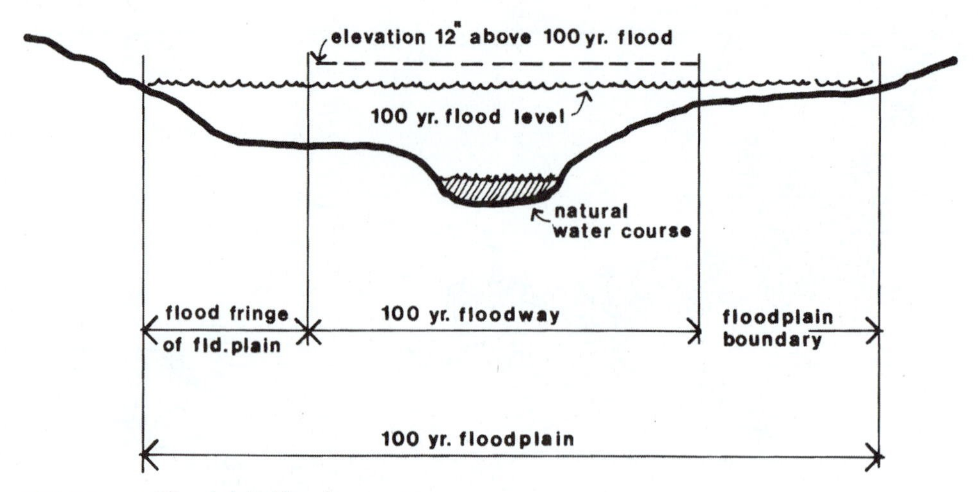

FIGURE 16-4. Floodplain/floodway section.

An additional quality of zoning is that it can be applied in river or coastal zones to regulate what uses may be permitted and under what set of conditions. Subdivision regulations are also to be employed to control the parameters under which development might occur in a floodplain.

Since a building code is limited by ordinance to the structure and site, codes can only focus on limiting the flood damage and property loss. A building code for construction in the floodplain will have established standards to prevent flotation of building by:

1. Specifying minimum anchorage systems or methods
2. Imposing material restrictions to exclude those which deteriorate when exposed to water
3. Imposing structural design standards which reflect anticipated water pressures and flood velocities[5]

Sanitary and water well codes are also ordinances that can be tailored to reflect the specific conditions of a flood-prone environment. In areas of high water table or that are susceptible to flooding, sanitary codes commonly prohibit the use of on-site waste disposal facilities in areas of high groundwater and flooding.

Good floodplain management is critical to the development health of many communities, and thus legislation adopted must reflect each community's unique physical, social, economic, and environmental conditions. The difficulty occurs in developing a balance between the public's interest and the environment, on the one hand, and private property rights and economics, on the other.

An excellent example of that balance was achieved in the development of The Woodlands community north of Houston, Texas. The 18,000-acre site was a serious challenge to the developer. On the positive side of the ledger, the existing pine–oak forest, streams, and abundant wildlife could be undeniably attractive to potential home buyers. On the negative side, approximately one-third of the site was within the 100-year floodplain of three major streams; the site was so flat that drainage patterns depended partially on prevailing wind direction, and much of the soil had a very slow percolation rate (Fig. 16-5, on page 178).

Conventional development practices would have eliminated much of what made the Woodlands such an attractive place to live. Prevailing development standards would have required the construction of an extensive artificial drainage system, complete with concrete drainage channels. The alternative was to analyze the existing ecology in an effort to discover what level of development the existing natural systems could support. The research was very productive. A feasible natural drainage system of existing swales, ponds, and pockets of permeable soil provided the basis for the plan to follow. Although the natural system needed some redirection and additional capacity, the developer's willingness to invest in the research without knowing the results was commendable.

The payoff was at two levels: one economic and one environmental. Utilizing the existing drainage patterns permitted the developer to reduce runoff, protect the natural vegetation, retard erosion and siltation, maintain water quality, enhance scenic amenities, recharge groundwater supply, and reduce maintenance. The ubiquitous economic "bottom line" is equally good. 1972 estimates of the engineering and construction costs for the artificial drainage system that would have been used under conventional circumstances totaled $18,679,300. By comparison, the alternative to supplement the natural drainage system to accomplish the same development objectives cost $4,200,400.[6] Making the process produce these types of dramatic results will not always be possible, but be assured that it always pays. Although the circumstances for alternative comparisons will never be the same, the experienced site planner

[5]Ibid., p. 9.

[6]Wallace McHarg Roberts & Todd, *The Woodlands New Community; Phase One: Land Planning and Design Principles* (Philadelphia: WMRT, December 1973), p. 6.

Table 16-1. Summary of Flood Data Classifications and Applicability

Classification	Determination of Three Hydraulic Factors Essential to Regulations			Desirable Areas of Application	Disadvantages	Advantages	Suggested Zoning Districts
	Profile	Floodway	Floodplain				
Class A: exact flood data based on hydraulic calculations	Determined initially from engineering study	Determined initially from engineering study	Determined initially from engineering study	Areas of high development for urban growth Areas of intense existing development Areas of high land values Areas where zoning is the only management tool	Very costly and sometimes takes many years before studies are complete	Provides sound legal base for zoning Expedites evaluation of proposed developments or amendments to adopted plans Minimizes hardships to applicant of proposed development Contributes to flood emergency preparedness plan	Floodway Flood fringe Basement
Class B: interpreted flood data based on known high-water marks	Extrapolation from past floods records	Normal depth analysis on a case-by-case basis	Location by elevations from extrapolated profile on topographical maps Street-sewer maps that show ground elevations or Field surveys	Small amount of existing development affected Little development pressure Strong sanitary subdivision controls and public policy Much land under public ownership Ongoing program to acquire class A data	Legal base questionable Requires technical assistance to evaluate floodway Applicant has greater burden to provide survey information	Low cost Discourages land speculation Valuable river basin Guide to public facilities and transportation Can be made available in short time to serve immediate need	General floodplain district
Class C: interpreted flood data based on nonhydraulic calculations	Determined case-by-case	Determined case-by-case	Experienced flood maps, aerial photo examination, or detailed soil maps that have been correlated with engineering studies on similar streams	Rural areas little potential for development Large portion of area under public ownership Land-easement or acquisition program Strong sanitary and subdivision regulations Ongoing program to acquire class A data	Frequency of mapped flood unknown All three hydraulic factors unknown Weak legal base Source of technical assistance is needed Burden to applicant to furnish surveys	Low cost Readily expected by local people Identifies pressure areas Discourages land speculation	General floodplain district

Source: Jon A. Kusler and Thomas M. Lee, "Regulations for Flood Plains," American Society of Planning Officials, Report No. 277, February 1972, p. 33.

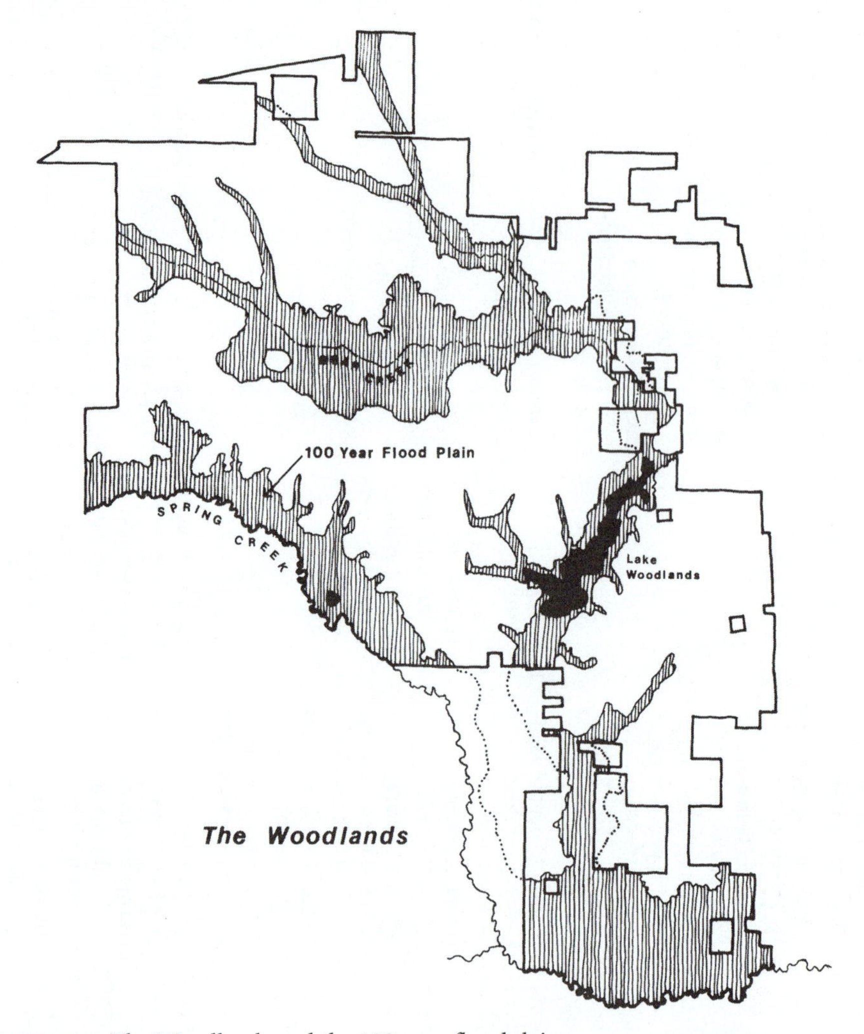

FIGURE 16-5. The Woodlands and the 100-year floodplain.

begins with the assumption that the natural environment is a collection of systems that can be integrated and positively supplement urban settlement.

STORM WATER RUNOFF ANALYSIS

Although most architects and landscape architects will relegate the responsibility of quantifying the amount of storm water to an engineer, it is, nonetheless, important to understand the basic concepts and techniques.

Rainfall is a natural phenomenon, and no two occurrences are the same. The data used in site analysis are based on historical documentation by the National Weather Service. Their data quantify three primary variables: amount (intensity), duration (for how long a time), and frequency (how often does this event repeat itself?). This information is ultimately translated into descriptions that generalize the real event.

There are various hydrological models for analyzing runoff. The selection of the model depends on a number of variables: the size of the area to be considered, its topography, level of development, the character of natural stream systems, existing storm sewers, aesthetic goals, and the level of detail desired, to name the most important. There are no rules relative to what is the best technique to use under every circumstance, so each site must be evaluated individually.

Four basic techniques, with the assistance of a high-speed computer, attempt to simulate rainfall:

1. Hydrological simulation models, which can provide a simulation of quantity (flows) and concentrations.
2. Unit hydrographs, which examine the implications of 1 in. of excess rain in a known drainage area over a limited period.
3. Regression models, which utilize statistical correlation to relate a cause such as rainfall in a watershed to an effect such as runoff volume or peak flow.
4. Empirical formulas. Although there are a variety of empirical formulas, the oldest and most commonly used is the rational method. This formula is widely used where a high level of accuracy is not necessary.[7]

[7]Ibid., pp. 27–28.

The rational method, $Q = CiA$, separates the numerous variables mentioned earlier into four numerical quantities:[8]

Q = peak runoff of the area in question, in cubic feet per second

C = runoff coefficient of the land in the drainage area; this factor can often be obtained from the local public works department

i = "average intensity of rainfall in inches per hour for a duration equal to the time of concentration, T, for a selected rainfall frequency"

T = "time in minutes after the beginning of rainfall for runoff to peak at the point under consideration"; although T is not a function in the formula, to identify the intensity of the rainfall (i), T must be determined

A = size of the drainage area, in acres

Using Fig. 16-6, begin by measuring the longest distance that the water must flow before discharge. The average slope that the water must traverse must also be determined. Although runoff coefficients can be obtained from local sources, Table 16-2 will be used as the source for this example. If the distance across the site is 720 ft and the slope is 3%, begin by reading from the "overland travel in distance" on the left side of Fig. 16-6 to the point at which the horizontal line intersects the slope line of 3%. Taken as an example, a site that is a 2-acre single-family residential land use, Fig. 16-6 indicates the C coefficient to be 0.30 to 0.50. Reading down on the graph from the intersection of the distance

[8] *Design and Construction of Sanitary and Storm Sewers,* American Society of Civil Engineers (ASCF) Manual of Practice, No. 37, 1970; notes revised by D. Earl Jones, Jr., p. 31.

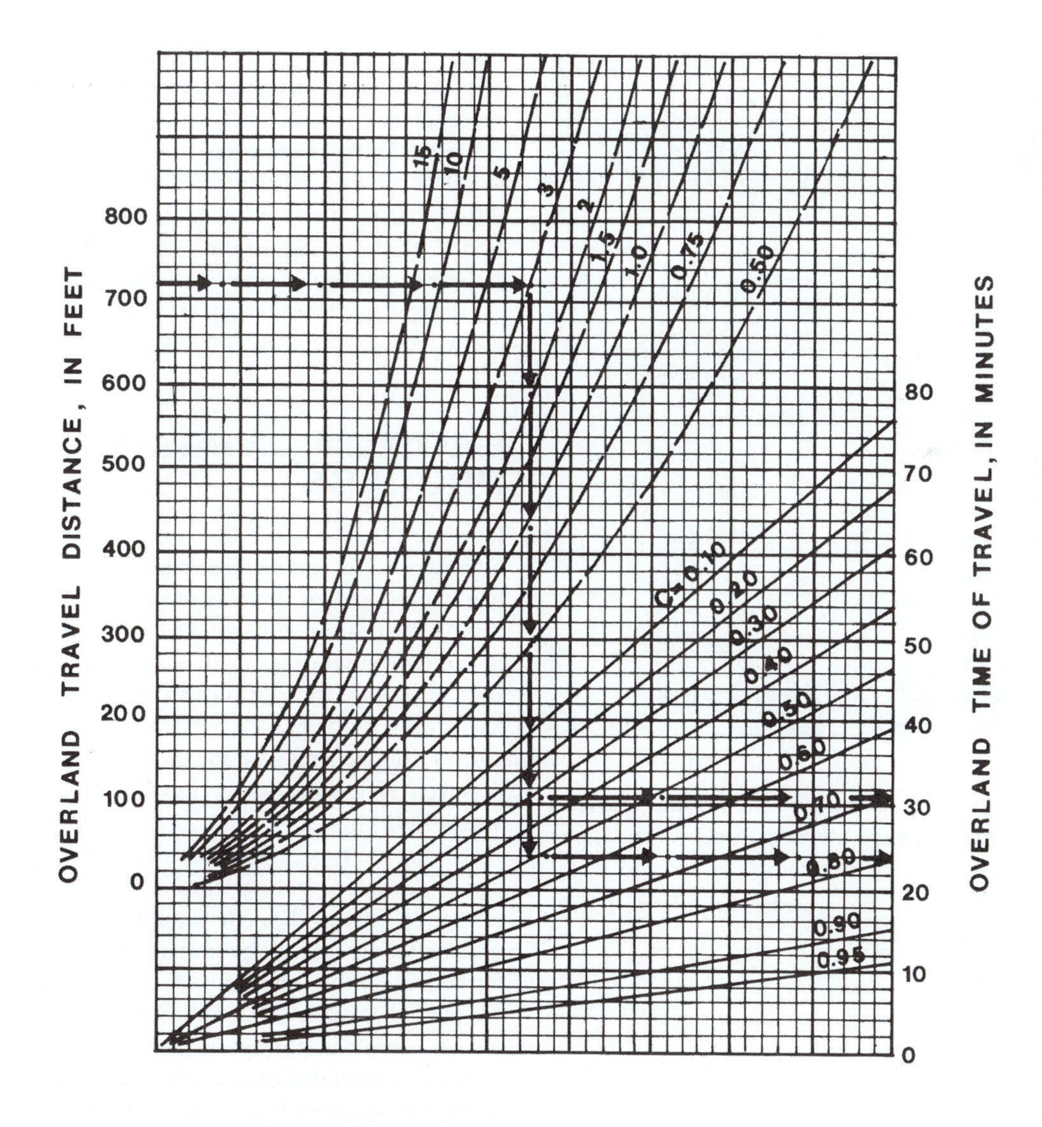

FIGURE 16-6. Graph of water distance flow. (After "Airport Drainage," Federal Aviation Agency, Department of Transportation, Advisory Circular, A/A 150-5320-5B, Washington, D.C., 1970)

Table 16-2. Runoff Coefficients

	Runoff Coefficients
Description of Area	
Business	
Downtown	0.70–0.95
Neighborhood	0.50–0.70
Residential	
Single family	0.30–0.50
Multiunits, detached	0.40–0.60
Multiunits, attached	0.60–0.75
Residential (suburban)	0.25–0.40
Apartment	0.50–0.70
Industrial	
Light	0.50–0.80
Heavy	0.60–0.90
Parks, cemeteries	0.10–0.25
Railroad yard	0.20–0.35
Unimproved	0.10–0.30
Character of Surface	
Pavement	
Asphalt or concrete	0.70–0.95
Brick	0.70–0.85
Roofs	0.70–0.95
Lawns, sandy soil	
Flat, 2%	0.05–0.10
Average, 2–7%	0.10–0.15
Steep, 7% or more	0.15–0.20
Lawns, heavy soil	
Flat, 2%	0.13–0.17
Average, 2–7%	0.18–0.22
Steep, 7% or more	0.25–0.35

and slope lines to the 0.30 and 0.50 curves, the overland time of travel is from 24 to 31 minutes. Time, *T*, can also be calculated with the following formula:

$$T = 1.8(1.1 - C)\frac{D}{S}$$

where:

D = distance across the site
S = average slope
T = time, in minutes

The value of T can now be used to identify I in the rational method equation ($Q = CIA$). To do that, either of the following references should be consulted:

1. Rainfall Frequency Atlas of the United States, Technical Paper 40, U.S. Department of Commerce, Weather Bureau
2. Five to Sixty Minute Precipitation Frequency for the Eastern and Central United States, NOAA Technical Memo N.W.S. Hydro-35, Silver Spring, Md., June 1977

These two resources provide the basis for the documentation of the amount of rainfall over a specific duration and with a specific frequency. Using Fig. 16-3, the 10-year 30-minute rainfall map for Ft. Worth, Texas, indicates that the city receives approximately 2.3" of rainfall in 30 minutes. This provides the remaining information for the formula. Although this process is rather general, the method is used extensively as a preliminary indication of storm runoff potential. The coefficients in these two references are applicable only for storms of 5- to 10-year return frequencies, and were originally developed when many streets were uncurbed and drainage was conveyed in roadside swales. For recurrence intervals longer than 10 years, the indicated runoff coefficients should be increased, assuming that nearly all the rainfall in excess of that expected from the 10-year recurrence interval rainfall will become runoff and should be accommodated by an increased runoff coefficient.

The runoff coefficients indicated for different soil conditions reflect runoff behavior shortly after initial construction. With the passage of time, the runoff behavior of sandy soil areas will tend to approach that of heavy soil areas. If the designer's interest is long term, the reduced response indicated for sandy soil areas should be disregarded.[9]

EROSION: TYPES AND PROCESS

One need only observe the condition of local roads and highways to see the evidence of erosion damage to embankments, graded slopes, and drainage ditches. Although this problem is sometimes a result of poor drainage design, it is often due to inadequate slope stabilization methods. Factors beyond the control of the site planner can wreak havoc on recently graded or planted soil. The weather is a perfect example and is all the more reason why care should be taken to ensure proper installation of soil-stabilizing vegetation.

There are two major types of erosion. The first is geological erosion or natural erosion. This is a process of soil formation change through natural processes. Most of our present topographical features in the rural landscape are formed under this long-term erosion process. The second is accelerated erosion. This is deterioration and loss of surface soil as a result of natural or human activities. Accelerated erosion is a brief process and dramatically affects the quality of the environment because the results are often severe in magnitude.

[9]Ibid., p. 30.

The erosion process involves (1) soil detachment, (2) transportation, and (3) sedimentation. The surface soil is first attacked by an external force such as wind or rain. After the bond is broken that binds the soil particles together, the detached soil particles are then carried by either medium (wind or water) to settle later when the energy source dies away.

Four types of erosion threaten slope integrity: sheet erosion, rill erosion, slump erosion, and wind erosion. Sheet erosion is just like it sounds; it is a thorough, somewhat even washing of water down a slope, like a sheet of water. To a point, this is the most controllable type of erosion.[10]

Rill erosion is the creation of rivulets or small channels that collect water and increasingly widen as they make their way down the slope. Some products are quite effective in resisting this type of erosion by providing a consistent deterrent to rivulet formation, thereby eliminating "weak spots" where this may occur. Rill erosion often occurs where fill has been necessary but has not had proper compaction or settlement.[11]

Slump erosion can be the most disastrous of all. It is the result of subsurface failure, where great areas of slope may suddenly sag or give way, necessitating regrading. This usually occurs due to extremely long or heavy rainfalls in which the soil becomes permeated with water, weakening subsurface bonding or compaction. Slump erosion may also be evidence of poor topsoil stabilization and inadequate surface vegetation or material to promote runoff.

Wind erosion most often occurs along ridges, where wind forces are greatest. Wind erosion can cause the loss of valuable nutrient-providing topsoil and mulch, exposing the roots of seedlings until seedlings are blown away, or simply causing poor growth and aiding water erosion.[12] Soil-stabilizing products deter wind erosion by holding mulch and topsoil in place until grass root systems are thoroughly developed.

Erosion control at its first stage should concentrate on the problems resulting from soil detachment. A concern about the loss of prime agricultural soil brought the issue of soil erosion into sharp focus. The effects of erosion on marine life and the associated flood hazard problems have required some site planners to treat erosion as an environmental problem as opposed to an agricultural issue. Studies of erosion control have concluded that surface water creates most of the erosion damage and that surface vegetation is often the most important factor in reducing erosion rates. Surface vegetation can absorb the energy of falling precipitation, reduce the soil detachment effect, and hold soil particles together. Vegetation also absorbs a fraction of the water through transpiration and as a component in the surface of the soil reduces the velocity of the water and the resulting scouring effect of runoff. Although the interception of precipitation by vegetation permits better absorption, impervious surfaces contribute to the volume of runoff and reduce any potential for ground water or aquifer recharge. When precipitation can be absorbed, the rate of infiltration slows down as the soil soaks up the water. When the rate of the rainfall exceeds the ability of the soil to absorb the water, the residual becomes runoff.

As water runs downhill, the runoff erodes soil as it moves. Excessive runoff will, first, flood the area it covers, and second, carry sediment that clogs flow channels, escalating the flood potential. The volume of runoff depends directly on:

1. The amount and rate of precipitation
2. The ability of the site's soils to absorb the water
3. The topography and intensity of ground cover

In urbanized areas, a substantial quantity of land is covered by impervious materials, which are, in most cases, connected to public storm sewer systems. The nature of storm sewer systems is such that they increase the capacity and speed of runoff. The removal of vegetation, soil disturbances, and other construction also increase the rate of runoff and erosion. Changes in the rural watershed due to urbanization and a decrease in the infiltration process result in:

1. An increase in the total volume of runoff
2. An increase in the peak discharge for any given storm, and as a result
3. An increase in the frequency of flood peaks
4. A reduction in the time an area goes from absorption to runoff
5. An increase in the amount and velocity of water in artificial channelization
6. A discernible increase in sediment during the construction process of urban development

Because erosion and runoff problems related to development extend far beyond the property lines where the development takes place, legal accountability of injury on other properties often becomes quite controversial. A majority of states rely on either the "common enemy" doctrine or the "civil law" doctrine to resolve direct conflicts. The "common enemy" doctrine holds that water is a "common enemy" and permits landowners to improve their properties regardless of resulting injury to other lands. The "civil law" doctrine, on the other hand, burdens the "upper" landowner with the responsibility for any damage created by an increase in downstream discharge.[13] Since following these two concepts to the letter has not been without problems, state laws have reflected various perspectives when applying these doctrines.

[10] *Handbook on the Principles of Hydrology*, Water Information Center, Inc., Water Research Building, Manhasset Isle, Port Washington, N.Y., p. X1.1.

[11] *Residential Erosion and Sediment Control* (Washington, D.C.: Urban Land Institute, American Society of Civil Engineers, and the National Association of Home Builders, 1978), p. 18.

[12] Ibid., p. 22.

[13] Ibid., p. 59.

Between the "civil law" and "common enemy" doctrines is the "reasonable use" theory. This is a compromise that permits a comparison of equities of each use in an individual case. In 1948, the Minnesota Supreme Court set down the following guideline for comparison:

> The rule is that in effecting a reasonable use of his land for a legitimate purpose a landowner, acting in good faith, may drain his land of surface water and cast them as a burden upon the land of another, although such drainage carries with it some water which would otherwise have never gone that way but would have remained on the land until they were absorbed by the soil or evaporated in the air if (a) there is a reasonable necessity for such drainage; (b) if reasonable care be taken to avoid unnecessary injury to the land receiving the burden; (c) if the utility or benefit accruing to the land drained reasonably outweighs the gravity of the harm resulting to the land receiving the burden; and (d) if, where practicable, it is accomplished by reasonably improving and aiding the normal and natural system of drainage according to its reasonable carrying capacity, or if, in the absence of a practicable natural drain, a reasonable and feasible artificial drainage system is adopted.[14]

The Soil and Water Conservation Districts Law provided one of the earliest public-sector erosion control programs. Because erosion was considered an agricultural problem early on, the Soil and Water Conservation Districts Law focuses on erosion in agriculture. This program was initiated by the federal government and was later adopted by state governments. It provided districts with authority to adopt land use regulations to control activities or practice that would result in soil erosion.

An indirect method to control erosion is through water pollution control ordinances. By regulating the amount of water pollution (among which sediment is the leader) that can be discharged, activities that involve large-scale erosion can be contained. Although the intent of water pollution control was not land use regulation, the relationship between the two is so strong that some regulation of land use practice is recognized as an integral part of water quality control.

EROSION AND SEDIMENTATION CONTROL

Unlike runoff control ordinances, erosion and sediment control ordinances are concerned principally with regulating the process of construction as it influences runoff and its quality. Most ordinances of this character center on two objectives: acceptable site-clearing methods and on-site detention during construction. Many communities have adopted policies that require developers to submit erosion and sedimentation control plans prior to receiving building permits. The rationale is easy to accept when communities are periodically faced with expending hundreds of thousands of dollars to remove sediment from their rivers and reservoirs. Many, in fact, have found the financial burden of removing the silt so expensive that they simply acquire new properties and build new reservoirs.[15]

The following are typical examples of erosion abatement methods:

1. Build sediment basins or traps to keep soil on the site. Stabilize cut-and-fill slopes with temporary diversions, berms, bench terraces, or dikes to intercept and divert storm runoff.
2. Leave vegetation on the site as long as possible. Plant temporary turf or ground cover promptly after grading.
3. Use jute matting or a similar stabilizer on slopes to protect seed or plants.
4. Preserve trees and shrubs on one side of the stream to provide shade and maintain wildlife habitat.
5. Reduce runoff velocity with grade stabilization structures, grassed waterways, or energy dissipaters.

Instead of controlling erosion problems through the restriction of land uses, erosion and runoff ordinances are more general and focus on performance plans or strategies. Ordinances often require engineering reports and development plans on clearing, grubbing, grading, vegetation preservation, erosion and sedimentation control, drainage improvements, maintenance of the intermediate regional floodplain, or maintenance of retention facilities. Since these plans are methods in which erosion might be abated or limited, they must be prepared, submitted, and approved before building permits are issued. The following is an example of common standards or policies:

1. The smallest amount of land area should be exposed at any time during development.
2. The period of exposure for land under development should be kept to the shortest practical.
3. Temporary vegetation or mulch should be used to stabilize critical areas where staging of development cannot avoid a time lag prior to permanent cover for the land.
4. Reasonable measures should be taken to prevent the carrying away of sediment by runoff water from land undergoing development.
5. Provisions should be made to accommodate the increased runoff caused by a change in surface conditions during and after development.
6. Permanent final vegetation and structures should be installed as soon as practical in the development.

[14]*Ballentine's Law Dictionary* (Rochester, N.Y.: The Lawyer's Co-operative Publishing Company, and San Francisco: Bancroft-Whitney Co., 1969), p. 1062.

[15]U.S. Dept. of Agriculture, Soil Conservation Service, *Sediment: It's Filling Harbors, Lakes and Roadside Ditches,* Agriculture Information Bulletin No. 325, p. 3.

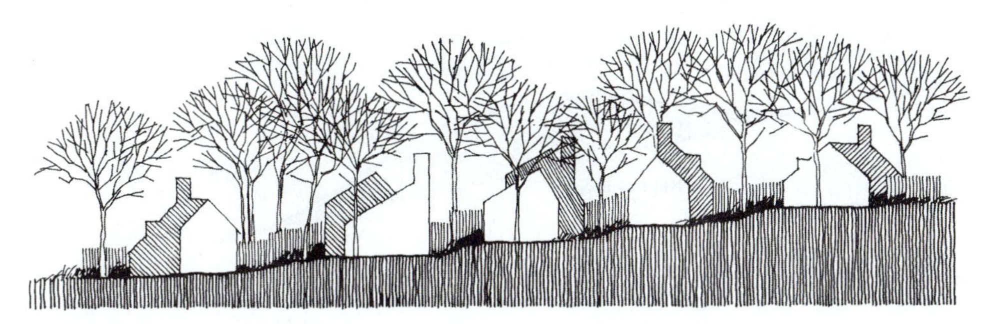

FIGURE 16-7. Bench terrace.

7. Development plans should fit the topography and soils, to minimize erosion potential.
8. Wherever feasible, natural vegetation should be retained and protected.[16]

Again, many of these policies are based on common sense and would be practiced by anyone concerned with the implications of development. The knowledgeable site planner must be conscious of the role and participation of various state and local agencies in the development process. Because of the nature of state laws and the fact that each state has an established code that can differ dramatically from its neighbor, each state's regulations must be recognized. Most of the time, these are adopted in local subdivision or development regulations and should state the reason for the regulation, specifics of the policies, and the process for application review and evaluation.

There are two conceptual methods of erosion and sediment controls: structural and natural, or vegetative. Natural, or vegetative, methods are those that employ the use of existing and/or new vegetation and mulches. Since the majority of sites are most vulnerable during construction, good advance planning can reduce the initial damage while facilitating the rejuvenation of permanent cover. When grass or other turf is used to stabilize a graded area, that new material should be planted parallel to the contours to reduce the flow of water. Vegetation used in this form are sometimes called filter strips. In those locales where clover or honeysuckle have a history of hardy growth, they have been used successfully in that capacity. Obviously, no construction or other traffic should be allowed into any area that has been planted as a vegetative buffer or filter strip.

Vegetative cover can be considered in either temporary or permanent terms. The value of temporary planting is that plants, such as annual rye grass, millet, and sudangrass, grow quickly and can successfully stabilize an area for a particular season when protection is critical.

Permanent cover requires some consideration to make certain that the plant material selected is well adapted to the site. Probably the most critical factor in the permanent cover selection process is maintenance. Although common sense would dictate otherwise, there are examples in every community where turf grasses requiring mowing have been planted on steep or inaccessible slopes. Recognizing the simple fact that plants will mature and that some will require maintenance should influence decisions that involve matching location, use, and species.[17]

Structural methods of erosion control are those which involve the use of landforms to retard or divert runoff. These techniques involve regrading or shaping the land as a method of slowing the velocity of the runoff or controlling the water's path. A bench terrace, as the term implies, is a flat terrace graded across the slope to slow the flow of runoff. In some cases where the area is large, bench terraces have been substantial enough to be used as building sites[18] (Fig. 16-7). In smaller areas the bench terrace has been used in conjunction with other cut and fill regrading to facilitate the stabilization of vegetation.

Berms are a common landform used in erosion control. Like the bench terrace, the berm can be used to intercept storm water runoff across a slope. The berm is essentially an earth mound and can be used to divert runoff into swales, gutters, or stable conditions below the berm (Fig. 16-8). Berms are used as both permanent and temporary methods of erosion control, but unlike other structural techniques can be modified into a permanent bench terrace form.

Diversion dikes or channels are also used to intercept potentially damaging runoff. A diversion channel is a trough cut in the slope to catch and divert runoff to a controlled outlet (Fig. 16-9). A diversion dike is a temporary ridge built parallel to contours of compacted soil. Both methods are used above critical slopes to reduce the potential for serious damage. Since both dike and diversion channels can concentrate significant amounts of surface water, before using either as an erosion control technique, some method must be developed for their integration into the adjacent landscape.

Outlets have been mentioned before as the disposal component of various structural techniques. Special attention

[16]*Residential Erosion and Sediment Control,* p. 32.

[17]Joseph DeChiara and Lee E. Koppelman, *Site Planning Standards* (New York: McGraw-Hill Book Company, 1978), p. 36.

[18]Ibid., p. 34.

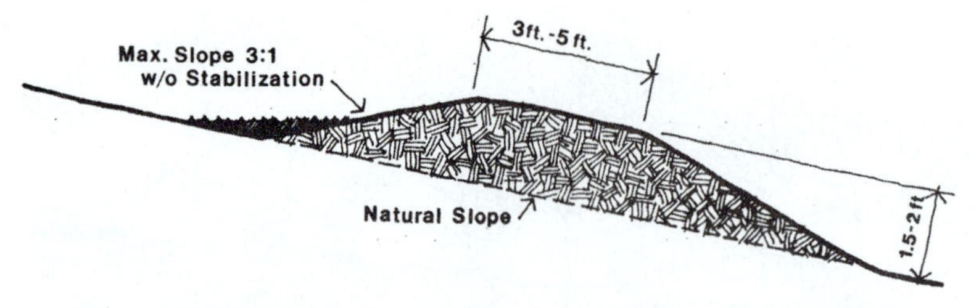

FIGURE 16-8. Filter berm.

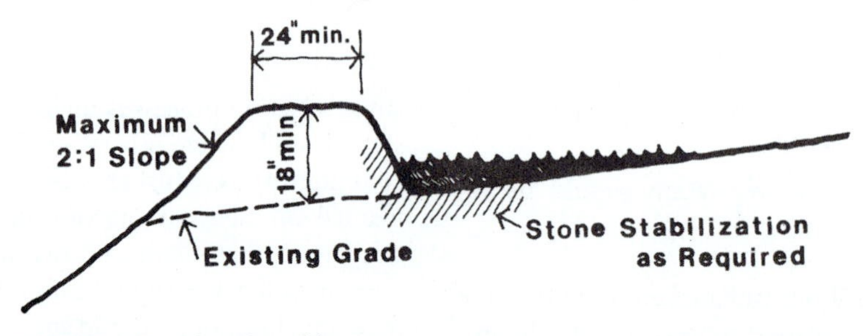

FIGURE 16-9. Diversion dike.

needs to be made in the selection of the appropriate grasses to reduce the velocity and scouring effect of the runoff. Although various techniques, both natural and artificial, can be used to protect channels and outlets, every effort should be made toward the maintenance and replenishment of the existing vegetation.

Permanent ponds or lakes are also common human-made erosion control methods. They are popular as amenities and often increase the value of a property while providing an important function in the site-planning process. The Woodlands developments near Houston, Texas, and Reston, Virginia, are examples of sites where permanent ponding has been used to enhance the aesthetic quality of development while functioning to retard erosion and siltation (Fig. 16-10). These two communities have enjoyed a quality of development planning that proved to be a major contributor to their later success. No small part of that planning was based on an extensive environmental analysis that seized opportunities to create amenities that were extensions of good land management concepts.

"Blue-green" storage areas are another method of storm water management that can often be employed as an

FIGURE 16-10. Permanent pond.

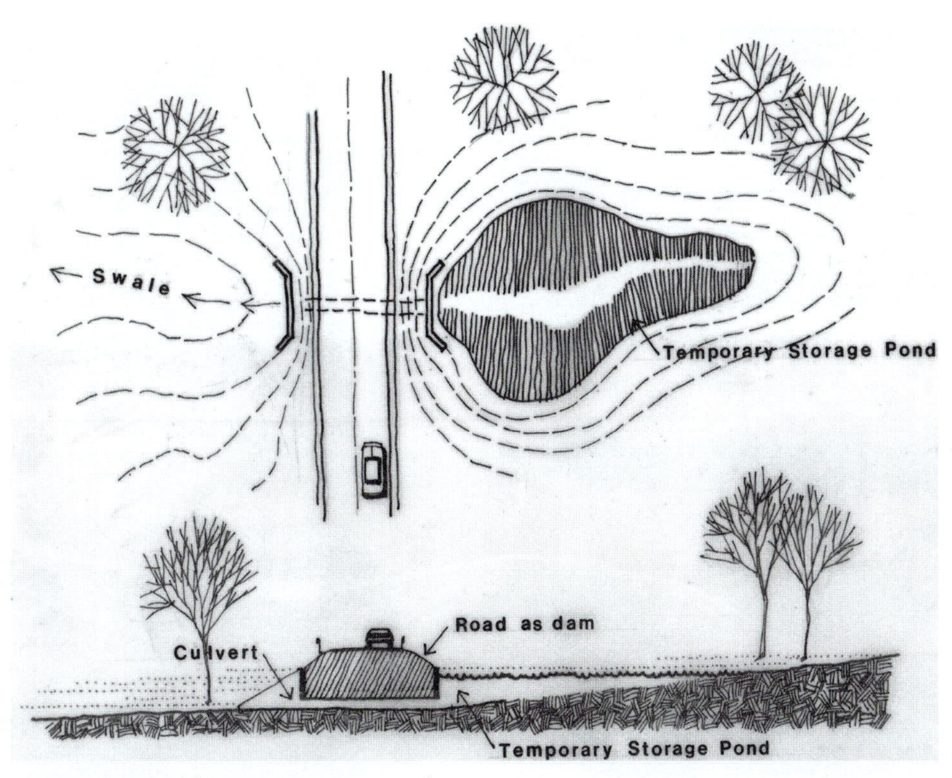

FIGURE 16-11. A blue-green storage pond.

amenity. The concept can be used in any situation where the water that naturally drains into a deep swale can be temporarily impounded. If a road needs to cross the swale, the grading to support the road becomes the bank, which retards the flow of water through the swale, changing the dry "green" area to a wet "blue" area until the water has drained through the culvert or pipe at the bottom of the swale (Fig. 16-11). This is a very simple yet positive mechanism for reducing on-site runoff and controlling the volume and rate of water being discharged from a site.[19]

Where possible, parking lots can be designed to hold water. This can be accomplished by designing the lot's slope, surface planes, and curbs to retain surface water and control its release. Some caution must be exercised in the design of the lot, to avoid deep ponding in areas in which people may need to walk. Since the speed of the auto in the parking lot will be slow, hydroplaning should not be a problem. The principal issues are the location of the area to retard the flow of water, the selection of the appropriate drain type, and the type of paving.

Erosion Reduction and Abatement Methods

There are numerous products available to assist in establishing proper erosion control on even the steepest grades. Although these products can sometimes be expensive to use, particularly over large areas, their savings in maintenance costs for severely eroded land can be significant.

These temporary soil-stabilizing products vary greatly in material and compositional makeup, but they are all used in a similar manner and toward the same end. Some must be installed after seeding and mulching; some provide their own mulch, and still others can be seeded after installation, thereby providing instant soil support. Most of the material is available in large rolls and is simply unrolled across the area to be stabilized and stapled to the earth. They are all designed, with varying degrees of success, to provide a beneficial microenvironment to protect and promote seedling growth and inhibit erosion forces until grass root systems are established enough to provide their own natural erosion deterrence.

Over time, many materials and products have been developed to deal with the problems of temporary soil stabilization and vegetation promotion. Although technological innovations are prevalent in various products, others rely principally on traditional materials and methods.

One product is a strong, durable, lightweight plastic netting for covering seeded and mulched areas. The netting can also be used to hold sod in place and disintegrates by sunlight as the turf is established (Fig. 16-12). Another product combines a mat of curled and seasoned wood excelsior with a layer of biodegradable plastic mesh. It is designed to hold soil in place, create conditions for

[19] *Residential Storm Water Management*, p. 35.

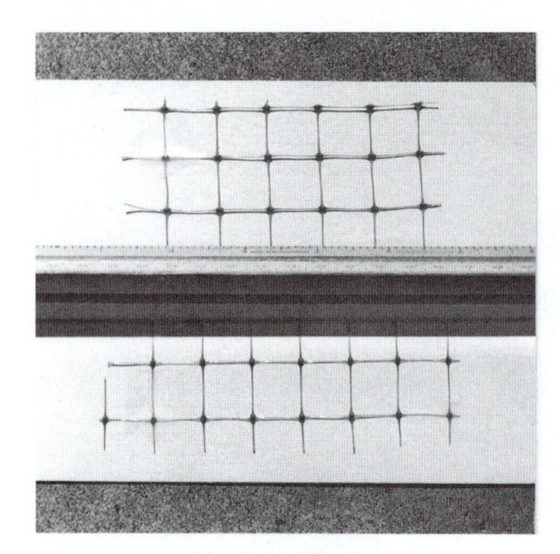

FIGURE 16-12. Plastic netting.

FIGURE 16-13. Wood excelsior.

FIGURE 16-14. Nylon monofilaments.

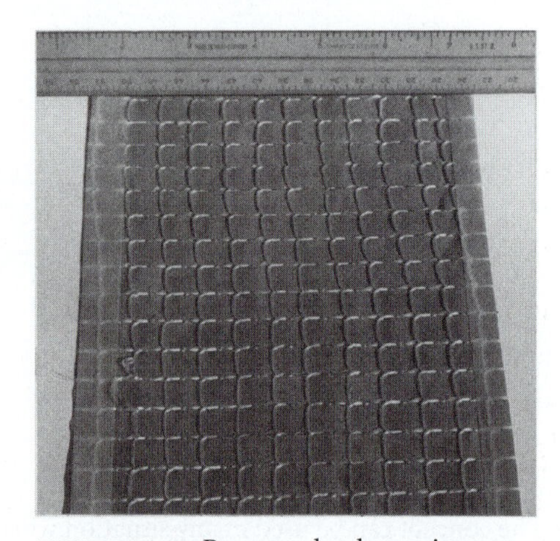

FIGURE 16-15. Paper and nylon twine.

FIGURE 16-16. Jute mesh.

successful grass seed germination, and protect seedlings after sprouting (Fig. 16-13).

Yet another product is a flexible soil reinforcement matting made from heavy nylon monofilaments fused together. Since 90% of its volume can be filled with soil or gravel, the mesh of wire provides a matrix for vegetative root growth and soil infiltration. The rigidity and open quality of this product make it a durable material with various erosion control uses (Fig. 16-14).

Another product is a combination of polypropylene yarn interwoven with strips of paper. The paper blocks the force of rain and acts as a mulch. When decomposed, the net holds the roots and soil in place as the vegetative mat develops (Fig. 16-15).

A jute mesh of natural fiber is a more traditional product and is ordinarily used to hold soil and seed intact on slopes and other areas of concentrated water flow. The fiber itself acts as a built-in mulch that will decompose in time and add organic matter to the soil (Fig. 16-16).

Many factors must be considered when choosing an erosion control product. Cost, availability, time, and specific applications are all important in making a selection. All products must be compared directly to the use of lay-in sod as a soil stabilizer: while the season, cost, and labor must be part of the considerations in selecting a material, when properly laid, sod is still the only alternative that offers nearly immediate living ground cover.

Open Paving Systems

A myriad of environmental problems associated with development have forced design professions to reevaluate the real costs of various methods and materials used for years without question. Although the simplicity and low cost of the asphalt or concrete parking lot was widely accepted, their aesthetic qualities left something to be desired. It is only recently, with the problems associated with on-site runoff and groundwater recharge, that the value of alternative systems became real.

At the turn of the century, many roads in American and European towns were paved with brick and cobblestones. At the time, these were simply the only readily available materials dense enough to be considered as paving surfaces. Considering the remedial suspension systems for horse-drawn carriages, these pavers were very coarse. Their aesthetic quality, however, was another matter. Their texture, color, and size permitted easy integration into the natural landscape. But with the onset of the automobile and advances in technology, bricks and cobblestones were replaced by more economical and smoother paving methods and materials. The cost of shipping, since kilns and quarries were often remote, and the labor-intensive character of the individual paver, coupled to keep their relative costs high. Compared to construction with monolithic slabs of concrete and asphalt, the construction of a road with individual pavers became exorbitant. Furthermore, the smooth, seamless quality of concrete and asphalt was a welcome change from a collection of masonry units.

Since the street was the first parking lot, the natural inclination was to apply the same technology to the parking lot. In a marginally urbanized society, naturally, little consideration was given to the surface runoff from parking lots. Although architects and landscape architects maintained a preference for the smaller masonry unit because of its aesthetic quality, its use has been limited to historical streets and motor courts of expensive hotels or corporate headquarters. However, problems associated with storm water management and erosion control have resulted in an examination of popular practice and development policies by various state and local governments. Since traditional concrete and asphalt densities accept negligible amounts of moisture, there emerged an obvious need for options to the usual paving standard.

The use of modern open pavers was developed in Europe. The Europeans hold their gardens in high esteem, and the problem of how to provide access for vehicles to their homes and parks is of major importance. The concept of open paving is as simple as garden flagstones. The open paver is designed to provide a solid surface only for the area necessary to support pedestrian traffic. The area that was unnecessary for support is left as open as possible, permitting plant growth. The green area is filled level to the top with soil to allow for cutting of the planting material when necessary. When in place, the solid surface becomes visually obscured by the grass. The open area accepts the runoff from the impervious surfaces and, ultimately, reduces the amount of runoff into the public sewer systems. Obviously, any reduction on one site can be multiplied by many sites, with the long-term implications becoming substantial.

Although there are numerous positive aspects associated with using open pavers, there are two negatives that must be recognized. It is unlikely that any system will cost less than the traditional poured-in-place concrete or asphalt paving compared as to cost per square foot. If the real economic value of the systems is to be considered at all, it must be considered in a comprehensive way and evaluated as part of the entire drainage system. The other negative has to do with women's fashions. Small, narrow-heeled shoes and open paving systems are not extremely compatible. Some prudence must be used in identifying the placement and pattern of the systems when pedestrians are anticipated users.

Several different manufacturers have created distinct open paving systems that can be separated into four general classifications: (1) individual precast units, (2) precast modular mats, (3) monolithic poured-in-place slabs, and (4) turf reinforcement.

The precast units employ two different conceptual design theories. One has been designed as a solid unit to be laid in place with an open joint between units. The open joint and porous underlayment permit surface water to filter through and recharge the groundwater supply. The second type of individual precast unit has been designed with an open void as an integral part of the unit. The voids are filled either with soil to support turf growth or a mixture of soil and gravel for a stable surface (Fig. 16-17). These open systems are often wire reinforced to accept the static and dynamic loads of vehicles. Although precast modular mats have a variety of configurations, the principles are essentially the same. The mats employ design concepts similar to

FIGURE 16-17. Precast open paving units.

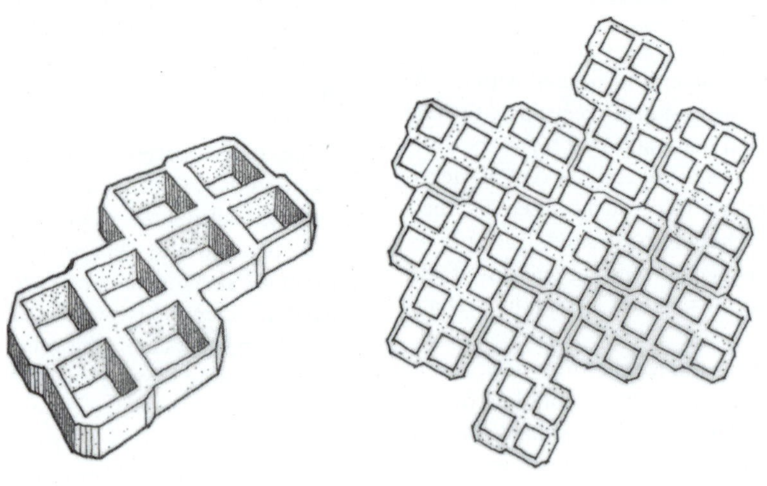

FIGURE 16-18. Precast open paving systems: mats.

those of individual open units but have been developed to respond to two principal issues. Designed as an integral interlocking system, the mats can accept higher vehicular loads than can the precast units (Fig. 16-18). Furthermore, because the mats are larger in area than the individual units, they might require less labor and have a lower total cost per square foot.

The poured-in-place slab techniques combine the technology of traditional concrete formwork with an open area of approximately 40% of the slab. These slabs have been designed with two-way reinforcing and, because of the nature of the poured-in-place techniques, can be modified to accommodate varying edge conditions and topography (Fig. 16-19).

The uses of open paving systems are as varied as the types. The predominant value of open paving systems has been one that resolves the onus of asphalt and concrete parking lots and driveways. Since every system is designed for a specific purpose and has inherent limitations, not every product is made with the capability of supporting firetrucks and heavy equipment. Some municipalities have become involved in the testing of open paving systems for use in fire lanes. Because criteria and standards vary widely from one municipality to another, a check with local building code officials is imperative.

Another prominent use of open paving systems is in erosion control. Open pavers have been used extensively on vulnerable slopes along highways and as stabilizers on stream and lake banks. Systems designed with interlocking characteristics work well across undulating surfaces. The monolithic quality of the poured-in-place systems, even in vertical planes, has effectively resisted the back swelling that commonly pushes out individual pavers along a shoreline.

Compared to concrete or asphalt surfaces, open paving systems can substantially reduce on-site runoff. In some cases, the percolation ability of a site can actually be improved if somewhat impervious soil is replaced by an open paving system with a permeable topsoil and gravel or sand base. When used in drainage swales or culverts, open systems not only permit better groundwater recharge, but the aesthetic quality of the ground cover is much preferred to the glaze of concrete channelization.

FIGURE 16-19. Poured-in-place open paving.

PART **FIVE**

STREETSCAPE AND SITE IMPROVEMENTS

- **CHAPTER 17:** Streets and Roads
- **CHAPTER 18:** Parking and Signage
- **CHAPTER 19:** Pedestrian and Bicycle Circulation
- **CHAPTER 20:** The Building and Energy
- **CHAPTER 21:** Land Use Controls

CHAPTER SEVENTEEN

STREETS AND ROADS

Roads and parking are two site-planning problems that architects and landscape architects approach with cautious optimism. These two aspects of the built environment are sources of intrusion and environmental degradation like nothing else in the client's program of requirements. While concerns and biases are justified, understanding more of the road's function and the precise criteria of parking can provide the basis for a more enlightened approach to the problem. The objective here is to understand the language, vocabulary, and forms of streets and roads as they relate to site planning. Details on parking design criteria are provided in Chapter 18.

STREET SYSTEMS

The scope of streets and roads must begin with a brief overview of street systems from a conceptual perspective. First, contrary to popular belief, street systems have a design capacity and as such have been designed to support land uses. A street's design capacity has been based on a design speed, a specific cross section, and a designated "level of service." Recognize that these three variables are linked and none can be changed without affecting the remaining two. This is critical to understanding the role of streets and roads as a component of land use planning. Freeways, parkways, expressways, beltways, thoroughfares, major and minor arterials, and neighborhood streets are all terms given to various road systems to serve a collection of land uses. The design of the system is one in which a balance is achieved between (1) public transit, (2) streets or thoroughfares, and (3) a community's land use densities. The common perception is that something is "wrong" with the street system when peak-hour traffic creates congestion on major thoroughfares. It often takes a significant amount of traffic and visibly deteriorating environmental quality for someone to question the zoning or integration between land use and transportation planning. Land use and circulation systems go hand-in-glove and to consider them as separate and independent factors invites a cycle of problems. There is probably at least one example of a road in your community in which the traffic generated by the road created a demand for rezoning (land use) changes, resulting in another demand for the "widening" or improvement of the original road.

The physical configuration of streets and roads at a larger scale has traditionally been separated into various models (Fig. 17-1): radial, orthogonal, linear, and so on. While understanding the nomenclature of a system can prove to be valuable, do not interpret a regions' physiography based solely on the form of its street system. As an example, in numerous communities in the Great Plains and Southwest, the orthogonal grid was an appropriate system. The grid was also used in the early planning of San Francisco, where the topographical relief is extraordinary (Fig. 17-2).

Transportation planning must begin with the identification of land uses that will be served by the system and recognition that each land use will generate a specific number of vehicle trips per day. Obviously, a neighborhood retail shopping center attracts (generates) a significantly larger number of auto/truck trips than does a single-family detached residence. The number and frequency of those trips have been researched extensively by the Institute of Transportation Engineers (ITE). Since national trends continue to change, the ITE publishes its research periodically. Although some communities have seen the need to conduct their own research on trip-generation factors, many others utilize the ITE's work in the critical step of the transportation planning process. Table 17-1 is an example of the ITE's latest report on residential land uses.

Without this base of data, the assignment of any numerical quantity of traffic to a road would be an arbitrary value judgment. Utilizing the data, the transportation planner can convert a land use type and its density into an average number of trips.

Levels of service are also part of the transportation planning process. Since each street's capacity is limited by the number of moving lanes and the planned design speed, the question of "how much is too much?" will always surface. Levels of service from A at the best to F at worst have been assigned for a qualitative evaluation. Since the ability of the system relies, like all systems, on the weakest link in the chain, that link is almost always at the intersections. An explanation of levels of service at an intersection follows:[1]

A: Conditions of free, unobstructed flow, no delays, and all signal phases sufficient in duration to clear all approaching vehicles.

B: Conditions of stable flow, very little delay; a few phases are unable to handle all approaching vehicles.

[1]Highway Research Board, *Highway Capacity Manual* (Washington, D.C.: National Academy of Sciences, 1965).

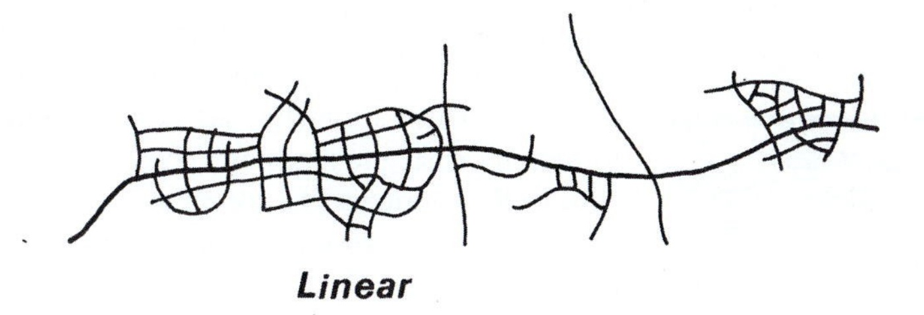

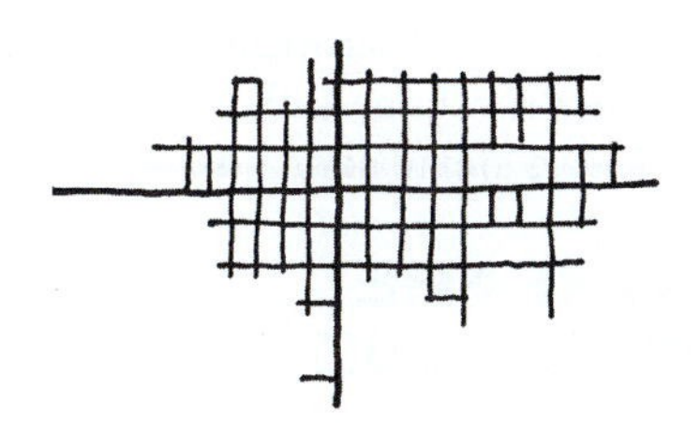

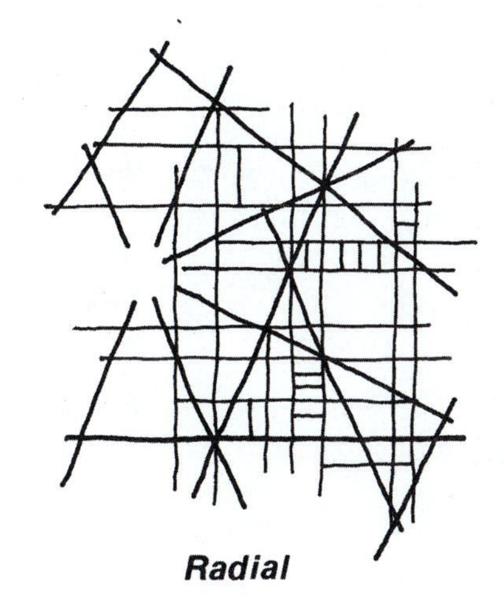

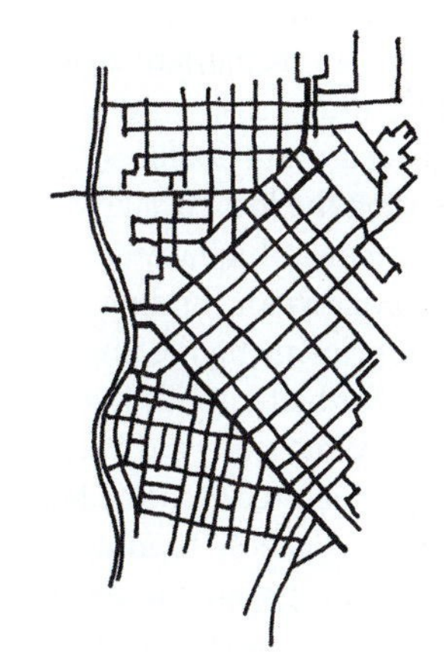

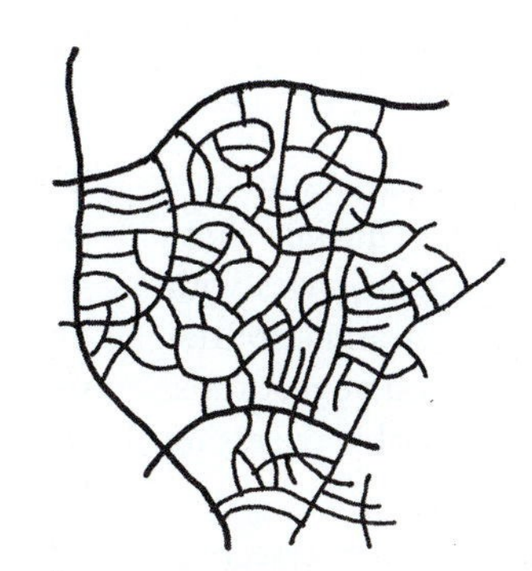

FIGURE 17-1. Conceptual diagrams of street systems.

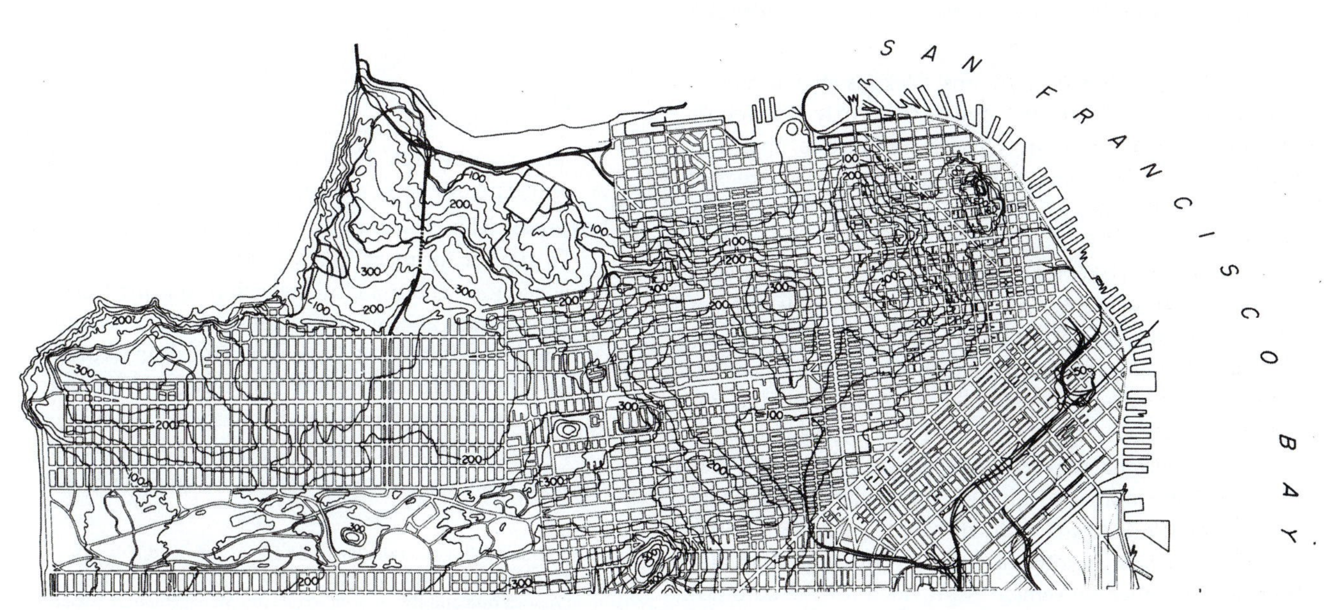

FIGURE 17-2. Map of San Francisco. (Photo courtesy of City Planning Department, City of San Francisco)

Table 17-1. Summary of Rate Tables of Various Types of Dwelling Units

Type of Dwelling Unit	Average Weekday Vehicle Trip Ends per Unit		
	Average	Maximum	Minimum
Single-family detached housing	9.57	21.85	4.31
Apartment, general	6.65	12.50	1.27
Low-rise apartment	6.59	9.24	5.10
High-rise apartment	4.20	6.45	3.00
Residential condominium/ townhouse	5.81	11.79	1.53
Mobile home	4.99	10.42	2.29
Senior adult -Adult housing detached	3.71	5.70	2.90
Residential planned unit development	7.50	14.38	5.79

Source: Institute of Transportation Engineers. *Trip Generation,* 8th ed. (Washington, D.C.: ITE, 2008, ITE, 1976), Section 200.

C: Represents stable operation with intermittent loading of the approaches. Drivers may have to wait through more than one red signal, and backups may develop behind turning vehicles. This is the level typically associated with urban design practice.

D: Approaches instability of flow, when delays to approaching vehicles may be substantial during short peaks within the peak period, but enough cycles with lower demand occur to allow periodic clearance of developing queues, thus preventing excessive backups. Level of service D is considered a reasonable operating level for urban intersections.

E: Represents the most vehicles that any intersection approach can accommodate. Long queues of vehicles develop, and delays are excessive (up to several signal cycles).

F: Represents jammed conditions. Backups from locations down-street or on the cross street may restrict or prevent movement of vehicles out of the approach under consideration, hence volumes carried are not predictable.

The selection of a road's cross section is based on supporting land uses, trip-generation factors, acceptable levels of service, and directional and peak-hours assignments. Although everything mentioned here relates principally to auto/truck vehicular systems, public transportation is an important ingredient in every community's transportation plan.

The important concept is to recognize that the automobile, with its insatiable appetite for parking and wider paving, can create havoc with any site. Experience reminds us that the pedestrian must be given a priority. Additionally, if every alternative movement is provided for the automobile, little of value will be left for the pedestrian.

As a part of the process, transportation planners continually seek to keep the system balanced and forecast those locations in which segments may be overloaded. As a method to identify what the loads are, checks are made periodically with traffic counters. The counting quantifies the number of vehicles that cross over a line on a street for a given period of time. The counters provide for 24-hour documentation through a sufficient number of weekly cycles that averages can be extrapolated and used in planning. These numbers are referred to as average daily trips (ADTs) and can be obtained from local and state highway departments. ADTs provide accurate identification of the volume, time, and direction of the flow of traffic on freeways and major roads.

THOROUGHFARE CLASSIFICATIONS AND CRITERIA

Just as each community adopts ordinances and laws that reflect its own values and historical precedents, each will also have adopted some general thoroughfare cross sections applicable to that particular municipality. Although there will be some similarities between municipalities, the site planner must begin by documenting the local design criteria.

In the early development of highways and road systems in the United States, the requests for funding of the programs required some method of prioritizing the improvements. Demands by rural legislators for better roads were challenged by their urban counterparts. What emerged was the collective agreement that the roads which received the lightest amount of travel would receive the smallest funding. State governments in the early 1930s also began the development of a classification system based on a road's function. Although the classification system varies from state to state, the National Committee on Urban Transportation recommends four functional classifications:[2]

Expressway. This classification includes similar terms, such as "freeways," "beltways," and "parkways." These are limited-access systems that rarely, if ever, have grade crossings. Expressways are designed to accommodate the longer intraregional trip, with the highest legal speed limits. These are multilane divided roads and generally carry over 40,000 trips per day.

Arterial. This classification also includes the terms "major" or "primary" road, as it connects major traffic generators within a city. Arterials also bring traffic to expressways and, in some cases, incorporate important rural routes. The system of arterials in a community should reflect a systematic form that integrates the major employment and activity centers with expressways and collectors. These streets will vary in design capacity from 25,000 to 40,000 trips per day.

[2]Frank So, Israel Stollman, Frank Beal, and David S. Arnold, Eds., *The Practice of Local Government Planning* (Washington, D.C.: American Planning Association and the International City Management Association, 1979), p. 216.

Collectors. The collector is also referred to as a "secondary" road and has two functions. The first is as a connector between the arterial system and local system. The second function is one of access and service to local land uses connecting major intracity functions. The collector is a local street serving an area within an arterial system zone and capable of supporting a traffic volume of 10,000 to 25,000 trips per day.

Local. Local systems are known by various terms, including "minor," "tertiary," and "neighborhood." This road classification is the one that generally represents the largest amount of street mileage in any community and the lowest design speed. The local street system is designed to provide access to adjacent land uses. Although local streets in a central business district (CBD) may have traffic counts in the thousands, neighborhood streets generally carry volumes below 1000 trips per day.[3]

Alan Voorhees, Walter Hansen, and Keith Gilbert, in their "Urban Transportation" contribution to *The Practice of Local Government Planning,* suggest that in new areas the development of a transportation system might have as many as nine classifications for a street system.[4] The nine-level classification system in Table 17-2 is more definitive concerning the site, function, and performance characteristics of various roads.

[3]Ibid.

[4]Alan Voorhees, W. G. Hansen, and A. K. Gilbert, "Urban Transportation," in *The Practice of Local Government Planning,* Frank So, Israel Stollman, Frank Beal, and David S. Arnold, eds. (Washington, D.C.: American Planning Association and the International City Management Association, 1979).

Every site-planning problem presents its own unique set of circumstances relative to land use and circulation systems. Any classification system can only serve as a guide to understanding how the variables of each street fit within the context of other streets. Since each community will have adopted its own standard sections, the site planner must match the municipality's requirements with the road and right-of-way classification most appropriate to the site and circulation problem.

INTERSECTIONS

The next level of road design issues of which the site planner should have some working knowledge is the configuration of intersections. The basic model for most communities is the 90° intersection, as it respects the layout of the numerous grid systems throughout the United States. The problem with this configuration is that there are eight points of potential conflict (Fig.17-3a). If either or both roads are one-way systems, the conflicts are reduced to four and one, respectively (Fig. 17-3b). The T intersection is widely used, as it essentially accomplishes a similar objective (Fig. 17-4). The T intersection works very successfully but should not be used in cases where the T of sets are closer than 120 ft (Fig. 17-5). Local ordinances may be more restrictive than this guideline and should be reviewed before final design decisions are made.

In the older sections of numerous cities and towns, conflicting street geometry often creates dangerous and difficult intersections that should not be duplicated (Fig. 17-6). Junctures of multiple street systems almost always result in at least one five-cornered intersection. It is very perplexing

Table 17-2. Urban Freeway and Thoroughfare Classification System[a]

Classification	Trips per Day	Moving Lanes	Width (ft)	Emergency Lanes (12 ft)	Park Lanes (8 ft)	Access: Residential	Access: Commercial	Design Speed (mph)	Paving Width (ft)
Freeway	40,000	4–8	12	2	□	□	□	60[b]	112–160
Major arterial	25,000–40,000	4–6	12	2	□	□	□	50	92–116
Arterial	10,000–25,000	4	12	2	□	⬓	■	40	68
Major collector	over 1500	2	12	□	■[c]	■	■	35	32–40
Major connector[d]	1,500	2	12	□	⬓	■	■	35	24
Minor collector[d]	1,500	2	10	□	■	■	■	30	28–36
Minor connectors[d]	1,500	2	10	□	□[e]	■	■	30	20
Loop streets[f] (cul-de-sac)	250	1	10	□	□	■	■	25	26
Parking connectors[f]	250	2	10	□	□	■	■	25	20

[a]□, *Inappropriate;* ⬓, *discouraged;* ■, *acceptable.*

[b]*Maximum legal speed 65 mph.*

[c]*One side or both sides.*

[d]*Collectors.*

[e]*One side in residential development.*

[f]*Locals.*

Source: Alan Voorhees, W. G. Hansen, and A. K. Gilbert, "Urban Transportation," in *The Practice of Local Government Planning,* Frank So, Israel Stollman, Frank Beal, and David S. Arnold, eds. (Washington, D.C.: American Planning Association and the International City Management Association, 1979), p. 220.

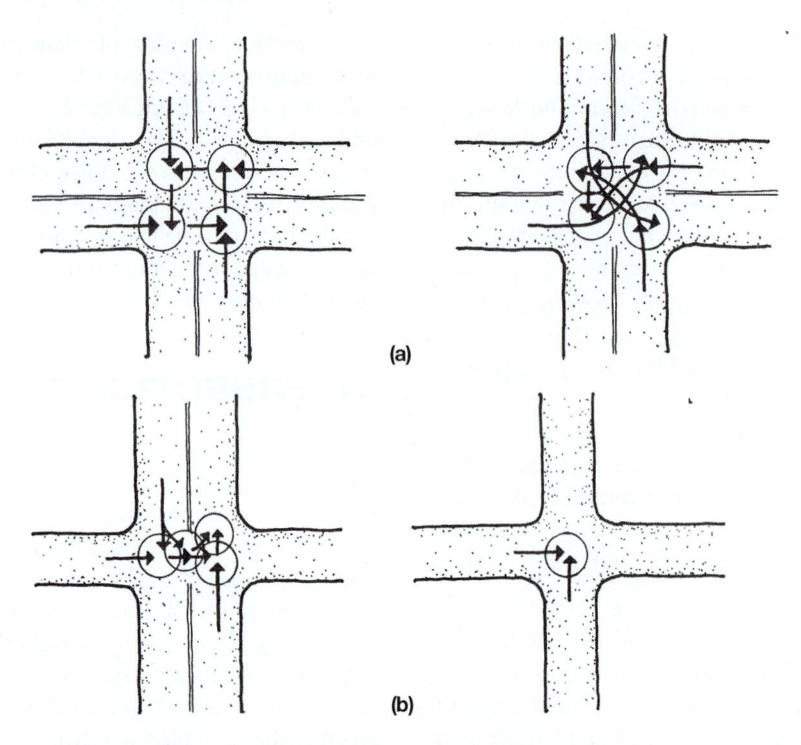

FIGURE 17-3. The 90° intersection. (a) Conflicts with two-way streets. (b) Conflicts with one-way streets.

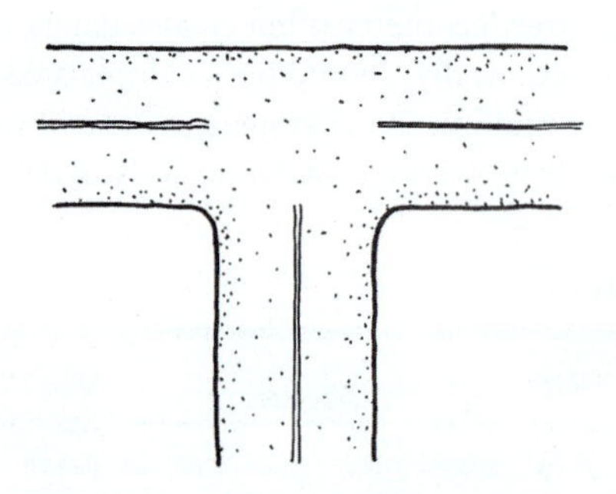

FIGURE 17-4. The T intersection.

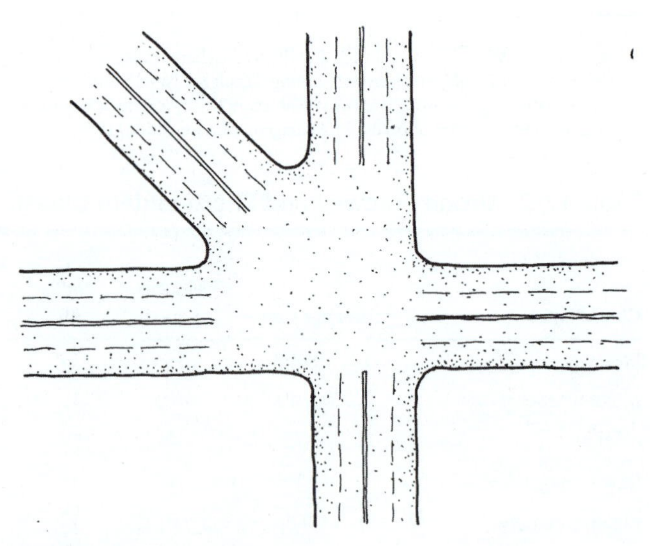

FIGURE 17-6. Five-corner intersection.

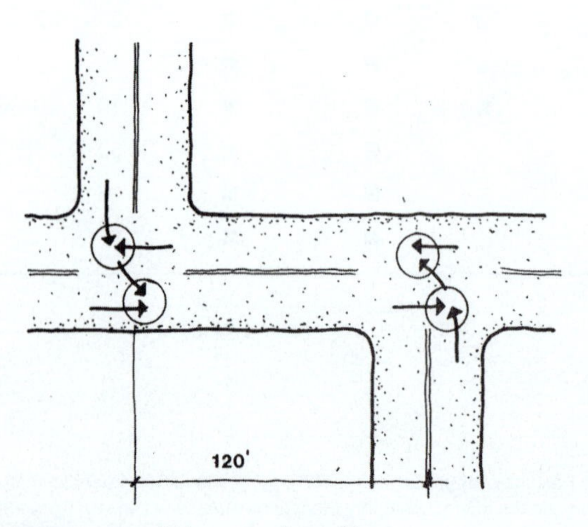

FIGURE 17-5. T-intersection offset.

to the driver who arrives at a corner and is faced with four options—one more than normal. With speed, the perplexed driver can become a moving hazard. The "red time" per signal is also longer because of the additional movements that need "green time" in the same cycle. Acute angles should also be carefully considered because the right-of-way that is necessary to accommodate the turning movements at the corner can become substantially larger than that required in

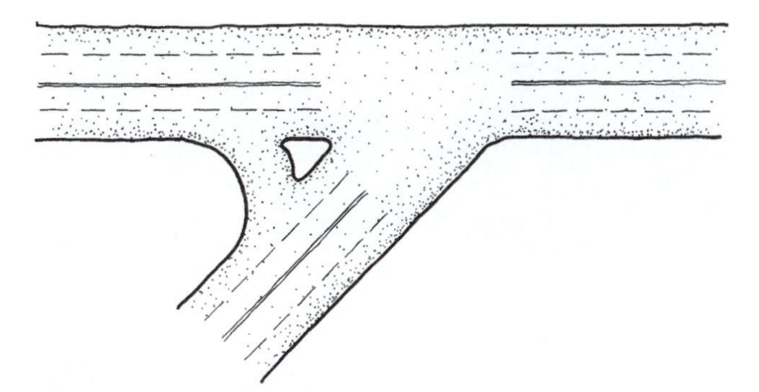

FIGURE 17-7. Obtuse-angle intersection.

the right-angle intersection. Furthermore, the 90° intersection is preferred because it does not favor one particular directional flow. Since the right-of-way is larger for any intersection that is other than 90°, configurations similar to those in Fig. 17-7 should be avoided.

Traffic Circles, Rotaries, and Modern Roundabouts

The circular intersection of roads has been used extensively throughout the history of city planning. This is an intersection where the paths of two roads do not intersect at 90° but follow the line of a concentric circle around an island at the center of what would have been the intersection of the two roads. The most notable of these circular junctions is the *Place de l'Etoile,* the setting for the *Arc de Triomphe* in Paris. The circle was a key part of the 1836 plan developed by Pierre L'Enfant for Washington, D.C. L'Enfant used the circle at strategic locations in his plan for the capital city, where he superimposed a new diagonal street system over the existing orthogonal street grid. Strategic junctures of the two street patterns were the sites for the new traffic circles to be punctuated by heroic sculpture, monuments, or other design features (DuPont Circle, Logan Circle, Thomas Circle, etc.) The *traffic circles* were not unique to Washington, D.C. More than half of the traffic circles in the world are in France.

In 1909, the British used the traffic circle or rotary as a traffic island for pedestrians in Letchworth Garden City. The circles used in U.S. cities were sufficiently large to accommodate classical fountains, monumental statues, or horticulture features (Fig. 17-8). But as the number of vehicles on streets and roads increased, the softer "let's-work-together" ethic of the traffic circle lost out to the strict instructions of the octagonal stop sign and tri-colored traffic lights. And assessing fault is certainly easier at a signaled intersection than in the circular merging lanes of a traffic circle.

Transportation planners' persistent pursuit of safer intersections continued to refine the circular intersection known as the traffic circle, or rotary. By the mid-1960s, British engineers had redesigned circular intersections in an effort to address the safety and capacity issues inherent with the older forms. Although the new configuration was about the same size as a signaled intersection, the smaller center islands encouraged slower speeds, resulting in safer modern roundabouts (Fig. 17-9). The success of the 1960s resulted in a renewed interest in the form from which a set of design criteria evolved. One of the criteria for the modern roundabout is that the traffic entering, or merging with the traffic, is burdened to provide the right-of-way to all other vehicles in the traffic circle. Exiting directly from the inner lanes of the modern roundabout is generally permitted. This is in contrast to exiting from a traditional traffic circle or rotary, which requires one to first achieve a position in the lane furthest from the center.

An important aspect of pedestrian and bike safety is that both are provided with walks, often along splinter islands, outside the vehicular traffic. Prohibiting pedestrian and bike traffic from merging with vehicles moving around the center island has, along with the other criteria mentioned here, helped to make the modern roundabout a statistically safer intersection compared to the un-signaled four-way stop. It is also important to note that the movement of vehicles around the traffic island is counterclockwise in those countries that drive on the right side of the road (Fig. 17-10). Since many of the roundabouts in the United States do not meet the criteria mentioned above, transportation planners are careful to use the term *modern roundabout* for the most current models and *traffic circle,* or *rotary,* when discussing an older design.

FIGURE 17-8. Columbus Circle, New York, circa 1907.

FIGURE 17-9 (a). Modern roundabout with cars. (Photo courtesy of Una Smith)

FIGURE 17-9 (b). Modern roundabout. (Photo courtesy of Andrew Bossi)

The modern roundabout has witnessed revived interest as a method to permit traffic to continue to move at a slower speed but without the need for electronic signals. Modern roundabouts have been used in conjunction with other strategies to achieve traffic calming in residential neighborhoods and urban centers. *Traffic calming* is a term that has been used to describe a program of methods to both reduce high-speed travel and increase the opportunity for a pedestrian-scale street to function in congested urban environments. The concept of traffic calming has been pursued by a variety of countries, using various methods. With the Dutch *Woonerf* street, the Netherlands was one of the first countries to advance the radical hypothesis that, under some circumstances, the auto and pedestrian can share the same street. The *Woonerf* street is surely a model example of traffic calming.

The Oregon Bicycle and Pedestrian Plan, published by the State of Oregon Department of Transportation,[3] states that some of the advantages of the roundabout include:

1. Keeping traffic moving through the intersection, thereby increasing the flow of vehicles and reducing the necessity for additional lanes
2. A reduction in the incidence of collision
3. A reduction in the severity of injury and vehicle damage due to the slower speeds through the roundabout
4. Lower capital improvements costs for the intersection due to the absence of electrical signals

It must be noted that the fact that although the size of the right-of-way required for the modern roundabout is somewhat larger, that fact is important only if the municipality will be required to purchase additional right-of-way.

Statistics and observations support using the modern roundabout as a strategy that can contribute to traffic calming. The change in the character of the streetscape gets a driver's attention and thereby leads to a reduction in speed. Motorists realize first that something different is evolving, and once into the roundabout they note that further caution and reduction in speed are warranted. The message here is to use the modern roundabout under those circumstances where traffic volumes are modest and can be a key part of a coordinated traffic calming program.

[3]Available at www.oregon.gov/ODOT/HWY/BIKEPED/docs/bp_plan_1.pdf

Give way sign at a roundabout in Australia (left-hand traffic)

Road sign before roundabout in the United States (right-hand traffic)

FIGURE 17-10. Roundabout directional signage.

CHAPTER EIGHTEEN

PARKING AND SIGNAGE

Parking is a difficult and thorny problem for the site planner. If all the time, energy, and financial resources to move vehicles onto a site were not enough, the storage of the devices adds insult to injury. When submitting projects for consideration by a jury for design awards, many site planners would prefer to avoid *any* photographs or drawings of areas allocated to parking. The site planner and client address parking, in many cases, as a necessary evil.

The user, however, has an entirely different attitude toward parking. Anyone involved in the rental or leasing of residential, retail, or commercial space will testify to the concern users have about parking. In suburban retail shopping areas, parking is critical to the ability of many merchants to survive economically. That same attitude is paramount to the residential user who prefers a "bedroom-to-bathroom" relationship regarding parking and dwelling. The reason is obvious: convenience. What the merchant often undervalues is the quality of the parking/pedestrian linkage. Although an adequate amount of spaces in a visible location is valuable, the aesthetic quality of the path is also important.

Parking as a design problem is compounded because of the inconsistencies associated with simple standards. One would think that with as much research and documentation as city planners, architects, landscape architects, and civil engineers have on the topic, a set of common standards would be easy to develop and adopt. Not so. The industry cannot agree on the size of a standard parking space, much less the curb-to-curb dimensions for a standard 90° parking lot. Although various publications will define the size, dimension, and criteria for site components, including parking, for disabled persons, it would be inappropriate for the site planner to assume that the specifics of any publication other than the Americans with Disabilities Act have become part of the design criteria for his or her community. If the concept of a "national standard" for a normal parking space and lot size is still some time away, it is easy to accept the fact that the size and percentage of "small" or "compact" car spaces on a national basis is further behind. Some municipalities require a certain percentage of small car spaces, but not all do.

Because of the inconsistencies between local governments, the site planner must consult his or her local zoning ordinances for specific parking criteria and requirements. It is recommended that some preliminary documentation of those requirements take place to ensure that the number of spaces required is accurate and that minimum sizes are properly recorded.

Always inquire as to the municipality's policy on spaces for disabled individuals and for small or compact cars. The sizes and arrangements illustrated here will be based on:

1. Normal parking space: 9 ft × 18 ft
2. Parking space for disabled persons: 13 ft × 18 ft, including ramp and loading access space
3. Compact/small car space: 8 ft × 16 ft

PARKING CONFIGURATIONS

There are generally four basic forms or configurations of parking:

1. The 90° two-way lot (Fig. 18-1) has the smallest amount of wasted space (in corners and aisles) and so is the most efficient. Some consider it to be the most difficult to negotiate.
2. The angled one-way lot has variations: 60° and 45° (Figs. 18-2 and 18-3). Since anything less than 90° creates two triangles of wasted area (Fig. 18-4), it is easy

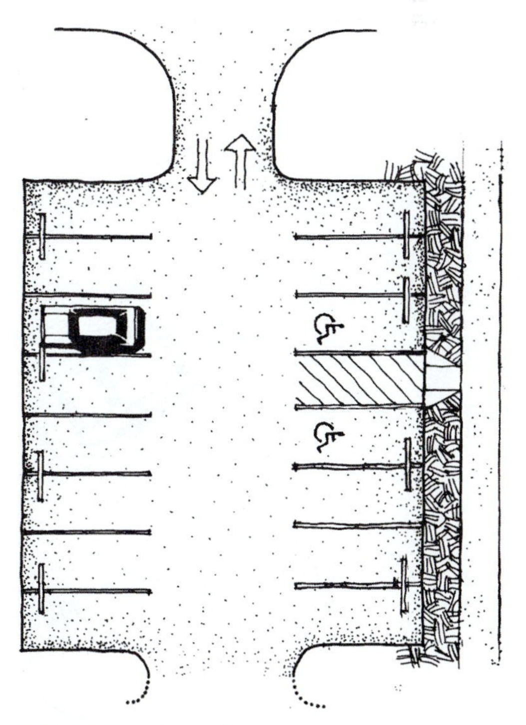

FIGURE 18-1. 90° parking.

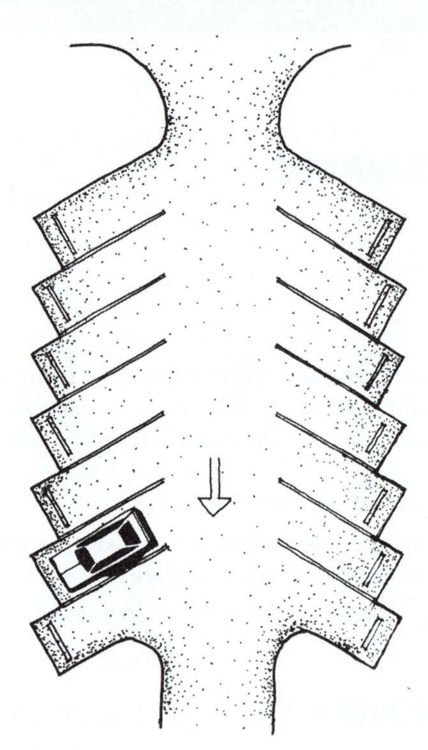

FIGURE 18-2. 60° parking.

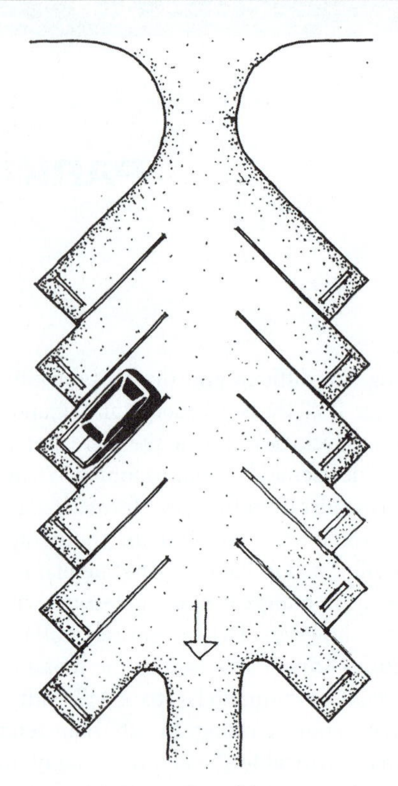

FIGURE 18-3. 45° parking.

to see why the efficiency goes down as the angle is increased. Since angled one-way parking can function with a narrower aisle width than any two-way system, it has obvious advantages in situations in which one-way systems are necessary (Fig. 18-5). A parking garage is a good example of a one-way system.

3. Two-way angled parking offers the advantage of easy ingress and egress. Suburban retail centers often use this form, as it accommodates two-way access down any aisle and has less than a perpendicular angle for those drivers who struggle with the perpendicular double-loaded lot.

4. Parallel parking is the final model. It is the most difficult for people to negotiate and the least efficient of the four models. One basic rule associated with parallel parking is that it should never be mixed with any

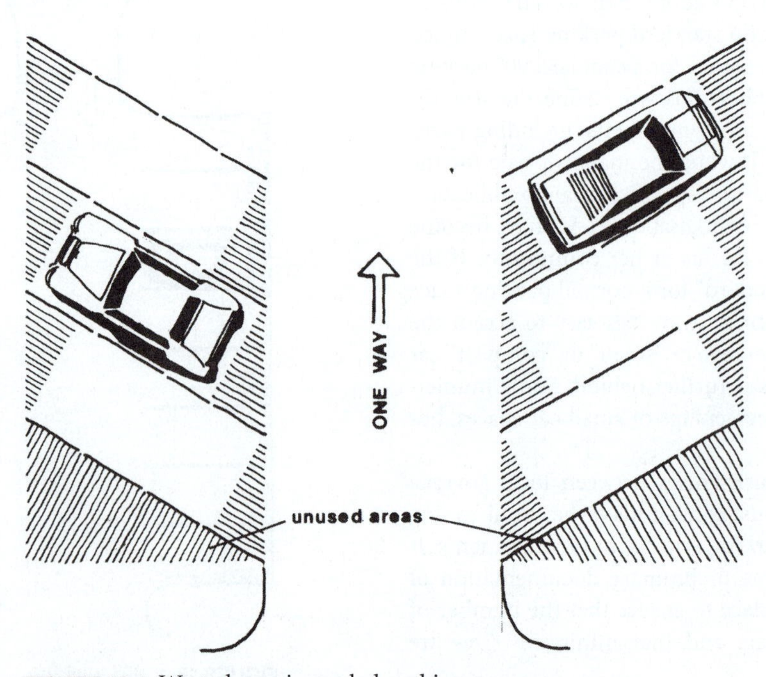

FIGURE 18-4. Wasted area in angled parking.

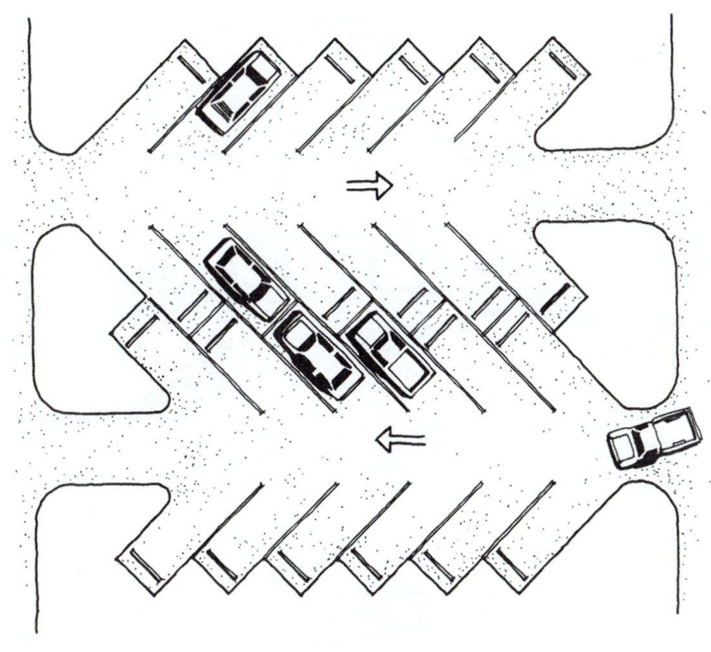

FIGURE 18-5. Angled parking: one-way.

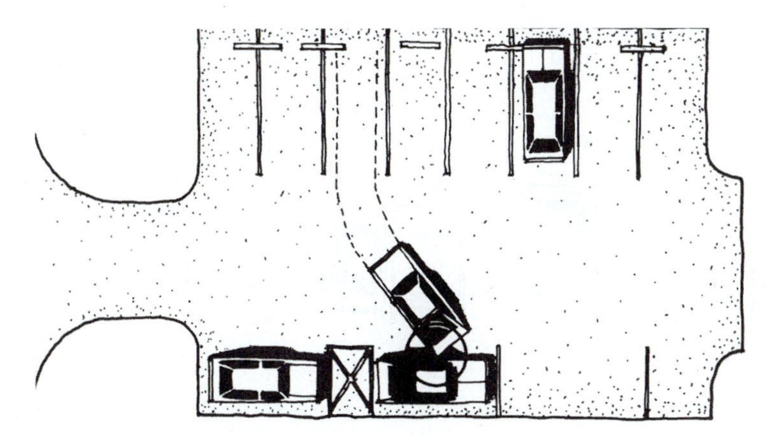

FIGURE 18-6. Mix of parallel and head-in parking.

other form of parking. Fig. 18-6 illustrates what not to do. The chances for a collision as a driver backs out of a space and into the side of a parked car are excellent!

PEDESTRIAN TRANSITION AND ACCESS FOR PEOPLE WITH DISABILITIES

On lots that receive a substantial amount of parking, it is often preferable to orient the parking such that people may walk either on paths between rows of cars or in the access lane (Fig. 18-7). Since the pedestrian is very vulnerable in a parking lot, the possibility of an injury can be reduced if the pedestrian is more visible. When parking is organized such that the pedestrian is forced to walk between cars, the possibility of someone stepping out from a line of parked cars into the path of a moving vehicle is ever present.

The ability to separate the pedestrian and the auto successfully is a challenge. Fire lanes and service accesses are also critical aspects of any circulation and parking plan. Since local ordinances will vary relative to a fire lane's paving, width, and structural characteristics, these criteria must be checked.

Single-loaded lots such as that described in Fig. 18-8 are inefficient. Although it is a common form for limited or restricted parking, it is obvious that the ratio of parked cars to paving is very low. Similarly, larger-than-necessary aisle widths consume development dollars, increase on-site runoff, and can be very confusing. Since there is little aesthetic value in most parking lots, increasing their size does not increase their contribution to a site plan.

When designing a parking lot that will be bounded on one edge by a building, make certain that one of two things

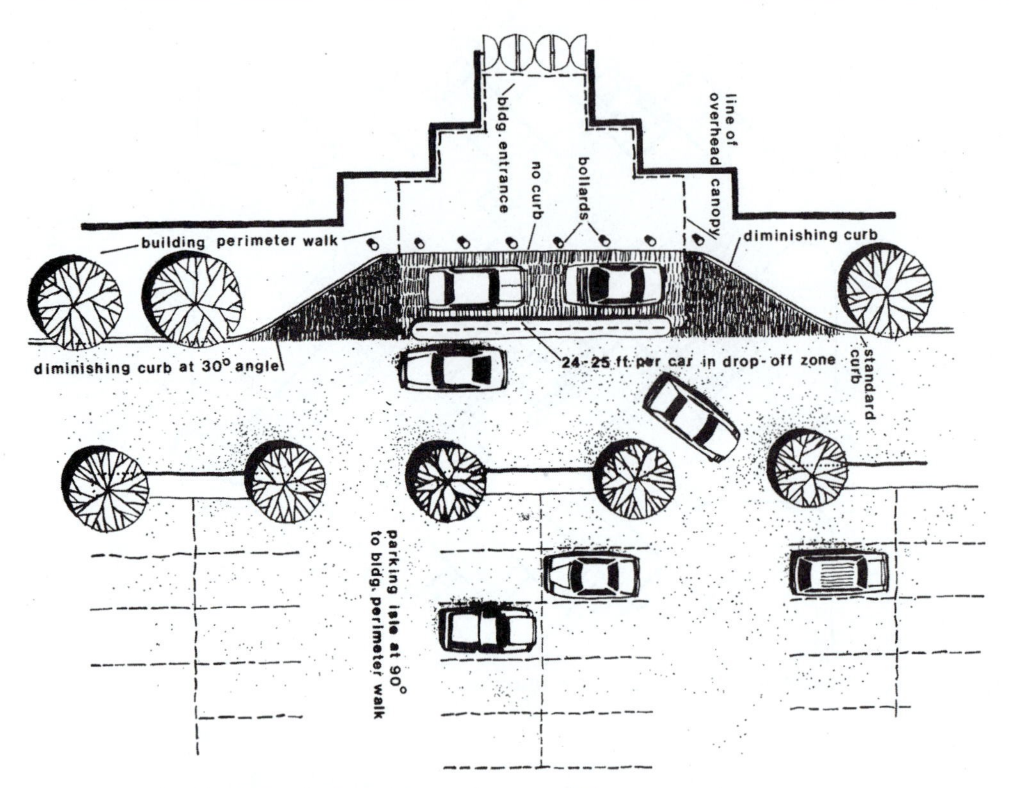

FIGURE 18-7. Parking organized so that people can walk in an access lane.

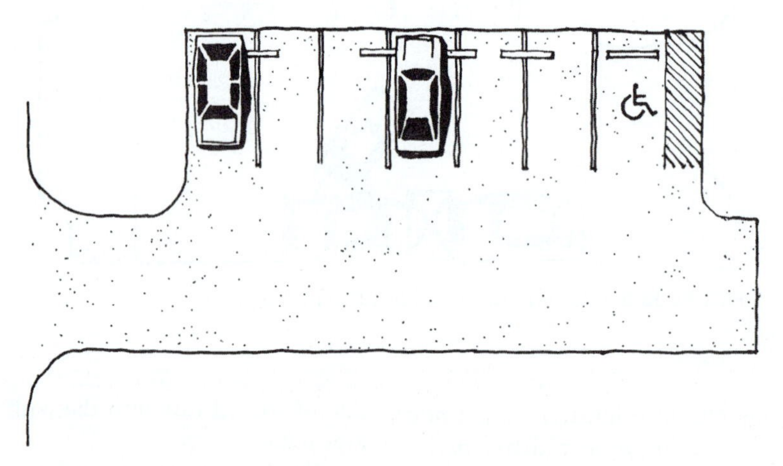

FIGURE 18-8. Single-loaded head-in lot.

takes place. First, a wheel stop is absolutely critical to keep auto bumpers away from a building wall (Fig. 18-9). Second, sometimes a screen or guard is necessary to protect walls or structures; they may be vulnerable B (Fig. 18-10).

On occasion, small lots present unique problems. One, in particular, confronts the site planner forced by limited area to a single-loaded lot parallel to the building. To the inexperienced, there may be a question as to whether the parking should be located next to or away from the building. Under normal circumstances the parking should always go next to the building (Fig. 18-11). On those occasions when either avoiding the intrusion of headlights on a residential building or the noise nuisance created by engines or exhaust is untenable, another response may be appropriate. Since the critical criterion is the user's safety, a question should always be raised where the car has priority over people.

As mentioned earlier, parking is a necessity as well as a nuisance. Any consideration given to reducing the level of that nuisance is valuable to any site plan. Screening methods include berms, fencing, walls, vegetation, and in some cases combinations of these basics, as vegetation always adds a softer quality to architectural elements. The concept of screening parking should also include the development of

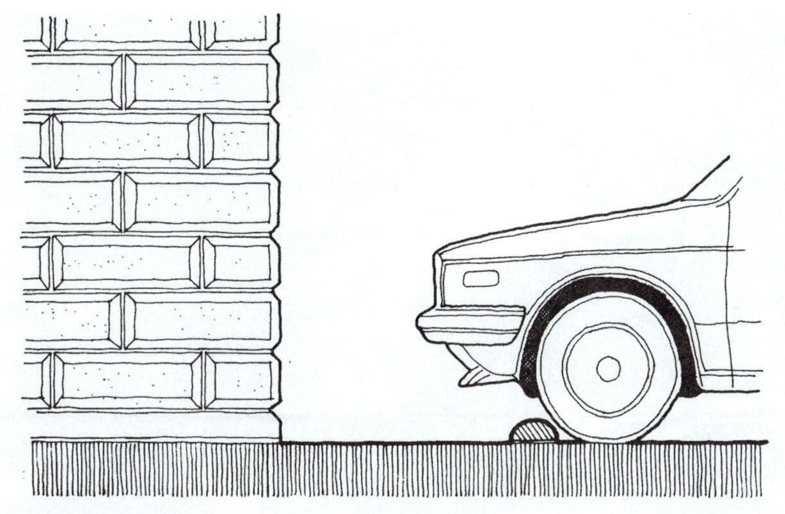

FIGURE 18-9. Wheel stop in front of building wall.

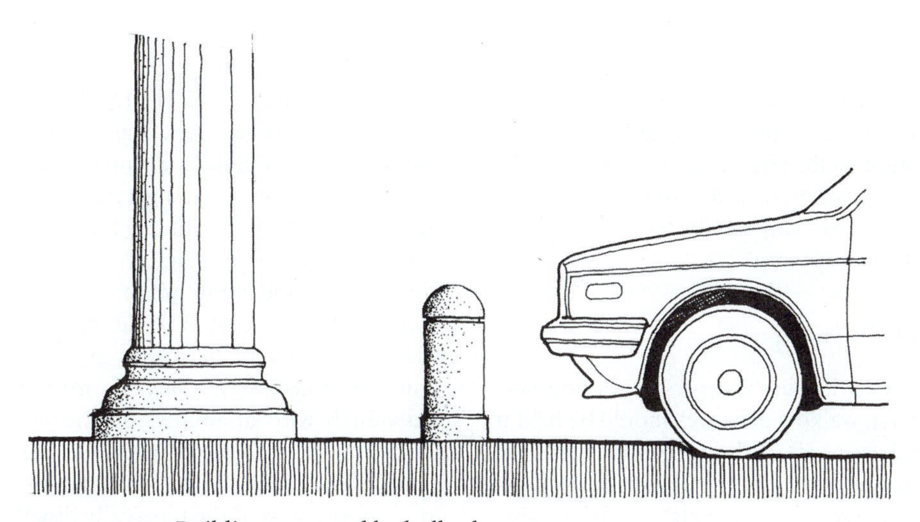

FIGURE 18-10. Building protected by bollard.

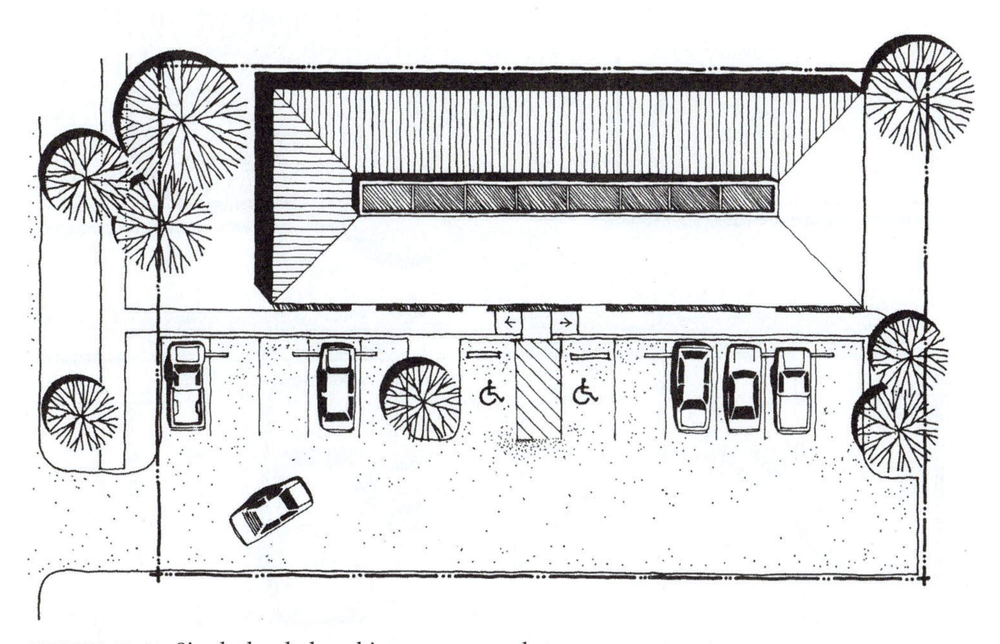

FIGURE 18-11. Single-loaded parking on narrow lot.

FIGURE 18-12. Parking buffer. (Photo by Patricia McCracken)

buffer areas along the edge of the street. The effort to reduce the visual impact of cars and impervious surfaces is always a positive contribution to the streetscape (Fig. 18-12). Numerous local governments have recognized the nuisance potential of parking lots, as many require fencing or screens as a condition of site plan approval.

Disabled Person Parking

The criteria associated with disabled person parking need special attention. First, walking distances should be held to a minimum. Second, every effort should be made to locate disabled person parking such that the physically impaired pedestrian can avoid crossing a circulation drive. Most ordinances require a wider (13 ft) space to permit wheelchair access between cars. Signs identifying disabled person parking should be located next to the wheelchair ramps. Common sense suggests that neither wheel stops nor drainage grates should be located in the direct path to the wheelchair ramp (Fig. 18-13). If the disabled person parking area and ramp are at the end of the parking lot, bollards should be used to define the edge of the pedestrian zone (Fig. 18-14). Signs and ramps should also be visible and accessible. There is more to accommodating individuals with disabilities on the site than we have described in these brief paragraphs. See Chapter 19 for a detailed review of dimensional requirements and criteria to support access of the physically disabled on the site.

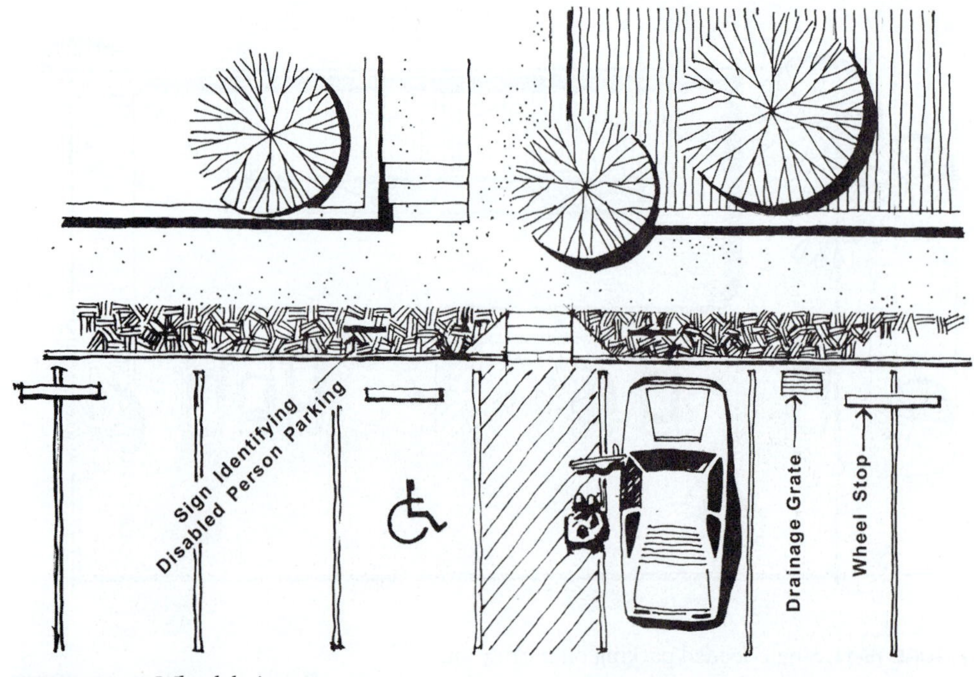

FIGURE 18-13. Wheelchair access.

FIGURE 18-14. Ramp at end of parking lot.

DROP-OFF ZONES

The interface between the pedestrian and the auto is one of the most difficult problems in the design of the parking lot. Conflicting perceptions as to who enjoys the preferential right-of-way creates a tenuous situation at best. Under some circumstances, zones that provide the auto momentary direct access to the building entrance in a "pedestrian only" environment is a very civilized alternative. Unlike the fire lane, which is based on a public safety issue, the drop-off zone is generally not a requirement and must be promulgated by the site planner. The drop-off zone should be clearly separated from the main stream of traffic. Although bollards may be used to demark the limits of the zone, no chains or other horizontal obstructions should be used. Drainage grates and other surface irregularities should also be excluded from this zone. When the curb is eliminated from a lengthy segment of the zone, the color of the paving should be in sharp contrast to that of the ramp.

In the drop-off zone, lighting is especially important. Where general pedestrian-level fixtures would be a property of the normal lighting plan, supplementary fixtures are necessary to provide a high level of illumination on ramps and curbs of the zone. Finally, some consideration should be given to shelter. To be successful, the structure need only provide shade and obstruction from the elements. The inclusion of a canopy or similar device can, functionally and aesthetically, enhance an important transition between parking and structures (Fig. 18-15).

FIGURE 18-15. Drop-off zone and shelter.

SIGNAGE

The objective of signs is to facilitate the use and accessibility of the site. While a visible and readable address is sufficient for a residential site, a school or shopping center is another type of problem. The four principal functions of signs are:

1. *To regulate or control:* "STOP," "Emergency Entrance," "No Swimming," "No Parking." These are the most common types, as they often involve health and safety issues (Fig. 18-16).
2. *To give directions:* "IN," "OUT," "To Interstate 20," "To Parking." These are often accompanied by directional graphics, as ambiguity or interpretation defeats the purpose of the sign (Fig. 18-17).
3. *To provide information:* "You are here"—the large-scale maps of parks and shopping centers, which identify the context of a campus or the alignment of a nature trail (Fig. 18-17).
4. *To give identification:* "Fine Arts Building," "Student Parking," "Ranger Headquarters." This is the last level of

FIGURE 18-16. Regulating or traffic control.

FIGURE 18-17. Directional and information.

FIGURE 18-18. Identification and amenity.

information in the hierarchy of signs: communication (Fig. 18-18).

It is likely that the majority of these four sign types will require review and/or coordination with a local governmental agency. This cooperation and coordination simply recognizes that there are numerous public- and private-sector entities involved in the successful development of a project. The larger and more complex an enterprise, the larger the circle of contributors.

Another type of signage that often requires review by municipalities is signs that display the names of subdivisions or commercial developments at or near an entrance to those locations (Fig. 18-19). Very often, these signs are considered

FIGURE 18-19. Monument sign.

to be "monument" signs. Monuments are free-standing signs located in the landscape areas away from buildings. Although the aesthetics of a monument sign can be an issue, most often the regulatory agency with the authority for review is concerned about the sign's location. All too often, the signs are located near existing or proposed utilities or in or near a public right-of-way. Therefore, to avoid creating a public hazard, coordination with the location of vehicular and pedestrian circulation is warranted.

It is common for signage regulations to vary widely from one jurisdiction to another. Aside from the aesthetic issues, one jurisdiction may regulate placement of signs in public street medians; others may not. Some may require signs be placed outside visibility triangles, regardless of the size of the sign. A *visibility triangle* is a triangle of a motorist's vision at an intersection. Its purpose is to indicate what objects in the triangle might block or encumber a motorist's view of oncoming or turning vehicles at an intersection. The length of each side of the visibility triangle is based on the speed limits of the roadway. That is, the higher the legal speed limit, the longer the legs of the visibility triangle. Fig. 18-20 illustrates that the triangle is formed by connecting the points along an imaginary line in the road to the apex of the visibility triangle considered to be the location of the motorist.

In many communities today, local jurisdictions may regulate the entire character of a sign, including size, location, material, color, and character and lighting, if any, well outside the issues of the visibility triangle. In general, sign ordinances are commonplace and have a history of established legal precedence. The knowledgeable site planner is aware of the potential for regulatory review and has acquired and followed all relevant sign ordinance specifications.

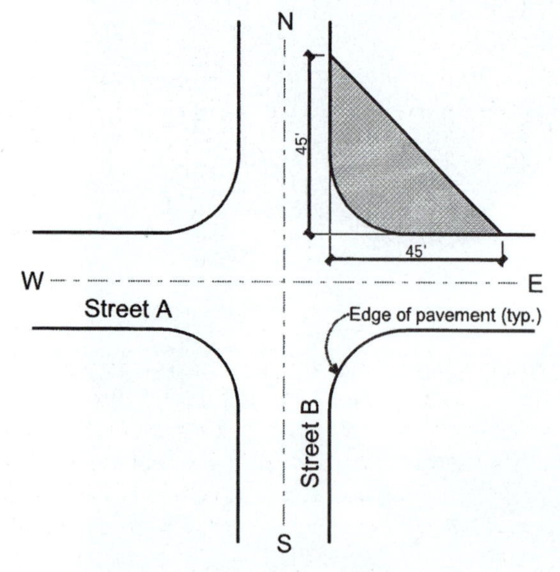

FIGURE 18-20. Visibility triangle diagram, northeast corner, Houston, Texas.

LIGHTING

Lighting on a site commonly functions to provide illumination for people to avoid obstacles, maintain security, and see adequately to perform certain tasks. Whereas one level and type of illumination may be necessary for general visibility in a parking lot, another is necessary for a residential driveway. As is obvious from the functions mentioned, each not only requires a different type of fixture but a different level of illumination as well.

Probably the first priority in the development of a lighting plan is safety. Although lighting alone does not create a crimeless environment, there is a very obvious cause–effect relationship between personal safety and lighting. Because of its integral role in the development of a safe urban environment, lighting may be the single most important component of pedestrian paths and open space to the elderly. For disabled persons, lighting is even more critical. It must be noted that if fixtures are either ill suited for a particular problem or poorly placed, there is little to be gained from increasing the illumination level. As an example, although overhead fixtures above 20 ft can provide an even level of illumination at a relatively low cost, they are often poorly suited for low-level, high-illumination conditions. Low levels of supplemental lighting often need to be provided for stairs, ramps, intersections, or other areas that would be hazardous if inadequately lit. This applies to pedestrian paths and parking lots as well as rest areas. In those conditions where low lighting is used, it should be below eye level. For the standing adult, eye level is below 5 ft 10 in.; for the wheelchair occupant, eye level is 3 ft 8 in.

Developing a lighting plan that matches the scale of the problem with an appropriately scaled fixture has more than the obvious benefits. A functional approach requires that the site planner recognize that lighting energy cannot only be wasted unnecessarily but that careful articulation of light levels and direction is necessary to keep lighting from becoming a nuisance to adjacent land uses. Furthermore, the range of illumination levels often required for a good lighting plan provides the opportunity for good nocturnal visual variety. There are five general types of site fixtures, three that have a pedestrian-related scale and two that support automobile functions. The first is the smallest in scale and provides illumination for a path, outdoor stair, driveway, or other specific function. Depending on the circumstances, this low level of lighting may be either a free-standing fixture or mounted to a wall. The fixture may be either incandescent or fluorescent and should be mounted within 24 in. of the ground plane.

The next fixture up in scale is that which provides lighting for public sidewalks or other major pedestrian spaces. This fixture should be 10 to 15 ft in height and may be fluorescent, or mercury vapor. There are numerous types of fixtures at this level which provide broad down-lighting, focused down-lighting, or general area lighting.

The third level in scale provides illumination for a special purpose. These include small-scale recreational uses (tennis, volleyball, etc.) and industrial, residential, or commercial activities (town square or amphitheater). The height of the fixture will vary between 20 and 30 ft above the site to achieve the desired level of illumination.

The last two types of fixtures support the operation and parking of automobiles. Lighting to accommodate parking lots and roads is between 30 and 50 ft in height. These fixtures are either mercury vapor or high-pressure sodium, as these lamp types provide high levels of illumination very cost-effectively. The "high mast" type is the largest in the scale of illumination and provides lighting for freeways, interchanges, and large parking lots. Similar to the fixture used for smaller-scale roads and parking lots, this fixture utilizes a mercury vapor or high-pressure sodium lamp. A note concerning these large-scale lighting types: They lack scale in lighting patterns. Additionally, the taller the fixture is from the ground plane, the more likelihood there is for the light to become a nuisance to adjacent land uses.

CHAPTER NINETEEN

PEDESTRIAN AND BICYCLE CIRCULATION

When faced with the problems of building location, stringent zoning, parking, and auto ingress and egress, it is easy to understand how accommodating pedestrians can become a low-level priority. Not only is the pedestrian system often considered as a residual space between buildings, but the design of the pedestrian movement as a kinetic experience demands the most of our creative skills. The challenge to mentally construct the reality of built form is particularly difficult when obligated to assimilate a sequence of diurnal and nocturnal experiences under varying climatic conditions. Issues of movement, direction, and orientation can become complex in an exterior environment where signs do little other than give names to places or streets that we do not know. Each of us has become lost or disoriented in a maze of walls, paths, and plants that left us wondering out loud: "Now where is my car?" This "lost feeling" creates not only a sense of insecurity but one of aggravation, neither of which is necessary or desirable. In *The Image of the City*, Kevin Lynch states that "the environmental image has its original function in permitting purposeful mobility," and "the terror of being lost comes from the necessity that a mobile organism be oriented in its surroundings."[1] Because a person's mobility is based on limited assumed images of the environment,[2] it is the site planner's role to provide reinforcement of those images, at least in terms of orientation.

PEDESTRIAN SYSTEMS

In this chapter the pedestrian system that functions as a link between buildings or activities will be considered separately from the system that is part of a community-wide, comprehensive circulation system. Although both involve pedestrian movement, the community-scale system has various aspects, which are common to those of bikeways and as such are addressed in that section of the chapter.

Pedestrian systems function in at least two general roles: one as a space to walk, another as a place to sit. As a place to walk, the conditions vary according to land uses served and environmental quality. In limited terms, many pedestrian linkages, sidewalks, and plazas are part of the public's domain. If the clients and users are the public, the criteria may be more stringent, as the range of alternatives and criteria are more limited than those that could be applied to a private residence.

Pedestrian systems in the public's domain, or those that will be dedicated to the public, need to reflect the design criteria established by the local government. Requirements for streets and roads, sidewalks, pedestrian paths, and accommodating individuals with disabilities will vary from one jurisdiction to another. The experienced site planner checks those requirements together with other development restrictions to make certain that there is complete agreement between the proposed plan and local regulations.

At the smallest scale of site planning, the planning objectives of the system should focus on:

1. The development of a system that functions as a connector while providing an enjoyable experience on its own merit
2. The design of a system sympathetic to the existing environmental context
3. The design of a system appropriate in scale to that of the proposed plan
4. The design of a path and succession of memorable images which create and reinforce the site's sense of place

Although the experienced site planner is cognizant of those objectives, the design of pedestrian systems is often the most primitive and the execution the most hostile. For whatever reasons, we forget that the most indelible impressions of an accommodating site plan begin with the person in the role of pedestrian. As pedestrians, we are more vulnerable and conscious of the quality of the environment. Our contact with light, color, sound, texture, built form, plants, and people is much more pronounced as a pedestrian than as an operator of or passenger in a motor vehicle. Our reliance on the auto for even the most menial trips has served to desensitize us to the genuine quality of the streetscape. To be effective, the design of pedestrian systems must reflect the mobility limits of pedestrians as well as the varying conditions within which the system might be used. As Robert Sommer has put it:

[1]Kevin Lynch, *The Image of the City* (Cambridge, Mass.: The MIT Press, 1960), pp. 124 and 125.

[2]Christian Norberg-Schulz, *Existence, Space and Architecture* (New York: Praeger Publishers, 1971), p. 35.

> It is curious that most of the concern with functionalism has been focused upon form rather than function.... Design professionals—city planners, landscape designers, architects . . .—would gain by adopting a functionalism based on user behavior.[3]

Connectors and Linkages

In general, pedestrian paths are designed to connect or link activities whose relationships are sufficiently interactive to warrant an articulated space. Paths are potentially significant contributors to creating an identifiable place, as the assembly of images alone can collectively produce an identity for the path itself. Although the functional obligations demand an enlightened response, the ability to design a system that contributes something to a site's sense of place requires an understanding of those objectives. Some are functional, others aesthetic, and others behavioral. In an attempt to analyze the direction that humanity takes in defining a path, Kurt Lewin employed the term "hodological space" or the "space of possible movement."[4] Lewin maintained that humans' paths are not necessarily those of straight lines but are those of preferred paths and represent the results of continual negotiations between shortest distance, pleasure, minimal effort, safety, and so on.

Designing a pedestrian system with the potential of creating a sense of place is no small challenge in a context where quality is often a marginal issue. Although some communities have adopted strict guidelines to maintain uniformity, site-specific pedestrian paths are often dictated by the road alignment and do little more than provide a path in the public right-of-way horizontally separated from the street. Unless the developer is willing to spend the time and money to supplement minimal requirements, pedestrian paths in the public domain, through most residential neighborhoods or commercial business districts, are remedial at best. At the smallest scale the path from a garage to a residence can be designed to accommodate the movement from vehicle to dwelling while articulating the transition from the quasi-public zone of the street to the privacy of the dwelling.

Pedestrian movement through a parking lot or in a parking garage traverses some of the most dangerous and hostile conditions in an urban context. Because of the normal change in roles from driver to pedestrian, and vice versa, the mix between people and cars is inevitable. Only in the rarest of circumstances can a pedestrian path through a parking area successfully transcend the problems associated with bodily injury and emerge as an enjoyable path. In many cases, the inability to separate people completely from vehicles creates an unreconcilable conflict. Many parking lots never attempt separation; the design of the facility is based strictly on its efficiency. A prevailing attitude appears to be: the larger the parking area, the less one needs to be concerned about the pedestrian. Sports stadiums and college or university parking lots are some of the obvious examples of this type of thinking.

FIGURE 19-1. Vehicle overhang of curb.

The pedestrian path through a parking lot is particularly difficult because the zone for the path needs to allow for the overhang of vehicles beyond the front or rear wheels (Fig. 19-1). The most common solution here is to pave the parking lot up to the line of the auto overhang and use precast concrete wheel stops to keep the auto bumper clear of the pedestrian path (Fig. 19-2). Because of the variety of automobile types and manufacturers, overhang distances vary widely and as such create some problems in maintaining clear distances. To reiterate the recommendation made in Chapter 18, when possible, orient the parking 90° to that of the desired pedestrian movement (Fig. 18-8). This will permit the design of a path network between the cars or at its worst, permit the pedestrian to walk in the parking lot circulation lanes. Prevailing wisdom is that there is less likelihood of an auto/pedestrian accident when the pedestrian is visible and walking in the lane with the auto.

If the pedestrian path is to become a contributor to the site's "sense of place," the system must be considered in a way that permits it to link the significant images of the site. After resolving the parking lot conflicts, the opportunities for successful transitional spaces occur in the pedestrian zones. But a word of caution: be reminded of the user's need to have some sense of orientation or direction. It is all too easy to assume that everyone enjoys the planner's familiarity with the site. In approaching the problem, the following

FIGURE 19-2. Paving to curb and wheel stop.

---

[3]Robert Sommer, *Personal Space: The Behavioral Basis of Design* (Englewood Cliffs, N.J.: Prentice Hall, Inc., 1969), p. 27.

[4]K. Lewin, "Der Richtungsbegriff in der Psychologie. Der spezielle und allgemeine hodologische Raum," *Psychologie Forschung*, 1934, p. 19.

issues may give some order to the considerations necessary in the design of the path.

1. *The location of the path:* Although other aspects of the process contribute to the quality of the path, the location is very important to its success. The expression "the path of least resistance" implies that people will find the most direct route given an open, unencumbered field. Successful pedestrian planning demands that one recognize human nature in a realistic appraisal of pedestrian "desire" lines. We have all seen examples where people have ignored paving in favor of using a dirt path through a lawn.
2. *The size, dimension, and shape of the path:* The configuration of the system is a group of decisions that are made in concert with the path's location. Since the location of a path is a response to its function and uses served, the physical character of the path will obviously require some adjustment. If a land use acts as a generator of pedestrian activity, the width of the path should reflect the larger demand. In many cases, local governments will stipulate either the size of the path or minimum building setbacks. As is often the case, these setbacks control both the size and location of the pedestrian path.
3. *The detailed quality of the path:* Every pedestrian corridor is a composition of elements (natural and human-made) that give a path character and vitality. These elements include paving material, texture, and color. The paving material and texture combine as a method not only to tell the pedestrian "this is the path" but also to signal "this is not meant to be walked on." People respond very well to physical barriers, whether paving textures, curbs, bollards, or vegetation (Fig. 19-3). However, practice suggests that changes in the physical character should not be subtle.

As an example, in *Urban Space for Pedestrians*, Boris Pushkarev and Jeffrey Zupan suggest that "pedestrians totally disregard any color patterns on a walkway."[5] The reasoning is that patterns are often annoying or do not relate to pedestrian flow, and are simply ignored. Although color can effectively differentiate one zone of activity or use from another, the color reinforces the real function. Pushkarev's and Zupan's observations invoke easy agreement, for the pedestrian is first, often one of many using the corridor and may not see the entire graphic, or second, has a purpose for a trip that may not correspond to the color.

As an objective in planning a system, scale seems simple enough at first, but the problem of producing a plan within a scale appropriate to a site is more elusive. Daniel Burnam's pronouncement to "make no little plans . . ." has misled more than one site planner. Some of the most successful plans were really based on simple ideas executed in a modest, sensitive way. Many misinterpreted "little plans" to mean little objects and, as such, equated the size of a system with its potential for success.

To be successful, it is not necessary for open-space systems and pedestrian networks to be either large or grandiose in physical section. Since large pedestrian plazas have a tendency toward monumentality, some consideration should be given to the introduction of elements that maintain a human scale. The experienced site planner approaches the development of pedestrian systems with restraint. Regardless of the scale of the plan, remember that the average pedestrian walks at speeds ranging from 260 to 275 ft/min across flat topography in neutral climatic conditions.

Large paved areas without people are uninviting and are avoided by many who equate people in a public space with security. While we recall examples of those grand and elegant plazas from history, we must also remember the pedestrian quality of the context as well as those supporting land uses adjacent to the plaza (Fig. 19-4).

William Wythe's publication *Social Life of Small Urban Spaces* is, without question, one of the most enlightening research efforts on people and their use of the exterior environment. Where Wythe's documentation on the use and the activities of people and the characteristics of urban open space may appear obvious, he also presents a lucid set of guidelines on the design and development of the urban space. Summarized, they are as follows:[6]

1. People tend to sit where there are places to sit.
2. Although people are attracted to sunny, wind-protected spaces in the winter, it is important to have a choice between shade and sun.
3. People are attracted to plazas where there are other people.
4. Trees are an important contribution to every site, especially where they are combined with sitting places.
5. Water is an immensely popular element in any public space, particularly where it is accessible.
6. Food service, at even the smallest scale (private vendor), attracts people, who attract more people.
7. The street can contain every quality that makes a successful plaza without a plaza.
8. For indoor plazas, retailing is very important to the pedestrian activity and life of the space.

A review of Wythe's publication is recommended, as his documentation and research for the conclusions mentioned here will provide a good basis for understanding how people use urban public open space.

Scale

The methods by which a pedestrian system is defined and articulated relative to other open-space elements contributes to its "scale." Vertical or horizontal separation of movement

[5]Boris Pushkarev and Jeffrey M. Zupan, *Urban Space for Pedestrians* (Cambridge, Mass.: The MIT Press, 1975), p. 156.

[6]William H. Wythe, *Social Life of Small Urban Spaces* (Washington, D.C.: The Conservation Foundation, 1980), pp. 28–94.

FIGURE 19–3. A clear, well-defined and embellished path.

FIGURE 19-4. A living street in Silver Spring, Maryland. (Photo by Daniel C. Brooks)

zones from other open-space components helps each to maintain their image and scale. Careful manipulation of grade changes also contributes to the definition of scale (Fig. 19-5). Grade change in the pedestrian path may be broad and gentle or direct and resolute, depending on the desired effect. The grade change has a tactile quality which can contribute to the aesthetic form of the path while giving it scale.

Bridges are an opportunity for the site planner to reinforce the human scale in a pedestrian system. Whereas the open path is visually part of its edges, the literal space of the bridge is more definitive. The bridge as a physical departure from two edges tells something of its duality. One is the special quality of the bridge as it absorbs the quality of the creek or street it traverses (Fig. 19-6). The second is the architectural character or linking function of the bridge. The structure as an image on the path reinforces the kinetic experience and the potential for defining "here" and "there." There are few components of a path network which have the potential for creating a sense of place compared to that of the bridge. The bridge is not only an event in the memory of the pedestrian but a piece of the network that contributes to the entire image of the pedestrian path.

Pedestrian and Living Streets

In the early 1970s, the Dutch began a process of reassessing the role and function of traffic, parking, and pedestrian systems in urban neighborhoods. Much of Europe, in the late 1920s, adopted transportation planning concepts developed in the United States. The American methods had good rationale and for that time were simple enough to apply. The component of transportation planning of particular interest here is the interface between pedestrians and vehicles. Based on the vulnerability of people, prevailing wisdom dictated that the pedestrian be horizontally separated from automobiles.

Few, however, forecasted the extent to which the auto would become a problem in urban neighborhoods. The indiscriminate use of the auto, unforeseen storage problems, and pedestrian conflicts created numerous marginally safe conditions. Although the pedestrian was horizontally separated

FIGURE 19–5. Grade change at pedestrian scale.

from the auto, on narrow streets, particularly in older historic districts, the safety of pedestrians had become a clear concern.

In Holland, the citizens of Delft petitioned the local government to close and prohibit several streets to auto traffic because of the hazards to pedestrians. Their request, however, was in direct opposition to those of the local public works and police departments. As a compromise, the municipality's planning department developed the concept of a residential street improved to accommodate the pedestrian with reduced automobile access. A plan emerged which designed the street in such a way as to give *no illusion* of separated systems. These *Woonerf* streets were residential streets to be occupied by cars as well as people, and as such were to undergo major changes. In many cases the streets were reduced to one lane, some as narrow as 3 meters (10 ft). In almost every case, alignments were never designed as straight paths. Maximum speed limits were reduced to 15 kph (9.5 mph), which are much more compatible with pedestrians and bicycles.[7] Not only are alignments circuitous and irregular, but design elements such as light standards, bollards, and plants are located so as to interrupt the visual plane of the street and reinforce the disconnected quality of the space (Figs. 19-7 and 19-8).

In some instances, additional parking has been introduced, which doubles as play areas when not in use. Although the idea emerged as a compromise (a word with very negative overtones in some design professions), the fact is that it works in various countries and communities in Europe. In France, this concept, known as "rues libres" (free streets), has worked

[7]George Wynne, ed., *Learning from Abroad. Traffic Restraints in Residential Neighborhoods* (New Brunswick, N.J.: Transaction Books, 1980), p. 3.

FIGURE 19–6(a). Bridge character contributes to historical C and O Canal in Washington D. C. (Photo by Daniel C. Brooks)

FIGURE 19–6(b). Bridge as an auto-pedestrian path and design feature.

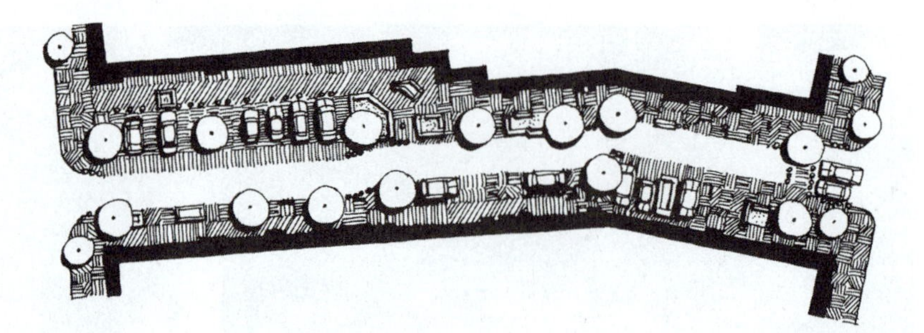

FIGURE 19–7. *Woonerf* street plan.

very successfully, particularly where shop owners have had difficulty with the creation of vehicle-free zones. Building on the success of the Dutch *Woonerf* and German *Verkehrsberuhingung* programs, the French made minor technical adjustments in the interval and length of free streets to reflect the character of existing urban neighborhoods.[8]

The Belgians and English have also employed the Woonerf concept to varying communities in their respective countries. The English considered the concept to be an alternative to the marginally successful total separation of vehicles and pedestrians at Radburn. Although the pedestrian-only zones at Radburn worked well, residential access roads also became burdened with quasi-pedestrian and bicycle activity. Application of the Woonerf concept was very successful, as it provided greater vehicular access to housing and commercial districts while giving the priority of use of the space to the pedestrian.[9]

It took some time for the concept of *Woonerf* streets to find willing takers in America. The idea was contrary to the traffic safety philosophy employed in the United States, where the emphasis on safety dictated separating, at least horizontally, motor cars from pedestrians and cyclists. But the *Woonerf* streets idea had real application potential and can also be found under the heading of "living streets" in the United States. The model is the same, though: pedestrians, skaters, and bicyclists all have the right-of-way in a living street (Fig. 19-9). Autos and buses have the burden of responsibility for movement through the street and are assumed to be at fault if they have an accident involving a pedestrian. In Boulder, Colorado, the living streets are called "access lanes" and follow the bow wave of planners eager to flesh out ways to calm traffic in cities. Traffic calming has been moderately successful as a method to introduce some sane urbanity to the streetscape of U.S. towns and cities. Additional techniques and details are available from a variety of sources, including the Federal Highway Administration (FHWA) and American Association of State Highway and Transportation Officials (AASHTO). The issue is no longer whether the strategy works but finding a willing constituency.

Accessibility

No material on site planning and the pedestrian would be complete without an overview of the guidelines necessary to provide access for individuals with disabilities. Although it is not our intent here to duplicate the efforts of texts that focus solely on the issues of barrier-free design, it is critical

[8]Ibid., p. 21.

[9]Ibid., p. 32.

FIGURE 19–8. *Woonerf* street section.

(a)

(b)

FIGURE 19–9. A living street in Washington, D.C. (Photos by Daniel C. Brooks)

to review the most common site-accessibility problems, design criteria, and supporting rationale.

The social consciousness of the United States in the late 1960s was the force behind numerous changes in our collective attitudes toward people and environment. Among those changes was recognition of the fact that the built environment had, for years, created unnecessary barriers that limited disabled individuals' access to public and private facilities. The publication of "Specifications for Making Buildings and Facilities Accessible to and Usable by the Physically Handicapped" in 1961 was based on the realization that individuals with disabilities accounted for 10% of the U.S. population and that attention to the issues of accessibility was long overdue.

Although the process was slow to start, by 1982, 38 states had adopted legislation based on American National Standards Institute (ANSI) standards. In that year, the Architectural and Transportation Barriers Compliance Board (ATCB) adopted and published standards for projects using federal financial assistance.

ADA Regulations

By 1994, the U.S. Department of Justice had published a detailed compilation of design criteria titled *ADA Standards for Accessible Design.* The document represents over three decades of research and analysis and is an excerpt from 28 CFR Part 36, subtitled *Nondiscrimination on the Basis of Disability by Public Accommodations and in Commercial Facilities.* These standards have become the cornerstone for other state and local governments, but all government regulations are subject to change. The discerning site planner will check the Web site www.ada.gov/reg to acquire the most recent amendments and updates to the Americans with Disabilities Act. And recognizing that all federal, state, and local regulations change and may vary, the design standards in this chapter should be used for general planning purposes only.

The obvious objective in every site plan should be to develop the site, paths, and walkways in such a way as to allow the greatest diversity of people access to buildings and facilities, safely and unhindered. Although the most common issues relate to topographical change and alignment, others associated with lighting and signs can be problematic, especially for individuals with disabilities.

Where building codes reflect design criteria associated with disabled person access to buildings, sites and site improvements are no more casual considerations. Paths, walkways, ramps, and signs must reflect current ADA standards.

The vast majority of site planners recognize that without the regulations, access for individuals with disabilities would not exist. Established criteria, functional guidelines, and sound research provide the basis for the progress toward barrier-free design found in ADA 1994.

Design Guidelines

Where each path should be designed to reflect an anticipated use, there are some minimal standards that must be mentioned. Although ADA 1994 requirements are different, the recommended two-way and one-way path widths to accommodate wheelchair traffic are 6 ft 0 in. and 4 ft 0 in., respectively[10] (Fig. 19-10). Concerning turning radii, it is possible to turn the wheelchair around in its own space, (one wheel moves forward while the second moves in reverse). Comfortable maneuvering distance, however, for the adult chair is 7 ft 0 in. × 8 ft 0 in. (Fig. 19-11).

Gradients. If a path has a grade of less than 5%, it is considered a walkway. Paths with a grade that exceeds 5% are considered ramps and have special design requirements. Paths with sustained grades above 4% should have brief 5 ft 0 in.

[10]American Society of Landscape Architects Foundation, *Barrier-Free Site Design* (McLean, Va.: ASLAF, 1977), p. 17.

FIGURE 19–10. Recommended path widths to accommodate wheelchairs.

flats interrupting the slope to permit a chairbound occupant to stop and rest. If possible, walks should not exceed a gradual 2% slope. Cross slopes should not exceed 1:50.

The maximum ramp lengths and slopes set out in ADA 1994 are illustrated in Fig. 19-13. In addition, 60 in. × 60 in. clear space should be provided at each end of the ramp to facilitate wheelchair movement.[11] Because of the incline, the ramp criteria are different from those of the walkway and path. The minimum one-way ramp widths should be no less than 3 ft 0 in., with two way ramps at a 5 ft 0 in. minimum.[12]

[11]Ibid.

[12]Ibid.

One aspect for accessibility is the design criteria for ramps from parking lots and across curbs.

Stairs are similar to ramps in some respects. Minimum one-way stair widths should be 3 ft 0 in. with two-way widths at 5 ft 0 in. Like the ramp, a topographical change in elevation is also a critical component of the design criteria (Fig. 19-13). Where the vertical change is in excess of 6 ft 0 in., an intermediate landing should be provided.[13] Since visual impairments often accompany physical problems, stairs and ramps should be in contrasting colors to those of adjoining walls or paving.

[13]American Society of Landscape Architects Foundation, p. 30.

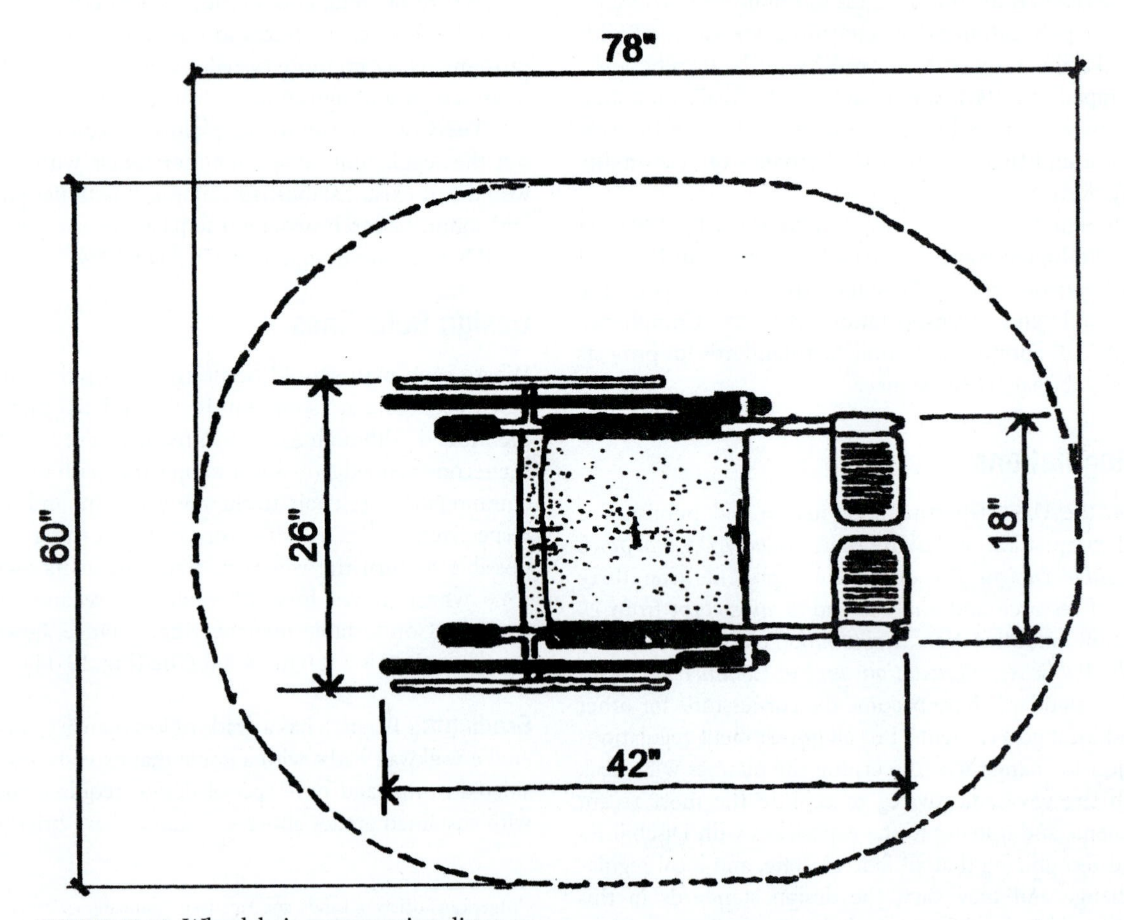

FIGURE 19–11. Wheelchair maneuvering distance.

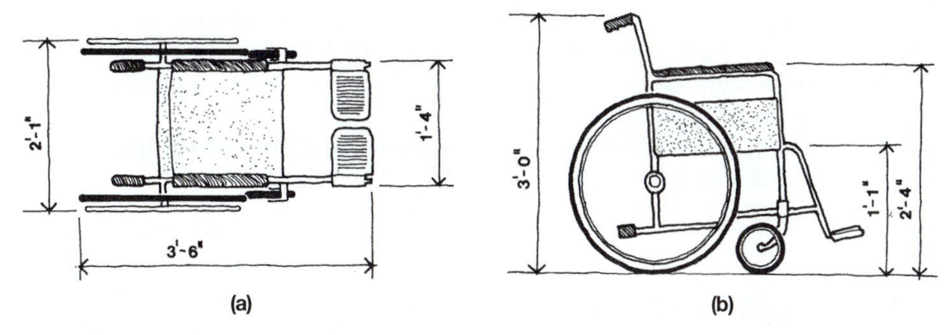

FIGURE 19–12. Wheelchair dimensions in plan and elevation.

Handrails for ramps and stairs are essential components of both ramps and stairs. The most comfortable height for ramp handrails is between 34 in. and 38 in. above ramp surfaces and 33 in. and 36 in. for stairs. A second handrail to accommodate children and wheelchairs is often mounted at 2 ft 4 in. In either case, the rail should extend $\frac{1}{2}$ in. beyond the ends of ramps and 6 in. beyond stairways.[14]

The surface is an important safety and aesthetic consideration in the design of paths, ramps, and stairs. Ramps and paths should be all-weather, nonslip, and free of surface irregularities such as large cracks and uneven edges. Although all joints cannot be eliminated in a paved surface, they should be minimized, with none larger than $\frac{1}{2}$ in.[15] All exterior surfaces should have a minimum slope of $\frac{1}{8}$ in. per foot to sustain positive drainage. When a ramp or stair is used as a part of a natural environment and its location on the path could become a hazard, a textural signal advising those with visual difficulties of the impending obstacle should be considered both above and below the ramp or stair.

The stability and functional quality of a surface is also dependent on its relative density. Harder surfaces are generally more stable and thus are preferred. Asphalt, brick in concrete, and concrete are considered to be both hard and smooth and to accommodate both pedestrians and wheelchairs very successfully. Rough surfaces, although dense, are generally less desirable unless special attention is given to reducing their inherent irregularities. Paths of flagstone, bricks laid in sand, or exposed aggregate concrete and wood decks could inhibit the mobility of those using crutches, canes, or wheelchairs. Furthermore, unlike with the smoother surfaces, snow and ice removal is much more difficult without inflicting damage to the surface, and when seasonal obstacles are not removed successfully, they can create obvious problems for every pedestrian.

Softer surfaces can create such accessibility problems that their use as a paving material, except under limited circumstances, is not recommended. The use of crushed rock, tanbark, earth/sand, or river rock should be avoided in the design of paths to accommodate those with mobility issues.

[14] Ibid., p. 31.

[15] American National Standards Institute, p. 9.

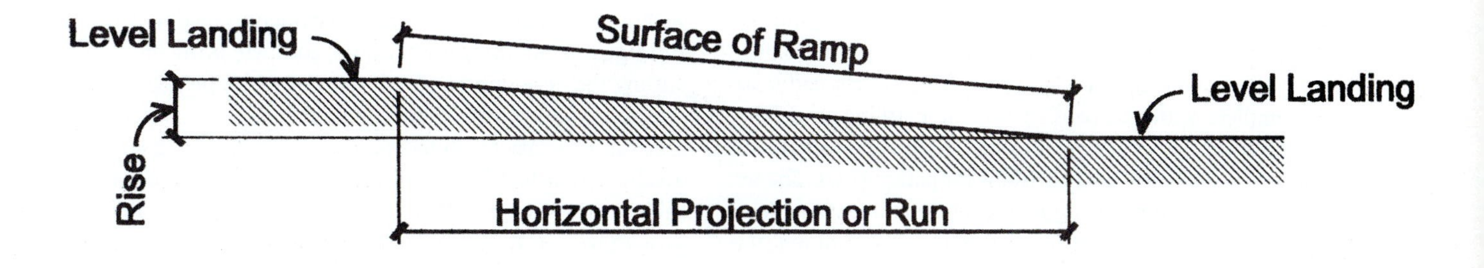

Slope	Maximum Rise		Maximum Horizontal Projection	
	In.	mm	Ft.	meters
1:12 to < 1:16	30	760	30	9
1:16 to < 1:20	30	760	40	12

FIGURE 19–13. Criteria for functional diversity access ramp length, slope, and height.

Because these materials are unstable, they are poorly suited for wheelchairs. In addition to the fact that they drain poorly and create orientation problems for individuals with visual impairments, they also generally involve high maintenance costs. Although their value as a paving surface is limited, discriminate use is appropriate in a natural setting in which their use does not present an access or safety hazard.

BICYCLES AND BIKEWAYS

The other general level of quasi-pedestrian activity is the bikeway. The cyclist in the United States is often relegated to paths that are hazardous to cyclists or to others. Two decades ago, the acceptance of the bicycle as a legitimate transportation system appeared to be some years away. The few bikeways that existed were predominantly limited to those conditions where the distances between housing and employment are relatively short, land costs and housing densities are high, and social pressures for two cars per household are minimal. That model still has currency. University campuses are excellent examples of these criteria.

Close proximity between housing and employment is particularly important if a system is to enjoy broad use. Although numerous cities have very high-density commercial or industrial employment centers, without adjacent residential land use, interest in the use of bicycles as an alternative to the automobile fails to materialize. The development of freeway systems has stretched the distances between employment and housing such that the relationships are often at marginally acceptable riding distances. When housing is relegated to strictly suburban locations, the impact of time and distance are often so high that only cycling enthusiasts are willing to brave the conflicts with motor vehicles and weather on a day-to-day basis.

The steady growth of cycling in America has led to cycling emerging as an acceptable alternative transport system. The concept of *alternative* is important to stress because cycling is not for everyone, nor could it be. It is important to remember that the cyclist depends on a system of paths or lanes, just as users of other vehicles require paving and rights-of-way. Like vehicular circulation systems, bicycle paths are not complete with the construction or marking of a lane. Where appropriate, parking, locking devices, and lockers are necessary, just as there are provisions for parking cars. The inclusion of all components is necessary to encourage people to utilize bicycles instead of motor vehicles. Exploiting the renewed attention in cycling is in the best interest of both cyclists and communities, as the cost of the supporting infrastructure is minute compared to that for other vehicles.

From the time when the bicycle was little more than a frame and two wheels propelled by the rider's feet, the bicycle has been an attraction. After 1866, when a patent was granted for a pedal drive, the use and success of the bicycle became dramatic. The "velocipede" enjoyed broad popularity in Europe and the United States during the heyday of the industrial revolution. The machine's economy, size, and speed provided the basis for a transport system that rivaled that of trains. Its broad social acceptability contributed to its use, and by 1899 the Cyclist Touring Club in Britain had 60,000 members, its largest membership ever.[16]

Bicycle Facilities Planning

It's unlikely that anyone anticipated the acceptance and growth of the bicycle in the land dominated by the automobile. In spite of the sprawling cities in the United States, interest in the bicycle has ballooned, from 10.6 million units in 1970 to over 44 million in 2008. Although that growth has not been straight-line upward, it has been solid. The reasons range from recreation and exercise alternative to employment commuting. While some growth can also be attributed to emerging forms of cycling as sport, the acceptance of the vehicle as a transportation alternative has become a legitimate component of transportation planning.

While not a design-related element, Toronto's recent commitment to funding public bicycles may be another bellwether of change of attitudes in North America. Recognizing that Toronto is not a U.S. city, it nonetheless now provides a model of how a public-supported bicycle system really works. The popular Canadian *BIXI* (a combination of bicycle and taxi) boasts all-new rugged cycles with lights, luggage carriers, and a comprehensive trail network with user-friendly facilities—all for a mere $5 per day. While the United States will likely rely on cars for years, the BIXI could provide still more evidence that a car is not essential to living in and enjoying the city. Nothing illuminates a new path like a successful green light.

In both the 1973 Federal Highway Act and the Surface Transportation Act of 1978, Congress supported the use of bicycles as an alternative transportation system, with additional federal financial assistance to be administered by the states. But in spite of that commitment, a 1980 report, "Bicycle Transportation for Energy Conservation," indicated that by-and-large, state governments had not allowed the funding support to local governments for bicycle programs at anything close to the levels authorized by Congress.[17] Although recalcitrant highway departments failed to provide funding for anything that would compete with highways, stifling the development of bicycle networks, people continued to demand better planning, bike lanes, and facilities in general. It is easy to understand some hesitation to support a vehicle that had been gathering dust in garages and basements everywhere. There was a view that the spike in growth was nothing more than a reaction to the cost of fossil fuels. But by the end of the first decade of the twenty-first century, bicycle planning was an active division in numerous state departments of transportation. California, Colorado, Oregon, Washington, and Wisconsin, to name but a few, have published bicycle facilities planning guidelines and design criteria, signage, and regulations detailing issues unique to their states.

---

[16]Mike Hudson et al., *Bicycle Planning: Policy and Practice* (London: Architectural Press Ltd., 1982), p. 1.

[17]Ibid., p. 8.

The continued interest in cycling, articulated in *Bicycle Planning* in the early 1980s, has never been more true than it is today:

1. From the public's viewpoint, it is a very cost-effective means of transport. The capital costs are modest relative to the potential uses.
2. The lanes and storage facilities needed are much smaller than those required for automobiles. This is particularly important in urban areas, where space is at a premium.
3. Walking and cycling are the two most energy-efficient forms of transportation. In a world of expanding population and finite limits of fossil fuels, every effort to conserve fuel should be made.
4. The public's cost of cycling is similar in scale to the private sector's costs. The machine is simple, and maintenance costs are minimal.
5. Cycling is a reliable and quick form of transport. Breakdowns are minimal, and door-to-door times for urban journeys between 2.5 and 4 miles (4 to 6.5 km) can be quicker on a bicycle than by any other means. Research in Tempe, Denver, Los Angeles, and Minneapolis/St. Paul indicates that the length of most bicycle trips is 5 miles or less.[18] In urban areas, walking or cycling can assist in improving the level of service on roads and thoroughfares.
6. Cycling is a benign means of transport for practically everyone. The machine is noiseless, pollution free, and can provide mobility for all age groups.

If there is any doubt about the acceptance of bicycles as a growing alternative to autos, the publication *A Guide for the Development of Bicycle Facilities* by AASHTO should dispel even the most contrarian voice. AASHTO's text is considered by many to be a first-choice resource for cycle facilities planning. While additional information can be viewed at the AASHTO Web site (www.transportation.org), it's recommended that the site planner first consult the relevant local government and state department of transportation for applicable design standards. Do not assume that the AASHTO standards will be acceptable to the jurisdiction where the site is located as some agencies have more restrictive criteria.

Planning Criteria

Bicycle and pedestrian systems have similar issues that must be addressed in their planning process. Although each aspect will have some variation, their commonalities suggest concurrent consideration.

Safety is the first priority for both systems. The fundamental concern is with the vulnerability of the cyclist or pedestrian relative to autos and trucks, but care must also be taken to avoid conflicts between cyclists and pedestrians. As the number of cyclists on paths and roadway shoulders grew and conflicts with autos escalated, a critical review of design standards for bicycle paths and motor vehicles became an imperative. Recognizing that there were different riders with different levels of skills and confidence, the FHWA initiated a reappraisal of bicycle facilities planning, beginning with an assessment of the cyclist's skill. With a focus on safety, the FHWA's report on design standards suggested that highway and bicycle planners use the following three categories of riders in determining the character of path/roadway types and conditions for cyclists:

- Advanced or experienced riders generally use their bicycles as they would motor vehicles. They are riding for convenience and speed and want direct access to destinations, with minimal detours or delays. They are typically comfortable riding with motor vehicle traffic; however, they need sufficient operating space on the traveled roadway or shoulder to eliminate the need for either themselves or a passing motor vehicle to shift position.
- Basic or less confident adult riders may also be using their bicycles for transportation purposes, such as to get to the store or to visit friends, but they prefer to avoid roads with fast and busy motor vehicle traffic unless there is ample roadway width to allow being overtaken by faster motor vehicles. Basic riders are comfortable riding on neighborhood streets or shared-use paths and prefer designated facilities such as bike lanes on busier streets.
- Children, riding on their own or with their parents, may not travel as fast as their adult counterparts but still require access to key destinations in their community, such as schools, convenience stores, and recreational facilities. Residential streets with low motor vehicle speeds, linked with shared-use paths with well-defined pavement markings between bicycles and motor vehicles, can accommodate children without encouraging them to ride in the travel lane of major arterials.

As an example of how extensively state highway departments have embraced the emerging role of bicycles, the Oregon Department of Transportation (DOT) adopted a bicycle and pedestrian plan. The plan recognizes that both cycling and pedestrian activities are expanding at a pace that requires definitive design criteria to assist municipalities in their planning. Where the FHWA's 1994 publication categorized bicyclists, the Oregon DOT plan found that establishing criteria for bikeway types would also facilitate consistency in the planning process. To begin, it defined a bikeway as a roadway used to accommodate bicycle travel on a road in the following four forms: shared roadway, shouldered bikeway, bike lane, and multiuse path. The multiuse path is different from other bikeway paths as the paving section for motor vehicles is separated and buffered from that used by pedestrians

[18] Nina Dougherty Rowe, "The Energy Crisis/Bicycles/Federal Government," *Planning, Design and Implementation of Bicycle and Pedestrian Facilities* (New York: American Society of Civil Engineers, 1975), p. 20.

and bicycles. Summarized below are the divisions and their functions:

- *Shared roadway:* Those facilities where bicyclists and motor vehicles share the same travel lanes without a striped bike lane are considered to be shared roadways. These are most common in neighborhood streets and low-speed rural roads (Fig. 19-14).
- *Shouldered bikeway:* Those facilities where paved shoulders to the motor vehicle lane and that are suitable for cycling are considered shouldered bikeways. These are most common in rural areas (Fig. 19-15).
- *Bike lane:* Those facilities where a portion of the roadway is designated, by striping or other visual means, for preferential use by cyclists are considered bike lanes. These are appropriate for use on urban arterials and major collectors and may be used in rural conditions where bicycle travel demand is substantial (Fig. 19-16).
- *Multiuse path:* Those facilities where all of the paving section for motor vehicles is separated from paving to accommodate bicyclists, pedestrians, runners, and skaters are considered to be multiuse paths. These are most commonly found in community trail plans and are appropriate in corridors not well served by major street systems (Fig. 19-17).

This classification system can be used as a working vocabulary to help avoid the confusion common in the planning process. The terminology is absent technical jargon, and the functional roles of the facilities should simplify communicating with other participants in the bike and pedestrian planning process.

Early cycle planning relied heavily on horizontal separation for safety; this is still the preferred option. But in balancing the growing need for bike lanes in built-up urban environments with limited right-of-way, real innovation

FIGURE 19–14. Shared bike lane.

FIGURE 19–15. Shouldered bikeway.

requires the need to address those conditions where cyclists must share the road with motor vehicles. It must be noted that the examples in Figs. 19-14 through 19-17 rely on clear pavement with marking, striping, and rumble strips.

In some cases, vertical separation has proved to be a legitimate method of reducing conflicts between pedestrians, cyclists, and vehicles. Horizontal separation is appropriate only in those conditions where the number of people and cars would create congestion and/or create serious safety problems (Figs. 19-18 and 19-19).

Pedestrian and cycle, security is also high on the list of priorities. Paths and storage facilities should be well lit. Visual obstructions of public spaces and conditions that provide concealment should be minimized. Closed-circuit television has become an increasingly common method of providing surveillance of building lobbies, transit terminals, and so on, where security is difficult because of the flow of people.

For pedestrian and cycle systems to function properly, they must be convenient and direct. Systems need to be designed to accommodate changes in modes (from auto to pedestrian). Minimizing the impact of mailboxes, light standards, newsstands, and other street furniture also facilitates the movement of people. Often, the most direct route is the path that encourages walking or cycling. Routes through midblock or connecting lobbies can often create important linkages for what would otherwise be a circuitous route (Fig. 19-19).

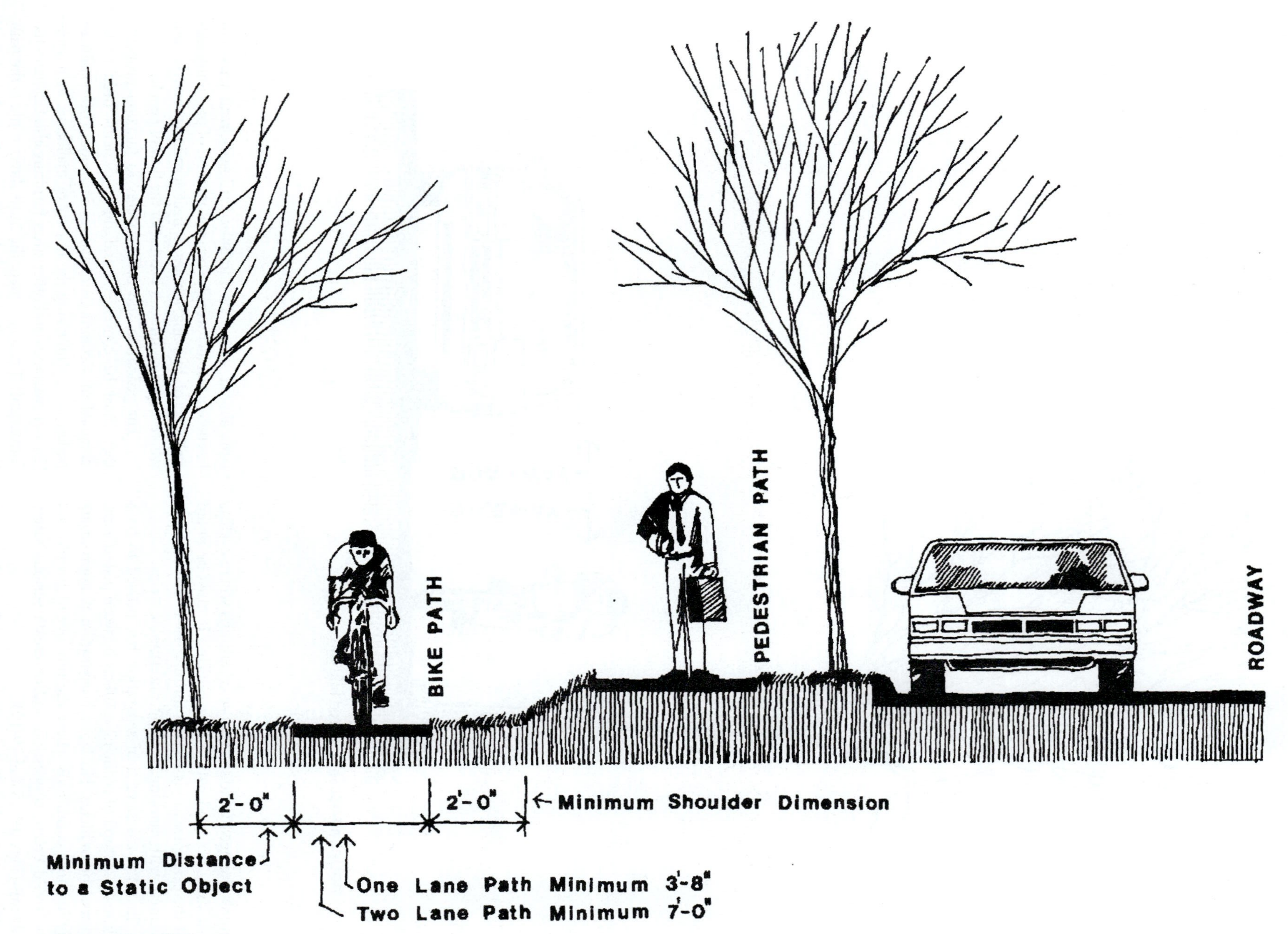

FIGURE 19–16 Bike lane.

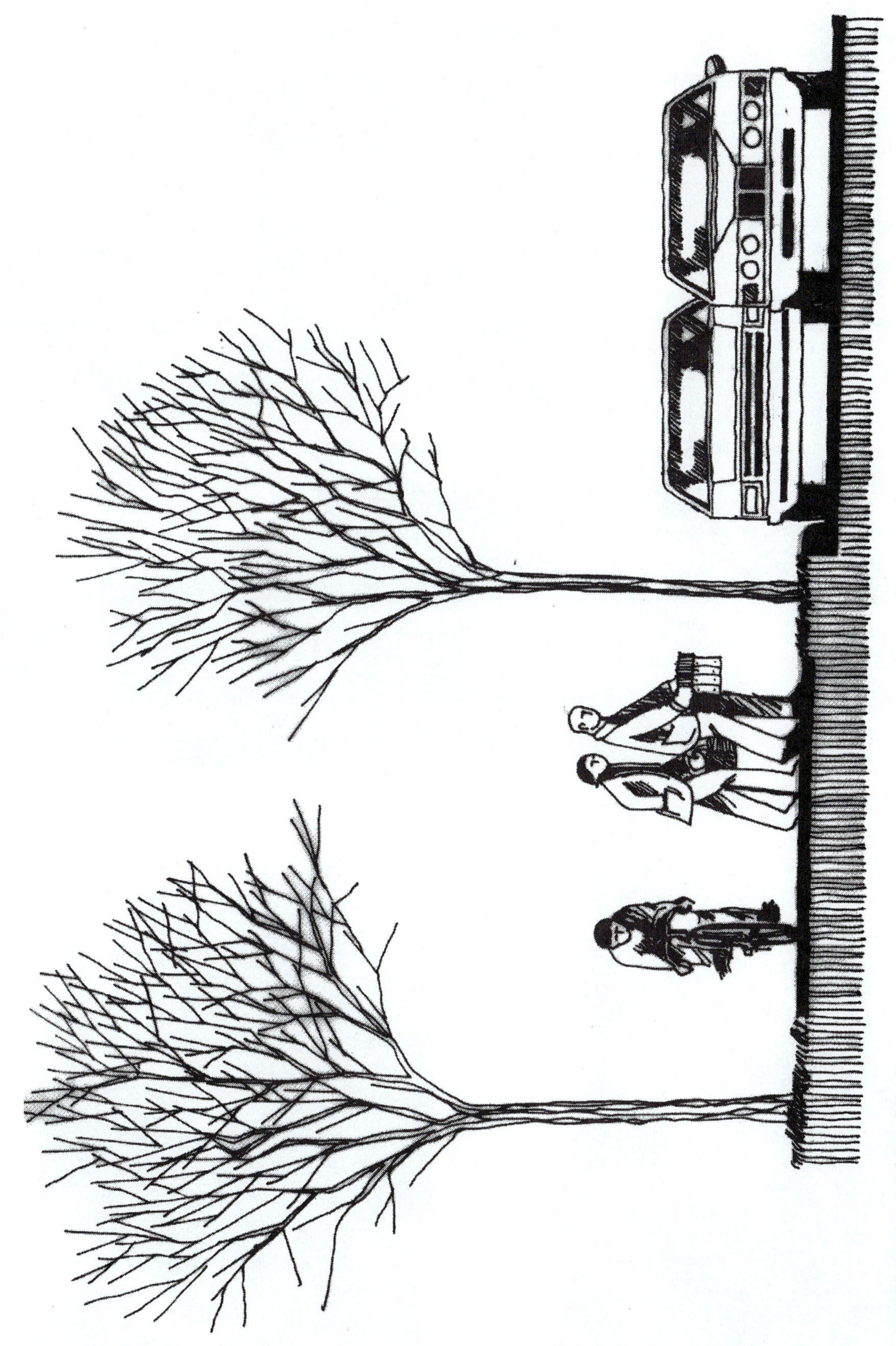

FIGURE 19–17(a). Multiuse bike lane.

FIGURE 19–17(b). Multiuse path.

FIGURE 19–18. Pedestrian overpass.

FIGURE 19–19. Midblock pedestrian connection.

If either a bicycle or pedestrian system does not maintain continuity throughout, the network will probably not be used. Convenience to the cyclist necessitates that locking and storage facilities exist at points of modal transfer. Shelter and seating also enhance and expand the use of both cycle and pedestrian networks.

On occasions, grade-separated linkages may be necessary to maintain the continuity of a network. Since the construction costs of underground systems are extremely high, these should be designed and built in conjunction with bridges or other structures. The underpass shown in Fig. 19-20 is a crucial link in the pedestrian system of Lake Anne Village at Reston, Virginia. But unless there are no alternatives, pedestrians are reluctant to walk up stairs to use elevated bridges. The safety hazard must be a genuinely perilous condition, as people would generally rather risk confrontation with vehicles than walk up stairs.

Clarity is essential, as every user must be able to understand the entire network. This is particularly crucial for cyclists at intersections, where the simultaneous movement of vehicles is always potentially hazardous. Good signs and directions are essential to provide the user with a sense of security and understanding as to a path's location, adjacent land uses, landmarks, and public facilities. In turn, understanding generates a higher use of the system and enhances its total cost-effectiveness.

FIGURE 19–20. Pedestrian underpass at Lake Anne. (Photo by Patricia McCracken)

CHAPTER TWENTY

THE BUILDING AND ENERGY

by George Truett James

Any change in the natural environment to accommodate a habitable structure should involve an analysis of the site with a focus on comfort and energy conservation. Historically, humans accepted the proposition that their welfare and nature were inextricably linked. Their survival often depended on their ability to adapt to climatic circumstances. The development of sophisticated, energy-intensive environmental control systems has freed us from the programmatic concerns of comfort and climate, and it has permitted the construction of habitable structures anywhere. However, recent energy supply disruptions and escalating utility costs have fostered an awareness of the finite nature of our resources and the importance of energy conservation in the built environment. Although our perception of the immediacy of our energy problems changes, our fundamental understanding of those problems should not. Because our resources are not limitless, design professionals must provide the leadership required to suppress the waste of conventional energy resources in our built environment.

The United States, with 5% of the world's population, uses 25% of the world's energy resources. Between 35 and 38% of U.S. energy consumption is used to heat, cool, light, and provide hot water for our buildings.[1] Our nation's buildings are therefore responsible for approximately one-ninth of annual world energy consumption. If these figures are not startling enough, consider that if our buildings and sites had been designed to reduce energy consumption, a conservative estimate of possible savings might be in the neighborhood of 20%.[2] This is simply a reduction of energy waste that has no effect on the standard of living. That wasted energy would be equivalent to the energy used for all purposes by the combined countries of Canada, Australia, New Zealand, and temperate Latin America. It is therefore important for the design professional to understand the implications of planning and design decisions on building energy use in order to reduce building energy waste. Site planning can profoundly influence the total energy demand of a building. An energy-conscious site-planning effort is simply one that directs the design decisions toward the concept of energy conservation and appropriate utilization of the site's natural energy resources. It has been estimated that proper site planning for a building can provide 30% energy savings without increasing development costs.[3] The site planner should therefore develop an understanding of the ways in which buildings use energy and the nature of the climatic forces acting on them. In response to these concerns, in this chapter we focus briefly on building energy use and how predesign analytic techniques can inform the site-planning process.

BUILDING METABOLISM

Buildings and their environmental control systems are affected by external thermal loads imposed by the climatic elements discussed in Chapter 11. Buildings are also affected by the thermal loads produced when heat is generated inside the building by people, lights, and equipment. This internal heat generation can be thought of as the building's metabolism. A building's metabolic rate, the rate at which internal thermal loads are generated, is dependent on and varies with building type, building function, and the way in which the building is used.

Without heating or air conditioning, the temperature of the air inside buildings with low metabolic rates, such as a house or warehouse, is primarily dependent on the external thermal loads produced by climate on the building's envelope or "skin." The rate of internal heat generation is insignificant compared to the rate of heat flow generated by climatic conditions. This type of building is referred to as a skin-load-dominated (SLD) building. Without the assistance of an interior climate control system (heater or air conditioner), inside air temperature is primarily a function of exterior climatic conditions. When it is cloudy and cold outside, it will be cold inside. When it is sunny and hot outside, it will be hot inside.

By contrast, in buildings with high metabolic rates, so much heat can be generated internally that the influences of climate on the building become insignificant by comparison (such as in a steel smelter). This type of building is referred to as an internal-load-dominated (ILD) building. When it is cloudy and cold outside, it may be perfectly comfortable inside without supplementary mechanical heating. It is not uncommon for buildings with high metabolic rates to require

[1] R. Stobaugh and D. Yergin, *Energy Future* (New York: Ballantine Books, Inc., 1979), p. 207.

[2] Ibid, p. 167; Yergin states that a 30 to 40% reduction of national energy use is possible with no effect on the standard of living.

[3] E. G. McPherson, ed., *Energy Conserving Site Design* (Washington, D.C.: American Society of Landscape Architects, 1984), p. 298, citing R. M. Byrne and L. Howland, *Background Information Summary* (Washington, D.C.: The Urban Land Institute, 1980).

air conditioning to maintain occupant comfort even during cold winter weather. There are instances of buildings where heating systems are rarely, if ever, activated because of their high internal heat generation rates.

When internal thermal loads are balanced by the rate of heat loss from the building, a thermal balance or equilibrium is achieved. This is a set of circumstances where neither mechanical heating nor cooling is required to maintain the inside temperature prescribed by the mechanical system's thermostat (the thermostat set point).

Disregarding the effects of solar radiation (cloudy conditions or nighttime), thermal balance is usually achieved in skin-load-dominated buildings when the outside temperature is only a few degrees Fahrenheit below the thermostat set point. For example, the internal heat generated in a single-family detached residence might produce a temperature of 72°F inside when the outside temperature is 65°F, a difference of 7°F. In other words, if the thermostat is set at 72°F, thermal balance will exist when it is 65°F outside. In order to maintain the 72°F prescribed by the thermostat setting, supplementary heating will be required when the outside temperature falls below 65°F, and supplementary cooling will be required above 65°F. Still disregarding solar radiation, the outside temperature at which the internal loads are balanced by the building's heat loss rate (65°F in this example) is referred to as the balance point temperature.

With a high rate of internal heat production, an ILD building will have a balance point temperature significantly lower than the earlier residential example. ILD buildings with very low balance point temperatures can, in a sense, eliminate winter heating loads, because internally produced heat can be sufficient to keep the inside temperature in the comfort zone during a cold winter day. Conversely, when summer climatic loads, including solar radiation, are added to these large internal loads, air-conditioning requirements can be quite substantial. Consequently, ILD buildings often require a great deal more auxiliary cooling energy than auxiliary heating energy.

The balance point temperature of a building can be estimated if enough programmatic information is available. As identified by Charles Benton and James Akridge in "Balance Point Exercise,"[4] balance point temperature is described by the following equation:

$$T_{bp} = T_{th} - \frac{Q_i}{UA + \text{infil}}$$

where:

T_{bp} = balance point temperature
T_{th} = thermostat set point
Q_i = average hourly internal loads (over a 24-hour period) (sum the number of Btu generated daily by people, lights, and equipment, and divide by 24 hours; the result is the average internal gain in Btu per hour)
UA = building's heat loss rate in Btu per hour per °F (Btu/hr-°F) calculated using standard steady-state heat transfer calculation procedures; this includes heat transfer due to conduction through the walls, windows, roofs, and floors
infil = infiltration rate in Btu per hour per °F

Following is an example of the calculation procedure for a small, residential structure with the thermostat set at 72°F:

House plan: 1,000 ft^2 (25 ft × 40 ft), 8-ft walls, volume = 8,000 ft^3

Floor area: 1,000 ft^2, avg. U-value = 0.1, UA = 100 Btu/hr-°F

Ceiling area: 1,000 ft^2, avg. U-value = 0.05, UA = 50 Btu/hr-°F

Wall/window area: 1,040 ft^2, avg. U-value = 0.077, UA = 80 Btu/hr-°F

Total UA = 230 Btu/hr-°F

Infiltration = air changes (cubic feet) per hour/ 60 minutes per hour = ft^3/min

Estimating one-half air change per hour:

[(8,000 ft^3 × $\frac{1}{2}$ air change per hour)/ 60 minutes per hour] × 1.08
(1.08 is the conversion factor for ft^3/min to Btu/hr-°F)

Infiltration = approximately 70 Btu per hour per °F (Btu/hr-°F)
UA + infil = 230 + 70 = 300 Btu/hr-°F

Internal loads:

4 people at 250 Btu (sensible heat) per hour	1,000 Btu/hr
Lights	300 Btu/hr
Equipment	800 Btu/hr
Total	2,100 Btu/hr

$$\begin{aligned} T_{bp} &= T_{th} - \frac{Q_i}{UA + \text{infil}} \\ &= 72 - \frac{2100}{230 + 70} \\ &= 72 - 7 \\ &= 65°F \end{aligned}$$

The balance point temperature will decrease if the thermostat set point is lowered, insulation levels are increased (lower U-values), or internal heat generation increases.[5]

When used in conjunction with a matrix describing daily temperature profiles for each month of the year, balance point temperature can provide a preliminary indication of shading requirements for a building. This can be useful to the site planner for preliminary building orientation, siting, and design decisions. Shading calendars are

[4]C. Benton and J. Akridge, "Balance Point Exercise," *Laboratory Exercises* (prepared for the U.S. Department of Energy Passive Solar Curriculum Project, *Teaching Passive Design in Architecture,* June 1981).

[5]There are other approaches that can be used to calculate balance point temperature; see G. Z. Brown, *Sun, Wind and Light* (New York: John Wiley & Sons, Inc., 1985), pp. 56–62; and D. Watson and R. Glover, *Solar Control Workbook* (prepared for the U.S. Department of Energy Passive Solar Curriculum Project, *Teaching Passive Design in Architecture,* November 1981), pp. 60–63 and Appendix E.

time	factor	J	F	M	A	M	J	J	A	S	O	N	D
12M	.222												
2am	.139												
4am	.056												
6am	0												
8am	.111												
10am	.583												
12 N	.861												
2 pm	1.0												
4 pm	.917												
6 pm	.694												
8 pm	.444												
10pm	.306												

FIGURE 20-1. Shading calendar.

developed using the balance point temperature and normal temperature data contained in *Local Climatological Data, Comparative Climatic Data,* or the *Airport Climatological Summary* (see "Information Sources and Interpretation," Chapter 11). Bruce Novell of the Alabama Solar Energy Center describes a series of adjustment factors that can be used to develop the daily temperature profiles from normal daily maximum and minimum temperature data.[6] These factors are shown along the left-hand side of the calendar, next to the hours of the day (Fig. 20-1). Intermediate temperatures during the course of the day are calculated by:

$$T_{\text{intermediate}} = [(T_{\max} - T_{\min}) \times \text{factor}] + T_{\min}$$

The balance point temperature (65°F for the residential example) is plotted as shown in Fig. 20-2. The shading calendar indicates that afternoon shading is necessary from

[6]B. J. Novell, "Passive Cooling Strategies," *ASHRAE Journal,* Vol. 25, No. 12, 1983, pp. 23–28.

time	factor	J	F	M	A	M	J	J	A	S	O	N	D
12M	.222	39	43	48	59	67	75	79	79	73	61	49	42
2am	.139	37	41	46	57	65	73	77	77	72	59	47	40
4am	.056	35	39	44	55	63	71	75	75	70	57	45	38
6am	0	34	38	43	54	62	70	74	74	69	56	44	37
8am	.111	42	49	51	63	69	78	78	82	74	62	51	42
10am	.583	47	51	57	67	74	82	87	87	81	69	58	50
12 N	.861	53	57	64	73	80	88	93	93	86	76	65	56
2 pm	1.0	56	60	67	76	83	91	96	96	89	79	68	59
4 pm	.917	54	58	65	74	81	89	94	94	87	77	66	57
6 pm	.694	49	53	60	69	77	85	89	89	83	72	61	52
8 pm	.444	44	48	54	64	71	79	84	84	78	66	55	47
10pm	.306	41	45	50	61	68	76	81	81	75	63	51	44

FIGURE 20-2. Shading calendar. Balance point temperature 65°F.

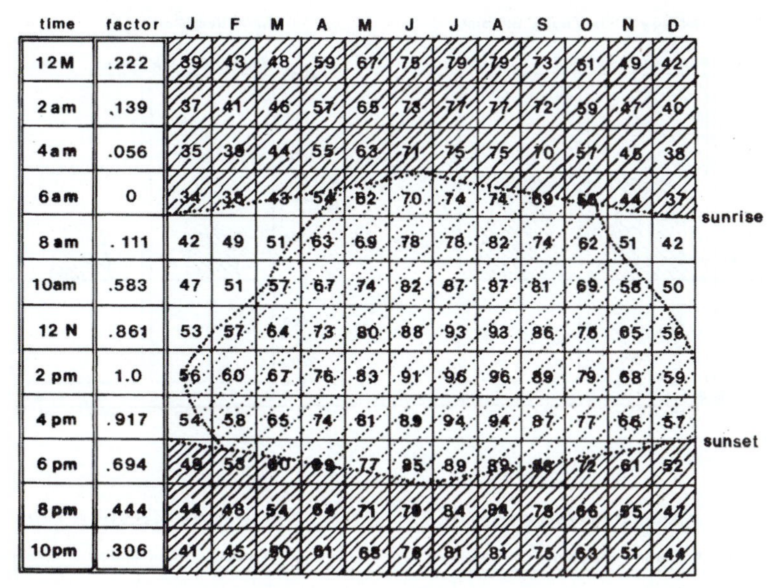

FIGURE 20-3. Shading calendar. Balance point temperature 55°F.

March through November for the house in this example. Complete shading is desirable from June through September. Afternoon solar gain is desirable in December, January, and February, and morning solar gain will be beneficial from October through April.

The shading calendar for a commercial building, which has higher internal loads and a balance point temperature of 55°F, is illustrated in Fig. 20-3. Here afternoon shading is suggested year-round, and early morning solar gain is useful for only a few months—November through March. This type of information can be valuable when making preliminary siting, massing, and orientation decisions.

PREDESIGN THERMAL LOAD AND BUILDING ENERGY ANALYSIS

With only limited, preliminary program information, there are methods available to the site planner to anticipate more precisely the nature of the thermal loading problems a building will experience when located on a site. Based on programmatic information similar to that used to calculate balance point temperature, a hypothetical base-case building can be subjected to analysis to identify thermal loading problems before the building is designed or even located on the site. This type of analysis will be discussed briefly.

One simple and extremely useful method for this type of analysis, described by G. Z. Brown in *Sun, Wind and Light,* is based on balance point temperature calculations.[7] The calculations can quickly be performed by hand, and the graphic depiction of results simplifies interpretation by the designer.

A hypothetical base-case building approach is used in the procedures presented in *Energy Graphics* by G. K. Hart, J. M. Kurtz, and W. I. Whiddon.[8] As with Brown's method, the calculation procedures are designed to be performed by hand. However, the calculations are tedious and time-consuming, which may inhibit the designer's investigation of alternatives. These calculation procedures are ideally suited for solution by computers, and many thermal load and energy analysis programs exist which can be used for the predesign analysis of a hypothetical base-case building.

A base-case building used in this type of analysis is hypothesized in the program information shown in Table 20-1. For this example the building is square (112 ft × 112 ft × 48 ft). The percentage of unshaded glass area for all four elevations is estimated to be 30%. In an effort to simplify the base analysis, all exterior walls will be considered as vertical planes, as opposed to being stepped or battered. Occupancy, lighting, and equipment schedules are developed from the building program, and the base-case building is "subjected" through analysis to these internal loads, as well as the external loads imposed by temperature, solar radiation, and humidity. The results of a computer analysis (using a microcomputer program called MPEG 1.0 developed by the author and based on *Energy Graphics* procedures) are illustrated in Fig. 20-4. Component loads (internal, solar, envelope, and ventilation loads) are graphed individually to facilitate comparative evaluations. "Total heat gain or loss" is obtained by summing the component loads. Loads below the "0" line on the "total heat gain or loss" graph indicate heating loads that will require auxiliary heating to maintain prescribed inside temperatures. Loads above the "0" line, but below the dashed "allowable heat

[7]Brown, *Sun, Wind and Light,* pp. 56–62.

[8]G. K. Hart, J. M. Kurtz, and W. I. Whiddon, *Energy Graphics* (Washington, D.C.: Booz, Allen & Hamilton, Inc., 1981).

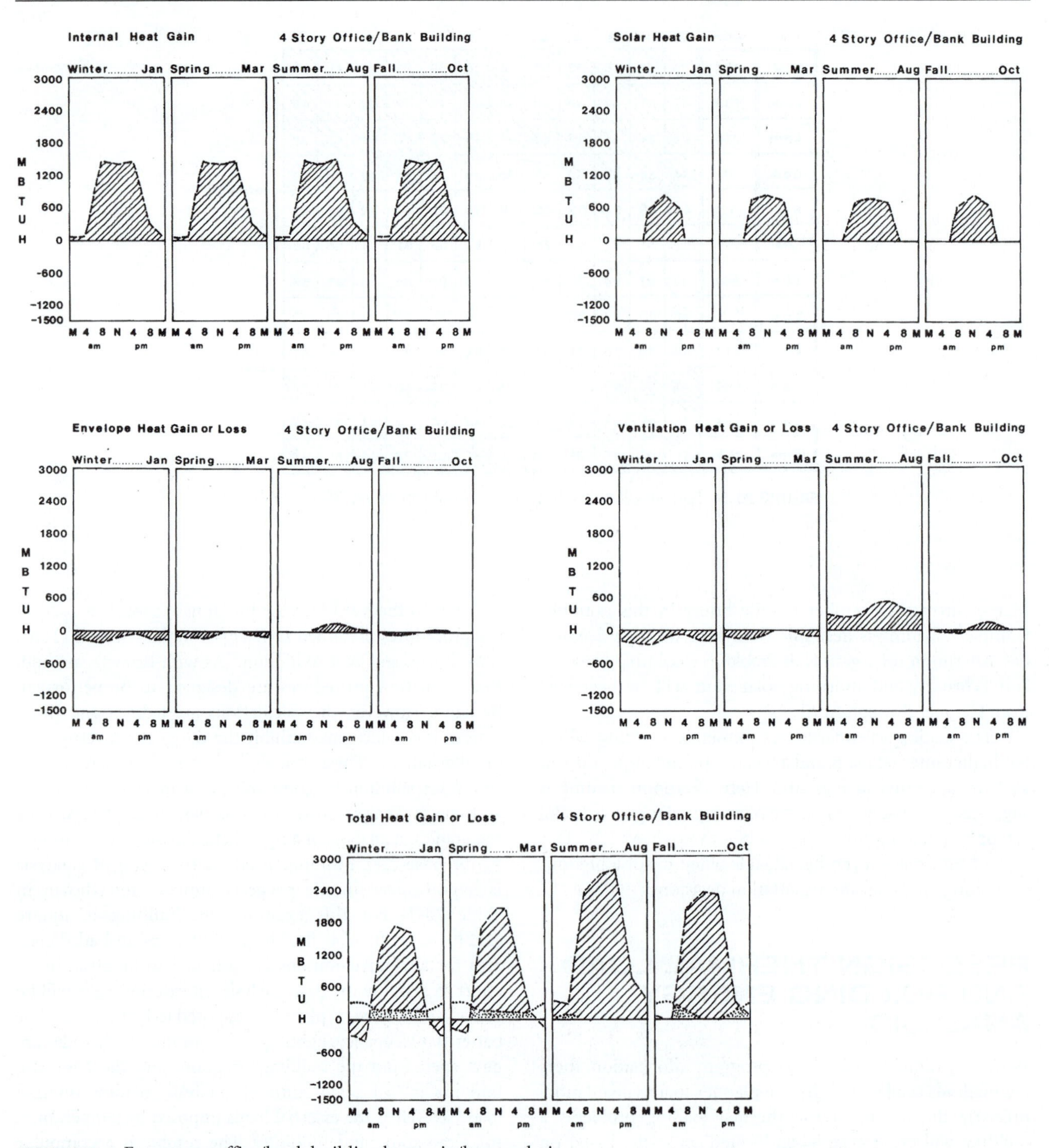

FIGURE 20-4. Four-story office/bank building heat gain/loss analysis.

gain" line, indicate heat gain which will not raise the inside temperature above the allowable limit. Loads above the dashed line will require mechanical cooling to mitigate their effect on interior temperature.

The base-case building that is being analyzed is a four-story building with a bank on the ground floor and three stories of office space above. The "total heat gain or loss" graph at the bottom of Fig. 20-4 indicates that on a normal January day in Dallas under clear skies, this building will experience excess heat gain and will require auxiliary cooling. This condition is exacerbated during warmer months. It is evident that "internal heat gain" from people, lights, and equipment is the dominant factor contributing to excess heat gain. "Solar heat gain" is also problematic, as well as "ventilation heat gain" in the summer, due primarily to latent heat gain. "Envelope heat gain or loss" is comparatively insignificant. With this type of thermal load profile, *Energy Graphics* recommends the following generalized design directives:[9]

[9]Ibid, pp. 56–59.

Table 20-1. Preliminary Building Program

1 Name of project analysis					Four-story bank/office building		
2 Floor area: volume					50,000		500,000
3 Roof area: roof *R*-value					12,500		10.0
4 South wall: area; % glass; shading coefficient					5,376	0.3	0.83
5 West wall: area; % glass; shading coefficient					5,376	0.3	0.83
6 North wall: area; % glass; shading coefficient					5,376	0.3	0.83
7 East wall: area; % glass; shading coefficient					5,376	0.3	0.83
8 *R*-value of walls; *U*-value of glass					7.00		0.58
9 Mechanical ventilation required; infiltration rate					10.00		1.0
10 Heat generated by occupants (Btu/hr/person) sensible; latent					250		200
	12M	4 am	8 am	12N	4 pm	8 pm	12M
11 Indoor temperature: upper limit	80	80	75	75	75	75	80
12 Indoor temperature: lower limit	60	60	65	65	65	65	60
13 Occupancy schedule	4	2	500	500	500	50	4
14 Internal heat from equipment	0.16	0.16	8.00	8.00	8.00	2.40	0.16
15 Internal heat from lighting	0.84	0.84	10.80	9.60	10.80	3.60	0.84
16 Heating equipment; cooling equipment efficiencies					0.70		2.00
17 Heating energy cost; cooling energy cost ($/MMBtu)					5.00		14.65
18 No changes to data							

1. Decrease internal loads.
2. Decrease solar heat gain.
3. Increase ventilation/infiltration heat loss (when outside air temperature and humidity are suitable).
4. Increase envelope heat loss.

The base building has a balance point temperature of 36°F, and the shading calendar, illustrated in Fig. 20-5, indicates that shading for windows is desirable for the entire year. This, along with the directives, would suggest to the site planner that building massing and orientation

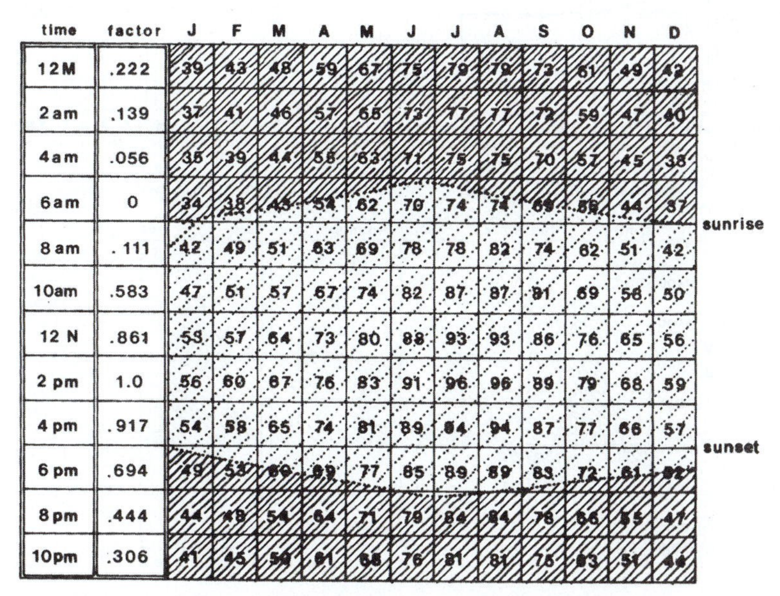

FIGURE 20-5. Shading calendar. Balance point temperature 36°F.

should be designed to reduce solar heat gain and take advantage of natural ventilation when appropriate. Artificial lighting is responsible for a significant portion of internal loads. This is primarily an architectural design problem—the utilization of daylighting to reduce artificial lighting requirements—but the site planner can respond by locating the building so as not to reduce access to daylight. The design directives suggest the following general architectural strategies:

1. Reduce electric lighting requirements through the use of daylighting.
2. Carefully shade all windows to reduce solar heat gain.
3. Utilize stack or natural ventilation if possible. Employ an economizer cycle in the air-conditioning system.

The office/bank building assumed in the base-case model is modified and re-evaluated as a structure that has a 4:1 length-to-width ratio. It is oriented along the east–west axis to facilitate daylighting and to simplify the shading of glass. Since east and west glass is difficult to shade, the glazing area of these two walls is reduced to 10% of the wall area. It is assumed that the narrower building with its emphasis on daylighting will permit a reduction in electric lighting by 50% during daylight hours. The effects of reorienting the building, carefully shading glass, and reducing electric lighting are shown in Fig. 20-6. The balance point

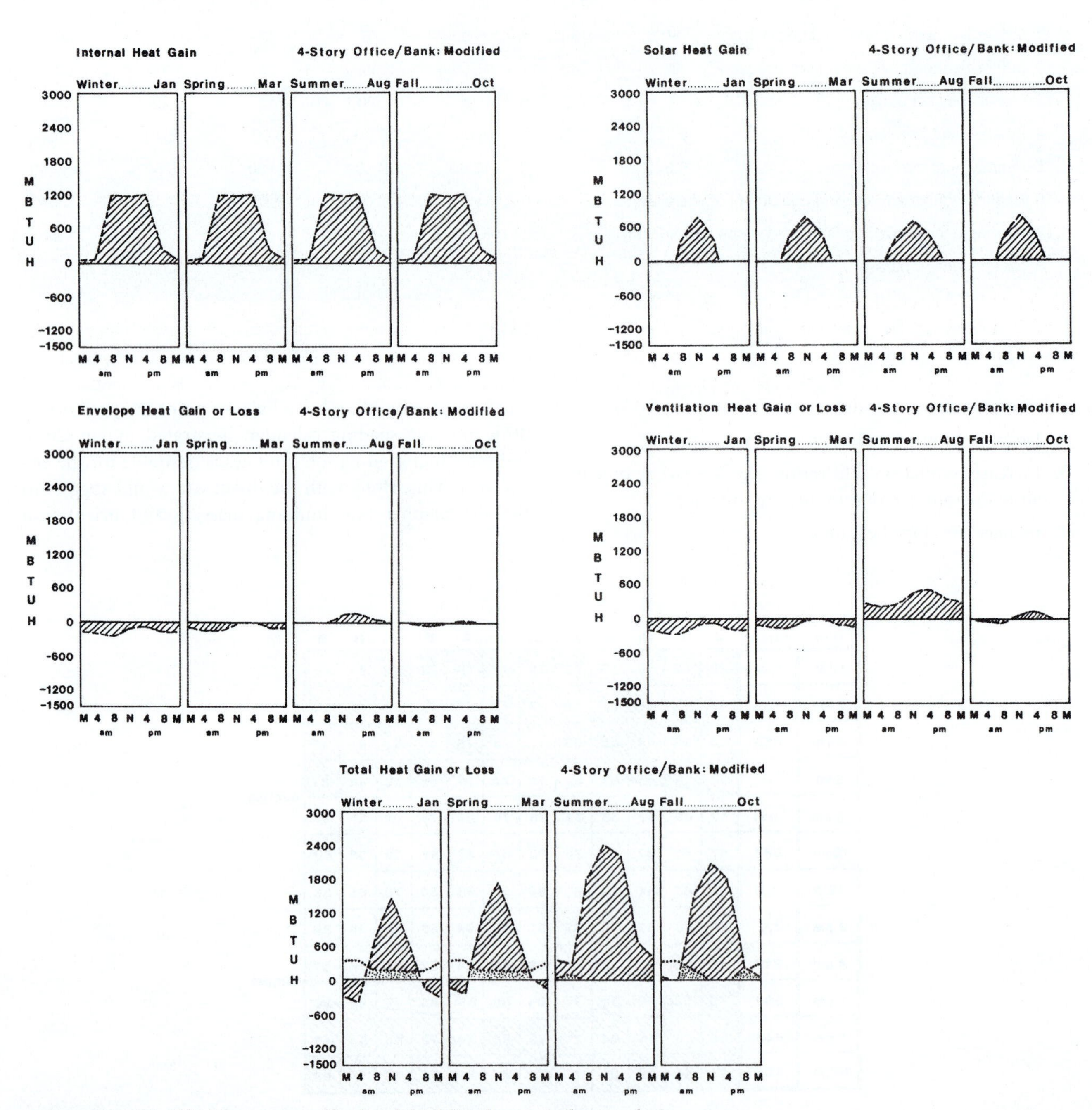

FIGURE 20-6. Modified four-story office/bank building heat gain/loss analysis.

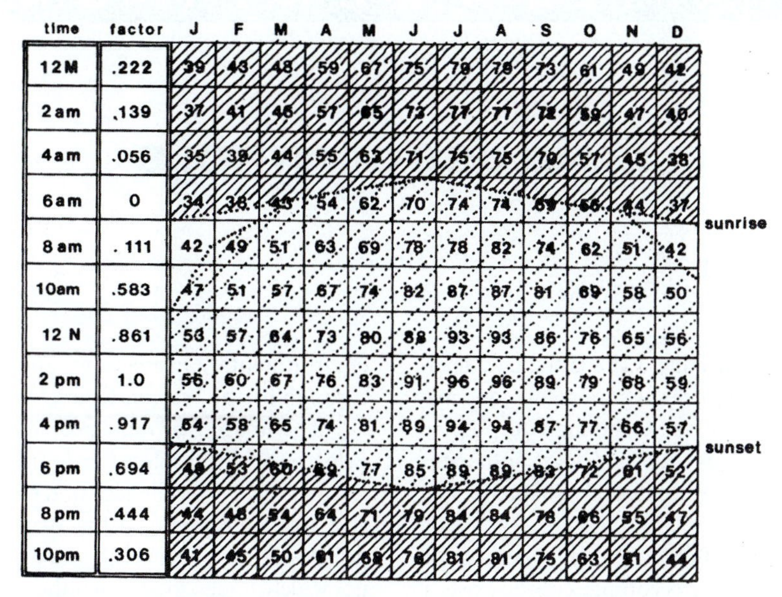

time	factor	J	F	M	A	M	J	J	A	S	O	N	D
12M	.222	39	43	48	59	67	75	78	78	73	61	49	42
2am	.139	37	41	46	57	65	73	77	77	72	59	47	40
4am	.056	35	39	44	55	63	71	75	75	70	57	45	38
6am	0	34	38	43	54	62	70	74	74	69	56	44	37
8am	.111	42	49	51	63	69	78	78	82	74	62	51	42
10am	.583	47	51	57	67	74	82	87	87	81	69	58	50
12 N	.861	53	57	64	73	80	88	93	93	86	76	65	56
2 pm	1.0	56	60	67	76	83	91	96	96	89	79	68	59
4 pm	.917	54	58	65	74	81	89	94	94	87	77	66	57
6 pm	.694	48	53	60	69	77	85	89	89	83	72	61	52
8 pm	.444	44	48	54	64	71	79	84	84	78	66	55	47
10pm	.306	41	45	50	61	68	76	81	81	75	63	51	44

FIGURE 20-7. Shading calendar. Balance point temperature 45°F.

temperature has been elevated to 45°F, and the effects on shading requirements are illustrated in Fig. 20-7. With the exception of early mornings in December and January, shading is still required for the entire year. The estimated annual load reduction produced by these simple strategies is 24%, or an annual air conditioning and lighting utility cost savings of 30%.

Again, this type of analysis can suggest strategies for preliminary site-planning decisions concerning buildings. Once building location and massing have been determined, the site microclimates should be remapped, using the mapping procedures outlined in Chapter 11 in the section "Site-Microclimate Mapping." The analysis of climate, site-microclimates, and the thermal loads on proposed buildings can inform the site-planning process, increase the opportunities for thermal comfort in exterior spaces, and reduce thermal loads on buildings and requisite energy use in buildings.

CHAPTER TWENTY-ONE

LAND USE CONTROLS

The least visible force affecting a site is the law. There are numerous public agencies whose regulations may affect the development of a site. Although the purview and jurisdictions of the various state and local authorities widely vary, the site-planning checklist in Chapter 7 will provide a start.

The most common condition where legal constraints materially influence site planning is zoning. Zoning exists at the city, town, county, and, at least in one case, state levels. Although state zoning is unusual, it has permitted the state of Hawaii to control the encroachment of urban land uses into prime agricultural resources and areas of scenic beauty. The island's landform may have induced Hawaii to adopt a concept of land use controls that has not been imitated by any mainland state since its adoption more than two decades ago.

But Hawaii is the exception. Zoning is basically a local governmental issue. What is done in a neighboring county, city, or other political jurisdiction does not necessarily mean that there are always commonalities. As with other conditions in site planning, the site planner should never assume anything relative to zoning and land use controls. Aside from a state's enabling legislation, which is common to all municipalities within a state, each will have adopted its own laws controlling land use controls.

What is most common from state to state and city to county is the process and the players. These will vary to some degree because one state may have given the power to zone only to its cities, another to its counties. Depending on the state, the nomenclature of local governments will vary: a city council, county council, county board of supervisors, county commissioners, and so on. Do not confuse your elected officials, who in most communities make the decisions concerning zoning, with those they appoint. These relationships and the process will be discussed later in this chapter. For an understanding of zoning and its influences, it helps to have some understanding of the evolution of the concept.

Misconceptions associated with individual rights abound. The popular belief is that each American has an inalienable right to do as he or she wishes with his or her land and that governmental control borders on an illegitimate exercise of police power. As Fred Bosselman, David Callies, and John Banta observed in *The Taking Issue*, "there is an opinion by many that foreign concepts like environmental protection and zoning were sneaked through by the Warren Court."[1]

The image that most people have of the colonist in the wilderness with absolute freedom from land use controls is far from the fact. A certain lack of regulation was not only acceptable but appropriate in the struggle to survive during the decades of the seventeenth and eighteenth centuries. What is not widely understood is that the colonists brought with them a set of legal concepts that were 400 years in the making.

HISTORY AND BACKGROUND

As a monarch, King John of England exercised broad power over the country's people, land, and treasury in the thirteenth century. He controlled the kingdom through noblemen on estates and levied taxes as he saw fit. John's abuses of his authority were so extensive that his noblemen collectively demanded that their rights be defined. This confrontation between noblemen and their king culminated in the development of the Magna Carta in 1215. John's reluctant acceptance of the Magna Carta is considered to be the keystone of individual rights in England and her colonies. Chapter 39 is of particular importance since it required the crown to respect "the law of the land," and consider the "judgment of peers" in the acquisition of a freeman's property:

> No freeman shall be arrested, or detained in prison, or deprived of his freehold, or in any way molested; and we will not set forth against him, unless by the lawful judgment of his peers and by the law of the land.[2]

Although there is some doubt as to whether subsequent monarchs respected the intent of the law, many continued to employ the crown's authority to regulate without pause. Elizabeth I, as an example, issued several proclamations associated with land use, density control, and building height and design in an effort to reduce the squalor of urban London.[3] Following the Great Fire of London in 1666, Parliament invoked and enforced strict building codes

[1]Fred Bosselman, David Callies, and John Banta, *The Taking Issue* (Washington, D.C.: U.S. Government Printing Office, 1973), p. 1.

[2]Translated from an original Latin text by Edward Coke (Milite), *The Second Part of the Institutes of the Laws of England* (London: printed for E. and R. Brouke, Bell Yard, New Temple Bar, and William S. McKechnie, 1797), p. 45.

[3]31 Elizabeth I C.7 (Statutes at Large, Vol. 6 at 409, *et seq.*).

which regulated roof lines, materials, and construction techniques.[4] It was from a tradition that respected the rights of the individual while acknowledging the important concept that property could be regulated for the public good that colonists set sail for the New World.

The New World immigrant had lived with land use controls, and as urban centers emerged, recognized the value of building codes and ordinances that restricted nuisances. Boston, New Amsterdam, and Philadelphia all had ordinances that sought to reduce the potential of fire or delineated specific design criteria for the landscaping of public streets and individual properties.[5]

The acceptance of the concept of the public good, or health, safety, and welfare, was also critical to the establishment of a German concept of land use regulation zoning. The notion that there was an appropriate arrangement or relationship between various land uses was based primarily on nuisance laws. It was not unreasonable to accept the premise that gunpowder factories, slaughterhouses, and stockyards should not be located adjacent to residential land uses. But although the concept of separating "incompatible" uses was reasonable, in the absence of the concept of zoning, the only recourse that one property owner had against a nuisance (another property owner) was a civil lawsuit. As the noted attorney Richard Babcock observed: "There is little evidence in the history of land development in the U.S. [in which] the private decision maker, left to his own devices, can be trusted to act in the public interest."[6]

Although land use regulation was broadly accepted, when the colonists sat down to draft the new law of the land, a fundamental difference relative to government and private property emerged. Where there was little disagreement as to the authority of the colonial legislature to take land for a public service, the concept of "just compensation" became a recurring text in various state constitutions. The principle appeared first in Vermont's first constitution in 1777,[7] to be followed by Massachusetts three years later. Madison's initial draft of the Bill of Rights included the following text:

> No person shall be subject, except in cases of impeachment, to more than one punishment or one trial for the same offense; nor shall be compelled to be a witness against himself; nor be deprived of life, liberty, or property without due process of law; *nor be obliged to relinquish his property, where it may be necessary for public use, without a just compensation* [emphasis added].[8]

The reasons for the "taking clause" text of the Bill of Rights have been debated by numerous legal scholars.[9] Among those considered are two plausible rationales for the distinct departure from the English tradition. The first was the simple fairness of a condition in which a property could be taken without compensation. Although this was an acceptable situation with a monarch, the government in the colonies was the people and, as such, was seen more as an extension of the authority of individuals than one ordained by birth or church. Second, the revolutionary army was apparently arbitrary and oppressive in the acquisition of supplies from the colonial farming community. Compensation was never discussed as the army consumed precious livestock and foodstuffs in the struggle against the British.[10] The propertied class was also somewhat apprehensive that those of lower "status" might one day administer the country and, as such, favored the "just compensation" clause.[11] Madison and his colleagues supported the principles of eminent domain inherited from the English, but they believed that the concept could lead to abuse without some controlling mechanism that maintained the rights of the individual property owner.

Land use controls in the United States conceptually separated into two areas: one as eminent domain, in which just compensation is mandatory; the second as police power or restrictions on land use, wherein no compensation is involved as long as the government (local, state, or federal) is acting in the interest of its citizenry. One of the earliest cases documenting the land use controls and police power in the United States is the *Brick Presbyterian Church* v. *The City of New York;* 1826.[12] The church had a history of permitting the bodies of deceased parishioners to be interred in the churchyard. The municipality had a real concern that the natural processes of decay would contaminate the municipality's groundwater supply and filed suit. The court held for the municipality, maintaining that the rationale for its decision to sustain the city's right to restrict land use was the health, safety, and welfare of the community.[13] The *Brick Presbyterian Church* case is an early model example of how a municipality's exercise of police power can control a property without "taking" the property.

One of the next critical issues that would need to be heard and tested would deal with the acceptable limits of the municipality's control. The inevitable argument would examine the case where the municipality's efforts to control a property's use result in a reduction in the property's value. In 1915 the California Supreme Court provided some clues as to the limits of property reduction when sustaining a

[4] 19 Ch. II C.3 (8 Stat. large), pp. 233–251.

[5] 2 Pennsylvania St. 66, Chap. 53.

[6] Richard F. Babcock, *The Zoning Game* (Madison, Wis.: The University of Wisconsin Press, 1966), p. 185.

[7] J. B. Wilbur, *Ira Allen: Founder of Vermont* (Boston, Mass.: Houghton-Mifflin Co., 1928), p. 96.

[8] *Annals of Congress*, pp. 451–452; Schwartz.

[9] Bosselman, Callies, and Banta, pp. 99–104.

[10] St. George Tucker, *Blackstone's Commentaries with Notes of Reference to the Constitution and Laws of the Federal Government of Virginia and Commonwealth of Virginia,* Book 1, Part 1, 1803, pp. 305–306.

[11] Oscar Handlin and Mary Handlin, eds., *The Popular Sources of Political Authority* (Cambridge, Mass.: Harvard University Press, 1966), p. 336.

[12] *The Brick Presbyterian Church* v. *City of New York,* 5 Cow (New York, 1826), p. 538.

[13] 7 Cow (New York, 1826), p. 595.

Los Angeles ordinance that prohibited the operation of a brick kiln in a residential area. In spite of the fact that the manufacturing function had been in place for some time and that its value would be reduced from $800,000 to $60,000, the court ruled that the ordinance was a reasonable restriction protecting the health and welfare of the community.[14] The guideline that delineates the difference between police power and eminent domain is that the state controls, through police power, the use of property for the public good, without compensation for diminution in value, when its use otherwise is harmful, whereas through eminent domain the state takes property, with compensation, when it is useful to the public.[15]

Although other cases provide insights into the changes in the court's attitude toward land use controls, few provide the conceptual examples presented in *Pennsylvania Coal* v. *Mahon* in 1921. At the turn of the century, northern Pennsylvania had extensive deposits of anthracite coal. Pennsylvania Coal, together with numerous other mining operations, had acquired the mineral rights to thousands of acres of land throughout the state and was actively involved in the extraction of the resource. In the exuberance of the industry to mine coal, many shafts and tunnels were excavated without proper shoring. The results were often surface collapse or subsidence. Since numerous tunnels and shafts transversed towns and communities throughout the state, many cities found the hazard of collapse to be a common occurrence.[16] Broken water mains, fires from broken gas lines, and the collapse of structures plagued numerous cities. The situation became so serious that the state adopted legislation (the Kohler Act) which prohibited subsurface mining that would cause subsidence and collapse of streets, buildings, and property damage.[17]

John and Mary Mahon owned and resided in Pittston, Pennsylvania, where, 20 years earlier, Pennsylvania Coal had owned land and mineral rights. Upon sale of the property, Pennsylvania Coal retained the mineral rights to its property as well as a waiver of any claim (against Pennsylvania Coal) for subsidence.[18] The Mahons were given written notice in September 1921 that Pennsylvania Coal was mining in their community and that the operation might create subsidence. If they were concerned about the safety of themselves or their household, they should evacuate the premises. Pennsylvania Coal reminded the Mahons that the mineral rights were theirs and that Pennsylvania Coal was not liable for any personal or property damage resulting from its mining.[19]

[14]*Hadacheck* v. *Sebastian*, 239 U.S. 394 (1915).

[15]*Words and Phrases*, Vol. 32A (St. Paul, Minn.: West Publishing Co., 1956), p. 404.

[16]Bosselman, Callies, and Banta, pp. 126–127.

[17]Public Law 1192.

[18]Brief on Behalf of the City of Scranton, Intervenor in the Supreme Court of the United States, October 1922, p. 1.

[19]Transcript of Record (29,099) filed August 17, 1922 in the Supreme Court of the United States, October 1922, p. 10.

The Mahons relied solely on the Kohler Act for protection and filed an injunction to prohibit Pennsylvania Coal from mining. Litigation ensued, and following disagreements between lower and appeals courts, the case was ultimately heard by the U.S. Supreme Court. The Court's decision, penned by Justice Oliver Wendell Holmes, held for Pennsylvania Coal and ruled the Kohler Act unconstitutional. In spite of the obvious health, safety, and welfare issues involved in the case, Chief Justice Holmes considered that "the right to the coal consists in the right to mine it."[20] The Court maintained that in the interest of the public's health, safety, and welfare, some rights may be taken with police power. But when the state attempted, through the Kohler Act and its police power, to take the rights to mine *any* of the coal, the state essentially attempted to acquire the property rights, which more appropriately should have been accomplished through eminent domain, where just compensation is required.[21] The implication was that although the state had appropriately tried to address an issue of "health, safety and welfare" through police power, where no obligation for just compensation was necessary, the results of the controlling legislation took *all of the rights* to the mineral (property). Although the Court maintained that taking all the rights was unconstitutional, they did not, nor have subsequent Supreme Court decisions, render any judgments that have given any guidelines as to "how many is too much."

The court's decision in *Pennsylvania Coal* v. *Mahon* established a clear line between the concepts of eminent domain and police power. As the nation grew, governments, at all levels, used the power of eminent domain to acquire property for a variety of uses—but always with compensation. The question as to the amount of just compensation is normally measured by the fair market value of the property at the time of the taking. Compensation also means:

> full indemnity or remuneration for the loss or damage of property taken or injured for public use; and just compensation means the full and perfect equivalent in money of the property taken, whereby the owner is put in as good a position pecuniarily as he would have occupied if his property had not been taken.[22]

Although there may be conditions in which "just" may be arguable, there is no question as to the importance of this legal mechanism in the operation of government. The police power segment of government's authority is equally important. This power to regulate and control the use of private property has been defined as:

> the inherent right of a government to restrict individual conduct or use of property to protect the public health, safety and welfare; it must follow due process of the law but, unlike eminent domain, does not carry the requirement of compensation for any alleged losses.[23]

[20]*Commonwealth* v. *Clearview Coal Company*, 256 Pennsylvania St. 328, 331.

[21]*Pennsylvania Coal Co.* v. *Mahon*, 260 U.S. 393 (1922).

[22]Corpus Juris Secundum, Sec. 96, p. 390.

[23]Diane Maddex, ed., *The Brown Book: A Directory of Preservation Information* (Washington, D.C.: Preservation Press, 1983), p. 52.

For years the concept of nuisance law worked to keep a needed separation between incompatible uses. In a time when land was cheap and life was simple, common sense prevailed and people kept some distance between the town's powder magazine and their homes. Although pigsties and rendering plants were also unwelcome in a residential atmosphere, in many cases the argument as to whether a use was, in fact, a nuisance was left to the civil courts to decide. The notion that there might be a variety of nuisance land uses that should be separated from residential activities emerged as a logical strategy through which local government might protect the public's interest.

In 1916, the City of New York became the first municipality to adopt a comprehensive zoning ordinance.[24] Although numerous communities throughout the United States later adopted and employed zoning ordinances, they were all challenged in the courts. State and district judges had failed to render any decision (one way or the other) with any argument of such profound logic that might mandate others to follow. The nature of the situation in the mid-1920s was quite unsettled. The dilemma was that although one might accept the concept that an industrial land use would be a nuisance to a residential neighborhood, questions concerning whether a grocery might be a nuisance to a sheet metal shop left the question open to debate. The basic legal haze that clouded the issue of control over private land now focused on the 14th Amendment. The salient section of the amendment provided that "no person be deprived of his property without due process of law." The question came to this: Even when state enabling legislation permitted zoning, was zoning a scheme of municipal regulation, structured to prohibit land uses that had never been considered a nuisance?

So it was not an unusual set of circumstances for the Village of Euclid, Ohio, to become engaged in litigation upon the denial of a request for the rezoning of a residential property to a commercial property. Ambler Realty, the plaintiff, maintained that the zoning ordinance was an unconstitutional taking of property rights and further, that the imposition of the zoning diminished the potential value of the property. After an unusual rehearing of the case, the high court overturned a federal trial court and sustained the concept that zoning was a legitimate function of police power.[25] The landmark decision favoring the Village of Euclid was one of the most significant in American land use law and established precedents that are still referenced by the courts. Although this was a very tumultuous time in the development of land use regulations in the United States, *Village of Euclid* v. *Ambler Realty Co.* became the cornerstone on which the system rests today.

[24]Frank So, Israel Stollman, Frank Beal, and David S. Arnold, eds., *The Practice of Local Government Planning* (Washington, D.C.: American Planning Association and the International City Management Association, 1979), p. 416.

[25]*City of Euclid* v. *Ambler Realty Co.*, 272 U.S. 365, 384 (1926).

Since zoning is a power passed from state to local governments, there are broad variations from state to state and from city to city. For further reading on the topic, the reader is referred to *American Land Planning* by Norman Williams, a five-volume text that examines the multiplicity of state laws in detail. Although efforts in 1922 and 1928 to develop a standard state enabling legislation was widely received, each municipality has interpreted and implemented laws to reflect local values and conditions. Although the court's ruling in 1926 closed one door, it opened many, as communities across the land struggled with the site-specific decisions associated with land use planning and zoning.

Some municipalities misread the Supreme Court's *Euclid* decision and found themselves before state supreme courts for unreasonable and/or capricious application of their ordinances. The courts maintained, among other things, that a reasonable use of the property was an essential consideration in the administration of an ordinance.

Eminent Domain and *Berman* v. *Parker*

Major changes in the legal theory regarding individual property rights did not appear until the case of *Berman* v. *Parker* in 1954. This case was a key precedent in establishing the elasticity of the "public use" phrase in the Fifth Amendment of the U.S. Constitution. The District of Columbia Redevelopment Act of 1945 resulted in the creation of a comprehensive plan of slum clearance and redevelopment. The plan was controversial as the eradication of substandard housing south of the capitol building was key to the creation of whole new neighborhoods. The redevelopment included communities of a mix of residential types, commercial retail shopping, and supporting parks and open space.

The Court's 8–0 decision held for Parker and the urban renewal efforts of the District of Columbia. Among the critical legal points growing from the Court's decision, to be repeated by numerous other governments in years to come, is that public ownership is not the sole method of promoting a public purpose. Recall that prior to this decision, eminent domain was used when a government needed to acquire property for public purpose, such as a school site or road right-of-way. However, in the case of the D.C. urban renewal effort, the government was developing a plan in which public and private entities would have ownership. The court, in this landmark case, decided that private property could be taken, with just compensation, based on a public purpose plan. The public need not become the ultimate owner of the subject property. Additionally, the Court's decision legitimized the efforts of numerous communities to pursue projects and redevelopment where beauty as well as sanitary issues were sound and worthy objectives.

More than three decades passed before the Supreme Court would consider another eminent domain case. It was not until 1984 that the Court heard a case that hinged on the question of whether a "public purpose" constitutes a "public use". Recall the Fifth Amendment's taking clause, "nor shall private property be taken for public use without just compensation." While the Court saw the elimination of slums and blight as a legitimate application of eminent domain in *Berman* v. *Parker*, the Court was silent as to other acceptable motives.[26]

Hawaii Housing Authority v. *Midkiff*

By the 1960s, urban renewal projects were common throughout the mainland United States, and the Court's decision in *Berman* v. *Parker* was the legal bedrock of the necessary acquisition of private property. Hawaii, however, had an entirely different kind of urban development issue. The Hawaii legislature had spent years completing a review of the disposition of the property ownership in the state. Among the facts that emerged from the exhaustive hearings was that the state and federal governments held almost 49% of the state's land area. As alarming to the legislature was the fact that of the remaining 51% of the property, 47% was held by a scant 72 private owners. On Oahu, the state's most urbanized island, 72% of the fee simple titles were held by 22 landowners. The legislature decided that the existing conditions resulted in an oligopoly, which was "skewing the state's residential fee simple market, inflating land prices, and injuring the public tranquility and welfare," and moved to rectify the imbalance.[27]

To accomplish the "redistribution" of property, the Hawaii legislature's decision was to condemn land in the oligopoly through eminent domain and sell the property to the existing leaseholders with the passage of the Land Reform Act of 1967. As there was clearly no "public use" that resulted from the eminent domain effort, the challenges to the 1967 legislation ended in 1984, when *Hawaii Housing Authority* v. *Midkiff* was heard by the U.S. Supreme Court. The Court's 8–0 decision reflected complete agreement with the state, underscoring the constitutionality of the legislature's eminent domain taking. The state's efforts to regulate the oligopoly was viewed as an exercise of police power (regulatory) that attempted to correct a free market's failure, thus satisfying the "public use" doctrine of the Fifth Amendment. The Court's ruling indicated that the critical aspect of an eminent domain taking was neither the method nor mechanics, but the taking's *purpose*.

Maher v. *The City of New Orleans* in 1975 provides some clues and an interpretation of the term "reasonable." The Vieux Carre is a nationally famous historic district and was one of the first in the United States to adopt a historic district ordinance. The ordinance was *very* restrictive, and when Mr. Maher's request for a demolition permit was denied, litigation ensued. The municipality maintained that the one-story apartment building was a reasonable use, providing the owner with a reasonable income as it existed. Although the city recognized that the structure was not of historic merit on its own, the property was part of the "tout ensemble" of the entire French Quarter. That is, the Historic District was not a landmark alone but a collection of structures. Change to any one part inflicted damage on the entire collection. Mr. Maher, however, considered the municipality's Historic District zoning an economic burden. He contended that the modest structure was not historically significant, and that the obligation to maintain a structure for the public benefit "combined with the rising costs of a more profitable development caused him 'irreparable harm' and amounted to a regulatory taking."[28]

The case ultimately came before the U.S. Supreme Court, and their decision held for the City of New Orleans. Basically on the grounds presented by the City of New Orleans, the high court maintained the existing use to be a *reasonable* use of the property.[29] The decision not only reinforced the city's historic zoning ordinance but on a national scale provided another example where the reduction in a property's value would not be an overriding issue in land use controls.

Penn Central v. *The City of New York*[30]

In 1978, the Supreme Court agreed to hear a case brought by Pennsylvania Central Railroad against the City of New York. The City of New York had refused an application by Penn Central to construct a high-rise office tower in the air rights over the historic Penn Central Railroad Station in Manhattan. The Penn Central Station was identified by the City's Historic Landmark Commission as a significantly important historic site and gave the structure historic landmark status. Since the historic landmark designation prohibited the structure from being razed for other development, the applicant requested permission to construct an office tower in the air rights space above the historic station. The city denied the application, stating that such a structure, if built, would be inconsistent with the character of the historic Penn Station.

The City of New York also noted that the city had provided the owners of the historic Penn Central structure the opportunity to use the municipality's transfer of development rights (TDR) program. This program provided the opportunity for the owners of historic properties to transfer the development rights from the sites upon which historic structures were located to other redevelopment sites in the city. Although the TDR program was not one that Penn Central owners wanted to pursue, it is interesting to note that the TDR strategy was subsequently used by numerous local governments throughout the United States for programs ranging from historic preservation to agricultural farm preservation.

The Court's 6–3 decision in favor of the City of New York acknowledged the city's efforts to soften the economic hardship while confirming the authority of government to pursue objectives that were other than economic. The decision was

[26] *Berman* v. *Parker*, 348 U.S. 26 (1954).

[27] *Hawaii Housing Authority* v. *Midkiff*, 467 U.S. 229 (1984).

[28] Complaint, *Morris Maher* v. *The City of New Orleans*, Civil Action No. 71-119, U.S. District Court, Eastern District of Louisiana, filed January 13, 1971.

[29] *Maher* v. *City of New Orleans*, 256 La. 131, 235 So. 2d 402 (1970).

[30] *Penn Central* v. *The City of New York*, 438 U.S. 104 (1978).

not a complete surprise, given the Court's decision in *Maher* v. *The City of New Orleans* in 1975. In those cases in which government purposes were legitimate and anchored by solid underlying regulation, the Court's review would likely be supportive.

Kelo v. *The City of New London*[31]

While the foundation for land use controls was jostled now and then, it moved little until *Kelo* v. *The City of New London* in 2005. In 1998 the Pfizer pharmaceutical company began construction of a new facility at the edge of the Fort Trumbull neighborhood in New London, Connecticut. The City of New London saw a potential rehabilitation opportunity at the edge of the new Pfizer plant. The city council reactivated the dormant New London Development Corporation, a private entity, and adopted a comprehensive plan to revitalize a depressed, not blighted area, of 115 lots. The city hoped the new plan would result in new development that would create new jobs and increase the tax base, along with other city revenues. To implement the plan, the city's strategy was to condemn the 115 lots, acquire the property through eminent domain, and resell the property to private entities.

Of the 115 lots to be acquired in the older Fort Trumbull neighborhood, 15 balked; the lead plaintiff, Susan Kelo, filed suit. After various lawsuits, decisions, and reversals, the case ended up before the U.S. Supreme Court. This case focused on the question of whether the City of New London could use the power of eminent domain to acquire property that the municipality was ultimately going to sell to private parties. The U.S. Supreme Court found that using the power of eminent domain for economic development did not violate the public use clause of the state and federal constitutions. Additionally, the Court sustained the right of the City of New London to delegate the power of eminent domain to a private entity—in this case, the New London Development Corporation. There was no slum clearance in the redevelopment plan, as there had been in the D.C. urban renewal plan. This was a plan whose objectives centered on economic development.

Probably no decision in which the power of eminent domain was an issue created as much angst as *Kelo* v. *The City of New London*. This was no surprise to anyone who followed the case in the courts as there were some 40 *amicus curiae* briefs filed for consideration by the Supreme Court in *Kelo*. To many, the *Kelo* decision turned the whole principle of eminent domain on its head. The public was again mystified as to the extent to which public use would be limited. Every decision by the Court seemed to imply an ever-widening definition of the term. The public's general perception of eminent domain was that it would be used to acquire rights-of-way for roads, school sites, and city halls—all simple, legitimate and straightforward public uses. But the decisions of Supreme Courts since the mid-1950s seemed to warp "public use" into "public purpose" in an ever-widening arc. *Kelo* v. *The City of New London* was more than some individual state legislators could tolerate, and a chorus of disapproval was heard from state senators of every stripe. The sentiment of lawmakers from Texas, Florida, Iowa, Michigan, and other states was succinctly reflected in the legislation introduced in New Hampshire, as it stated:

> No part of a person's property shall be taken by eminent domain and transferred, directly or indirectly, to another person if the taking is for the purpose of private development or other private use of the property.[32]

The Court's decision in *Kelo* was an affront to many lawmakers, regardless of their political persuasions. Where some states have remained silent regarding the interpretations of the high Court, others have taken a more proactive stance and sought to head off decisions they consider inconsistent with regionally acceptable standards of public use. It's unlikely that the last legal word has been written on this topic.

Conclusion and Counsel

No decision to date has resolved the question of "how much" (control) is considered "too much" and "taking" without compensation. What has been observed is that the courts maintain that a segregation of incompatible land uses, prevention of congestion, and an ordinance based on a plan are all components of sound ordinances. The problems that have developed since 1926 are built around several *assumptions* associated with zoning:

1. That segregated land uses would result in a high-quality environment
2. That a zoning map, once prepared, would exist for all time
3. That local and state government would not change the rules
4. That nonconforming uses would, under pressure of redevelopment, go away
5. That zoning could be used as a method to cure problems that were *not* restricted to physical land use[33]

The problem is that none of the points listed above were correct assumptions. This is not an attack on the concept of zoning, but one must acknowledge that there were perceptions (of success) associated with zoning that left many with second thoughts. Euclidean zoning was an appropriate start; it just could not resolve all problems.

[31] *Kelo* v. *The City of New London*, 545 U.S. 469 (2005).

[32] www.sos.nh.gov/con/con-2006. CACR 30 Article 12-a (power to take property limited).

[33] Ira M. Heyman, "Legal Assaults—On Municipal Regulation," *Management and Control of Growth* (Washington, D.C.: Urban Land Institute, 1975), pp. 187–188.

Second, the concept has not been skillfully administered in every community. Any tool, poorly applied, will produce questionable results. This is an enduring misunderstanding by many communities as to the function of zoning. Zoning is a tool used to develop a community plan. Without a plan, the community and its local government have little guidance. Without a well-conceived and specific set of implementation strategies, including zoning, any plan is little more than a wish.

While it is unlikely that many architects, landscape architects, or engineers will find themselves in court, defending the property rights of a client or the eminent domain authority of a local government, understanding the history and legal precedents is valuable. But knowledge of the theory and history of the law, unless you are also an attorney, does not provide you with the skills to provide legal advice on the taking issue. In any case in which a client has misgivings regarding an eminent domain taking or a *quid-pro-quo* proffered by a state or local government, the site planner should recommend that the client seek the counsel of an attorney with expertise in property rights and land use regulation. Just as hiring a registered professional land surveyor to execute a property line survey is essential to establishing a property's perimeter boundary line, so too is seeking the counsel of an attorney when questions arise concerning the legality of a taking or the constraints of a zoning ordinance.

EUCLIDEAN ZONING: THE BASIC ORDINANCE

As the variety of local governments in the United States is extensive, the effort here will not cover every known set of circumstances but will outline in general terms the components and their influence on the development of a site plan. Some variations will be explored, but strict application cannot be made in all cases without some interpretation for local laws, process, and nomenclature. Since the Euclidean zone is the foundation of land use controls in the United States, the language of the zoning text and its meaning warrant closer inspection.

Apparently, the term "Euclidean zoning" has two origins. Some are of the opinion that "Euclidean" refers to the geometric patterns of zones. Others suggest that the nomenclature refers to the origin of the landmark zoning case: Euclid, Ohio. But regardless of opinions as to the term's origin, there appears to be no disagreement relative to the meaning of the term "Euclidean zoning." It is that traditional prescriptive zoning which, in very specific terms, identifies land use and density, with a rubric of performance standards. Euclidean zones have been the backbone of all zoning ordinances. The performance standards, or design controls, are both loved and hated by the developer, the community, and citizens. Since there is no variation in the standards, everyone knows, within certain limits, what the building mass and location on a site will be. The performance standards establish front, rear and side-yard setbacks, building height limits, and, in some cases, building coverage. These performance standards also include parking requirements. All too often the site planner applies the requirements established in the zoning ordinance and finds little flexibility in the alternative location for building and parking. The problem is that Euclidean zones are structured so that they can be applied to any site without deviation, regardless of context or topographical relief.

Euclidean zones treat every site as though there is no encumbrance to development in the form of slopes, vegetation, hydrology, geology, and so on. The notion that "one size fits all" or one set of standards can be applied to any site has some rueful consequences. Although the administration of Euclidean zones is simple, "what you see is what you get," there are obviously circumstances in which all agree that the building envelope and its location on a site need adjustment.

Many ordinances begin with a "purpose and intent" paragraph which spells out the reason for that particular section of the ordinance. This opening section in many cases capsulates the uses and general densities specified in the text. The section of the ordinance that prescribes permitted uses has three characteristics. One is normally a laundry list of "uses permitted by right." This is an alphabetical catalog of land uses for which a property may be developed. In a review of the range of permitted uses in a residential zoning classification, the difference between land use and zoning will become obvious.

Cumulative Zoning

The second characteristic of "uses permitted" is the reference to other zoning categories in the ordinance. This is a method utilized by some communities to avoid duplication within the text of the ordinance. Although it does reduce the volume of text, uses can get lost in the translation. It is very important to keep an accurate record of the references so that the real development potential of a zone can be clearly identified and understood.

The most potentially negative aspect of this form of referencing is the inclusion of uses within a zoning category that are incompatible with other uses in the zone. This concept is called cumulative zoning. Some zoning ordinances are structured to restrict residential land uses to residential zones as it should. In a cumulative zoning structure residential uses are not limited to residential zones, and that is where the conflict occurs. Cumulative zoning is organized to keep commercial and industrial activities out of residential zones but does not prohibit residential *land uses* in a commercial or industrial zoning category (Fig. 21-1).

Because of the variety of land uses permitted by right in the ordinance that is cumulative, the development community is known to have provided broad support for the concept. But imagine the problems created by a situation in which a developer builds a housing project in a commercial zone, then, clearly within his or her rights, builds a shopping center adjacent to the housing. Anyone can understand the

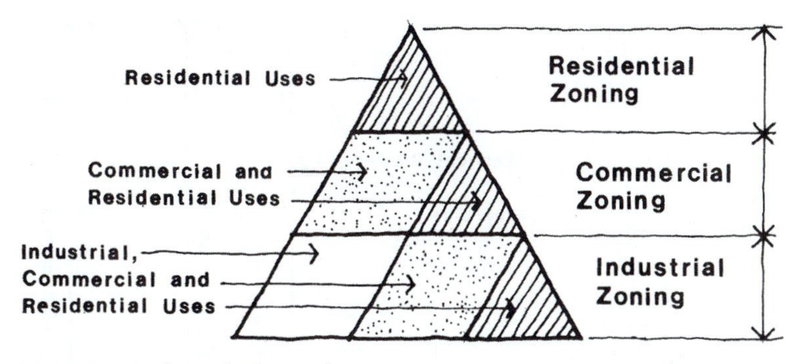

FIGURE 21-1. Cumulative zoning.

anger and frustration of a person who has purchased a home assuming it to be a property zoned as residential but finds that the land use does not imply zoning, and worse yet, a shopping center as a neighbor is within the law. Although many communities have structured their Euclidean zones to reduce potential conflicts by limiting residential uses to residential zones, the experienced site planner reads the fine print.

Special Exceptions and Special Uses

The third characteristic of the permitted uses section references those uses permitted by special exception or special use permit. This is another category of uses that are permitted *but* only under certain circumstances. The inclusion of this section in an ordinance is a recognition by the municipality that there are circumstances in which a particular segment of uses may be compatible with uses permitted by right in this zoning category. Because of the range of variables involved (access, topography, adjacent land uses, existing zoning, public facilities, etc.) these uses must be evaluated on a case-by-case basis. Obviously, the municipality wants to monitor and control carefully the location of uses listed under "special exception."

Performance Standards

This section provides the development guidelines for the uses prescribed in the preceding section. In most ordinances this section will, as a minimum, establish:

1. Minimal front-, rear-, and side-yard setbacks
2. Maximum height limits (Fig. 21-2)
3. Off-street parking requirements

These minimums and maximums provide the basis for the maximum building envelope on a site. Sometimes, these two requirements are coupled together in more restrictive language. As an example, the ordinance might state "the minimum front yard dimension will be 10 ft for the first 12 ft in vertical height of the building. After that point, the setback will be 1 ft back for each 2 ft in building height to a maximum building height of 48 ft 0 in." (Fig. 21-3). This language evidences strong control over the shape and location of any building that exceeds the dimensions established in the minimum setback. These statements need to be separated and graphically analyzed for their site-specific implications. This section may also embody restrictions on building forms which vary from one zoning category to the

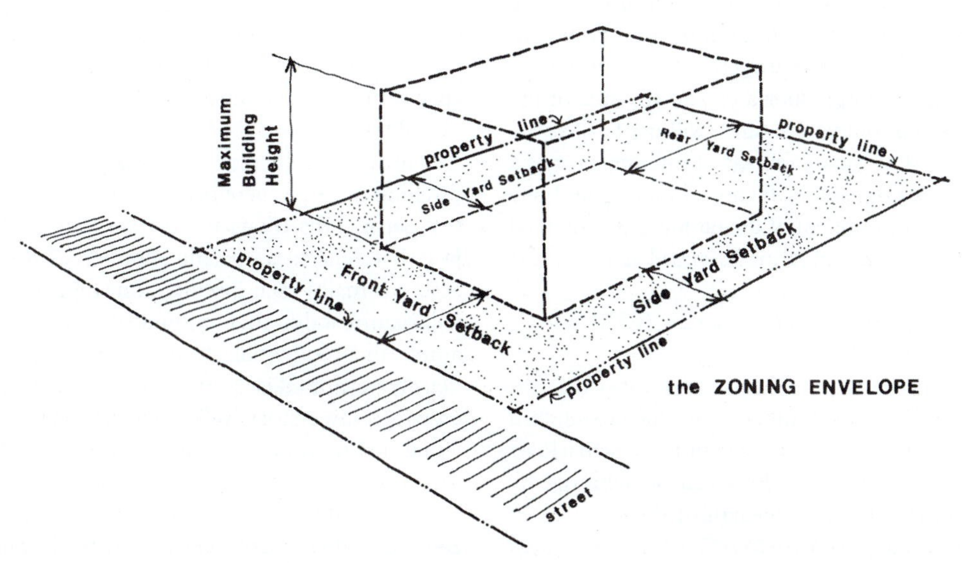

FIGURE 21-2. Euclidean zoning setbacks.

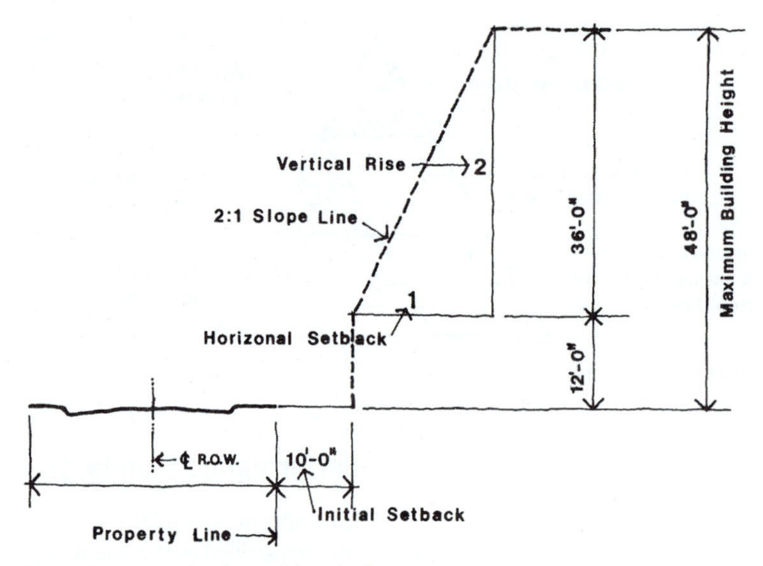

FIGURE 21-3. Setbacks and height limits.

next. For example, the setback of a building in a residential category may, logically, be more restrictive than a setback in an industrial zone.

Building coverage is another restriction common to Euclidean zones. The word "coverage" has different interpretations in different communities. Some ordinances consider "coverage" only as it refers to the building; in other ordinances, the term includes building and parking. In the first case the concept of coverage is a response to the method used to control a building's mass. In the interpretation that includes parking areas, the method employed is designed to keep runoff and human-made surfaces to a minimum. In either case, the effect on the site plan can be significant.

Whereas all of the performance standards noted above can be the subject of a request for a variance, density normally is not. Recall that density, like land use, is a decision that is normally made by the local legislative authority.

Density requirements take on two general formats. The first is the expression of density in the form of a floor area ratio (FAR). This is a term that is given to the numerical relationship between a building's floor area and the area of the site. FAR is a term the zoning ordinance will use to quantify and limit the amount of square footage of building area as a function of the building site. With both building and site areas in square feet, the quotient of the building area divided by the site area is the floor area ratio. Expressed as a formula:

$$\frac{\text{Building area}\,(\text{ft}^2)}{\text{Site area}\,(\text{ft}^2)} = \text{Floor area ratio (FAR)}$$

It must be emphasized that the floor area ratio has absolutely nothing to do with *any* other performance standard in the sense that it does not specify height limit, setback, or coverage. To the contrary, a 1.0 FAR can be achieved by building a two-story building over half of the site area or a four-story building over one-fourth of the site area (Fig. 21-4). FAR is strictly a method to control or limit the amount of building floor area that can be built on a site. It is also important to mention that some ordinances use percentage in lieu of the decimal equivalent. For example, 100% would equal 1.0 FAR. In any case, a check with the zoning administrator is imperative.

By and large, the concept of FAR as a density control mechanism is used predominantly to restrict commercial retail, office, and industrial land uses. It is easy to see how the concept works when a building's use and economic potential are functions of the building's gross floor area and parking requirements.

For residential land uses, FAR is not a popular density control device. In many zoning ordinances, the maximum density permitted is expressed simply in terms of a number of dwelling units per acre. For example, a maximum density of four dwelling units per acre will commonly result in single-family detached housing. The lot size will probably range from 8,000 to 9,000 square feet, depending on the amount of public right-of-way or open space required by the subdivision process. Other performance standards influence control over a building's location and form on an individual lot.

In residential zones of higher densities, the number of dwelling units per acre becomes more critical. Density control permits the client and the site planner to design a project that reflects a specific number of dwelling units per acre. Normally, each dwelling unit type in a housing project will require a different amount of square footage. As an example, in a residential zoning category where the maximum density permitted is 20 units per acre (a nominal density for a two- to three-story walk-up apartment) one might expect a one-bedroom unit to range in size from 700 to 800 ft^2. In the same project, a three-bedroom unit might range from 1,200 to 1,400 ft^2. It is easy to understand why in residential zones density control in terms of dwelling units per acre is more effective and equitable than FAR.

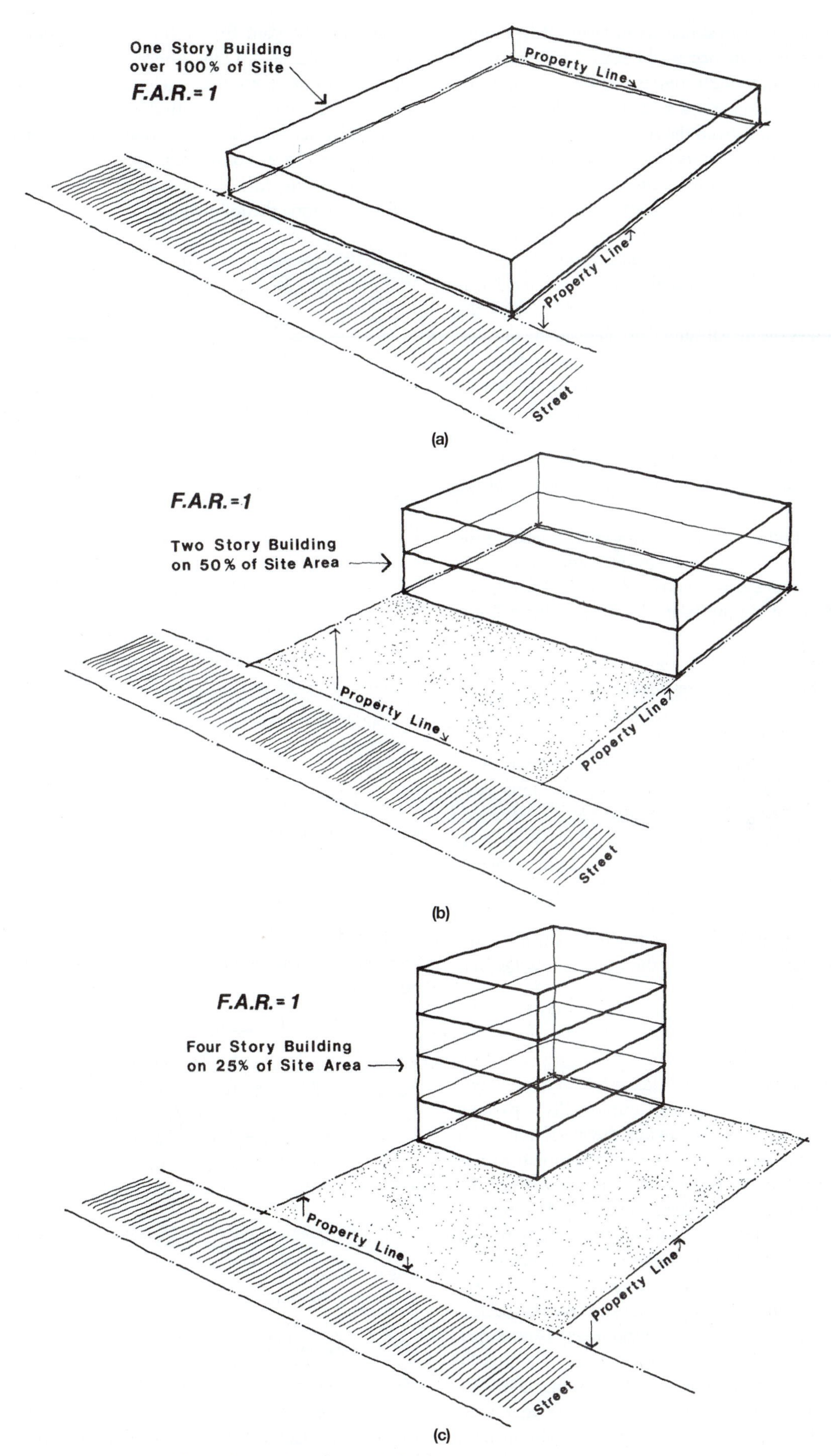

FIGURE 21-4. Three alternatives for an FAR of 1.

Some ordinances express density in terms of land area. As an example, the ordinance might read: "the maximum permitted density in the R.20 zone is one dwelling unit for each 2,000 $ft.^2$ of net land area assigned to residential development." The net results from the mathematics that follows might range from 19 to 21 units per acre, depending on the way "net land area assigned to residential development" is defined in the zoning ordinance. Many communities have found that there is a certain beauty in the simplicity of a zoning ordinance which states: "The maximum density permitted in the R.20 zone is 20 dwelling units per acre." It is hard to misunderstand that type of directive. Obviously, the community knows the density and can calculate the number of trips that will be generated by a residential zone of 20 units per acre. Similarly, the developer knows how density can affect and dictate an economic pro forma and dwelling unit mix. The site planner, then, interprets the program and resolves any emerging conflicts between the permitted density, performance standards, parking requirements, and the site's ecology.

Another method of density control involves tying residential density to an open-space or recreational area requirement. This concept is a variation on the "bonus" or "incentive" theme of zoning strategies. Essentially, it holds that residential densities should be a function of passive and active recreational facilities. This method is largely mathematical and is not limited to residential land uses. Similar "density for open-space areas" are, in many urban areas, linked to office and retail activities.

In 1926 no one imagined the future proliferation of density control mechanisms or stratification of zoning categories. The three ordinary zoning categories—residential, commercial, and industrial—have multiplied like junk mail. In urban areas, the array of Euclidean zones is phenomenal, as each community struggles to delineate finer lines between highway commercial zones, regional retail centers, and mixed-use commercial-office zones. Regardless of the community, the practice of developing the right zoning text has become a high art.

The Parking Schedule

The final aspect of the zoning ordinance, which has a major impact on site planning, is the municipality's parking schedule. The parking schedule provides the site planner with an index of parking requirements for each land use. Just as there is no uniform national agreement on the size of the parking space, there is no agreement on parking demands for any specific land use. There is so little commonality on the way the schedules function that it is not worth discussing.

What is important is that, in some cases, the parking requirement can become a surrogate density control mechanism. There is a basic rule in the development community which suggests that as long as the price of structured parking (per square foot) exceeds the cost of additional land (per square foot), the parking requirement will be met with parking "at grade." A theorem of real estate development is that the cost of structured parking, either above or below grade, must be paid for by the rents generated from the principal land use. Only under unusual circumstances, where a corporation's goals include the development of an image that is incongruous with the "least cost" method, does the site planner find that an alternative is plausible. Often in the case of office, commercial, and retail land uses, the process to define the holding capacity of a site is simple. For example, assume the following:

1. A 100 ft × 200 ft site
2. A zoning ordinance that permits:
   a. An FAR of 1.5
   b. A maximum building height of 60 ft or six stories
   c. Parking: 1 space for every 350 ft^2 of gross floor area of office buildings

Assuming that the cost of the land is significantly below the cost per square foot of structured parking, the following would provide an outline of the site's development potential:

$$100 \text{ ft} \times 200 \text{ ft} \times 1.5 \text{ (FAR)} = 30{,}000 \text{ ft}^2 \text{ maximum building floor area permitted by zoning}$$

As a method for preliminary identification of the area of the site that will be occupied by the building, assume that the total building area permitted in zoning is achievable:

$$30{,}000 \text{ ft}^2 \text{ divided by six stories} = 5{,}000 \text{ ft}^2 \text{ per floor}$$

Then:

20,000 ft^2	site
− 5,000 ft^2	building area
− 1,000 ft^2	building/parking transition
14,000 ft^2	gross site area remaining for parking

Assume that for normal and compact car spaces the average area (parking and circulation) required for each space will be 300 ft^2:

$$14{,}000 \text{ ft}^2 \text{ divided by } 300 \text{ ft}^2 = 46 \text{ parking spaces}$$

Recall from the example zoning requirements that one parking space will be required for every 350 ft^2 of gross building area. The amount of gross building area that could be achieved on the site is the product of the number of parking spaces conceptually possible (46) and the maximum building area permitted for each parking space (350 ft^2):

$$46 \text{ spaces} \times 350 \text{ ft}^2 = 16{,}100 \text{ ft}^2 \text{ building area}$$

It is interesting to note that just over 53% of the 30,000 ft^2 permitted "by right" can actually be realized, given:

1. Surface parking only.
2. A requirement of one parking space for every 350 ft^2 of building area.
3. The total land area for parking is not to exceed 14,000 ft^2 and 300 ft^2 per car.

Since the height limit is six stories, an alternative is to raise the building and minimize the amount of the site that must be allocated to the building's access. Although this modification will provide for some additional building square footage, it is obvious that the 30,000 ft^2 of building floor area permitted by the zoning can be achieved on this site only with some form of structured parking. This is an example of a condition wherein the ordinance implies that the development potential is almost twice that which can realistically be achieved, given the realities of the marketplace. The experienced site planner is cognizant of these quirks in the ordinances and takes the time to evaluate the implications before a property is purchased.

THE DECISION MAKERS

Sometimes, understanding the difference between who makes a recommendation and who makes decisions can be the difference between success and starting over. Because of the process and the distinct obligations of each, the responsibilities for a review of requests have generally been divided among three entities: the local legislative authority, the planning and zoning commission, and the board of appeals or adjustment. Although the nomenclature will vary, each will be more identifiable given its role in the process.

The Local Legislative Authority

With a few exceptions, the local legislative authority makes the final decision concerning zoning in a community.[34] These are the people elected to serve as the principal governmental authority at the local level. Depending on whether the local government is a city or county, the elected officials will have various titles. These include, but are not limited to, city councils, county councils, county board of supervisors, and county commissioners. To emphasize the importance of the elected officials in the process, in most communities they *appoint* the members to serve on both the planning commission and the board of zoning appeals (Fig. 21-5).

[34]So, Stollman, Beal, and Arnold, p. 434.

Planning and Zoning Commissions

Although the name for this agency does not vary widely, the organization and its authority might. Common variations are (1) a separate, but appointed, planning commission and zoning commission; and (2) an elected planning and zoning commission. This separation between planning and zoning may appear to be paradoxical, as it institutionalizes a division between the planning and implementation functions. Although it can create some friction between the appointed bodies, they appear to work together more effectively in circumstances in which a member of each group serves as an ex-officio member on the other commission.

There are some municipalities whose enabling legislation has given the power to make zoning decisions to elected planning commissions. Although this structure is unusual, a review of the authority of each entity in the process should readily clarify the functions of each. In most cases, the operative word for the planning and zoning commission is "recommendation," as they *recommend* changes to the legislative authority. Under some circumstances, the

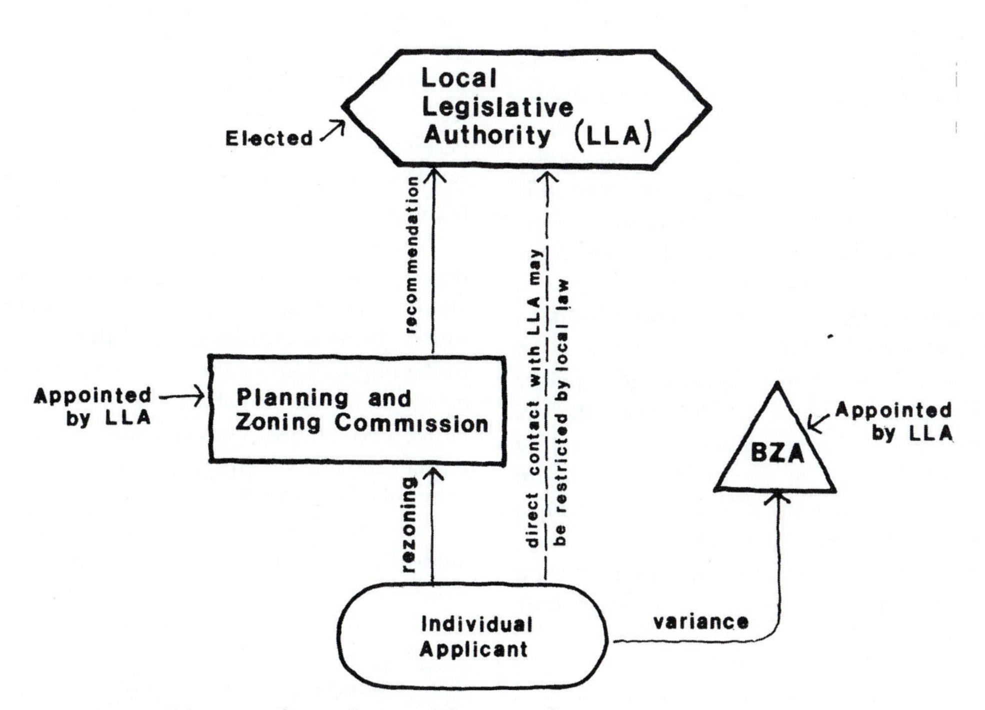

FIGURE 21-5. Diagram of rezoning participants and process.

planning commission may have final approval of the site plan in a two-stage performance zone. Recall, however, that the initial decision concerning land use and density was made by the local legislature. This is critical to understanding the process. It is highly recommended that prior to making a decision to rezone a property, one attend at least one public hearing of both the legislative authority (city council, county council, etc.) and the planning commission. The public hearing permits an examination and evaluation of a council's and commission's values, biases, and sophistication.

The Board of Zoning Appeals or Adjustment

If the performance standards thwart the achievement of the existing zoning or create a hardship, one should consider a request for a waiver from the board of zoning appeals (or local equivalent). In most cases, this process is extremely short and does not require an extensive amount of preparation or exposure, and rightfully so. Communities have adopted various terms for this function in local government, the most common two being the board of zoning appeals or adjustment (BZA) and the zoning board of adjustment or appeals (ZBA). One of the functions of this board is to review cases or appeals in which the performance standards of the ordinance, when applied to a site, create a unique condition and unusual (economic) burden or hardship. The text from Robert M. Anderson's *American Law of Zoning* sets down two generally recognized rules. "Where an applicant for a variance can demonstrate that owing to the size and shape of his land, he cannot make any reasonable use of it unless the literal application of the zoning regulation is varied, he is entitled to a variance" and "in most jurisdictions, an applicant whose problems in the use of his land are caused by his own conduct, rather than by circumstances which are peculiarly related to the land, is not entitled to a variance."[35]

In most communities, the BZA can only consider waivers of performance standards, not land use or density. Land use and density are generally the purview of the legislative authority of the community (city council, county council, board of supervisors, county commissioner, etc.). These authorities appoint their respective BZAs and in most communities the BZA's decision is *final.* If an applicant makes a request for a variance and is refused, an appeal to that decision must be heard before the local courts. Depending on the way the BZA/ZBA has been charged, in many communities they also hear special exceptions or special use permits. The point here is that the BZA has an extraordinary amount of control, and the effectiveness of the zoning ordinance in many ways rests with the way the BZA interprets "unique and unusual economic burden or hardship."

Many BZAs are very strict, and any request for variation will be scrutinized extensively. Others are somewhat more liberal and interpret their role in a different way. The following are examples of the behavior of two different BZAs.

The first is an example of the Board of Adjustment for the city of Dallas, Texas in which it reviewed a case for eight variances at the junction of two major freeways in the city. The applicant's request for variances was not recommended by the planning staff but was approved by the Board of Adjustment. The site plan and schedule of variances for the proposed development are as follows:

Schedule of Variances for Dallas Galleria

1. Setback variance of 228 ft for building 10 (shopping mall)
2. Setback variance of 664 ft for building 11 (hotel)
3. Setback variance of 10 ft 6 in. for buildings 2, 3, 5, and 6 (office)
4. Setback variance of 10 ft 0 in. for building 1 and 7 (parking)
5. Height variance of 128 ft for buildings 5 and 6
6. Variance to reduce the width of a parking space from the required 9 ft 0 in. to 8 ft. 6 in.
7. Variance to reduce the parking required by 3,637 total spaces (14,959 required; 11,322 approved reduction)[36] (Fig. 21-6)

If variances of this magnitude were commonplace, they could create a degree of cynicism within a community relative to its government and zoning. Since performance standards may be the single aspect of zoning important to many citizens, variances should be scrupulously evaluated. What must be recognized is that the application of "unique hardship or burden" prescribed by Anderson is obviously not universally accepted.

As a contrasting example, the Montgomery County, Maryland, ZBA received a request from a friendly foreign government in 1975 to convert an existing parochial school to a school for the children of embassy personnel. The physical changes to the site and facility were none: no new roads, parking, or buildings, and no environmental degradation. But since there was a proposed "use" change, the ordinance required a review by the zoning board of adjustment. The ZBA took their responsibility for thorough review very seriously. A public hearing on the request permitted proponents and opposition alike to express their opinions. Lengthy testimony by various witnesses and thorough examination as to the implications of the "new" school became part of the extensive record on which the ZBA's final approval was based. Two different municipalities in two different regions reflect dramatically different views of the role of the board of zoning adjustments.

As a footnote on variances, in some communities the authority for minor waivers is invested in the zoning administrator's office.[37] Since neither the office nor the power to

[35] Robert M. Anderson, *American Law of Zoning* (Rochester, N.Y.: The Lawyer's Cooperative Publishing Co., 1968).

[36] City of Dallas, BDA Request 80-152/5161.

[37] So, Stollman, Beal, and Arnold, p. 434.

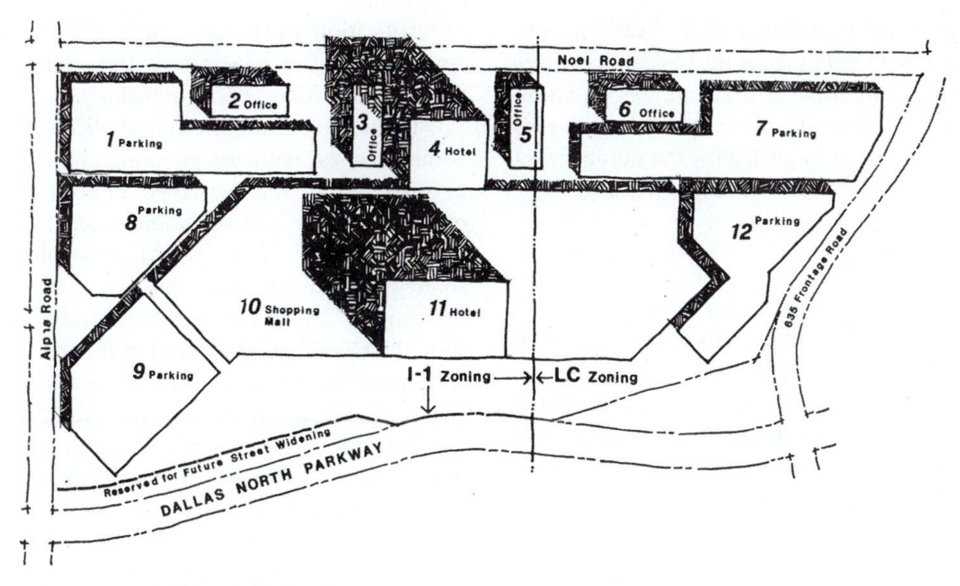

FIGURE 21-6. Dallas Galleria plan.

give a variance is common to every community, the rule is to call and inquire. It could be a much less time consuming process than the BZA hearing.

The Hearing Examiner

The employment of a hearing examiner is one of the variations in the traditional process that is growing in use throughout the United States. This format involves a review of the rezoning application prior to consideration of the request by the local legislative authority (Fig. 21-7). This review is executed by a person referred to as a hearing examiner. In many cases this individual is an attorney who has been hired by the local government to review, analyze, and conduct his or her own hearing on a rezoning request. Often the hearing examiner is empowered to call and cross-examine witnesses and take testimony. This quasi-judicial forum provides the basis on which the hearing examiner develops a case record. This record will include testimony from the planning and zoning commission, professional planning staff, and those on both sides of the rezoning request. The intent of this process is to develop an objective opinion, insulated from political pressure and influence, based solely on the facts presented in the record.

The hearing examiner's opinion is then forwarded to the local legislative authority and becomes, together with the planning commission's recommendations, part of the case record. Although the recommendations of the hearing examiner are not binding, they do provide an additional layer of testimony and documentation that is very important for a fair and equitable evaluation in a process that is often emotionally charged.

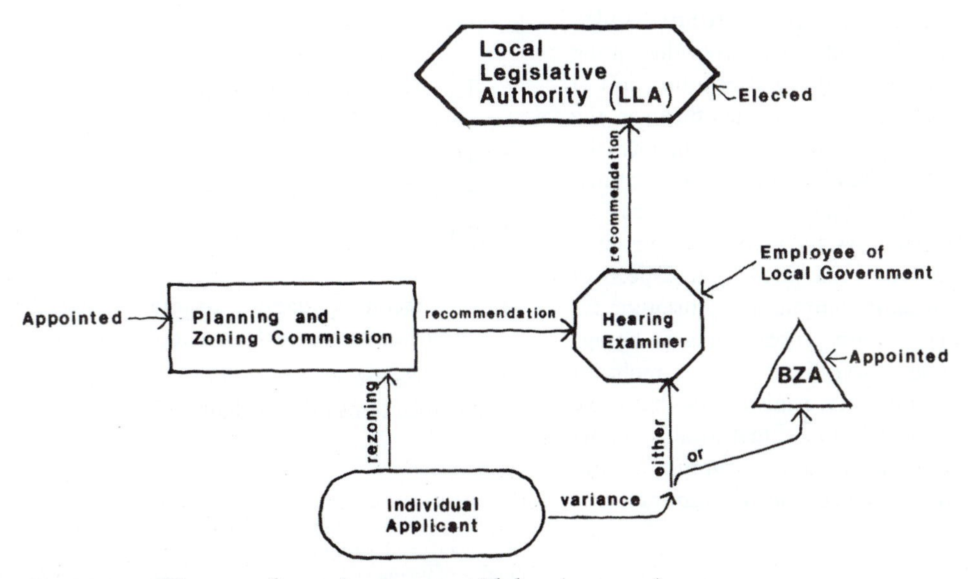

FIGURE 21-7. Diagram of rezoning process with hearing examiner.

In some communities the function of the hearing examiner is also less formal. Depending on the obligations of the position, the hearing examiner may also consider variances. Although the additional layer of review and evaluation may not be appropriate for every municipality, the number that find the process beneficial has been steadily increasing.

Neighborhood Zoning

The increase in public participation over the past few years has emerged to become a significant force in the zoning process. Almost every community has watched the success of a few politically savvy citizens' organizations affect the decision-making process of the planning commission and local legislative authorities. Some have mustered enough votes to change local ordinances and establish neighborhood zoning authority. As with the zoning process, the influence and working relationship with the local legislature will vary. In some cases the role of citizens is sought and valued. In others, citizens are tolerated as an obligatory segment of public hearings. Whereas many design professionals avoid discussing a rezoning request with neighborhood groups, others consider it an opportunity to present the case for a proposal before stepping into an intemperate public hearing. Since there seems to be little reason for the demise of the neighborhood zoning authority, taking the time to understand the concerns of citizens can provide a basis for reasonable and successful compromise.

THE REZONING PROCESS

The rezoning process begins when a person decides that the existing zoning on a property will not permit the desired development program. This decision may be necessary because (1) the land use or densities desired are not permitted in the existing zoning, or (2) the performance standards within the existing zoning will not permit the development program to be achieved. Recognizing that there will be local variations, a rezoning request for a Euclidean zone will follow a fairly common format. The rezoning request is filed with those responsible for the administration of the zoning ordinance. In most communities, this will be a function of the planning staff, zoning administrator's office, or the city/county manager's office and will be accompanied by their recommendation. The planning and zoning commission will review the rezoning request at a public hearing. Depending on local and state laws, the planning and zoning commission may be required to examine specific aspects of each rezoning request. In some communities a rezoning request is required to meet certain proofs if it is to be approved. These proofs might necessitate that an applicant prove that the existing zoning was a mistake or that there had been a change in the neighborhood that would support rezoning.[38] Compliance with the adopted municipal land use plan may also be a requirement. Obviously, some rationale or justification for the rezoning is necessary or every request could be based on new criteria and attract a charge of arbitrary and capricious decision making.

In some communities an applicant may discuss a rezoning request with the planning and zoning commission or members of the legislative authority. Although this is often the custom, in other municipalities discussion of a zoning request with an elected or appointed official is absolutely prohibited. This is referred to as an "ex parte communication" and is based on the concept that the rezoning should be based on the record of the public hearing, and as such, *no* communication with any elected or appointed official outside that hearing is appropriate. The opposing opinion holds that there is no harm in a person-to-person discussion of the rezoning request. Whatever the case, know and respect the local practice and ethics codes, as a moderate infraction could jeopardize the success of an application.

Following the planning commission's public hearing and recommendation, in most communities the request is forwarded to the local legislative authority for its decision. Assuming that the request is a Euclidean zone, the local legislative authority will consider the request at one of its meetings. This is an important phase of the process, as the meetings are public hearings and permit the local legislative authority and citizens to discuss the application and its implications. This phase of the process is one of the very rare, ongoing examples of local government interacting with its constituency. Of all the obscure and fragmented examples of government of, by, and for the people, a public hearing on a controversial rezoning request gives a new meaning to the word "democracy." There will always be circumstances in which a good proposal is dismembered by citizens while one of obscure socially redeeming value goes unchallenged. The process does not ensure against regrettable decisions but does permit changing the leadership if stupidity seems to be the continual "bill of fare." The local legislative authority's decision will be based on a number of elements. These cannot be listed in order of importance, because there is no way of knowing what is important to every local government, but the list should include the following:

1. The compatibility of the rezoning request with the existing neighborhood
2. The land use and zoning recommended in the municipality's adopted land use plan[39]
3. The recommendations of the planning and zoning commission
4. The recommendations of the municipality's zoning administrator or professional planning staff
5. The responses (opposition or support) of citizens' organizations in the neighborhood

[38] Ibid.

[39] Do not confuse the zoning map with the adopted land use plan. The land use plan is just that—it identifies land uses, densities, and defines zones only. If you check one of the residential zones you will probably find a number of permitted uses that are *not* residential. The land use map and zoning map are *not* the same.

6. The impact of the request on the health, safety, morals, and welfare of the community
7. The relative success of the applicant's presentation, which would be based on:
   a. A rationale for the request
   b. The request's ability to accomplish a public good (school site dedication, etc.)
   c. A respect for the neighborhood's concerns reflected in the rezoning request prior to filing the application
   d. A respect for the municipality's adopted plan or the intent of the plan
8. The record of the hearing

In the event of a denial, the range of alternatives open to the applicant depends on the municipality. Each community will have some restriction limiting the resubmittal. As a last resort, the courts are always available to redress a grievance.

Rezoning Alternatives

The first, and most traditional, request is one to change the Euclidean zone classification. This is the simplest request and probably accounts for 95% of rezoning cases in the United States. The second, a variation on the first, is a request for a change in Euclidean zones with "conditions." This strategy is somewhat more unusual and in some states and local governments is absolutely prohibited. Conditional zoning can take on other variations. One of the more prevalent involves the application for a Euclidean zone, but with certain restricting conditions attached. For example, restricting conditions may take the form of a height limit or increased setback, but whatever the case, can be an acceptable strategy to "tailor" the zone to fit a set of unusual contextual conditions. It is imperative that a check be made with local government and its attorney to make certain that conditional zoning is legally acceptable. This needs to be discussed with the municipality's attorney on a case-by-case basis. Land use laws are constantly changing and the directions of state supreme courts today may differ greatly with those of 10 years ago.

Performance Zones

The third alternative is a request for a performance zone. These zoning forms are site specific; that is, the zoning category usually requires a site plan that delineates precisely what the applicant plans to build. Even with the latitude available in variances, the site planner presented with a moderately large site (20 to 100 acres) often has a problem compounded by its size. Until the late 1960s, there were no alternatives to the Euclidean zone. Large acreages throughout the United States were subdivided into small tracts following the specifications outlined in the ordinance. The results of the 1949 Housing Act and Euclidean zoning sponsored the mindless conversion of millions of acres of land into low-density subdivisions of $\frac{1}{4}$-acre lots. The notion of separating incompatible uses in some communities meant that each housing type needed to be separated from the others. That is, detached houses should be segregated from attached homes (town houses), which should be separated from multifamily housing.

Although the concept of segregating housing types is prevalent in many communities today, to some it thwarted the potential of developing communities that could offer a variety of housing types for a variety of lifestyles. As long as the zoning ordinance dictated the physical outcome of any site of any size, the results were going to be stolid and unimaginative. Furthermore, any possibility to save any unique ecology or open space was jeopardized by the necessity to make certain that the density permitted by the zoning was achieved. The concept of planned unit development (PUD) emerged as a direct response to the prescriptive qualities of Euclidean zoning. The PUD is different in various ways, but most important, it was the first form of *performance zoning*. Simply stated, performance zoning is based on the site plan prepared by the applicant. This site plan must not be confused with the site plan that many developers commission their architects or landscape architects to prepare for a hearing on a Euclidean rezoning. In most communities the illustrative site plans prepared for a Euclidean rezoning have no legal status and are prepared to beguile the citizens or neophyte planning commission member.

The planned unit development text can be found in various forms and descriptions in zoning ordinances throughout the United States. The nomenclature may be planned development (PD), residential or planned residential development (PRD), planned unit development (PUD), or other. The variations are limited only by the ability of planning commissions and their staffs to find another name for a rose. But regardless of title, the concept is generally the same.

Performance zones traditionally require a site-specific plan as a part of the rezoning request. These characteristics are not nationally uniform, but there are some generally common qualities. Most PUDs have very liberal setback and height-limit controls. The idea is that the site planner needs the flexibility to do good design work and as such needs some release from the arbitrary dictum of Euclidean zones. In many cases a maximum density might be established with setbacks respecting adjacent properties. The prevailing policy is flexibility, with an emphasis on the quality of the plan. Often, the PUD plan requirements are very detailed and require the location of all buildings, parking lots, pedestrian circulation systems, open space, and recreational facilities. Although this may appear to be a burden, it is also not uncommon for a municipality to grant higher densities as an incentive to a developer to plan development comprehensively and utilize a PUD ordinance.

Unlike the Euclidean zone, the PUD often permits a mix of housing types. Some municipalities go so far as to provide for a mix of uses, that is, residential and commercial. This is again an important departure from traditional zoning, but it must be emphasized that this varies across the United States. The mix of housing types, for example, is easy to accept, but the mix in land use types might appear to fly in the face of the rationale for zoning in the first place. This departure from the traditional zoning concepts puts the rationale for the zoning squarely on the quality of the plan, which is where it always should have been. To reiterate, the applicant prepares and submits the PUD plan for approval to the same authority as one would for a Euclidean zone. A word of caution: In most communities, when a PUD plan is approved, the site must be built just as the plan describes. Since the PUD plan is considered the zoning, it is recorded as the basis on which building permits will be reviewed. Although the PUD has some very definite advantages, in many communities the PUD plan is a definitive commitment of the developer's intent.

Why, with the flexibility of the PUD, is the zoning concept not used more extensively? There are at least three reasons. The first two can be attributed to the developer. The PUD requires the development of a plan, and that means some front-end costs for consultants that most entrepreneurs would like to avoid. Furthermore, the plan must be followed regardless of the owner. An undeveloped property locked into a specific PUD plan can certainly restrict its speculative value.

The third issue that has retarded wide acceptance or implementation of performance zones has its origin in the strict interpretation of zoning and the role of the local legislative authority. This more conservative perspective suggests that performance zones are a form of "conditional" zoning and in some states are simply illegal. The same philosophical viewpoint contends that in the evaluation of a performance zone, the local legislative authority is often forced to rely on the advice of its professional staff. Under normal conditions that would appear to be appropriate and an intelligent way to utilize a community's professional planners. But the conservative position contends that if the PUD is not successful, local governmental officials would attempt to shift the responsibility for failure to their professional advisors.

Similarly, some planning commissions and councils do not permit their professional planning staffs to make any recommendation on a zoning case. That position maintains that the planning commission and council should not be influenced by any recommendation that might cloud their objectivity. So regardless of political philosophy, there are some encumbrances to what is unquestionably a proven feasible alternative to Euclidean zoning. There is at least one variation on the theme of PUDs that is prevalent enough to mention and remember. There are some local governments that have adopted two-stage PUD ordinances. The first stage obligates the applicant to develop a plan that identifies land uses, density, and major circulation systems. If the first stage of the applicant's plan is approved by the local legislative body, the applicant is then permitted to develop a plan of the more specific elements: buildings, parking lots, recreational facilities, and so on. More often than not, the second stage is only a phase of an entire development plan.

As with most other innovations, this PUD variation has both positive and negative effects. The two-stage process has some benefits for both developer and the municipality. Identifying the land use, density, and primary circulation plan permits the developer and the municipality to focus on those issues alone. At the public hearing there is no discourse in this first stage on the location of a building or the orientation of a parking lot. Additionally, the limited focus of this conceptual planning in the first stage is simply easier to deal with at all levels. If the first-stage plan is approved, the applicant has at least some assurance that if the property must be sold, it will have some basis for economic value without the adoption of the second, more restrictive phase. If sold, the new owner could develop the second-stage plan to reflect his or her own preferences in the context of the adopted first stage. The problem is that unless the municipality has drafted some very specific policy statements relevant to open space, recreational facilities, storm water management, and so on, it could have adopted a plan of dubious conclusion. But since the zoning does not provide the developer with a building permit, the process still provides the community with an opportunity to negotiate an acceptable site-specific plan.

Although the variations on the PUD are extensive, the commonalities are flexibility, land use mix, and the opportunity for creativity. Check your local ordinance for the alternatives and the process for application and review.

This section on performance zones would not be complete without mention of another variety of "performance zones." The term "performance zoning" has been used here in the generic sense. Another form of performance zoning emerged in the mid-1970s as "Performance zoning," with a capital *P*. This classification refers to a zoning method that defines performance standards and establishes guidelines associated with on-site runoff, trip generation, open space, recreational facilities, and water and sanitary sewer requirements. Although these requirements would obviously vary from project to project, the concept was based on the hypothesis that each land use generates certain demands on a community's public facilities. Since those demands can be documented, they could, in turn, become the rationale on which land use decisions (zoning) could be made. This concept, also termed "fiscal zoning" or "impact zoning," was an important step toward a structure or format that attempted to qualify the rationale for land use decisions at the local level. The problem is that this method is not simple to administer and in most cases requires a substantial amount of review time on the part of the community. Also, the preparation of the documentation is no simple task, and many developers would simply prefer to take their chances with simpler mechanisms.

This is by no means the last word on performance zoning. There are numerous other variations as municipalities continue to explore alternatives to the straitjacket restrictions of Euclidean zoning.

Bonus and incentive zoning have enjoyed general popularity in the past, with some ordinances more successful then others. This was a form of PUD in which the developer could achieve certain bonuses, normally an increase in density, if certain public amenities were incorporated in the plan. The value here again is in the flexibility and the process to develop a plan that respects a specific context. Transfer of development rights (TDR) is also a variation on land use controls. Similar to zoning, TDR is a mechanism adopted by a municipality, on its own volition, to control the density of development. Transfer of development rights is based on the concept that since the municipality gives development rights (land use and density) away in the form of zoning, it also has the power to facilitate the transfer of those rights from one area of the community to another. TDR has been particularly effective in circumstances in which public facilities are strained beyond capacities or preservation and conservation are valued community objectives.

It must be noted that TDR is not an easy strategy as it necessitates the development of a comprehensive plan that defines those areas in a community where development rights can be "sent" or relocated to areas appropriate for "receiving" additional development capacity. The development of the plan would logically include a revision to the community's comprehensive master plan. Planning and zoning have been an effort to balance the health, safety, and welfare of citizens with the growth objectives of the development community, yet minimizing the bureaucratic review processes—not an easy task.

The final alternative to a request for an existing Euclidean zone or performance zone is to consider the development of a text for a new Euclidean zone designed to respond to a specific set of conditions. Zoning laws change, as they must, to reflect changes in technology and society. Ordinances written 20 years ago may not reflect the densities and performance standards appropriate for the varying types of development today. As an example, historic landmark ordinances were uncommon components of zoning ordinances in 1970. That situation has changed dramatically over the past decade as communities, citizens, and developers have worked to conserve these special resources.

A zoning ordinance can be amended and under some circumstances may be the only remaining feasible alternative. This option should be pursued as a last resort, as it is probably the most time consuming and expensive. The need for an additional zoning classification must obviously go beyond the immediate needs of your client or otherwise will appear to be totally self-serving. It goes without saying that a new zoning classification must have the support of the local zoning administrator and professional planning staff. Without visible evidence of the staff's favor, a request to expand an ordinance will probably be met with suspicion and recalcitrance by local elected officials.

Down-Zoning and Back-Zoning

When a local legislative authority believes that the densities allocated by a previous action to a specific property are incongruous with the community's health, safety, and welfare, the action of changing a zoning classification to one of a lower, permissible, or more restrictive performance standard is commonly called down-zoning or back-zoning. The reduction of a permitted residential density from 20 dwelling units per acre to 10 dwelling units per acre would be an example of down-zoning. Contrary to popular belief, it is unusual for a property owner to enjoy any vested rights to a zoning classification on a vacant parcel unless some on-site construction has taken place.[40] Although the application for a building permit is also a common companion to establishing a vested right, state laws vary widely, and legal counsel on this matter is imperative. Although it is not a common procedure, it does occur often enough to warrant understanding the process. The courts have generally sustained the concept that the local legislative authority that has the power to rezone property to a higher use has the power to rezone to a lower density as well. Because the legal authority exists does not necessarily mean that the municipality does not have certain obligations with respect to process and criteria in down-zoning. Although legislative action to down-zone a property must not be arbitrary, capricious, unreasonable, or discriminatory, the property owner should be reminded that the burden of proof rests with those challenging the decision of the municipal legislative authority. Since the vast majority of those whose property is to be down-zoned seek private legal counsel, the process can be very excruciating to both the public and private sectors. Chapter 15, written by Richard Babcock, of So and colleagues' *The Practice of Local Government Planning*, is a good general resource that sets out in greater detail some of the interrelationships between the comprehensive community planning and the administration of the zoning ordinance.

[40] So, Stollman, Beal, and Arnold, p. 429.

PART SIX

APPLYING THE THEORY

- **CHAPTER 22:** The Site-Planning Process

THE SITE-PLANNING PROCESS

Paramount to understanding site planning is understanding that it is a process. By definition, a *process* is a sequence of interrelated events or decisions that results in a product or conclusion. This characteristic of site planning must be fully comprehended, as attempts to short-circuit the process can produce unnecessary and unfortunate consequences. There is always a cadre of those who believe that process planning subverts the creative, esoteric aspect of the art. But this is not true. As will become evident, there are various stages in the process that require creative thinking other than that required to locate a building on the site.

It is of principal importance for the site planner to grasp the difference between the objective and subjective issues of the process. Clear objectivity is imperative for various stages of the procedure and anything less than an unbiased evaluation of critical issues could jeopardize a successful resolution of the problem. There is a place in the process for the integration of subjective issues. These "softer" issues can often be identified as functions of design preference, style, or taste. These vary widely and will depend directly on the preferences of those participating in the process. This is where "I want . . ." and "I believe . . ." appear. Although these components are strictly a matter of personal preference, they are essential to transforming a collection of facts or data into built form. It must also be understood that the building and the site must be considered as an integral design problem. Neither should be developed without the other, as the success of the ultimate union is then risky, at best.

Recognizing that site planning is a process directed toward problem solving also accepts the participation of others in the process. Just as every creative designer abhors a situation in which the client takes over the role as designer, the same holds true for the designer who usurps the role of client/user. Recall that each design problem has a collection of identities to be recognized. If developed properly, the process permits these identities to materialize.

For the most part, site planning in the United States is practiced in a reactionary way; the architect develops a site plan as a response to design criteria based on short-term economic parameters, the zoning ordinance, and current architectural form vocabulary. Although the practice of site planning is not the purview of one profession, the notion that it may be interdisciplinary in scope is not broadly accepted. Historically, it was the obligation of the architect. The process today, however, often becomes a perfunctory exercise for the civil engineer, the architect, and the landscape architect, as each alternates in the role of site planner. Too frequently, we forget that coalescing the focus of these disciplines into one set of design decisions can occur only when there is a complete understanding of the process and the range of issues to be considered.

Site planning begins by assuming that we are ignorant of the site. This is not to say that one is oblivious to the process or components, but that, in accepting lack of knowledge of a site, the planner proceeds with objectivity and without preconceived notions as to the conclusions. Initially, these early stages of research, data collection, documentation, and synthesis are an analytical procedure and provide the foundation of information and data for future design decisions.

To facilitate an understanding of that process, two hypothetical sites with separate development objectives are used as examples. Site R is a residential land use; site C is a commercial/retail use. Each site is used to explore the application of various site-planning issues discussed in previous chapters. To avoid confusion in discussing sites R and C simultaneously, the consideration of each site will be separated beginning with the "Inventory" section.

GOALS AND OBJECTIVES

The initial stage of the site-planning process begins with a qualified acceptance of the client's development program. The emphasis is on "qualified" because subsequent research and analysis may not generate complete support for the client's program on the site. The professional seeks to develop a working relationship in which a candid dialogue of both good news and bad news can be held. What is important here is to recognize that accepting a given site and given program, without caveat, is hazardous at best. If the site planner and client approach matching a program to a site with specific goals, the test of the match can be made through research and analysis.

At this stage it is very important that the planner glean from the client any expression of other goals or images that he or she wishes to objectify in the project. Making a genuine effort to understand the client's preferences at this stage will greatly encourage participation in the process, while forcing any latent or obscure objectives to surface. This is a critical part of the process as there can be a serious disconnect between the client's desires and reality.

Site R (Fig. 22-1) is a wooded 1-acre lot in a residential subdivision at the edge of a small creek. The client wants to consider subdividing the acreage into four lots, permitted by the existing zoning. Because of the topographical relief and the floodplain created by the creek, it is unknown whether the four $\frac{1}{4}$-acre lots could be developed within the minimum areas and dimensions prescribed in the zoning ordinance. Whatever the circumstances, the client's principal objective is to build four single-family detached dwellings on the site, selling three and retaining one for personal use.

Site C (Fig. 22-2) is a 5.07-acre parcel zoned for commercial office and retail land uses. It is adjacent to a residential PUD on the west and a suburban shopping center on the east. Richardson Blvd. exists on the north and a new road (Sullivan) is under construction to the south. A local developer has acquired the site and with it a promise from a local bank to occupy 12,500 ft^2 of ground floor space. The developer would like to supplement the bank's function as a generator with 10,000 ft^2 of retail (commercial) space. The remaining floor area allowed by the ordinance would be "built out" as speculative office space. The new owner wants to achieve the 0.7 floor area ratio currently permitted under the municipality's zoning ordinance.

RESEARCH AND DATA COLLECTION

Good site planning begins with good research. There is no substitute for getting the right information properly documented the first time. Even when the client is from the private sector, the public sector is always a participant in the site-planning process. Although the client may provide the site planner with a program and development criteria, various public entities will have jurisdictional control over the development of a site.

It is more than appropriate to check the local zoning ordinance at this time to make certain that the client's proposed use for the property is permitted by the current zoning. If rezoning is anticipated, it is important to recognize that requirement as a part of the process.

Following an environmental analysis, the site needs to be evaluated for alternative locations of the building and parking. Since each land use has its own criteria, there are likely to be specific areas of the site uniquely suited for each use. What is important is to make certain that the site is appropriate for each use in the client's program. For example, surface parking should not be located on any slope exceeding 7%; buildings are generally more adaptive to sloping topography. Structured parking can be developed on slopes in excess of 7%, but the cost of surface parking must also be considered in a total evaluation. Environmental maladies such as floodplains, swales, easements, and other dubious areas obviously should be avoided.

The zoning ordinance needs to be consulted again to identify two variables associated with the development of the site. One is the "zoning envelope," or buildable volume created by the performance standards of the zoning. This volume is described by the density, height limits, setbacks, and area coverage in the text of most Euclidean ordinances. Although ordinances will vary widely, delineating the correct profile of the zoning restrictions is imperative. The other controlling element in the zoning ordinance is the parking requirement. The owner's development objectives must now be interpreted in the form of general floor areas and parking requirements. It is important at this point to know the limits of development prescribed in the zoning ordinance. No knowledgeable site planner pursues the design process further without having researched the ordinance. Make no assumptions concerning the ordinance; any questions should be answered by the zoning administrator and preliminaries confirmed.

INVENTORY

An inventory of the site and its resources requires the collection and documentation of various site components. The first level of data collection requires documentation of the legal description of the site and would include, but not be limited to:

1. Plat or boundary-line survey and description
2. Acreage
3. Restrictive covenants or deed restrictions
4. Easements or other encumbrance
5. Topography at 1- or 2-ft intervals, depending on the amount of slope and irregularity

This research would also document the public's control, which would include:

1. Zoning (existing and proposed) (if rezoning is contemplated)
2. Public right-of-way (existing and proposed)
3. Street standards and design criteria
4. Utilities (existing and proposed)
5. All special permits or inspections required by local, state, or federal agencies

Data collection requires documentation of the site's existing natural and human-made conditions as well as its off-site contextual qualities. In addition to the checklist in Chapter 7, the following characteristics should be noted in the site reconnaissance phase.

- Views and vistas
- Rock outcroppings
- Context (built and natural)
- Vegetation (species, variety, and maturity)

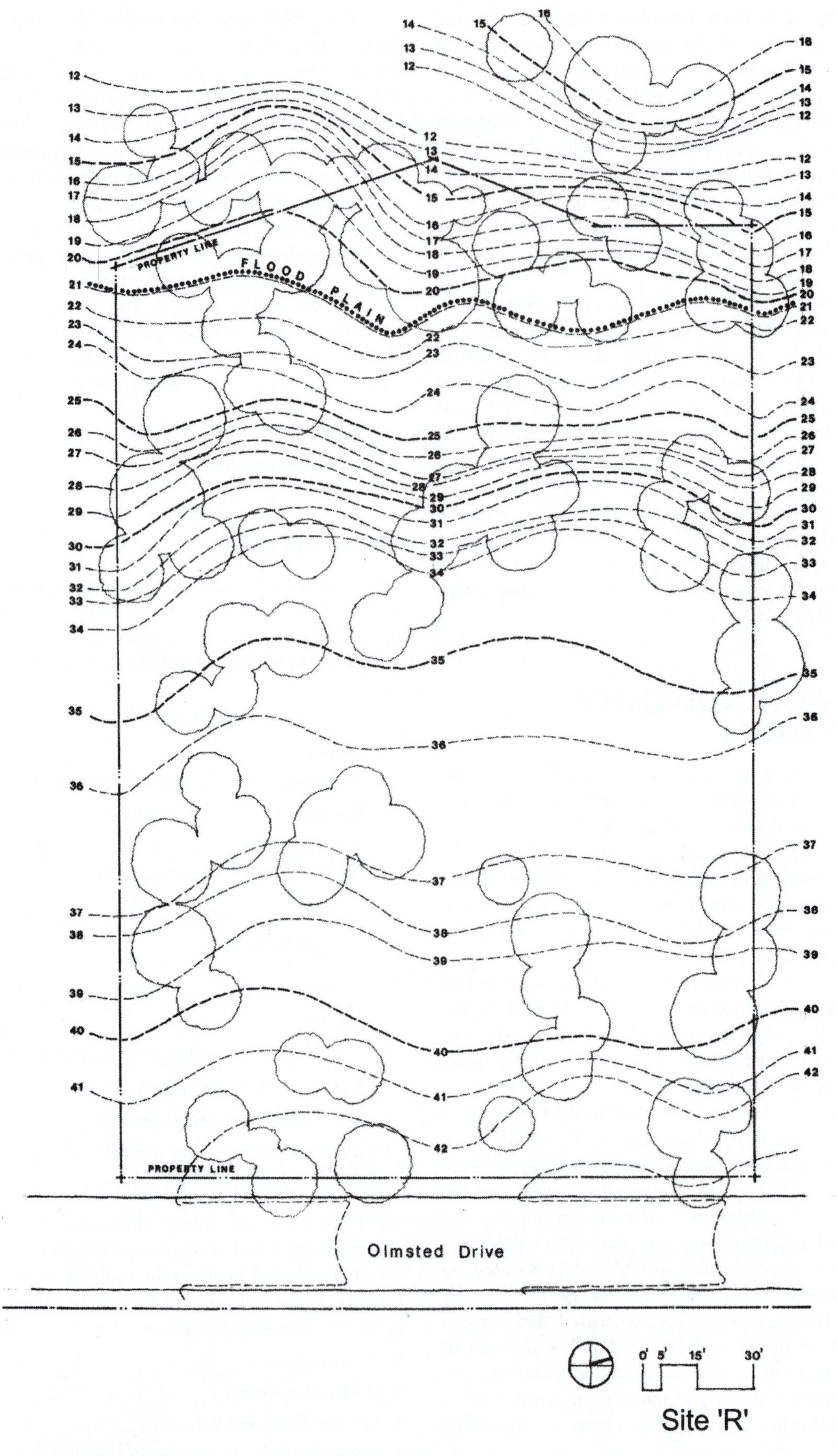

FIGURE 22-1. Plan of site R.

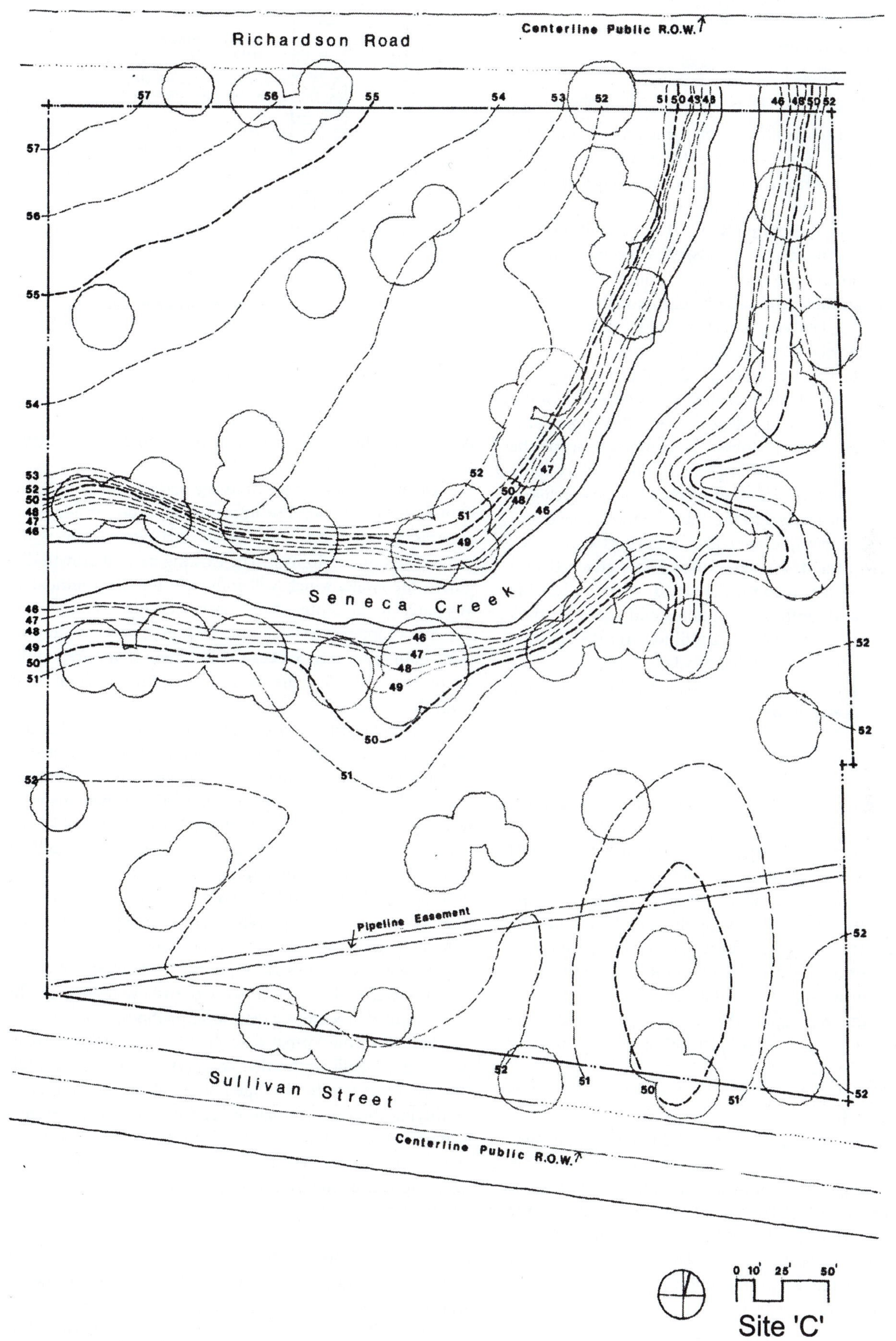

FIGURE 22-2. Plan of site C.

- Abandoned roads or cemeteries
- Streams or water courses
- Recent construction
- Oil and gas transmission lines, electrical and telephone lines

While the data collection and mapping process continue, the experienced site planner makes a visit to the site. There is no substitute for a personal walking reconnaissance. If either time or budget will not permit the expense of a site visit, today's technology will permit a surrogate to walk the site with a camera/phone in your place. Slides and photographs cannot communicate the experience of being there. Look for things that might not necessarily be on the maps, and note any discrepancies between the information that is documented and that which actually exists on a site.

This inventory process is essential, as it provides the site planner and client with an unqualified understanding of the site. The combination of photographs and field notes provides the site planner with physical documentation of the site's character. This personal inspection provides the opportunity to assess the quantifiable elements of the research and evaluate their contribution to the entire context. This documentation is also an important reference during the process.

Figs. 22-4(a) and (b) show color prints attached to a foam core board that is in turn attached to the wall of the site planner's studio/office. The photos and corresponding notes are physical reminders of conditions on the site that are instantly visible—all simultaneously.

The same photos can be downloaded as PDFs and used in large group presentations.

Site R

A visit to site R has revealed waterborne debris on the site at a level which appears higher than that indicated by the most recent floodplain maps. The higher floodplain level may be the result of additional development above the site since the documentation of the 100-year floodplain. Further research indicates that there is also a rock outcropping in the area of the site. Since the rock outcropping can be the precursor of expensive excavation and foundation work, a soil testing lab is contacted immediately. Test borings always clarify the geology of the site and provide the site planner with a more accurate "picture" of the site's physiographic past (Fig. 22-3).

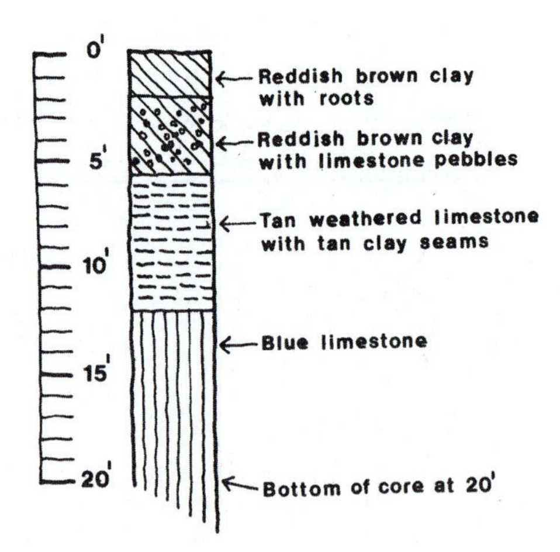

FIGURE 22-3. Test boring.

Analysis: Site R

The overlays for site R provide two illuminating facts. One is the site has a zone of steeper slopes than first appeared in the site visit (Fig. 22-5). This is a reminder of how deceptive land forms can be. The second revealing aspect of the analysis is that the floodplain is, in fact, higher than the older drawings indicated. The height of water in Rock Creek has increased due to the amount of development in the watershed, and contact with the Army Corps of Engineers confirms that their latest hydrograph establishes a new 100-year floodplain at the (5)21 foot contour. Other than the isolated rock outcropping, the soils and subsurface geology for site R indicate that there are no major problems associated with the soils.

Documentation of the drainage patterns for both R and C provide another level of information to be considered in analysis. Site R has a major swale that almost bisects the property from east to west. The existing natural growth disguised the swale, which will influence the development of any parcel in the center of the 1-acre site (Fig. 22-6).

Site C appears to be simple. The north and south portions both drain into Seneca Creek. There is one area to the southeast that is a depression and appears to drain away from the creek.

Climatological analysis might be considered as the next stage in the inventory. Of the various methods presented in Chapter 11, it appears that microclimate mapping for these sites would have genuine applicability. Microclimate mapping will delineate those zones of both the R and C sites which are either sunny and windy, sunny and calm, shaded and windy, or shaded and calm in both summer and winter. This information is particularly valuable for residential land uses, as it identifies those areas of an open site that are inherently suited for outdoor activities as well as those locations where a structure might enjoy maximum solar exposure. Remember that a residential use is a "skin"-loaded structure whose heating can be supplemented by various passive solar methods.

Following the format described for microclimate mapping, Fig. 22-7 illustrates that character of site R. The addition of structures to the open site will obviously require modifying the microclimate map. Since structures create a human-made shelter from the wind, knowing those zones where an outdoor seating area is most accommodating would influence the location of a terrace for either a residence or a restaurant.

Problem Statement: Site R

Although it is not the condition with the example sites C and R, it is not uncommon find that the client has chosen the wrong site for a use that may be a legitimate contribution to a specific area of the community. Although clients often engage the counsel of others with regard to market

(a)

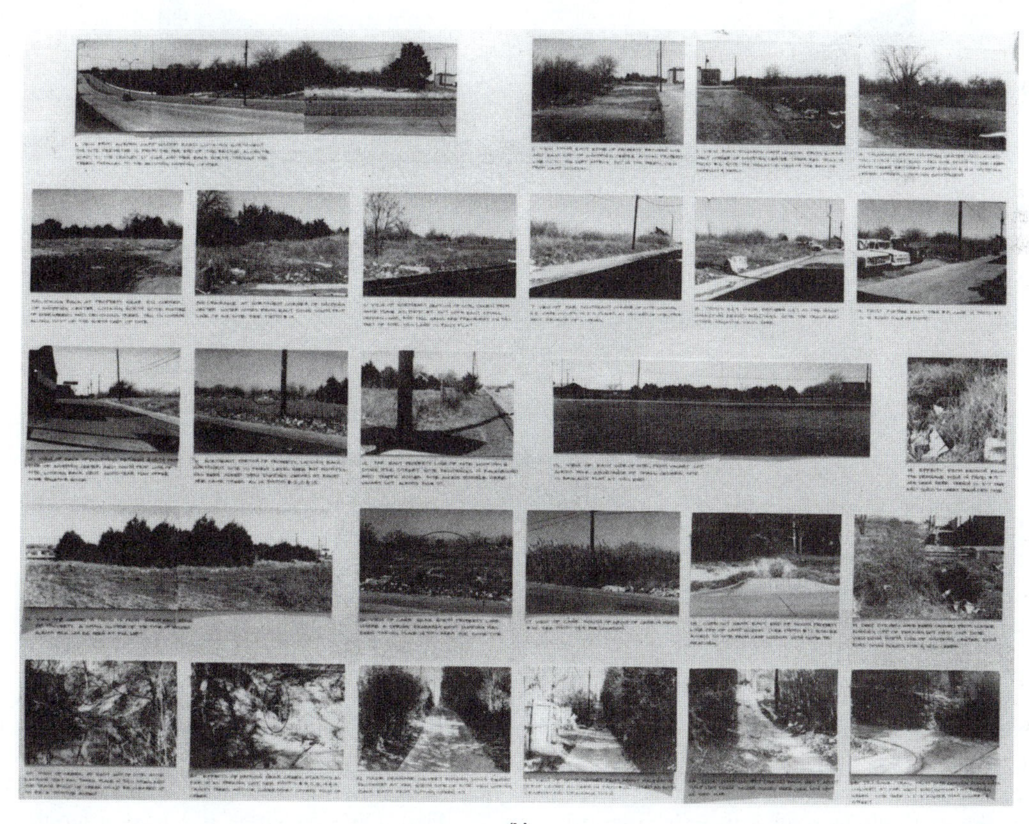

(b)

FIGURE 22-4. Photos on board with notes.

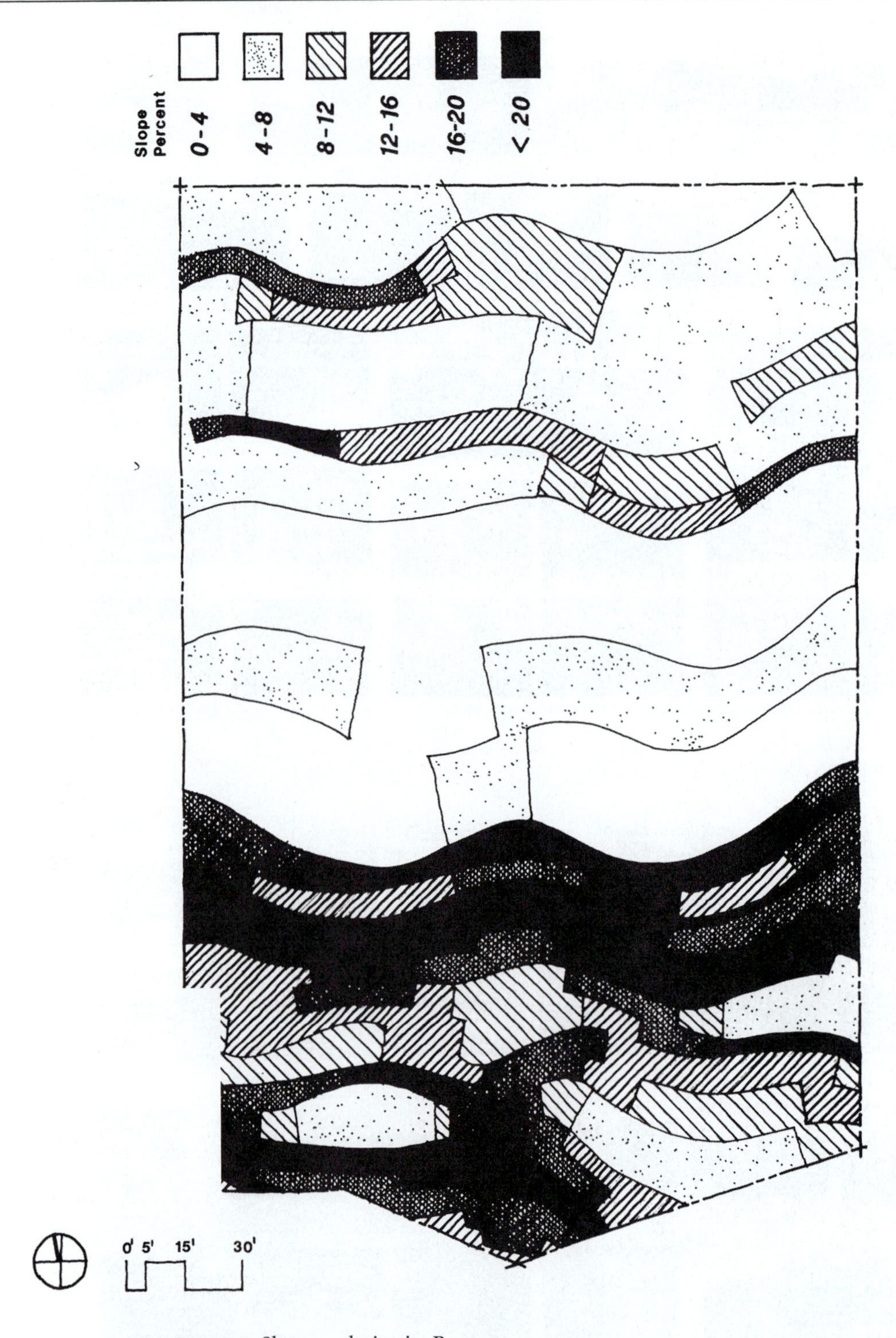

FIGURE 22–5. Slope analysis, site R.

analysis and economic feasibility, the site-selection effort is often a decision based primarily on land cost, often without benefit of the site planner's critique or opinion.

It is important to determine, in precise terms, where the conflicts exist, the methods to resolve their incongruities, and the implications of that resolution. This stage in the process permits the client to review the circumstances and make the next, more crucial set of decisions.

For site R, the total acreage must be evaluated relative to the subdivision of the property for four lots. Traditionally, the property would be subdivided so that each lot has equal frontage on the street (Fig. 22-8). Following that method would create four lots of approximately 42 ft × 260 ft. Although that formula has some advantages, it has some fairly obvious disadvantages.

First, Olmsted Drive is a busy arterial. While turning into a driveway off of the street is simple enough, exiting from any of the lots forces the driver to back into an often heavy flow of traffic. Second, between the side-yard setbacks (minimum 6 ft for the first 8 ft; thereafter 1 ft back for every 3 ft in vertical height) and the existing vegetation, the shape of the structure and location of private outdoor open space

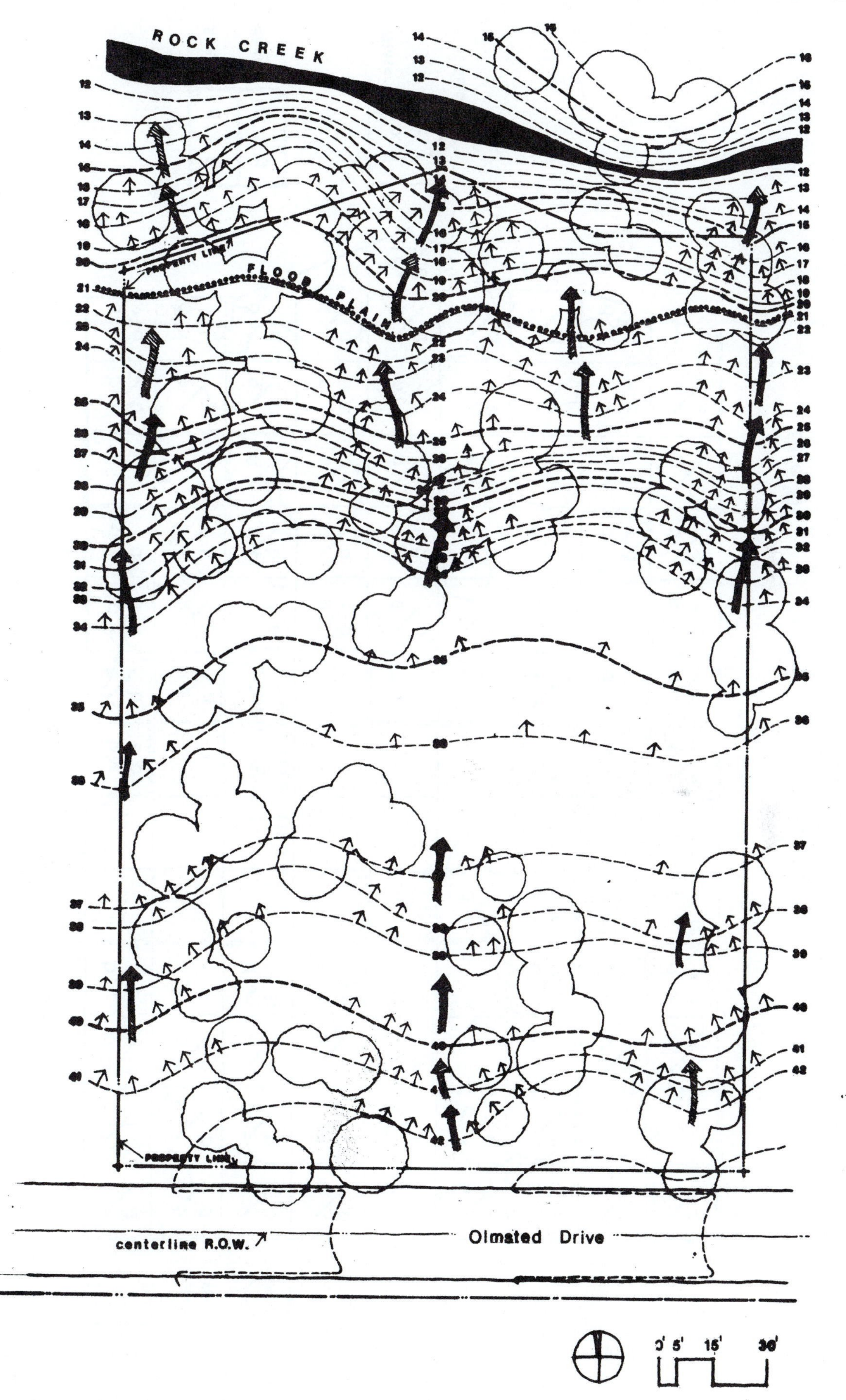

FIGURE 22-6. Drainage analysis, site R.

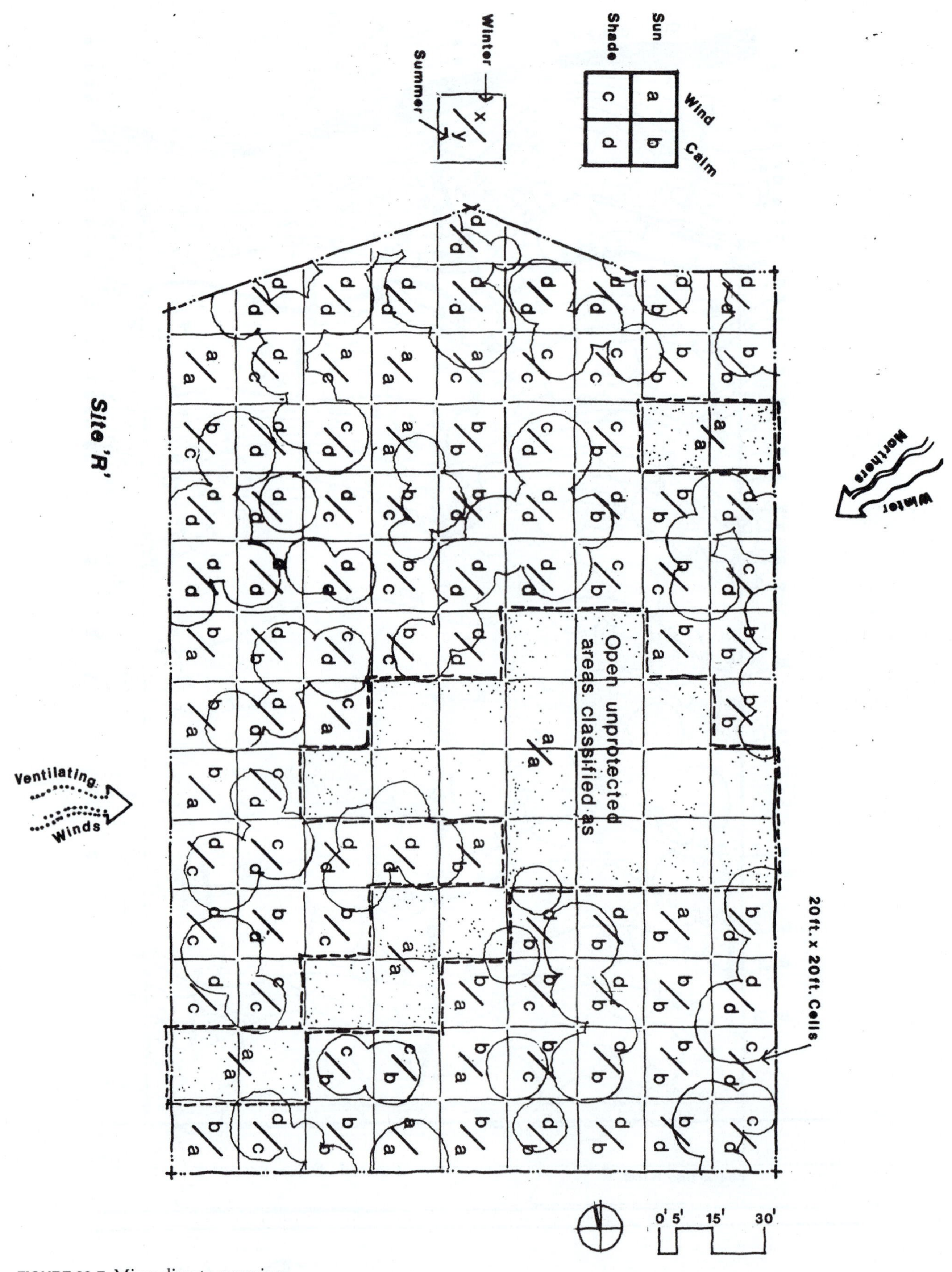

FIGURE 22-7. Microclimate mapping.

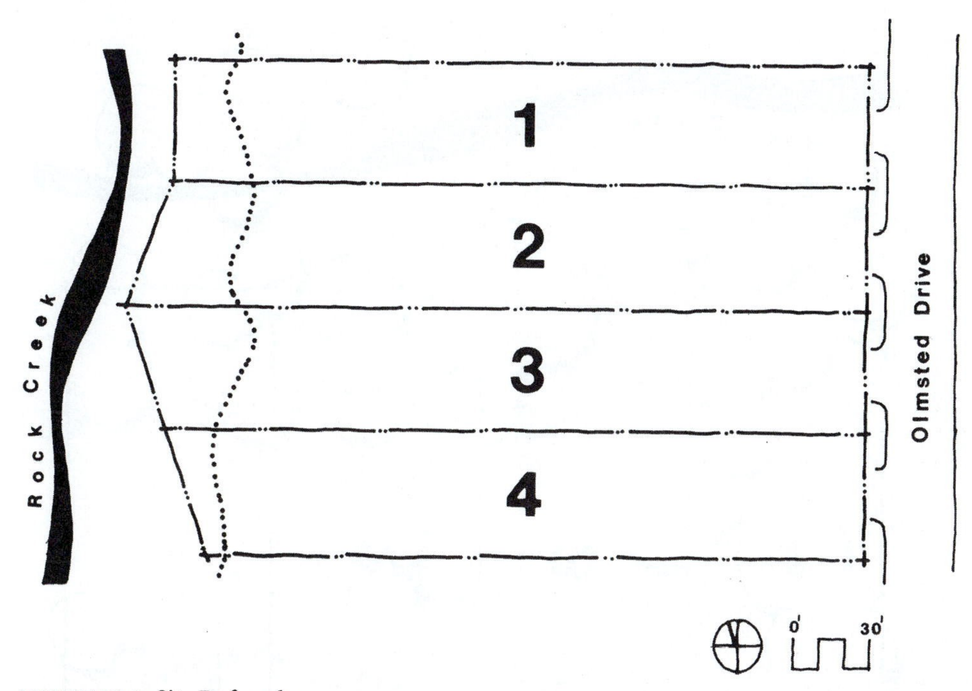

FIGURE 22-8. Site R: four lots.

are very restrictive (Fig. 22-9).Third, although the lots could be oriented with their long sides almost due north–south, the design flexibility necessary to keep one structure from obscuring the southern winter sun from another is marginal at best (Fig. 22-10). Fourth, the drainage swale that traverses the property will also affect the development of either lot 2 or lot 3 and possibly both. Finally, the opportunity to create a sense of place for four dwellings along the edge of a thoroughfare is extremely limited. The object building model restricts the site to a condition in which each building has little opportunity to contribute something to the adjoining structure. As with the traditional residential subdivision, the identity of each residence becomes relegated to an address and zip code.

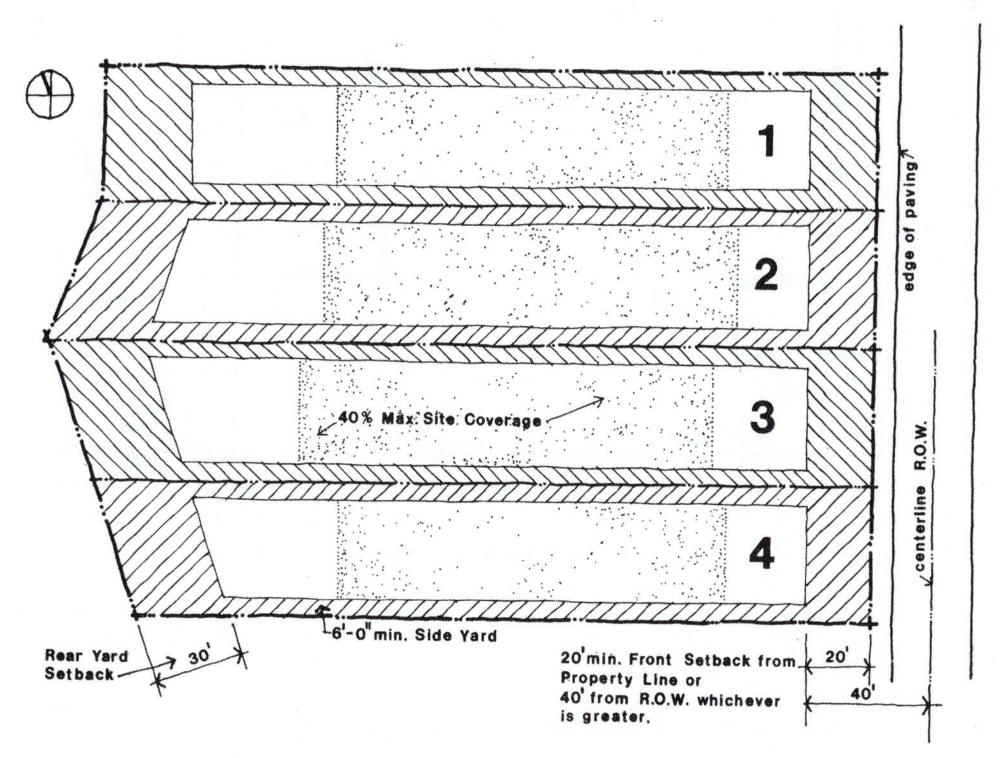

FIGURE 22-9. Site R: four lots with setbacks.

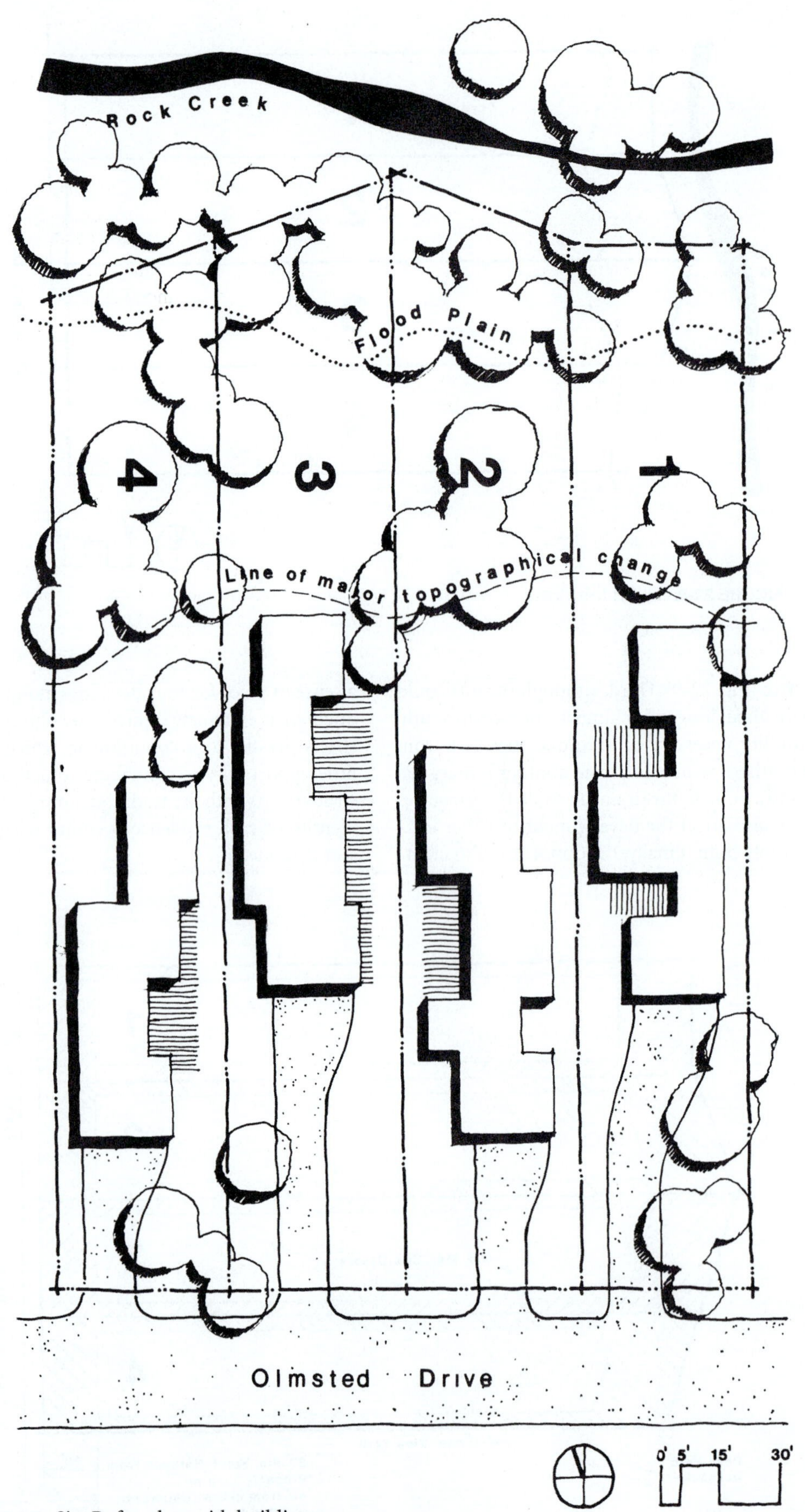

FIGURE 22-10. Site R: four lots with buildings.

Alternatives: Site R

Following the earlier review of site R's size, topography, drainage patterns, zoning, and floodplain, the developable limits of the property are now clear. One conceptual alternative to the four parallel lots is subdivision of the property with two lots about 100 ft deep fronting on Olmsted. This would permit the design of a road to reinforce the natural drainage of the existing swale as well as the development of the site with four less radically shaped lots (Fig. 22-11). Although this appears, at first glance, to be a reasonable method of subdivision, the city's minimum right-of-way for a public street is 40 ft 0 in., with an accompanying 60 ft 0 in.-diameter cul-de-sac. Although the impact of these criteria on the site is substantial, it would provide access to the rear lots.

An alternative would be to serve the two rear lots (2 and 3) with a private road built in an easement across lots 1 and 4 (Fig. 22-12(a)). The easement could be granted in perpetuity and "run" with the deeds of all four lots.

The problem with the second alternative is that the municipality has a rule which prohibits the subdivision of any property that does not have access to a public street. The third alternative is a variation on the second. If lots 2 and 3 have a 10 ft 0 in. strip of land that connects the "body" of the lot to Olmsted (public street) (Fig. 22-12(b)), the two 10-ft strips are sufficiently wide enough to build a *private two-way* road complete with cul-de-sac. The configurations of lots 2 and 3 are unconventional and are sometimes referred to as "pipe stem" or "flag" lots. The design flexibility now possible with all four lots is significantly better than the traditional format. The configuration of the four lots not only provides a broader range of site-planning and architectural solutions, but the ingress/egress problems are substantially better, as all four lots can now use the private drive.

The private drive can also be designed with a concave section and open paving. This will imitate the natural drainage characteristics of the site as well as reduce the amount of runoff that would normally occur following development. Another benefit of the private drive is that it permits the orientation and design of the four dwellings into an identifiable residential cluster, as opposed to the parallel lots facing the street. The potential for the development of four independent, yet cohesive units becomes more plausible when the scale of the street is broken and a hierarchy becomes a visible and functional part of the site plan.

The architectural solutions to lots 1 and 4 will be significantly different from those of 2 and 3. The two westerly lots should have units that have been designed to accommodate the steeper slopes. Since the slopes on part of the site considered for residential development are in excess of 20%, a review of the surface soil conditions and test borings is in order. Since the sewer line that will be built in the private drive to serve all four units will be at elevations ranging between 31 and 39 ft, any uses in the dwellings on lots 2 and 3 below elevation 31.0 ft which require a sanitary sewer connection will also require a sump pump to lift the waste to the gravity flow sewer. This is not an uncommon occurrence, but it emphasizes the need to consider each site and building as an integral design problem. Although overlooking the need for a sump pump is not itself an insurmountable problem, it is embarrassing.

Structures for all four lots can now be oriented as need be with considerable flexibility. The small private street can also be designed as a pedestrian street. As with the Dutch *Woonerf* streets, the residents are conscious of its multifunctional role. Because of the lot configurations, private road and smaller front/rear yards, the use of a performance zone (PUD) should be requested. While this is an additional "cost," recognize too that the scale of the small neighborhood cluster would have been extremely difficult to achieve without it (Fig. 22-13).

Reconciling Conflicts: Site R

Structures on site R have been located in those areas that provide maximum solar exposure with minimal environmental degradation. The long sides (north and south) have

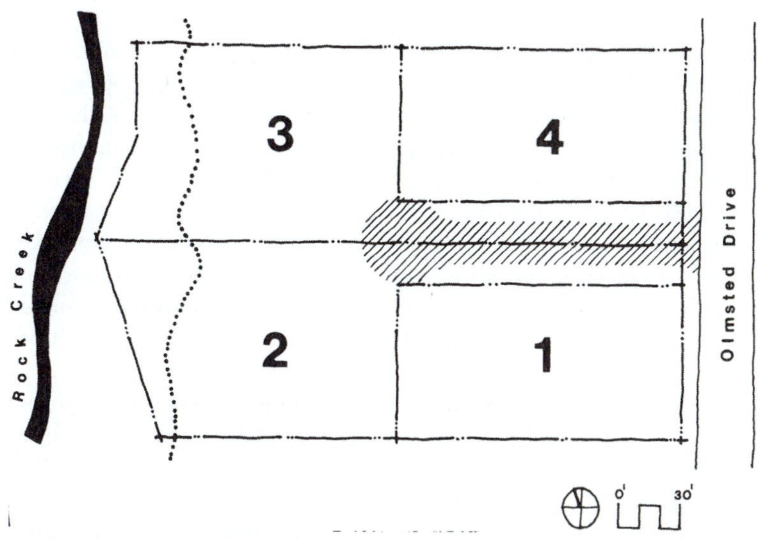

FIGURE 22-11. Site R: alternative plan.

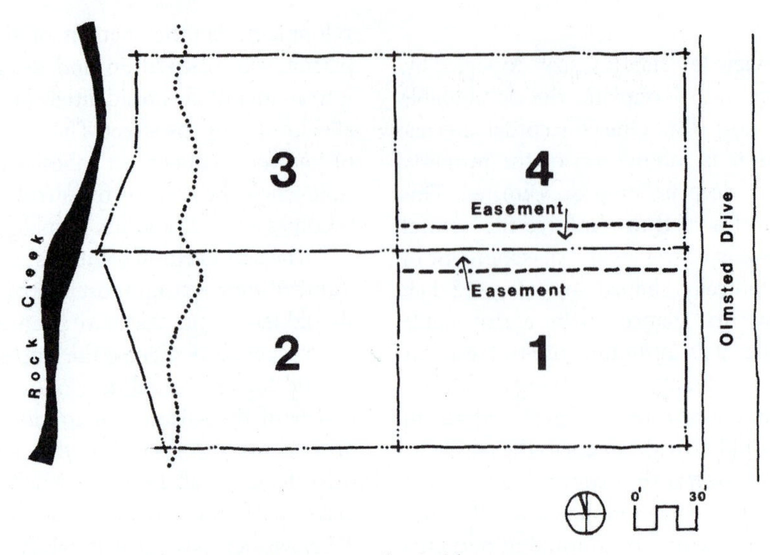

FIGURE 22-12(a). Site R: alternative plan with easement.

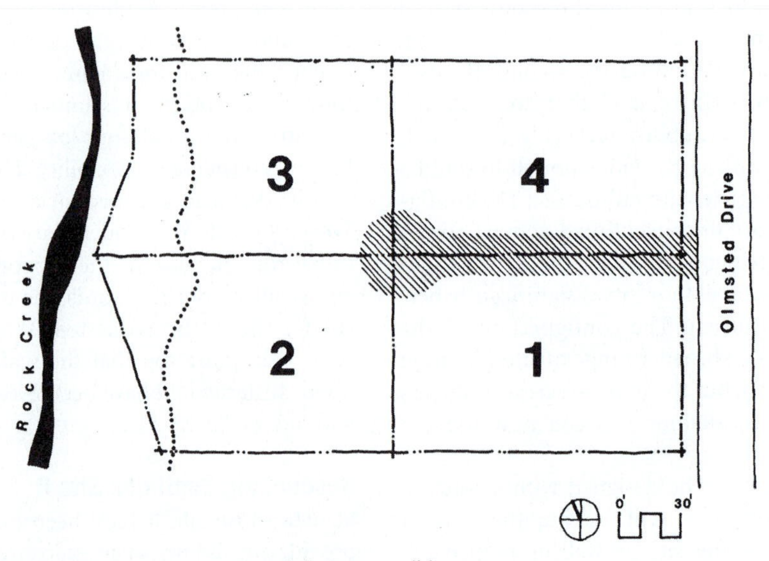

FIGURE 22-12(b). Site R: with "pipe stem" lots.

been oriented to enjoy maximum solar exposure and direct gain for winter heating. The narrower elevations have been oriented toward the east and west, minimizing the potential heat gain during the summer months. Since the topography on site R generally slopes from east to west, structures on all four lots will probably run perpendicular to the contours. Since the topographical change across the length of the slab is rather substantial, "breaking" or separating the slab into two elevations would be an appropriate consideration. This articulated response permits locating the eastern portion of the slab sufficiently high to provide the vertical distance adequate for the design of a swale. An additional advantage is that the length of the swale along the east side of each building can be modest because the length of the building parallel to the contours is relatively short. Fig. 22-14 illustrates some of the primary spot elevations necessary to accommodate structures on lots 1 and 4.

Although the westernmost lots have substantially steeper slopes, the principles applied to the two east end lots (1 and 4) still apply. Obviously, the floor levels should reflect that of the topography, and in this case the structure can become two stories (Fig. 22-15). Were it not for a very stable soil condition, this would be a questionable solution. Furthermore, the potential for more floor levels makes the view of the creek available to more of the house. Fortunately, the grove of deciduous trees alongside the creek is sufficiently dense as to screen out much of the intense heat from the summer sun.

Similar to lots 1 and 4, the water that flows from east to west must be carried around the building. The problem is accentuated on lots 2 and 3, as the amount of runoff increases with the increase in the parking area. Care needs to be taken here to make certain that any additional water is either directed away from the building with grading or with a

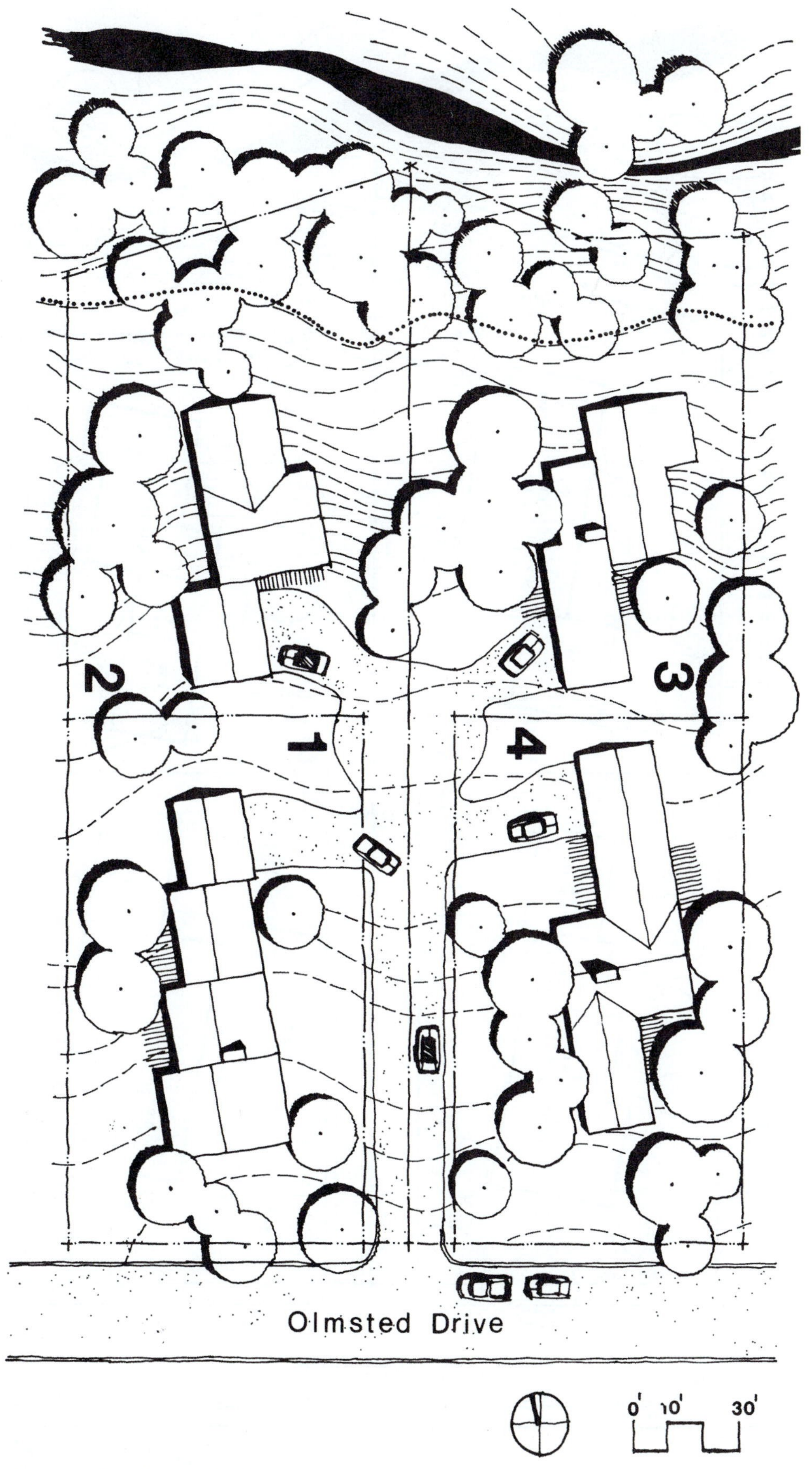

FIGURE 22-13. Site plan for site R.

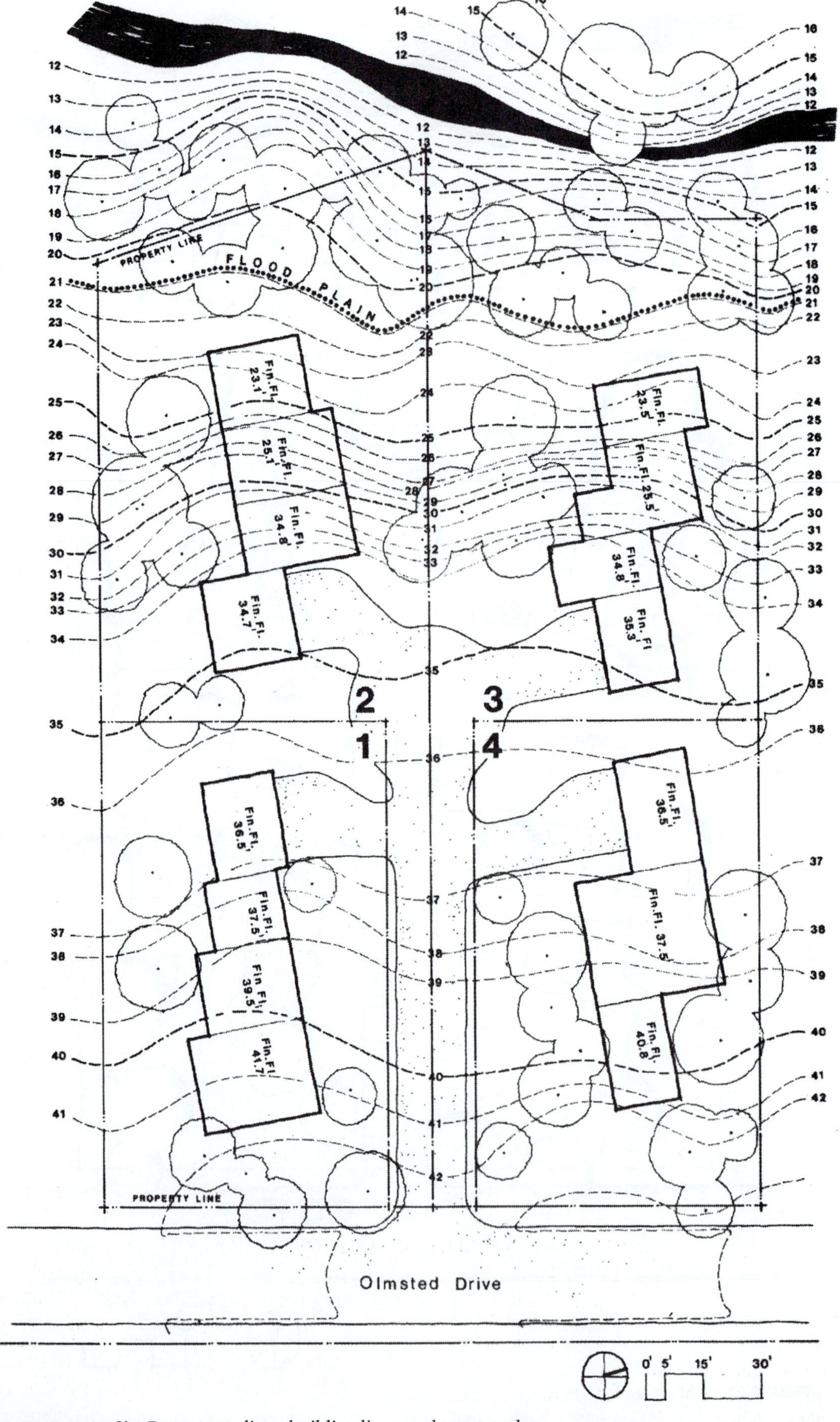

FIGURE 22-14. Site R: property lines, building lines, and topography.

FIGURE 22.15 Site R: section through buildings.

closed drainage system or a combination of both. Any time a parking area is topographically higher than a structure, the potential exists for a drainage problem.

Inventory and Analysis: Site C

The initial visit to site C produces no visual surprises. Trees are confined to the creek's edge and are scattered sparsely throughout the remainder of the site. Although signs identifying a pipeline easement confirm the indication of that encumbrance, there is something puzzling about the alignment of the easement (Fig. 22-16). The drawing describes one thing; reconnaissance implies something else. Under these circumstances, a request is made of the owner to engage the services of a registered engineer to clarify the question.

A review of the Soil Conservation Service's research for site C suggests that the permeability of the dominant soil type north of Seneca Creek is moderate at best. This comes as no surprise, since soils close to creeks and streams are often borderline sites. A topographical depression on the south side of the site is predominantly sandy loam and might be considered as an on-site detention pond if necessary.

A microclimate review of site C with pending commercial and retail activities will be quite different from the residential uses of site R. The structures on site C are internally loaded. That is, the thermal loads for the heating, ventilating, and air conditioning (HVAC) systems are created by the heat generated from the occupants and lights inside the structure. It is very unlikely that supplementary heat will be required and more than likely that cooling will be necessary even in the winter. In this circumstance, where the interior is already a heat load, any additional solar heat gain only aggravates the problem.

Whereas the intent is to limit the amount of solar gain at all times to site C, the residential sites should be scrutinized carefully to identify those areas where there is maximum solar exposure in the winter and maximum shading in the summer. Again, this preferential location is not unique to a particular region but has general applicability.

For site C, the parking component of the zoning ordinance is somewhat more complex. The bank and retail uses are permitted without caveat, but the setback and height limits are somewhat more confining. It seems that the zoning ordinance has a very restrictive setback for any retail use that adjoins a multifamily residential category. Fig. 22-17 illustrates various physical conditions of the site as well as the basic controlling performance standards of the zoning ordi-

FIGURE 22-16. Pipeline easement warning.

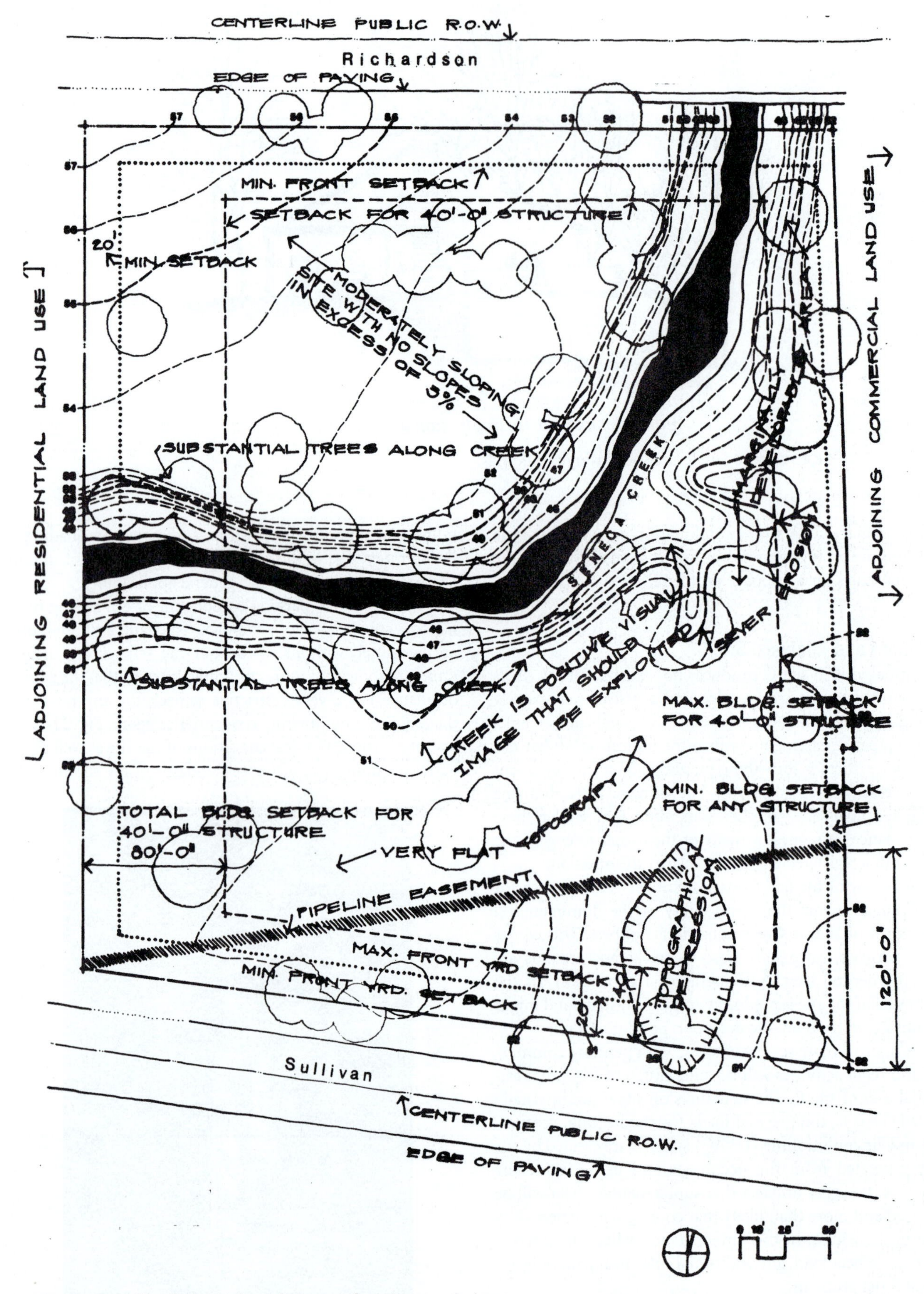

FIGURE 22-17. Site C physical conditions and zoning restrictions.

nance. Additional development controls include a 10% open-space requirement. It must be noted that the zoning ordinance's definition of "open space" excludes slopes in excess of 25% and water courses.

The parking requirement will also require interpretation at this point. A realistic appraisal of the impact of parking must be delineated since miscalculations can have a deleterious effect on the site plan. "Realistic" needs to be emphasized since many people have a bias toward structured parking. Recall that the simplest way to conceptualize the feasibility of structured parking in open-market conditions is to compare the price of unimproved property with that of the cost of unimproved property plus a parking structure. In most cases clients will favor the alternative that provides the parking required by ordinance at the lowest cost per square foot. This analysis is very straightforward, but it must recognize that there can be circumstances in which a decision will not be made on economic issues alone.

Careful attention has also been given to the parking requirement, particularly since there are three uses: retail commercial, banking, and office space. The 10,000 ft^2 of retail space has a parking requirement of one car for every 300 ft^2 of retail space. The banking function has a requirement of one space for every 250 ft^2 of gross building area. The remaining allowable building area, planned office activities, has a requirement of one space for every 350 ft^2 of gross building area. The ordinance further stipulates that every parking lot for office, retail, or banking activity will have at least 2 spaces for individuals with disabilities for lots up to 50 cars. Above that size, spaces for individuals with disabilities are required at a ratio of 1 for every 30 cars. Fig. 22-18 describes the numerical extension of the parking requirements in addition to the spatial allocations of site functions.

A review of the client's development objectives reveals he anticipated more density than the site can accommodate. The 0.7 FAR permitted by the ordinance would require 380 parking spaces! Since committing that much area to parking alone has some serious implications relative to the developer's objectives, the client is contacted for a preliminary review. First, an extension of the parking spaces in terms of site area indicates that most of the site must be committed to parking if the permitted FAR could even be approached. This might be ameliorated if the use considered for the site was speculative office space alone. But given the fact that one potential user is retail commercial and the second a bank, eliminating the entire ground floor of a functional use would create major problems with the project's economic pro forma. The developer concedes that the commitment from the bank and retail activities have more priority, and instructions are to continue development of the site plan by reducing the floor area of speculative office space. This dialogue between the site planner and client is common, and conveying your best professional opinion is critical.

On site C, a review of the utilities and their alignments confirms their common location parallel to the roads. The petroleum pipeline easement was a discrepancy and has been corrected on the final base map.

Design criteria form a critical aspect of the design process and should be delineated in written as well as visual form. Both public and private sectors will have criteria, both specific and implied. The criteria should be articulated, as they can become a development constraint that will have to be resolved before legitimate design alternatives can be considered. On occasions, the site planner may be confronted with the necessity to change the criteria. On others, however, exceeding recommended standards would be inappropriate and invite serious consequences. As an example, topographical fill that exceeds 4:1 slope without stabilization considerations could result in serious erosion or even slumping.

Some criteria have already been established. Zoning is a form of criteria that will ultimately dictate parking and general building locations on both sites.

As to site C, the community has also been faced with major property damage resulting from flooding. Since Seneca Creek is dry most of the time, it has not yet created a problem. But the city council has adopted a new development policy which stipulates that any new construction must demonstrate a reduction in on-site runoff by 30% to that of comparable land uses. Earlier considerations associated with on-site detention and/or open paving systems must now become an integral part of the program.

Problem Statement: Site C

Early resolution of the differences between the perceived and realistic development potential diffuses a problem that makes all others pale by comparison. Fig. 22-19 describes the revised development program and the areas necessary to be allocated to parking. When this area is distributed in equitable percentages to both north and south sectors, the residual land area becomes apparent. Although nothing prohibits locating all the uses in one structure on one side of the creek, it makes little sense considering the implications of walking distances alone. Since the site is bordered on the north and south by major roads, providing access to both zones divided by the creek is largely a matter of building location and the configuration of parking and circulation. Although a vehicular bridge connecting both north and south zones has been considered and discounted as unnecessary, it appears that a pedestrian connection of some type is required.

The soils, slopes, and drainage patterns present few problems. The site's topography, parking lots, and paving systems will be evaluated collectively as methods to resolve the municipality's runoff policies. Although the creek is a problem in the sense that it has divided the site, it should also be considered an asset, as it contains and supports most of the larger trees on the site.

Since the office, retail, and bank uses create an "internally" loaded building, reducing the additional heat load of the summer sun is a primary objective in the siting and orientation of the buildings. Furthermore, the use of any reflective material on both buildings will be carefully evaluated. Glare and reflected heat can damage surrounding vegetation as well as create additional heat loads to adjacent structures. This poses an issue that must become part of the architectural problem

Land-Use Computations

1. North parcel	1.63 acres × 43,560 ft² =	71,002.8 ft²
2. South parcel	2.37 acres × 43,560 ft² =	103,237.2 ft²
3. Creek	1.06 acres	
4. Total	5.06 acres	174,424.4 ft² (total buildable area)

5. 174,424 ft² × 0.7 floor area ratio (FAR) = 121,968.0 ft² of building area permitted by zoning ordinance

6. 121,968.0 ft²	total building area
7. − 10,000.0 ft²	of retail activity
8. − 12,500.0 ft²	of bank
9. 99,468.0 ft²	potential area for speculative office space

Parking Requirements

10. 10,000.0 ft²	of retail requires 1 space/300 ft² of gross floor area = 34 parking spaces
11. 12,500.0 ft²	of banking use requires 1 space/250 ft² of gross floor area = 50 parking spaces
12. 99,468.0 ft²	of office space requires 1 space/350 ft² of gross floor area = 284 parking spaces

13. Total parking required: 368 parking spaces

14. 368 × 350 ft²/space = 128,800 ft² of site area
368 × 400 ft²/space = 147,200 ft² of site area

Allocation of Site Functions

15. 174,240.0 ft²	(line 1 + line 2)
16. − 15,000.0 ft²	marginally developable area in south parcel between Seneca Creek and property line
17. − 17,424.0 ft²	10% of site area for open space
18. − 10,000.0 ft²	of retail space on ground floor
19. − 12,500.0 ft²	of banking activity on ground floor
20. − 2,250.0 ft²	(10% of lines 18 and 19) for pedestrian circulation
21. 117,066.0 ft²	of site area remaining for parking and auto circulation

Summary: Since area required for parking based on 0.7 FAR ranges from 128,800 to 147,200 ft² (line 14), some downward adjustment on the realistic development potential must take place.

FIGURE 22-18. Site C use allocations and computations

statement. A strategy must be developed to permit exploitation of views of the creek without increasing the direct solar gain.

Alternatives: Site C

Since the revised development objectives are more compatible with the zoning, proposed land uses, and parking requirements, the question now evolves to the distribution of uses, allocation of parking, and the method by which the site can be developed as a cohesive whole. The cost of connecting the northern and southern segments with a road and bridge would be extremely expensive and provide marginal utility. Thus it appears that at least one alternative should examine the development of the site with two buildings connected by a pedestrian bridge. As two separate buildings, both bank and retail uses could enjoy individual identities, while the pedestrian bridge facilitates a good integration of uses. Where the bridge helps to tie two commercial uses together, it is also easy to see how important it would be to a residential and commercial mix.

The pedestrian bridge and path also contribute to the development of an identifiable place. Without the bridge the plan is dependent on the design of two object buildings. The bridge also presents an exceptional opportunity to bring human scale into an auto-dominated environment. Where

Parking Computations

North Parcel

Retail area	10,000 ft^2	requires:	34 parking spaces
Office space	17,500 ft^2	requires:	50 parking spaces
		Total:	84 parking spaces
First floor	Retail:	10,000 ft^2	
Second floor	Office:	10,000 ft^2	
Third floor	Office:	7,500 ft^2	

South Parcel

Bank area	12,500 ft^2	requires:	50 parking spaces
Office space	37,500 ft^2	requires:	107 parking spaces
		Total:	157 parking spaces
First floor	Bank:	12,500 ft^2	
Second, third, and fourth floors	Office:	12,500 ft^2 each	

SUMMARY:

Total north parcel:	27,500 ft^2
Total south parcel:	50,000 ft^2
Site total:	77,500 ft^2 of building area

Approximately 0.45 FAR reflected in site C development plan, shown in Fig. 22-20.

FIGURE 22-19. Revised development program for site C.

pedestrian systems are so often minimal components of the site plan, this site and its uses require a well-articulated system.

Where the north–south linkage is imperative for site C, it appears that there may be some value in extending a path to the western edge of the property. A path that parallels the creek and provides direct access from the adjacent residential land uses could provide a valued exposure for both bank and retail commercial uses. Although the functional issues must be scrutinized before a final decision can be made, an additional east–west path would clearly be a welcomed amenity for residents, employees, and retail customers alike.

Following the decision to locate the bank to the south of Seneca Creek and the retail activity to the north, the next task was to achieve some balance as to the site's area and required parking. To some degree this phase is "trial and error," as estimates must be adjusted until an acceptable relationship between building, parking, and open space can be found. Just as the example portrayed in Chapter 21 on the parking schedule, the parking requirement had a direct relationship to the ultimate achievable office and/or retail area.

The next phase is to locate the buildings and support parking so as to provide a visible and efficient auto and pedestrian system that integrates and exploits the residual open space. The development of a pedestrian bridge between the north and south sections of the site, terminating at each building, resolves both functional and aesthetic issues. The bridge can become a physical unifying element while creating its own set of views and images. The geometry of the human-made path, buildings, and parking also evolve to contrast sharply with the natural sinew of the creek (Fig. 22-20).

The grading plan for site C is fairly uncomplicated. Although the north side of the creek drains toward the creek and the building, coordinating the relationships between the finish floor elevations, peripheral walks, and drives will permit the design of a positive surface drainage plan. Recall the process of establishing spot elevations first to determine the necessary topographical changes. As before, some of the issues on the north side of the creek also exist on the south. Although a large area of the parking in the south will drain naturally toward the creek, another area drains toward the topographical depression near Sullivan Street. Retaining the trees in this area and directing some of the runoff into this zone will enhance the aesthetic as well as the functional qualities of the plan.

Ingress to and egress from the site are reasonably straightforward. The parking lots are oriented such that the pedestrian might use the drive as the path. A drop-off zone has been integrated into the fire lane, which is accessible to the three sides of the building. Parking for individuals with disabilities has been located on two sides of the building in an effort to keep walking distances to a minimum.

Entrance drives have been located so as to communicate a clear understanding of the circulation and parking system on both sides of the creek. Two-way drives and head-in 90° parking have been used to keep the area of impervious surfaces to a minimum. Since storm water runoff is also a specific problem with site C, the use of an open paving system has a high priority.

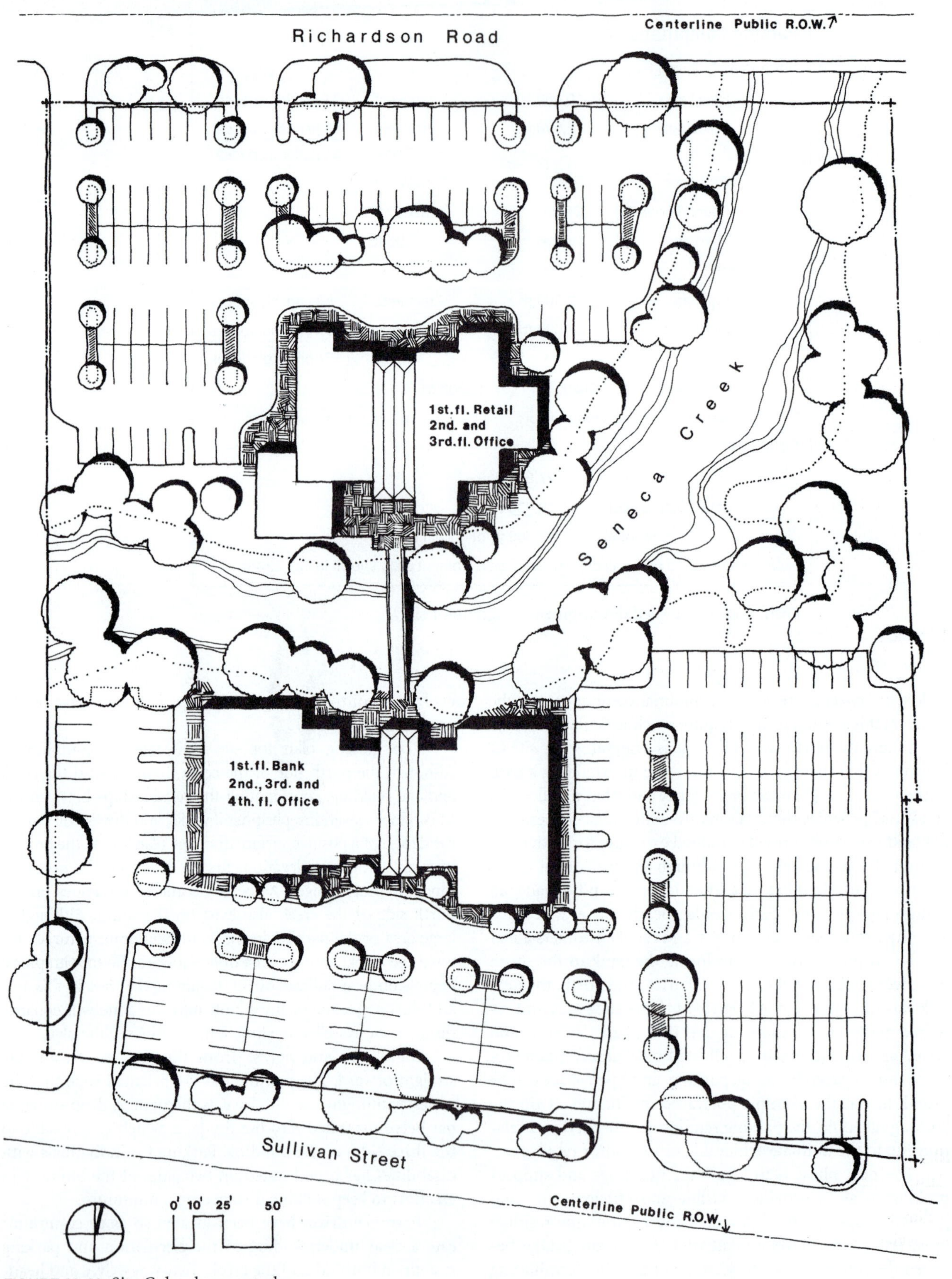

FIGURE 22-20. Site C development plan.

Appendix A

STATIONS PROVIDING LOCAL CLIMATOLOGICAL DATA*

Alabama
Birmingham C.O., Birmingham AP, Huntsville, Mobile, Montgomery

Alaska
Anchorage, Annette, Barrow, Barter Is., Bethel, Bettles, Big Delta, Cold Bay, Fairbanks, Gulkana, Homer, Juneau, King Salmon, Kodiak, Kotzebue, McGrath, Nome, St. Paul Is., Talkeetna, Unalakleet, Valdez, Yakutat

Arizona
Flagstaff, Phoenix, Tucson, Winslow, Yuma

Arkansas
Fort Smith, Little Rock, North Little Rock

California
Bakersfield, Bishop, Blue Canyon, Eureka, Fresno, Long Beach, Los Angeles AP, Los Angeles C.O., Mount Shasta, Red Bluff, Sacramento, San Diego, San Francisco AP, San Francisco C.O., Santa Maria, Stockton

Colorado
Alamosa, Colorado Springs, Denver, Grand Junction, Pueblo

Connecticut
Bridgeport, Hartford

Delaware
Wilmington

District of Columbia
Washington Dulles AP, Washington National AP

Florida
Apalachicola, Daytona Beach, Fort Myers, Jacksonville, Key West, Miami, Orlando, Pensacola, Tallahassee, Tampa, West Palm Beach

Georgia
Athens, Atlanta, Augusta, Columbus, Macon, Savannah

Hawaii
Hilo, Honolulu, Kahului, Lihue

Idaho
Boise, Lewiston, Pocatello

Illinois
Cairo, Chicago, Moline, Peoria, Rockford, Springfield

Indiana
Evansville, Fort Wayne, Indianapolis, South Bend

Iowa
Des Moines, Dubuque, Sioux City, Waterloo

Kansas
Concordia, Dodge City, Goodland, Topeka, Wichita

Kentucky
Jackson, Lexington, Louisville

Louisiana
Baton Rouge, Lake Charles, New Orleans, Shreveport

Maine
Caribou, Portland

Maryland
Baltimore

Massachusetts
Blue Hill, Boston, Worcester

Michigan
Alpena, Detroit, Flint, Grand Rapids, Houghton Lake, Lansing, Marquette, Muskegon, Sault Ste. Marie

Minnesota
Duluth, International Falls, Minneapolis–St. Paul, Rochester, Saint Cloud

Mississippi
Jackson, Meridian

Missouri
Columbia, Kansas City, Kansas City Downtown AP, St. Louis, Springfield

Montana
Billings, Glasgow, Great Falls, Havre, Helena, Kalispell, Miles City, Missoula

Nebraska
Grand Island, Lincoln, Norfolk, North Platte, Omaha Eppley AP, Omaha (North), Scottsbluff, Valentine

Nevada
Elko, Ely, Las Vegas, Reno, Winnemucca

New Hampshire
Concord, Mount Washington

New Jersey
Atlantic City AP, Atlantic City C.O., Newark

New Mexico
Albuquerque, Clayton, Roswell

*Through the National Climatic Data Center in Asheville, North Carolina.

New York
Albany, Binghamton, Buffalo, New York C. Park, New York Kennedy AP, New York LaGuardia AP, Rochester, Syracuse

North Carolina
Asheville, Cape Hatteras, Charlotte, Greensboro (High Point), Raleigh, Wilmington

North Dakota
Bismarck, Fargo, Williston

Ohio
Akron, Greater Cincinnati AP, Cleveland, Columbus, Dayton, Mansfield, Toledo, Youngstown

Oklahoma
Oklahoma City, Tulsa

Oregon
Astoria, Bums, Eugene, Medford, Pendleton, Portland, Salem, Sexton Summit

Pacific Islands
Guam, Johnston Island, Koror, Marshall Islands (Kwajalein, Majuro), American Samoa (Pago Pago), E. Caroline Islands (Ponape, Truk), W. Caroline Islands (Yap), Wake Is.

Pennsylvania
Allentown, Avoca, Erie, Harrisburg, Philadelphia, Pittsburgh, Williamsport

Puerto Rico
San Juan

Rhode Island
Block Island, Providence

South Carolina
Charleston AP, Columbia, Greenville–Spartanburg AP

South Dakota
Aberdeen, Huron, Rapid City, Sioux Falls

Tennessee
Bristol–Johnson City–Kingsport, Chattanooga, Knoxville, Memphis, Nashville, Oak Ridge

Texas
Abilene, Amarillo, Austin, Brownsville, Corpus Christi, Dallas–Fort Worth, Del Rio, El Paso, Galveston, Houston, Lubbock, Midland–Odessa, Port Arthur, San Angelo, San Antonio, Victoria, Waco, Wichita Falls

Utah
Milford, Salt Lake City

Vermont
Burlington

Virginia
Lynchburg, Norfolk, Richmond, Roanoke, Wallops Island

Washington
Olympia, Quillayute, Seattle C.O., Seattle Sea-Tac AP, Spokane, Stampede Pass, Walla Walla, Yakima

West Virginia
Beckley, Charleston, Elkins, Huntington

Wisconsin
Green Bay, La Crosse, Madison, Milwaukee

Wyoming
Casper, Cheyenne, Lander, Sheridan

SUN PEGS FOR VARIOUS LATITUDES

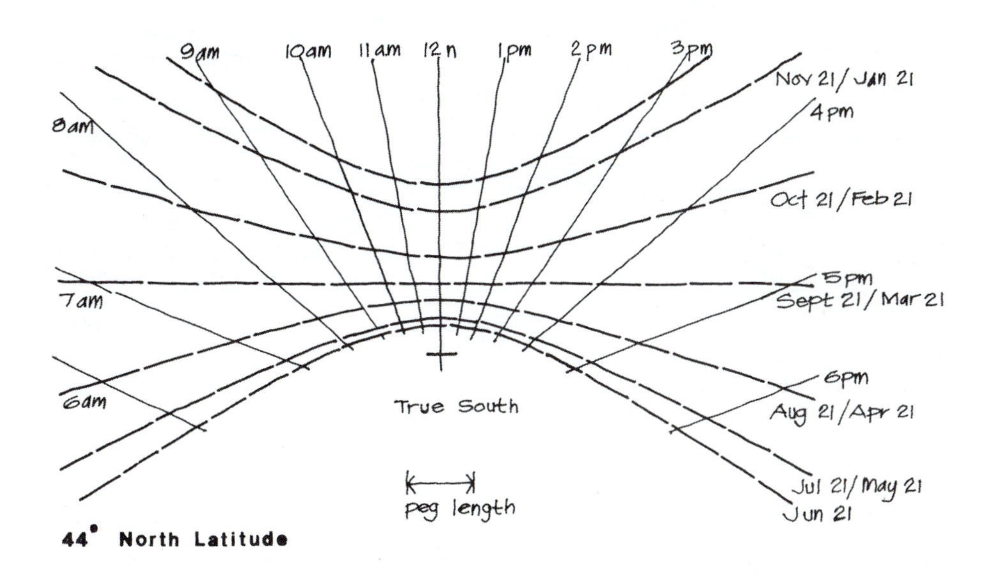

44° North Latitude

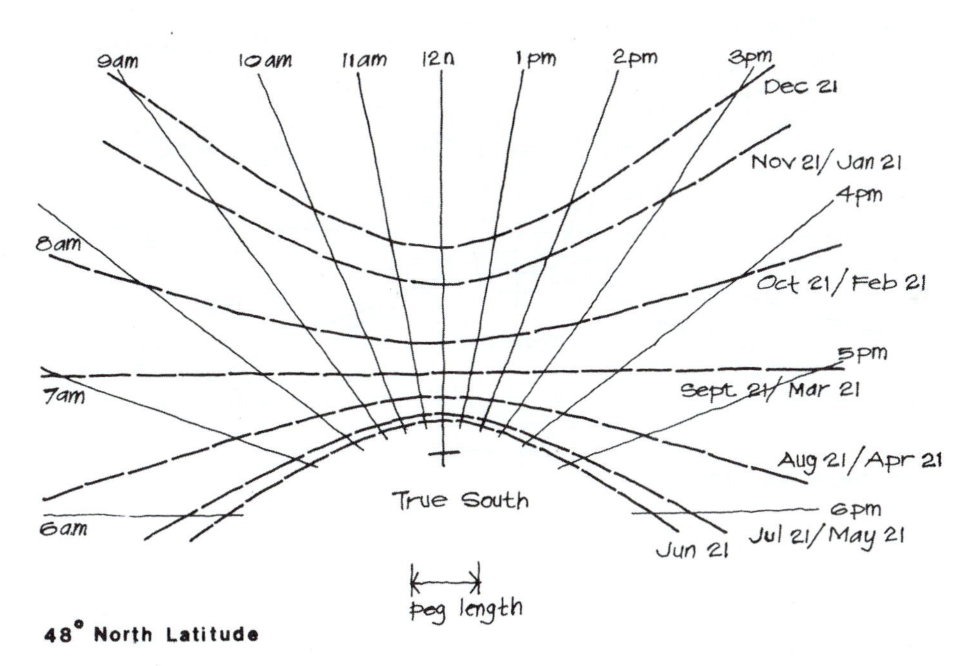

48° North Latitude

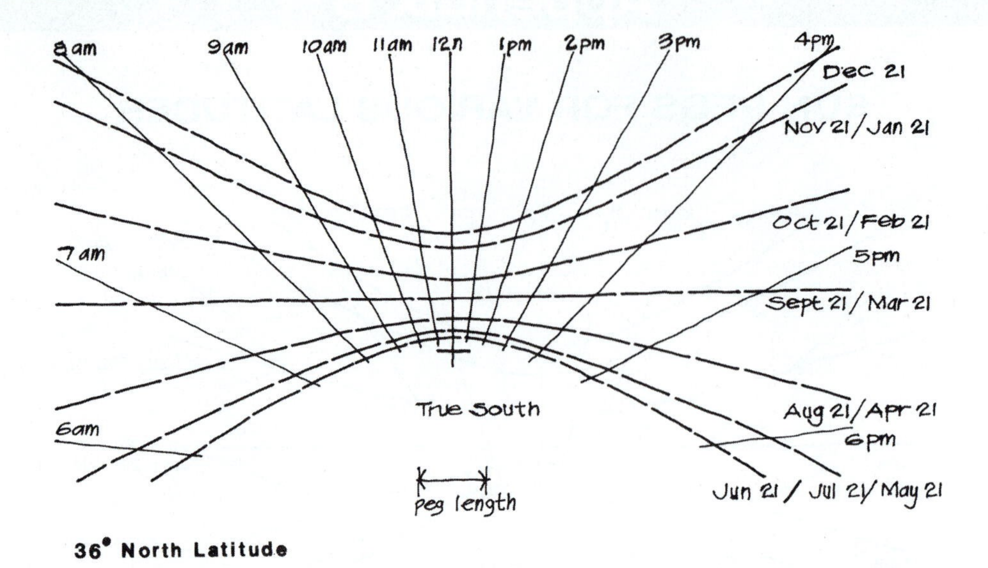

36° North Latitude

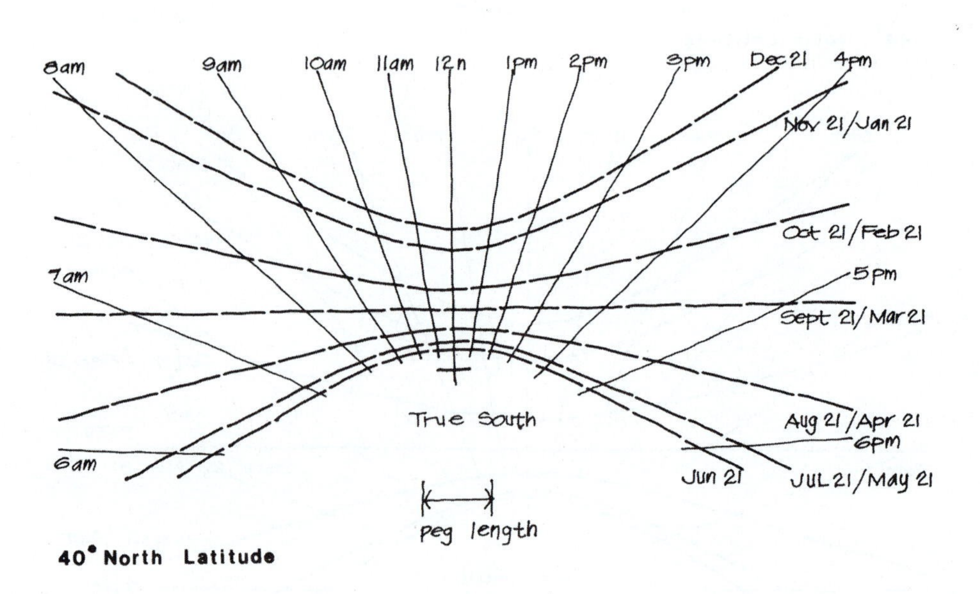

40° North Latitude

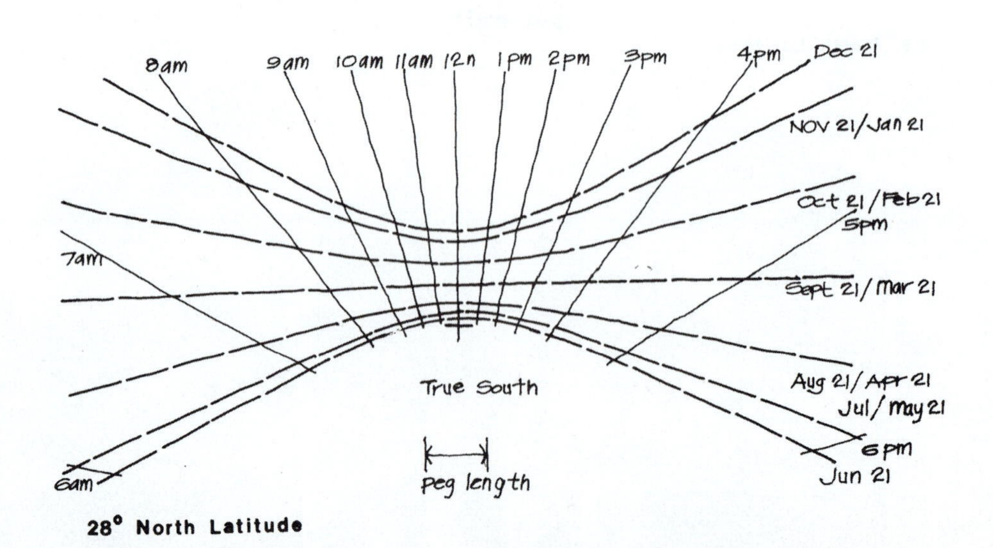

28° North Latitude

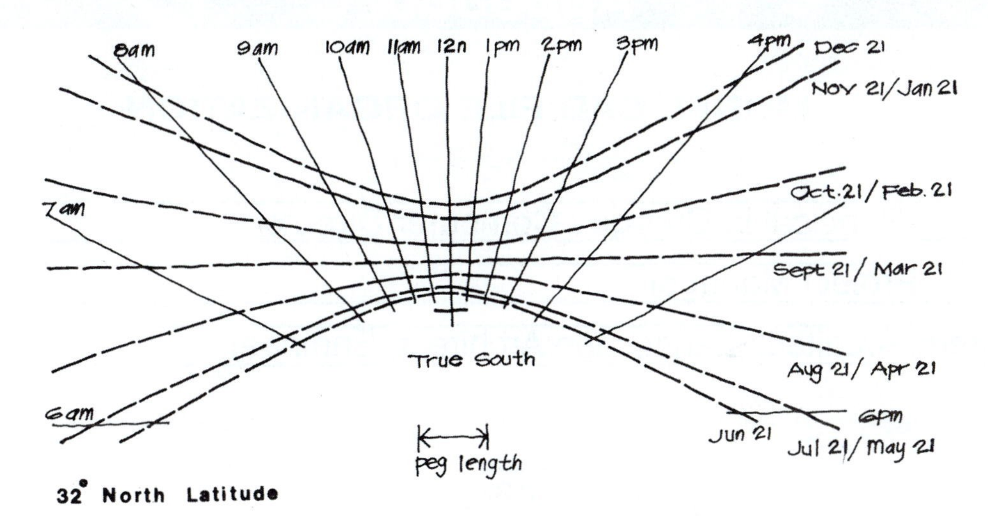

32° North Latitude

Appendix C

MODEL CAD FILE ORGANIZATION

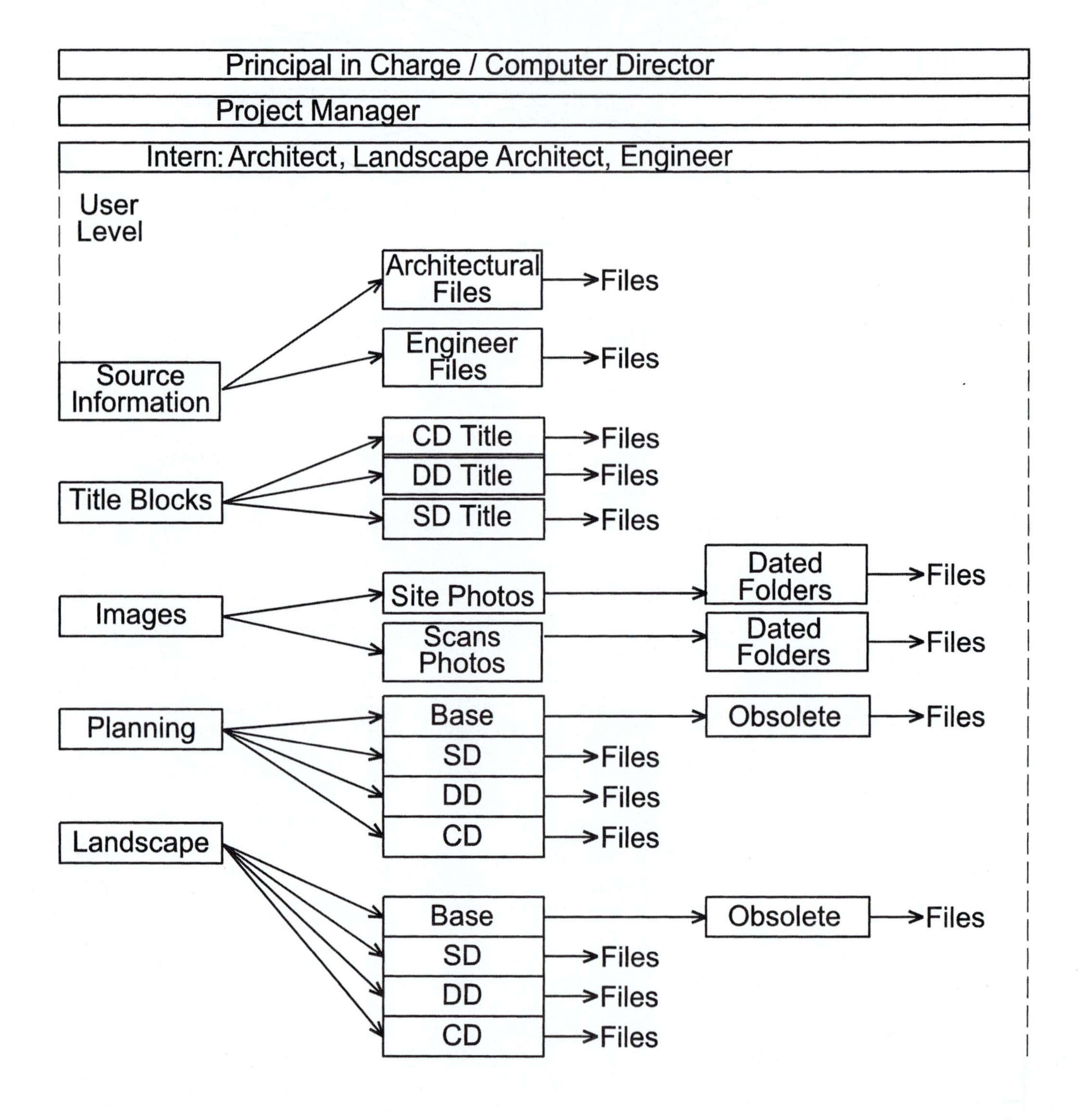

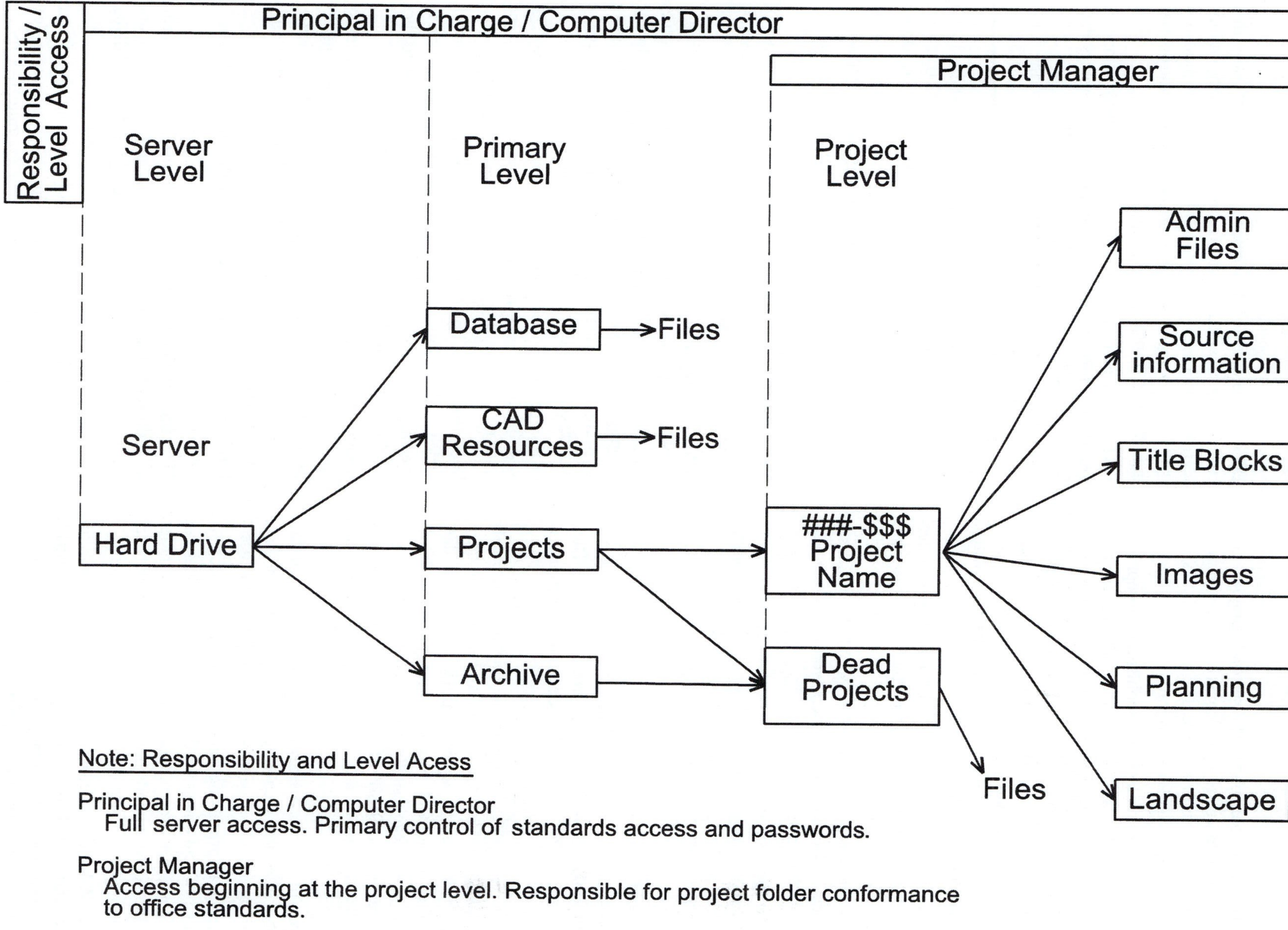

Note: Responsibility and Level Acess

Principal in Charge / Computer Director
Full server access. Primary control of standards access and passwords.

Project Manager
Access beginning at the project level. Responsible for project folder conformance to office standards.

Intern Architect / Planner
Access beginning at the user level. End user responsible for maintaining conformance to project standards in working with 'active' files.

Glossary of Terms*

AML (Arc Macro Language) The software language used by programmers to create special queries or programs to be executed in the ArcInfo software. The term AML is sometimes used as a synonym for an ArcInfo program.

ArcInfo The most complex and sophisticated proprietary GIS software of ESRI available at the time of this publication.

ArcView Proprietary software of Environmental Systems Research Institute (ESRI), one of the most user-friendly GIS software products on the market. The software was designed to permit a user to organize, maintain, visualize, and create maps and spatial information based on a database, with a minimum amount of training.

ASCII American Standard Code for Information Interchange, pronounced "askee." The base binary code for data that is used on most computers.

AutoCAD Proprietary software of Autodesk, Inc., used by architects, engineers, and planners world-wide to create digital 2D and 3D graphics files.

CAD (computer-aided drafting) A system using DOS, Windows, or Macintosh platform PCs or UNIX workstations and software to create 2D and 3D graphics on monitors or hard copy plotters. Also known as computer-aided drafting and design (CADD).

CADD *See* CAD.

card A small device that plugs into any of the open slots in the main (motherboard) printed circuit of a PC. These cards permit a PC to execute additional functions or processes.

CD-ROM (compact disc–read-only memory) A media format used to hold text, graphics, and sound. It's like a music CD but uses a different track format for data. Although a music CD player cannot play CD-ROMs, CD-ROM players can usually play music CDs. CD-ROMs hold in excess of 600 MB of data, which is equivalent to about 250,000 pages of text or 20,000 medium-resolution images.

coverage In ArcInfo, a set of thematically associated data considered as a unit. A coverage usually represents a single theme, or layer, such as roads, soils, hydrology, or land use.

CPU (central processing unit) The microprocessor, or brains, of a PC and one of the three basic parts that make up the PC. The other two components of the PC workstation include the keyboard and mouse, which are *input* devices, and the monitor and printers, which are *output* devices.

database A logical collection of interrelated information, managed and stored as a unit. A GIS database usually includes data about the spatial location and shape of geographic features recorded as points, lines, areas, or grid cells attributes.

DBMS (data base management system) Software that controls the organization storage, retrieval, security, and integrity of data in a database. A DBMS accepts requests for data from the application program and instructs the operating system to transfer the appropriate data.

DXF (data exchange format) One of the most common graphic file formats, used to transfer digital graphics data from one software to another. While there are some protocols that restrict the transfer of some lettering fonts and graphic patterns, line-work usually moves without complications.

dumb map A map or graphic created and retained as a digital file but prepared without support of a GIS database. All maps and graphics prepared in AutoCAD are dumb maps.

ESRI (Environmental Systems Research Institute) The corporation responsible for the development and maintenance of ArcInfo and ArcView.

giga Billion. It often refers to the precise value of 1,073,741,824 since computer specifications are usually based on binary numbers. Abbreviated G.

GIS (geographic information system) In a digital environment, a system in which polygons describing geographies are attached to a database of information unique to the geographies. Queries of the database are processed using a software which permits responses to the queries to be portrayed on both monitors and plotters.

GUI (graphical user interface) Pronounced "gooey". A graphics-based user interface that incorporates icons, and pull-down menus that are accessible via a mouse.

hypertext Text linked to information. A system through which a programmer links data or expanded information to specific words in digital text. A user can use a mouse to click on the selected, highlighted hypertext word.

inkjet A printer mechanism that sprays one or more colors of ink onto paper and produces high-quality printing similar to that of a laser printer. The HP DesignJet plotters used by the planning department at MRO use this technology.

interactive Involving back-and-forth dialog between a user and a computer.

IT (information technology) An acronym used to describe an industry that employs digital hardware and software and human resources to support the storage and use of digital information.

k An abbreviation for kilo, meaning thousand. It refers to the more precise value of 1,024 since computer specifications are usually binary numbers. For example, 64k means there are 65,536 bytes of storage.

*The sources for many definitions in the glossary are *The Electronic Computer Glossary* by Alan Freedman, published by the Computer Language Company of Point Pleasant, Penn., and *Understanding GIS*, published by Environmental Systems Research Institute, Inc., of Redlands, Calif.

LAN (local area network) A network of personal computers within a confined geographical area that is made up of servers, workstations, a network operating system, and a communications link.

layer A set of thematic data, such as soils, property lines, streams, etc., described and stored in a map library. CAD technicians refer to *layers* of line work in AutoCAD and *coverages* of line work in ArcInfo.

laser printer A printer that uses the electrophotographic method used in copy machines. A laser paints the dots of light onto a photographic drum or belt. The toner is applied to the drum or belt and then transferred onto the paper.

Map Maker ESRI software, developed in ArcInfo, that permits the creation of hard copy maps by even nontechnical users. The menus permit a variety of alternative coverages to be selected in the process of creating a unique map.

meg An abbreviation for the term megabyte, or 1 million bytes. Also written as MB, Mbyte, and M-byte.

multimedia Communicating information in more than one form, such as text, audio, graphics, animated graphics, and full-motion video.

multitasking Running two or more programs on one computer at the same time. The number of programs that can be effectively multitasked depends on the amount of memory available, CPU speed, capacity and speeds of peripheral resources, and the operating system.

network In communications, the transmission channels and supporting hardware and software.

pathname The route to a file or directory location on a disk.

PC (personal computer) A machine that conforms to the IBM PC or PS/2 Standard. PCs are used as stand-alone computers or as file servers in a local area network.

pixel A contraction of the words *picture element.* The smallest unit of information in an image or raster map. Referred to as a *cell* in an image.

planimetric map A map of line work drawn from aerial photography depicting the physical features of a landscape. Examples of the physical features on a planimetric map are structures (buildings), road paving, parking lots, bridges, railroads, rivers, lakes, creeks, golf courses, cemeteries, and wooded areas. Although our planimetric maps include topography, topo is not a common component of all planimetric maps.

plot file A digital file instructing our HP DesignJet *plotters* to convert line work data in a file to a graphic image on paper. Plot files in AutoCAD and ArcInfo can be retained for future use. Neither ArcView nor Map Maker create separate plot files in the process of creating a hardcopy graphic.

plotter *See* inkjet.

polygon A multisided area on a map. In ArcInfo, the polygon contains one label point inside its boundary to which data can be assigned. In ArcInfo, a polygon has attributes that describe the geographical feature it represents.

query The process of eliciting information from ArcView or ArcInfo by asking questions of the geographical data. A query can be spatial or logical. A spatial query involves spatially selecting features (e.g., selecting all features within 100 ft of a road intersection). A logical query involves selecting features whose attributes meet logical criteria (e.g., selecting all properties zoned R-60 with a land area greater than 5 acres).

RAM (random access memory) A computer's primary workspace. Although true of most memory chips, *random* means that the contents of each byte can be directly accessed without regard to the bytes before or after it.

raster image In computer graphics, a technique for representing a picture image as a matrix of dots. It is the digital counterpart of the analog method used in creating an image on TV. Unlike with TV, however, which has one standard, there are many raster graphics standards.

real-time Immediate response. Any electronic operation that is performed in the same time-frame as its real-world counterpart.

ROM (read-only memory) A memory chip that permanently stores instructions and data.

scan A file created when a scanner reads a hard copy image and electronically converts the lines and patterns in the image into a bitmapped raster file. Scanned images can be manipulated in software for either display or publication-size production.

site license A license to use software within a facility. A license provides authorization to make copies and distribute them within a specific jurisdiction.

smart map A graphic or map prepared and stored as a file, using a database.

spreadsheet Software that simulates a paper spreadsheet, or worksheet, in which columns and rows create a matrix of cells. Usually, each row represents a separate account or constituent described by the data in the columns, or fields. While accounts vary in size and geography, all have common fields of data describing the account.

tile The spatial unit by which a geography is organized, subdivided, and stored in a (digital) map library. To provide base mapping coverage of 4,000 ft × 6,000 ft per tile, at a scale of 1 in. = 200 ft, a tiling pattern of 664 map sheets were required to cover the geography of Montgomery County.

UNIX A multiuser, multitasking operating system that runs on a wide variety of computer systems, from micro processions to mainframes. UNIX is written to operate in C, a computer language designed for system-level programming.

vector graphics In computer graphics, a technique for representing a picture as points, lines connecting points, and polygons. AutoCAD and ArcInfo graphics are constructed as vector graphics. This technique is in contrast to raster graphics, which are composed of a series of unconnected pixels.

WAN (wide area network) In communications, a network that interconnects geographical boundaries such as cities and states.

Windows A graphics-based operating system from Microsoft that coexists with DOS. It provides a desktop operating system similar to that of the Macintosh. Different applications, or multiples of the same application, are kept alive in windows that can be resized and located on the screen. Users can switch back and forth between active windows.

workstation (1) A high-performance, single-user microcomputer or minicomputer that has been specialized for graphics, CAD, or scientific applications. (2) In a LAN, a personal computer that serves a single user, in contrast with a file server that serves all users on the network. (3) Any terminal or personal computer.

WYSIWYG An acronym for "what you see is what you get." This expression is used to convey to the user that the hard copy print, requested by the user, will duplicate the image on the screen.

Index

Note: Page numbers followed by "f" refer to figures. Page numbers followed by "t" refer to tables.

A

Accelerated erosion, 180
Acropolis, 8–9, 9f
ADA Standards for Accessible Design, 215
ADTs. *See* Average daily trips (ADTs)
Agoras
 at Corinth, 8, 10f
 at Miletus, 8, 10f
AIA. *See* American Institute of Architects (AIA)
Air movement, 97–102
 control and modification of, 99–102
 described, 97–99
 effects of, 99
Airport Climatological Summary—(CLIM90), 104–105, 107t, 110
Air temperature, 94–96
 control and modification of, 95
 effects of, 95
Albedo rates, of ground surfaces, 91, 92t
Alberti, Leon Battista, 2
Alexander the Great, 11
Altis at Olympia, 9, 10f
Altitude angles, 102
American Institute of Architects (AIA), 36
AML (Arc Macro Language), 282
Anaxogoras of Clazomene, 78
Angled parking, 198
Angle of repose, 72, 72t
Aquifers, 80
 confined, 80
 unconfined, 80
ArchInfo software, 38–49, 282
Architectural elements, for control and modification of, 91–92
 air movement, 101–102
 air temperature, 95
 relative humidity, 97
ArcInfo maps, 39
ArcView software, 38, 39, 282
Area drains, 140, 140f
Arterial classification, 192
ASCII, 282
AutoCAD, 33–34, 36, 282
 about, 38
 drafting software, 38
 simple graphics, 38
AutoLISP, 33
Average daily trips (ADTs), 192
Azimuth angles, 102–103

B

Back-zoning, 251
Baily, Francis, 13
Balance point temperature, of buildings, estimating, 227–229
Basemaps, 58
Bearing capacities, of rocks and soils, 70
Benchmarks, 122
 for surveys, 164
Bench terraces, 183
Berman v. *Parker*, 237–238
Berms, 183, 184f
Bicycles and bikeways, 218–225
 facilities planning, 218–219
 planning criteria, 219–225
BIM. *See* Building information modeling (BIM)
Bioclimatic analysis, 107–110
Blue-green storage areas, as storm water management method, 184–185, 185f
Board of Zoning Appeals (BZA), 246–247
Bogs, 80
Bonus zoning, 250–251
Brick Presbyterian Church v. *The City of New York*, 235
Bridges, pedestrian systems and, 212
Building codes, construction in floodplains and, 176
Building information modeling (BIM), 49–50
Buildings
 balance point temperature of, estimating, 227–229
 energy use and, 226
 estimating thermal loading problems of, 229–233
 internal-load-dominated, 226–227
 metabolism of, 226–229
 skin-load-dominated, 226

C

CAD. *See* Computer-aided design (CAD)
Canopy trees, 73
Capillarity, 72
Card, 282
Catch basins, 140, 140f
Caution, concept of, 52
CD-ROMs, 35, 282
Central processing unit (CPU), 282
Chains, surveying, 13
Chinese geomancy, 6
Circle, as Roman model, 5
Circular intersections, 195–196
Cities
 influences of forms on, 2
 plans for ideal, 2, 2f
Citizens' organizations, zoning process and, 248
City Beautiful movement, 14
City planning, site planning and, 2
Civil law doctrine, 181–182
Clean Water Act (CWA) (1972), 55, 83
Clean Water Act (CWA) (1977), 85
Climatic analysis
 general, 107–110
 publications for, 104–107
Climatic elements, 87–102
 air movement, 97–102
 air temperature, 94–96
 information sources for, 102–107
 relative humidity, 96–97
 solar radiation, 88–94
Climatic regimes
 macroclimate, 87
 mesoclimate, 87
 microclimate, 87–88
Collector classification, 193
Comfort, 86–87
 planning for, 107–110
Comfort zone, 86
Common enemy doctrine, 181–182
Commoner, Barry, 22
Compaction, 71
Comparative Climatic Data for the United States, 104, 106t
Compressibility, soil, 71
Computer-aided design (CAD), 26–27, 282
 creating standards, 36–37
 model file organization, 280–281
 organization, 36–37
 perceptions and expectations of, 29
 vector graphics and, 33–34
Computer education, design schools and, 27
Computer platforms, selecting, 32
Computers, site planning and, 26
Conditional zoning, 249
Confined aquifers, 80
Construction Specifications Institute (CSI), 36
Contour intervals, 119
Contour lines, 118–120, 120f, 121
 formula for identifying location of whole-number, 169
Corbusier, Le, 14–15
Corinth, agora at, 8, 10f
Coverage, 282
Cover bondage, 73
CPU. *See* Central processing unit (CPU)
Creek, 128, 129f
CSI. *See* Construction Specifications Institute (CSI)
Cumulative zoning, 240–241
Cut and fill, estimating, 148–152
Cycling. *See* Bicycles and bikeways

D

Database, 282
Data base management system (DBMS), 282

Data exchange format (DXF), 282
"Deadman" orientation, 157
Deciduous vegetation, 89
Deepwater Horizon disaster, 20
Deflection, wind control and, 75
Delphi, 4
Department of Defense (DOD), 36
Depressions, 122
Design schools, computer education and, 27–28
Diffuse radiation, 88
Digital drawings, creating, 37–38
Dinocrates, 11
Direction, as mechanism for development of geometric typologies, 6
Direct radiation, 88
Disabled persons
  parking and, 202
  pedestrian systems and, 214–218
Discipline designators, 37
Dissolved load, 82
Diversion channels, 183
Diversion dikes, 183, 184f
Doctrine of prior appropriation, 81
Documentation, 58–66
  basemaps for, 58
  government agencies involved in, 58–65
Downcutting erosion, 83
Down-zoning, 251
Drainage, 118
  considerations, 138
  grading and, 136–138
  subsurface systems and structures for, 140–141
Drainage basins, 82
Drainage plans, guidelines for designing, 138–139
Drawings
  creating digital, 37–38
  simple, 38
Drip line, 130, 130f
Drop-off zones, 203
Drought-tolerant plants, 76
Due diligence, 52–53
Dumb map, 282
Dumps, site inventory of, 57
DVDs, 35

E

Earth Observation Satellite Company (EOSC), 174
Earthworks, 118. *See also* Grading
Ecology, planting design and, 73
Elasticity, soil, 71
E-mail, 35
Eminent domain
  Berman v. *Parker* and, 237–238
  land use controls and, 235–238
Endangered Species Act (ESA) (1973), 57
Energy conservation, 75
Environment, laws of man's relationships with, 22
Environmental Impact Statements (EISs), 23
Environmental Protection Agency (EPA) (1969), 83, 170
Environmental Systems Research Institute (ESRI), 282
Erosion, 83
  abatement methods, 182
  accelerated, 180
  geological, 180
  grading/drainage plans and, 138
  natural, 180
  open paving systems for, 187–188, 187f, 188f
  ordinances, 182–183
  process of, 181
  reduction and abatement methods, 185–186
  rill, 181
  runoff and, 181
  sedimentation control and, 182–185
  sheet, 181
  slope integrity and, 181
  slump, 181
  types of, 180
  urbanization and, 181
  wind, 181
Erosion control, 75, 118, 181
  structural methods of, 183–185
Euclidean zoning, 239–245
  cumulative zoning, 240–241
  performance standards, 241–244
  rezoning process and, 249
  special exceptions and uses, 241
Evergreen vegetation, 89
  for control and modification of air movement, 100
Expressway classification, 192

F

Federal Emergency Management Agency (FEMA), 66, 171
Feng shui, 6
File management, 35–36
File security, 35
File sharing, 35
File storage systems
  organizing, 35–36
  standardized, 35
Filing, office, 35–36
Filing schemes, creating, 36
Filtration, wind control and, 75
First-order streams, 82
Fish and Wildlife Service (FWS), 57
Flood discharge, 172
Flood frequency, 172
Flood fringe, 172
Flood Insurance Rate Maps (FIRM), 66, 171
Floodplains, 172–178
  building codes for construction in, 176
  defined, 174
  management of, and community development, 176–178
  sanitary and water well codes for, 176
  site planning and, 174
  summary of flood data classifications and applicability, 174, 177t
Floods, 171–172
  strategies for controlling, 171
  zoning and land use controls and, 174
Floodway, 172
  defined, 174
Food bondage, 73
FTP sites, 35
Funnel, as method to control runoff, 129, 129f

G

Garden Cities, 14
Geddes, Sir Patrick, 14
Geiger, Rudolph, 88
Genius loci, 6
Geographic information system (GIS), 38, 282
  defined, 39
Geological erosion, 180
Geomancy, 4
  Chinese, 6
Geomorphology, site inventory of, 53
Giga, 282
GIS. *See* Geographic information system (GIS)
GIS maps, 39
"Going green," 24
Google Earth, 34
Gordis, Robert, 18
Grading, 118. *See also* Earthworks
  concepts and vocabulary, 118–127
  considerations, 138
  drainage and, 136–138
  estimating cut and fill, 148–152
  principals, 136–138
  process, 131–136
  purposes of, 118
  for streets and roads, 141–148
  topographical forms employed in, 128–131
Grading plans, guidelines for designing, 138–139
Graphical user interfaces (GUIs), 32–33, 282
Greenhouse effect, 88
Green site planning, 24
Green sites, 24
Grid
  in European and colonial settlements, 12–14
  orthogonal, 11, 13
Ground covers, 73

Groundwater, 70
Groundwater underflow, 78
Gunter, Edmund, 13

H

Habitats, 75
Hachures, 118f, 119, 119f
Hammer, 164, 167f
Hardin, Garrett, 21
Hardscape
 articulated, 129, 129f
 for control and modification of air temperature, 95
 for control and modification of solar radiation, 92–94
Hawaii Housing Authority v. *Midkiff*, 238
Headward erosion, 83
Hearing examiners, zoning, 247–248
Hippodamus, 11
Housing Act (1949), 16
Howard, Ebenezer, 14
Humidity, relative, 96–97, 107–109
 control and modification of, 97
 effects of, 96–97
 low, buildings and, 96
Hydrological cycle, 78–81
Hydrology, site inventory of, 53
Hypertext, defined, 282

I

Imperial Forum, 11
Incentive zoning, 250–251
Infiltration, of water, 78
Information technology (IT), 282
Inkjet, defined, 282
Institute of Transportation Engineers, 190
Interactive, defined, 282
Internal-load-dominated (ILD) buildings, 226–227
 thermal balance and, 227
Internet, 34
Interns
 CAD skills and, 29
 defined, 26
 graphic skills of, 27
 implications of technology on roles of, 26–27
 professional expectations, 30
Intersections, 193–195
 circular, 195–196
Inventory. *See* site inventory

J

Jacobs, Jane, 15–16
Josephus, 18
Judeo-Christian ethic, in New World, 17–21

K

Ka'aba, 4
Kelo v. *The City of New London*, 239
k (kilo), defined, 282
Knibshuah, 18
Kudzu vine, 76

L

"Land as commodity" philosophy, 20
Landforms
 for control and modification of air movement, 102
 for control and modification of air temperature, 95
 for control and modification of solar radiation, 92–94
Land ownership
 in Europe, 17
 in New World, 17–18
Landscape character, 73–74
Landschaft, 12
Land use controls
 about, 234
 Berman v. *Parker*, 237–238
 conclusions, 239–240
 eminent domain and, 235–238
 Euclidean zoning, 239–245
 Hawaii Housing Authority v. *Midkiff*, 238
 history and background of, 234–237
 Kelo v. *The City of New London*, 239
 Penn Central v. *The City of New York*, 238–239
Laser printer, 283
Laws of man's relationships, with environment, 22
Layer, defined, 283
Layering Guidelines (AIA), 36, 37
Layer names, AIA guidelines for, 37
Legislative authorities, local, as zoning decision makers, 245
L'Enfant, Pierre, 195
Leopold, Aldo, 21
Le Plan de Paris, 15
Lighting, 206–207
Liquid limit, 72
Living streets, 214, 215f
Local area networks (LANs), 35, 283
Local climatological data, states providing, 275–276
Local Climatological Data—Annual Summary with Comparative Data, 104, 105t, 109, 110
Local systems classifications, 193
Longwave radiation, 88

M

Macroclimate, 87
Maher v. *The City of New Orleans*, 238
Major groups, 37
"Man dominant" position, 21–22
Manifest Destiny, 19–20
Map Maker, 38, 283
Mapping, 52–53
Marcus Vitruvius Pollio, 88
Marshes, site inventory, 53–57
Mechanically stabilized earth (MSE) walls, 161–163
Meg, defined, 283
Mesoclimate, 87
Metabolism, building, 226–229
Microclimates
 as climatic regime, 87–88
 site inventory of, 57–58
MicroStation, 33–34
 about, 38
Miletus
 agora at, 8, 10f
 plan of, 11, 11f
Minor groups, 37
Modern roundabouts, 195–196, 196f
Mt. Olympus, 4
Multimedia, defined, 283
Multitasking, defined, 283
Mumford, Lewis, 14

N

Nash, Roderick, 21
National Climatic Data Center (NCDC), 102
National Environmental Policy Act (NEPA) (1969), 23
 purpose and policies of, 23–24
National Oceanic and Atmospheric Administration (NOAA), 57, 102
National Pollution Discharge Elimination System (NPDES), 55
National Stream Quality Accounting Network (NASQAN), 66
Native plants, 76
Natural erosion, 180
Nature, sacred places and, 3–5
Netting, for erosion reduction, 185–186, 186f
Network, defined, 283
New World
 grid system in, 12–14
 Judeo-Christian ethic in, 17–21
 land ownership in, 17–18
North American Wetlands Conservation Act (1989), 85
Nuisance law, 237

O

Obstruction, wind control and, 75
Office filing, 35–36
Offices. *See* Site-planning offices
Olympia, Altis at, 9, 10f
On-site retention ponds, 138, 139f
Open paving systems, 187–188, 187f, 188f
Orientation, as mechanism for development of geometric typologies, 6
Orthogonal grid, 11, 13

P

Parallel parking, 198–199
Paris
 Le Plan de, 15
 Voisin plan for, 15

Parking, 197
configurations, 197–199
for disabled persons, 202
drop-off zones and, 203
pedestrian transition and, 199–202
sizes and arrangements, 197
Parking lots
pedestrian movement through, 209–210
as storm water management method, 185
Parking schedules, 244–245
Passive soil pressure, 154
Passive solar heating, 88
Pathname, defined, 283
Pedestrian systems, 208–218. *See also* Streets and roads
accessibility for individuals with disabilities and, 214–218
bridges and, 212
connectors and linkages, 209–210
living streets and, 212–214
scale and, 210–212
Pedestrian transition, parking and, 199–202
Pennsylvania Coal v. *Mahon*, 236
People, sites and, 6
Performance zones, 249–251
Performance zoning, 250
Permeability, 72
Personal computers (PCs), 283
Photoshop, 33
Pins, 164, 167f
Pixels, 283
Planimetric line work, 49
Planimetric maps, 283
Planned development (PD), 249
Planned residential development (PRD), 249
Planned unit development (PUD) ordinances, two-stage, 250
Planned unit developments (PUDs), 249–251
Planners. *See* Site planners
Planning commissions, 245–246
Planting design, ecology and, 73
Planting plans, 75–76
Plant materials
divisions of, 73–74
functional roles of, 74–76
Plants
drought-tolerant, 76
native, 76
Plasticity, soil, 71
Plot files, 283
Plotters. *See* Inkjet
Polygon, 283
Ponds, permanent, as erosion control method, 184
Portable document formats (PDFs), 35
Precipitation, 82
hydrological cycle and, 78
site inventory of, 58
Premises address data, 49
Primitives, 33
Principals, 27
CAD perceptions and expectations of, 29
defined, 26
Prior appropriation, doctrine of, 81
Prior fillings, site inventory of, 57
Professionals
CAD literacy and, 28–29
CAD perceptions and expectations of, 29
defined, 26
implications of technology on roles of, 26–27
professional expectations, 30
required computer skills for, 27
Property base maps
creating, 49
methods for preparing, 49

Q

Query, defined, 283
Quid-pro-quo relationship, 21

R

Radiant City, 14–15, 15f
Radiation
diffuse, 88
direct, 88
longwave, 88
reflected, 88
shortwave, 88
solar, 88–94
Rainfall
hydrological cycle and, 78
site inventory of, 58
Random access memory (RAM), 283
Raster graphics, 33
Raster image, defined, 283
Read-only memory (ROM), 283
Real-time, defined, 283
Reasonable use theory, 182
Reference files (xrefs), 36
Reflected radiation, 88
Reinforced Earth, 161
Relative humidity
control and modification of, 97
effects of, 96–97
low, buildings and, 96
"Resource" philosophy, 20
Restraint, concept of, 52
Retaining walls
alignment and, 156
concrete masonry units for, 155–156
economy and, 156
masonry for, 155
materials for, 155
mechanically stabilized earth (MSE), 161
precast concrete systems, 158–161
railroad ties for, 156–157
in situ soil conditions and, 154–155
structural design review, 153–158
Rezoning process, 248–251. *See also* Zoning
alternatives for, 249
Rill erosion, 181
Riparian doctrine, 81
Rivers and Harbors Act (1899), 83–85
Roads. *See* Streets and roads
Rock
bearing capacities of, 70, 70t
existence of, site inventory of, 53
Rod readings, 164–167, 167f
ROM. *See* Read-only memory (ROM)
Roman Forum of Trajan, 9–11, 11f
Rotaries, 195–196
Roundabouts, 195–196, 196
Runoff, 82
erosion and, 181
factors for volume of, 181
ordinances, 182–183
topographical forms employed in, 128
urbanization and, 181

S

Sacred places, nature and, 3–5
Sanitary codes, floodplains and, 176
Scan, defined, 283
Secondary road classification, 193
Second-order streams, 82
Section 404, of Water Quality Act (1977), 83
Sedimentation controls
erosion and, 182–185
methods, 183
Sediments, 83
Shading calendars, 227–229, 228f, 229f
Shearing strength, 71
Sheet erosion, 181
Shelterbelts, 99–101
Shortwave radiation, 88
Shrinkage, 71–72
Sideways erosion, 83
Signage, 204–206
Simple drawings, 38. *See also* Drawings
Site inventory, method of, 53–58
Site license, defined, 283
Site-microclimate mapping, 110–115
Site planners, 27
thinking with pencil and, 30–31
Site planning
changing views of, 24
checklist and review, 66–69
city planning and, 2
common oversights of, 16
Egyptian model of, 7
four basic models of, 6–7, 7f
Hellenic Greece and, 8–9
scale and authorities controlling, 2–3

Site-planning offices, 28
Site-planning process, 254–274
 about, 254
 goals and objectives, 254–255
 inventory of site, 255–274
 research and data collection, 255
Site planning profession
 intern expectation of, 30
 lay expectations of, 29–30
 professional expectations of, 30
Sites
 people and, 6
 topsoil and, 131
Sketching, computer *vs.* hand-drawn, 26
SketchUp, 33
Skin-load-dominated (SLD) buildings, 226
Slope analysis, 122–127
Slope integrity, erosion and, 181
Slope retreat, 83
Sloping plane, 28, 128f
Slump erosion, 181
Smart maps, 283
Smith, Adam, 19, 21
Soil and Water Conservation Districts Law, 182
Soil compressibility, 71
Soil Conservation Service (SCS), 65
Soil elasticity, 71
Soil plasticity, 71
Soils
 bearing capacities of, 70, 70t
 classification chart, 71t
 classification systems, 72
 life-support capacity of, 72
 performance criteria, 72
 properties of, 70
 site inventory of, 53
 structural and drainage characteristics of, 71–72
 structural suitability of, 72
 textures of, 70–71
Soil surveys, 65–66
Solar geometry, 102–104, 110
Solar radiation, 88–94
 architectural elements for control and modification of, 91–92
 control strategies for, 89
 landforms and hardscape for control and modification of, 92–94
 vegetation for control and modification of, 89–91
Solar transits, 103, 104f
Spot elevations, 122
Spread sheets, 283
Steel tape, 164, 167f
Stein, Clarence, 14
Stilgoe, 17, 18
Stone, Christopher, 21
Stonehenge, 5, 5f
Storm water management, 170–171
 methods of, 183–184
Storm water pollution prevention plan (SWPPP), 170
Storm water runoff analysis, 179–180
Streams
 first-order, 82
 second-order, 82
Streets and roads. *See also* Pedestrian systems
 grading for, 141–148
 intersections, 193–196
 living, 214, 215f
 pedestrians and, 212–214
 systems, 190–192
 thoroughfare classifications and criteria, 192–193
Subsurface collection, 140–141
Subsurface geology, 53, 70
Subsurface water, 78
 aquifers and, 80
Summary of Meteorological Observations, Surface (SMOS), 107, 108t, 110
Summits, 122
Sun angle calculator, 102
Sun pegs, 102, 110
 for various latitudes, 277–279
Surface geology, 70–71
Surface water, 78, 81–82
 as process, 82–83
Surveying chains, 13
Suspended-sediment load, 82
Sustainable communities, 20–21
Swale, 128, 129f
Swamps, 80
Swell, 71–72

T

Tao Te Ching, 6
Thermal balance, buildings and, 227
Thermal comfort, 86–87
Thermal loading problems, of buildings, estimating, 229–233
Thomas Aquinas, Saint, 18
Thoroughfare classifications and criteria, 192–193
3D modeling, 33–34
Tile, defined, 283
Topographical surveys, 164–169
Topography
 grading and, 120–122
 site inventory of, 53
Topsoil, 131
Traffic calming, 196
Traffic circles, 195–196, 195f
Transfer of development rights (TDR), 250–251
Transit and rod, 164, 167f
Transportation planning, 190
 levels of service and, 190–192
Trees, foliation periods for, 89–90, 90t
Trench drains, 140, 141f
Tri-Service Plotting Guidelines (DOD), 36
True south, 102
Two-stage planned unit development (PUD) ordinances, 250

U

Unconfined aquifers, 80
Understory, 73
Uniform Drawing System (CSI), 36
UNIX, defined, 283
Upslope, 83
Urbanization, runoff/erosion and, 181
U.S. Geological Survey (USGS), 66

V

Valley, as method to control runoff, 128, 129f
Values, 21–22
Vector graphics, 33, 283
Vegetation
 for control and modification of air movement, 99–101
 for control and modification of air temperature, 95
 for control and modification of relative humidity, 97
 for control and modification of solar radiation, 89–91
 deciduous, 89
 evaluating existing, 73–74
 evaluation standards for, 74
 evergreen, 89, 100
Vegetation, site inventory of, 57
Vidal, Henri, 161
Village of Euclid v. *Ambler Realty Co.*, 237
Voisin plan for Paris, 15

W

Water Quality Act (1977), 83
Water resource management, 81
Water rights, 80–81
Water tables, 70
Water well codes, floodplains and, 176
Wetlands, 80, 83–85
 criteria for determining existence of, 85
Wetlands, site inventory of, 53–57
Wide area networks (WANs), 284
Wildlife, site inventory of, 57
Wildlife habitat, 75
WIMP (window, icon, menu, pointing device)-style interfaces, 32–33
Wind, 97–102
Windbreaks, 99–101
Wind control, 75
Wind erosion, 181
Windows, Microsoft, 284
Woonerf streets, 196, 213–214, 214f
Workstations, 284
WYSIWYG, defined, 284

X

Xeriscape, 76
Xeriscaping, 76
Xrefs (reference files), 36

Z

Zoning. *See also* Rezoning process
- back-, 251
- bonus, 250–251
- conditional, 249
- cumulative, 240–241
- down-, 251
- Euclidean, 239–245, 249
- for flood-prone areas, 174–176
- incentive, 250–251
- performance, 249–251

Zoning Appeals, Board of, 246–247
Zoning commissions, 245–246
Zoning decision makers, 245–248
- Board of Zoning Appeals, 246–247
- citizens' organizations, 248
- hearing examiners, 247–248
- local legislative authorities, 245
- planning and zoning commissions, 245–246

Zoning ordinances, 237
- parking schedules and, 244–245